VOLUME SEVENTY FIVE

ADVANCES IN MARINE BIOLOGY

Mediterranean Marine Mammal Ecology and Conservation

ADVANCES IN MARINE BIOLOGY

VOLUME SEVENTY FIVE

ADVANCES IN MARINE BIOLOGY

Mediterranean Marine Mammal Ecology and Conservation

Edited by

GIUSEPPE NOTARBARTOLO DI SCIARA
Tethys Research Institute
Acquario Civico
Milano, Italy

MICHELA PODESTÀ
Museum of Natural History of Milan
Milano, Italy

BARBARA E. CURRY
Physiological Ecology and Bioenergetics Laboratory
Conservation Biology Program
University of Central Florida
Orlando, FL 32816
United States

ELSEVIER

AMSTERDAM • BOSTON • HEIDELBERG • LONDON
NEW YORK • OXFORD • PARIS • SAN DIEGO
SAN FRANCISCO • SINGAPORE • SYDNEY • TOKYO
Academic Press is an imprint of Elsevier

AP

Academic Press is an imprint of Elsevier
125 London Wall, London, EC2Y 5AS, United Kingdom
The Boulevard, Langford Lane, Kidlington, Oxford OX5 1GB, United Kingdom
525 B Street, Suite 1800, San Diego, CA 92101-4495, United States
50 Hampshire Street, 5th Floor, Cambridge, MA 02139, United States

First edition 2016

Notices
Knowledge and best practice in this field are constantly changing. As new research and experience broaden our understanding, changes in research methods, professional practices, or medical treatment may become necessary.

Practitioners and researchers must always rely on their own experience and knowledge in evaluating and using any information, methods, compounds, or experiments described herein. In using such information or methods they should be mindful of their own safety and the safety of others, including parties for whom they have a professional responsibility.

To the fullest extent of the law, neither the Publisher nor the authors, contributors, or editors, assume any liability for any injury and/or damage to persons or property as a matter of products liability, negligence or otherwise, or from any use or operation of any methods, products, instructions, or ideas contained in the material herein.

ISBN: 978-0-12-805152-8
ISSN: 0065-2881

Publisher: Zoe Kruze
Acquisition Editor: Alex White
Editorial Project Manager: Helene Kabes
Production Project Manager: Surya Narayanan Jayachandran
Cover Designer: Victoria Pearson

Typeset by SPi Global, India

CONTRIBUTORS TO VOLUME 75

S. Airoldi
Tethys Research Institute, Milano, Italy

D. Alarcón
CIRCE (Conservation, Information and Research on Cetaceans), Pelayo-Algeciras, Cádiz, Spain

A. Azzellino
Politecnico di Milano, University of Technology; Tethys Research Institute, Milano, Italy

G. Bearzi
Dolphin Biology and Conservation, Oria, Italy; OceanCare, Wädenswil, Switzerland; Texas A&M University at Galveston, Galveston, TX, United States

S. Bonizzoni
Dolphin Biology and Conservation, Oria, Italy; OceanCare, Wädenswil, Switzerland; Texas A&M University at Galveston, Galveston, TX, United States

A. Cañadas
ALNILAM Research and Conservation, Navacerrada, Madrid, Spain

M. Castellote
National Marine Mammal Laboratory, Alaska Fisheries Science Center/NOAA, Seattle, WA, United States

R. de Stephanis
CIRCE (Conservation, Information and Research on Cetaceans), Pelayo-Algeciras, Cádiz, Spain

J.-N. Druon
European Commission, DG Joint Research Centre, Directorate D—Sustainable Resources, Unit D.02 Water and Marine Resources, Ispra, Italy

L. Eddy
Dolphin Biology and Conservation, Oria, Italy; OceanCare, Wädenswil, Switzerland

M. Elasar
School of Marine Sciences, University of Haifa; Israel Marine Mammal Research & Assistance Center (IMMRAC), Haifa, Israel

R. Esteban
CIRCE (Conservation, Information and Research on Cetaceans), Pelayo-Algeciras, Cádiz, Spain

M.C. Fontaine
Groningen Institute for Evolutionary Life Sciences (GELIFES), University of Groningen, Groningen, The Netherlands

A.D. Foote
CMPG, Institute of Ecology and Evolution, University of Bern, Bern, Switzerland

M.C. Fossi
University of Siena, Siena, Italy

A. Frantzis
Pelagos Cetacean Research Institute, Vouliagmeni, Greece

N.B. Furey
Dolphin Biology and Conservation, Oria, Italy; University of British Columbia, Vancouver, BC, Canada

S. Gaspari
National Research Council (CNR), Institute of Marine Sciences (ISMAR), Ancona, Italy

P. Gauffier
CIRCE (Conservation, Information and Research on Cetaceans), Pelayo-Algeciras, Cádiz, Spain

O. Gimenez
Centre d'Ecologie Fonctionnelle et Evolutive, Montpellier, France

J. Giménez
CIRCE (Conservation, Information and Research on Cetaceans), Pelayo-Algeciras, Cádiz; Estación Biológica de Doñana (EBD-CSIC), Sevilla, Spain

O. Goffman
School of Marine Sciences, University of Haifa; Israel Marine Mammal Research & Assistance Center (IMMRAC), Haifa, Israel

J. Gonzalvo
Tethys Research Institute, Milan, Italy

N. Hadar
Israel Marine Mammal Research & Assistance Center (IMMRAC), Haifa, Israel

P.S. Hammond
Sea Mammal Research Unit, Gatty Marine Laboratory, University of St Andrews, Fife, Scotland, United Kingdom

D. Kerem
School of Marine Sciences, University of Haifa; Israel Marine Mammal Research & Assistance Center (IMMRAC), Haifa, Israel

S. Kotomatas
WWF Greece, Athens, Greece

C. Lanfredi
Politecnico di Milano, University of Technology; Tethys Research Institute, Milano, Italy

G. Lauriano
Institute for Environmental Protection and Research (ISPRA), Roma, Italy

T. Lewis
North Atlantic & Mediterranean Sperm Whale Catalogue (NAMSC), London, United Kingdom

L. Marsili
University of Siena, Siena, Italy

A. Moulins
CIMA Research Foundation, Savona, Italy

A. Natoli
UAE Dolphin Project, Dubai, United Arab Emirates

G. Notarbartolo di Sciara
Tethys Research Institute, Acquario Civico, Milano, Italy

S. Panigada
Tethys Research Institute, Acquario Civico, Milano, Italy

M. Podestà
Museum of Natural History of Milan, Milano, Italy

L. Rendell
Sea Mammal Research Unit, University of St Andrews, St Andrews, Fife, United Kingdom

M. Rosso
CIMA Research Foundation, Savona, Italy

J.M. Salazar-Sierra
CIRCE (Conservation, Information and Research on Cetaceans), Pelayo-Algeciras, Cádiz, Spain

N.L. Santostasi
Dolphin Biology and Conservation, Oria, Italy; Centre d'Ecologie Fonctionnelle et Evolutive, Montpellier, France

A. Scheinin
School of Marine Sciences, University of Haifa; Israel Marine Mammal Research & Assistance Center (IMMRAC), Haifa, Israel

T. Scovazzi
Dipartimento giuridico delle istituzioni nazionali ed europee, Università degli Studi di Milano-Bicocca, Milan, Italy

P. Tepsich
CIMA Research Foundation, Savona; University of Genoa, Genoa, Italy

V.D. Valavanis
Marine Geographic Information Systems Lab, Hellenic Centre for Marine Research, Heraklion, Greece

P. Verborgh
CIRCE (Conservation, Information and Research on Cetaceans), Pelayo-Algeciras, Cádiz, Spain

K.A. Viaud-Martinez
Illumina, Inc., San Diego, CA, United States

CONTENTS

SERIES CONTENTS FOR LAST FIFTEEN YEARS*

Volume 38, 2000.
Blaxter, J. H. S. The enhancement of marine fish stocks. pp. 1–54.
Bergström, B. I. The biology of Pandalus. pp. 55–245.

Volume 39, 2001.
Peterson, C. H. The "Exxon Valdez" oil spill in Alaska: acute indirect and chronic effects on the ecosystem. pp. 1–103.
Johnson, W. S., Stevens, M. and Watling, L. Reproduction and development of marine peracaridans. pp. 105–260.
Rodhouse, P. G., Elvidge, C. D. and Trathan, P. N. Remote sensing of the global light-fishing fleet: an analysis of interactions with oceanography, other fisheries and predators. pp. 261–303.

Volume 40, 2001.
Hemmingsen, W. and MacKenzie, K. The parasite fauna of the Atlantic cod, Gadus morhua L. pp. 1–80.
Kathiresan, K. and Bingham, B. L. Biology of mangroves and mangrove ecosystems. pp. 81–251.
Zaccone, G., Kapoor, B. G., Fasulo, S. and Ainis, L. Structural, histochemical and functional aspects of the epidermis of fishes. pp. 253–348.

Volume 41, 2001.
Whitfield, M. Interactions between phytoplankton and trace metals in the ocean. pp. 1–128.
Hamel, J.-F., Conand, C., Pawson, D. L. and Mercier, A. The sea cucumber Holothuria scabra (Holothuroidea: Echinodermata): its biology and exploitation as beche-de-Mer. pp. 129–223.

Volume 42, 2002.
Zardus, J. D. Protobranch bivalves. pp. 1–65.
Mikkelsen, P. M. Shelled opisthobranchs. pp. 67–136.
Reynolds, P. D. The Scaphopoda. pp. 137–236.
Harasewych, M. G. Pleurotomarioidean gastropods. pp. 237–294.

*The full list of contents for volumes 1–37 can be found in volume 38

Volume 43, 2002.
Rohde, K. Ecology and biogeography of marine parasites. pp. 1–86.
Ramirez Llodra, E. Fecundity and life-history strategies in marine invertebrates. pp. 87–170.
Brierley, A. S. and Thomas, D. N. Ecology of southern ocean pack ice. pp. 171–276.
Hedley, J. D. and Mumby, P. J. Biological and remote sensing perspectives of pigmentation in coral reef organisms. pp. 277–317.

Volume 44, 2003.
Hirst, A. G., Roff, J. C. and Lampitt, R. S. A synthesis of growth rates in epipelagic invertebrate zooplankton. pp. 3–142.
Boletzky, S. von. Biology of early life stages in cephalopod molluscs. pp. 143–203.
Pittman, S. J. and McAlpine, C. A. Movements of marine fish and decapod crustaceans: process, theory and application. pp. 205–294.
Cutts, C. J. Culture of harpacticoid copepods: potential as live feed for rearing marine fish. pp. 295–315.

Volume 45, 2003.
Cumulative Taxonomic and Subject Index.

Volume 46, 2003.
Gooday, A. J. Benthic foraminifera (Protista) as tools in deep-water palaeoceanography: environmental influences on faunal characteristics. pp. 1–90.
Subramoniam, T. and Gunamalai, V. Breeding biology of the intertidal sand crab, Emerita (Decapoda: Anomura). pp. 91–182.
Coles, S. L. and Brown, B. E. Coral bleaching—capacity for acclimatization and adaptation. pp. 183–223.
Dalsgaard J., St. John M., Kattner G., Müller-Navarra D. and Hagen W. Fatty acid trophic markers in the pelagic marine environment. pp. 225–340.

Volume 47, 2004.
Southward, A. J., Langmead, O., Hardman-Mountford, N. J., Aiken, J., Boalch, G. T., Dando, P. R., Genner, M. J., Joint, I., Kendall, M. A., Halliday, N. C., Harris, R. P., Leaper, R., Mieszkowska, N., Pingree, R. D., Richardson, A. J., Sims, D.W., Smith, T., Walne, A. W. and Hawkins, S. J. Long-term oceanographic and ecological research in the western English Channel. pp. 1–105.

Queiroga, H. and Blanton, J. Interactions between behaviour and physical forcing in the control of horizontal transport of decapod crustacean larvae. pp. 107–214.

Braithwaite, R. A. and McEvoy, L. A. Marine biofouling on fish farms and its remediation. pp. 215–252.

Frangoulis, C., Christou, E. D. and Hecq, J. H. Comparison of marine copepod outfluxes: nature, rate, fate and role in the carbon and nitrogen cycles. pp. 253–309.

Volume 48, 2005.

Canfield, D. E., Kristensen, E. and Thamdrup, B. Aquatic Geomicrobiology. pp. 1–599.

Volume 49, 2005.

Bell, J. D., Rothlisberg, P. C., Munro, J. L., Loneragan, N. R., Nash, W. J., Ward, R. D. and Andrew, N. L. Restocking and stock enhancement of marine invertebrate fisheries. pp. 1–358.

Volume 50, 2006.

Lewis, J. B. Biology and ecology of the hydrocoral Millepora on coral reefs. pp. 1–55.

Harborne, A. R., Mumby, P. J., Micheli, F., Perry, C. T., Dahlgren, C. P., Holmes, K. E., and Brumbaugh, D. R. The functional value of Caribbean coral reef, seagrass and mangrove habitats to ecosystem processes. pp. 57–189.

Collins, M. A. and Rodhouse, P. G. K. Southern ocean cephalopods. pp. 191–265.

Tarasov, V. G. Effects of shallow-water hydrothermal venting on biological communities of coastal marine ecosystems of the western Pacific. pp. 267–410.

Volume 51, 2006.

Elena Guijarro Garcia. The fishery for Iceland scallop (Chlamys islandica) in the Northeast Atlantic. pp. 1–55.

Jeffrey, M. Leis. Are larvae of demersal fishes plankton or nekton? pp. 57–141.

John C. Montgomery, Andrew Jeffs, Stephen D. Simpson, Mark Meekan and Chris Tindle. Sound as an orientation cue for the pelagic larvae of reef fishes and decapod crustaceans. pp. 143–196.

Carolin E. Arndt and Kerrie M. Swadling. Crustacea in Arctic and Antarctic sea ice: Distribution, diet and life history strategies. pp. 197–315.

Volume 52, 2007.
Leys, S. P., Mackie, G. O. and Reiswig, H. M. The Biology of Glass Sponges. pp. 1–145.
Garcia E. G. The Northern Shrimp (Pandalus borealis) Offshore Fishery in the Northeast Atlantic. pp. 147–266.
Fraser K. P. P. and Rogers A. D. Protein Metabolism in Marine Animals: The Underlying Mechanism of Growth. pp. 267–362.

Volume 53, 2008.
Dustin J. Marshall and Michael J. Keough. The Evolutionary Ecology of Offspring Size in Marine Invertebrates. pp. 1–60.
Kerry A. Naish, Joseph E. Taylor III, Phillip S. Levin, Thomas P. Quinn, James R. Winton, Daniel Huppert, and Ray Hilborn. An Evaluation of the Effects of Conservation and Fishery Enhancement Hatcheries on Wild Populations of Salmon. pp. 61–194.
Shannon Gowans, Bernd Würsig, and Leszek Karczmarski. The Social Structure and Strategies of Delphinids: Predictions Based on an Ecological Framework. pp. 195–294.

Volume 54, 2008.
Bridget S. Green. Maternal Effects in Fish Populations. pp. 1–105.
Victoria J. Wearmouth and David W. Sims. Sexual Segregation in Marine Fish, Reptiles, Birds and Mammals: Behaviour Patterns, Mechanisms and Conservation Implications. pp. 107–170.
David W. Sims. Sieving a Living: A Review of the Biology, Ecology and Conservation Status of the Plankton-Feeding Basking Shark *Cetorhinus Maximus*. pp. 171–220.
Charles H. Peterson, Kenneth W. Able, Christin Frieswyk DeJong, Michael F. Piehler, Charles A. Simenstad, and Joy B. Zedler. Practical Proxies for Tidal Marsh Ecosystem Services: Application to Injury and Restoration. pp. 221–266.

Volume 55, 2008.
Annie Mercier and Jean-Francois Hamel. Introduction. pp. 1–6.
Annie Mercier and Jean-Francois Hamel. Gametogenesis. pp. 7–72.
Annie Mercier and Jean-Francois Hamel. Spawning. pp. 73–168.
Annie Mercier and Jean-Francois Hamel. Discussion. pp. 169–194.

Volume 56, 2009.

Philip C. Reid, Astrid C. Fischer, Emily Lewis-Brown, Michael P. Meredith, Mike Sparrow, Andreas J. Andersson, Avan Antia, Nicholas R. Bates, Ulrich Bathmann, Gregory Beaugrand, Holger Brix, Stephen Dye, Martin Edwards, Tore Furevik, Reidun Gangst, Hjalmar Hatun, Russell R. Hopcroft, Mike Kendall, Sabine Kasten, Ralph Keeling, Corinne Le Quere, Fred T. Mackenzie, Gill Malin, Cecilie Mauritzen, Jon Olafsson, Charlie Paull, Eric Rignot, Koji Shimada, Meike Vogt, Craig Wallace, Zhaomin Wang and Richard Washington. Impacts of the Oceans on Climate Change. pp. 1–150.

Elvira S. Poloczanska, Colin J. Limpus and Graeme C. Hays. Vulnerability of Marine Turtles to Climate Change. pp. 151–212.

Nova Mieszkowska, Martin J. Genner, Stephen J. Hawkins and David W. Sims. Effects of Climate Change and Commercial Fishing on Atlantic Cod *Gadus morhua*. pp. 213–274.

Iain C. Field, Mark G. Meekan, Rik C. Buckworth and Corey J. A. Bradshaw. Susceptibility of Sharks, Rays and Chimaeras to Global Extinction. pp. 275–364.

Milagros Penela-Arenaz, Juan Bellas and Elsa Vazquez. Effects of the *Prestige* Oil Spill on the Biota of NW Spain: 5 Years of Learning. pp. 365–396.

Volume 57, 2010.

Geraint A. Tarling, Natalie S. Ensor, Torsten Fregin, William P. Good-all-Copestake and Peter Fretwell. An Introduction to the Biology of Northern Krill (*Meganyctiphanes norvegica* Sars). pp. 1–40.

Tomaso Patarnello, Chiara Papetti and Lorenzo Zane. Genetics of Northern Krill (*Meganyctiphanes norvegica* Sars). pp. 41–58.

Geraint A. Tarling. Population Dynamics of Northern Krill (*Meganyctiphanes norvegica* Sars). pp. 59–90.

John I. Spicer and Reinhard Saborowski. Physiology and Metabolism of Northern Krill (*Meganyctiphanes norvegica* Sars). pp. 91–126.

Katrin Schmidt. Food and Feeding in Northern Krill (*Meganyctiphanes norvegica* Sars). pp. 127–172.

Friedrich Buchholz and Cornelia Buchholz. Growth and Moulting in Northern Krill (*Meganyctiphanes norvegica* Sars). pp. 173–198.

Janine Cuzin-Roudy. Reproduction in Northern Krill. pp. 199–230.

Edward Gaten, Konrad Wiese and Magnus L. Johnson. Laboratory-Based Observations of Behaviour in Northern Krill (*Meganyctiphanes norvegica* Sars). pp. 231–254.

Stein Kaartvedt. Diel Vertical Migration Behaviour of the Northern Krill (*Meganyctiphanes norvegica* Sars). pp. 255–276.

Yvan Simard and Michel Harvey. Predation on Northern Krill (*Meganyctiphanes norvegica* Sars). pp. 277–306.

Volume 58, 2010.

A. G. Glover, A. J. Gooday, D. M. Bailey, D. S. M. Billett, P. Chevaldonné, A. Colaço, J. Copley, D. Cuvelier, D. Desbruyères, V. Kalogeropoulou, M. Klages, N. Lampadariou, C. Lejeusne, N. C. Mestre, G. L. J. Paterson, T. Perez, H. Ruhl, J. Sarrazin, T. Soltwedel, E. H. Soto, S. Thatje, A. Tselepides, S. Van Gaever, and A. Vanreusel. Temporal Change in Deep-Sea Benthic Ecosystems: A Review of the Evidence From Recent Time-Series Studies. pp. 1–96.

Hilario Murua. The Biology and Fisheries of European Hake, *Merluccius merluccius*, in the North-East Atlantic. pp. 97–154.

Jacopo Aguzzi and Joan B. Company. Chronobiology of Deep-Water Decapod Crustaceans on Continental Margins. pp. 155–226.

Martin A. Collins, Paul Brickle, Judith Brown, and Mark Belchier. The Patagonian Toothfish: Biology, Ecology and Fishery. pp. 227–300.

Volume 59, 2011.

Charles W. Walker, Rebecca J. Van Beneden, Annette F. Muttray, S. Anne Böttger, Melissa L. Kelley, Abraham E. Tucker, and W. Kelley Thomas. p53 Superfamily Proteins in Marine Bivalve Cancer and Stress Biology. pp 1–36.

Martin Wahl, Veijo Jormalainen, Britas Klemens Eriksson, James A. Coyer, Markus Molis, Hendrik Schubert, Megan Dethier, Anneli Ehlers, Rolf Karez, Inken Kruse, Mark Lenz, Gareth Pearson, Sven Rohde, Sofia A. Wikström, and Jeanine L. Olsen. Stress Ecology in *Fucus*: Abiotic, Biotic and Genetic Interactions. pp. 37–106.

Steven R. Dudgeon and Janet E. Kübler. Hydrozoans and the Shape of Things to Come. pp. 107–144.

Miles Lamare, David Burritt, and Kathryn Lister. Ultraviolet Radiation and Echinoderms: Past, Present and Future Perspectives. pp. 145–187.

Volume 60, 2011.

Tatiana A. Rynearson and Brian Palenik. Learning to Read the Oceans: Genomics of Marine Phytoplankton. pp. 1–40.

Les Watling, Scott C. France, Eric Pante and Anne Simpson. Biology of Deep-Water Octocorals. pp. 41–122.

Cristián J. Monaco and Brian Helmuth. Tipping Points, Thresholds and the Keystone Role of Physiology in Marine Climate Change Research. pp. 123–160.

David A. Ritz, Alistair J. Hobday, John C. Montgomery and Ashley J.W. Ward. Social Aggregation in the Pelagic Zone with Special Reference to Fish and Invertebrates. pp. 161–228.

Volume 61, 2012.

Gert Wörheide, Martin Dohrmann, Dirk Erpenbeck, Claire Larroux, Manuel Maldonado, Oliver Voigt, Carole Borchiellini and Denis Lavrov. Deep Phylogeny and Evolution of Sponges (Phylum Porifera). pp. 1–78.

Paco Cárdenas, Thierry Pérez and Nicole Boury-Esnault. Sponge Systematics Facing New Challenges. pp. 79–210.

Klaus Rützler. The Role of Sponges in the Mesoamerican Barrier-Reef Ecosystem, Belize. pp. 211–272.

Janie Wulff. Ecological Interactions and the Distribution, Abundance, and Diversity of Sponges. pp. 273–344.

Maria J. Uriz and Xavier Turon. Sponge Ecology in the Molecular Era. pp. 345–410.

Volume 62, 2012.

Sally P. Leys and April Hill. The Physiology and Molecular Biology of Sponge Tissues. pp. 1–56.

Robert W. Thacker and Christopher J. Freeman. Sponge–Microbe Symbioses: Recent Advances and New Directions. pp. 57–112.

Manuel Maldonado, Marta Ribes and Fleur C. van Duyl. Nutrient Fluxes Through Sponges: Biology, Budgets, and Ecological Implications. pp. 113–182.

Grégory Genta-Jouve and Olivier P. Thomas. Sponge Chemical Diversity: From Biosynthetic Pathways to Ecological Roles. pp. 183–230.

Xiaohong Wang, Heinz C. Schröder, Matthias Wiens, Ute Schloßmacher and Werner E. G. Müller. Biosilica: Molecular Biology, Biochemistry and Function in Demosponges as well as its Applied Aspects for Tissue Engineering. pp. 231–272.

Klaske J. Schippers, Detmer Sipkema, Ronald Osinga, Hauke Smidt, Shirley A. Pomponi, Dirk E. Martens and René H. Wijffels. Cultivation of Sponges, Sponge Cells and Symbionts: Achievements and Future Prospects. pp. 273–338.

Volume 63, 2012.
Michael Stat, Andrew C. Baker, David G. Bourne, Adrienne M. S. Correa, Zac Forsman, Megan J. Huggett, Xavier Pochon, Derek Skillings, Robert J. Toonen, Madeleine J. H. van Oppen, and Ruth D. Gates. Molecular Delineation of Species in the Coral Holobiont. pp. 1–66.
Daniel Wagner, Daniel G. Luck, and Robert J. Toonen. The Biology and Ecology of Black Corals (Cnidaria: Anthozoa: Hexacorallia: Antipatharia). pp. 67–132.
Cathy H. Lucas, William M. Graham, and Chad Widmer. Jellyfish Life Histories: Role of Polyps in Forming and Maintaining Scyphomedusa Populations. pp. 133–196.
T. Aran Mooney, Maya Yamato, and Brian K. Branstetter. Hearing in Cetaceans: From Natural History to Experimental Biology. pp. 197–246.

Volume 64, 2013.
Dale Tshudy. Systematics and Position of *Nephrops* Among the Lobsters. pp. 1–26.
Mark P. Johnson, Colm Lordan, and Anne Marie Power. Habitat and Ecology of *Nephrops norvegicus*. pp. 27–64.
Emi Katoh, Valerio Sbragaglia, Jacopo Aguzzi, and Thomas Breithaupt. Sensory Biology and Behaviour of *Nephrops norvegicus*. pp. 65–106.
Edward Gaten, Steve Moss, and Magnus L. Johnson. The Reniform Reflecting Superposition Compound Eyes of *Nephrops norvegicus*: Optics, Susceptibility to Light-Induced Damage, Electrophysiology and a Ray Tracing Model. pp. 107–148.
Susanne P. Eriksson, Bodil Hernroth, and Susanne P. Baden. Stress Biology and Immunology in *Nephrops norvegicus*. pp. 149–200.
Adam Powell and Susanne P. Eriksson. Reproduction: Life Cycle, Larvae and Larviculture. pp. 201–246.
Anette Ungfors, Ewen Bell, Magnus L. Johnson, Daniel Cowing, Nicola C. Dobson, Ralf Bublitz, and Jane Sandell. *Nephrops* Fisheries in European Waters. pp. 247–314.

Volume 65, 2013.
Isobel S.M. Bloor, Martin J. Attrill, and Emma L. Jackson. A Review of the Factors Influencing Spawning, Early Life Stage Survival and Recruitment Variability in the Common Cuttlefish (*Sepia officinalis*). pp. 1–66.
Dianna K. Padilla and Monique M. Savedo. A Systematic Review of Phenotypic Plasticity in Marine Invertebrate and Plant Systems. pp. 67–120.

Leif K. Rasmuson. The Biology, Ecology and Fishery of the Dungeness crab, *Cancer magister*. pp. 121–174.

Volume 66, 2013.

Lisa-ann Gershwin, Anthony J. Richardson, Kenneth D. Winkel, Peter J. Fenner, John Lippmann, Russell Hore, Griselda Avila-Soria, David Brewer, Rudy J. Kloser, Andy Steven, and Scott Condie. Biology and Ecology of Irukandji Jellyfish (Cnidaria: Cubozoa). pp. 1–86.

April M. H. Blakeslee, Amy E. Fowler, and Carolyn L. Keogh. Marine Invasions and Parasite Escape: Updates and New Perspectives. pp. 87–170.

Michael P. Russell. Echinoderm Responses to Variation in Salinity. pp. 171–212.

Daniela M. Ceccarelli, A. David McKinnon, Serge Andréfouët, Valerie Allain, Jock Young, Daniel C. Gledhill, Adrian Flynn, Nicholas J. Bax, Robin Beaman, Philippe Borsa, Richard Brinkman, Rodrigo H. Bustamante, Robert Campbell, Mike Cappo, Sophie Cravatte, Stéphanie D'Agata, Catherine M. Dichmont, Piers K. Dunstan, Cécile Dupouy, Graham Edgar, Richard Farman, Miles Furnas, Claire Garrigue, Trevor Hutton, Michel Kulbicki, Yves Letourneur, Dhugal Lindsay, Christophe Menkes, David Mouillot, Valeriano Parravicini, Claude Payri, Bernard Pelletier, Bertrand Richer de Forges, Ken Ridgway, Martine Rodier, Sarah Samadi, David Schoeman, Tim Skewes, Steven Swearer, Laurent Vigliola, Laurent Wantiez, Alan Williams, Ashley Williams, and Anthony J. Richardson. The Coral Sea: Physical Environment, Ecosystem Status and Biodiversity Assets. pp. 213–290.

Volume 67, 2014.

Erica A.G. Vidal, Roger Villanueva, José P. Andrade, Ian G. Gleadall, José Iglesias, Noussithé Koueta, Carlos Rosas, Susumu Segawa, Bret Grasse, Rita M. Franco-Santos, Caroline B. Albertin, Claudia Caamal-Monsreal, Maria E. Chimal, Eric Edsinger-Gonzales, Pedro Gallardo, Charles Le Pabic, Cristina Pascual, Katina Roumbedakis, and James Wood. Cephalopod Culture: Current Status of Main Biological Models and Research Priorities. pp. 1–98.

Paul G.K. Rodhouse, Graham J. Pierce, Owen C. Nichols, Warwick H.H. Sauer, Alexander I. Arkhipkin, Vladimir V. Laptikhovsky, Marek R. Lipiński, Jorge E. Ramos, Michaël Gras, Hideaki Kidokoro, Kazuhiro Sadayasu, João Pereira, Evgenia Lefkaditou, Cristina Pita, Maria Gasalla, Manuel Haimovici, Mitsuo Sakai, and Nicola Downey. Environmental

Effects on Cephalopod Population Dynamics: Implications for Management of Fisheries. pp. 99–234.
Henk-Jan T. Hoving, José A.A. Perez, Kathrin Bolstad, Heather Braid, Aaron B. Evans, Dirk Fuchs, Heather Judkins, Jesse T. Kelly, José E.A.R. Marian, Ryuta Nakajima, Uwe Piatkowski, Amanda Reid, Michael Vecchione, and José C.C. Xavier. The Study of Deep-Sea Cephalopods. pp. 235–362.
Jean-Paul Robin, Michael Roberts, Lou Zeidberg, Isobel Bloor, Almendra Rodriguez, Felipe Briceño, Nicola Downey, Maite Mascaró, Mike Navarro, Angel Guerra, Jennifer Hofmeister, Diogo D. Barcellos, Silvia A.P. Lourenço, Clyde F.E. Roper, Natalie A. Moltschaniwskyj, Corey P. Green, and Jennifer Mather. Transitions During Cephalopod Life History: The Role of Habitat, Environment, Functional Morphology and Behaviour. pp. 363–440.

Volume 68, 2014.
Paul K.S. Shin, Siu Gin Cheung, Tsui Yun Tsang, and Ho Yin Wai. Ecology of Artificial Reefs in the Subtropics. pp. 1–64.
Hrafnkell Eiríksson. Reproductive Biology of Female Norway Lobster, *Nephrops norvegicus* (Linnaeus, 1758) Leach, in Icelandic Waters During the Period 1960–2010: Comparative Overview of Distribution Areas in the Northeast Atlantic and the Mediterranean. pp. 65–210.

Volume 69, 2014.
Ray Hilborn. Introduction to Marine Managed Areas. pp. 1–14.
Philip N. Trathan, Martin A. Collins, Susie M. Grant, Mark Belchier, David K.A. Barnes, Judith Brown, and Iain J. Staniland. The South Georgia and the South Sandwich Islands MPA: Protecting A Biodiverse Oceanic Island Chain Situated in the Flow of the Antarctic Circumpolar Current. pp. 15–78.
Richard P. Dunne, Nicholas V.C. Polunin, Peter H. Sand, and Magnus L. Johnson. The Creation of the Chagos Marine Protected Area: A Fisheries Perspective. pp. 79–128.
Michelle T. Schärer-Umpierre, Daniel Mateos-Molina, Richard Appeldoorn, Ivonne Bejarano, Edwin A. Hernández-Delgado, Richard S. Nemeth, Michael I. Nemeth, Manuel Valdés-Pizzini, and Tyler B. Smith. Marine Managed Areas and Associated Fisheries in the US Caribbean. pp. 129–152.

Alan M. Friedlander, Kostantinos A. Stamoulis, John N. Kittinger, Jeffrey C. Drazen, and Brian N. Tissot. Understanding the Scale of Marine Protection in Hawai'i: From Community-Based Management to the Remote Northwestern Hawaiian Islands. pp. 153–204.

Louis W. Botsford, J. Wilson White, Mark H. Carr, and Jennifer E. Caselle. Marine Protected Area Networks in California, USA. pp. 205–252.

Bob Kearney and Graham Farebrother. Inadequate Evaluation and Management of Threats in Australia's Marine Parks, Including the Great Barrier Reef, Misdirect Marine Conservation. pp. 253–288.

Randi Rotjan, Regen Jamieson, Ben Carr, Les Kaufman, Sangeeta Mangubhai, David Obura, Ray Pierce, Betarim Rimon, Bud Ris, Stuart Sandin, Peter Shelley, U. Rashid Sumaila, Sue Taei, Heather Tausig, Tukabu Teroroko, Simon Thorrold, Brooke Wikgren, Teuea Toatu, and Greg Stone. Establishment, Management, and Maintenance of the Phoenix Islands Protected Area. pp. 289–324.

Alex J. Caveen, Clare Fitzsimmons, Margherita Pieraccini, Euan Dunn, Christopher J. Sweeting, Magnus L. Johnson, Helen Bloomfield, Estelle V. Jones, Paula Lightfoot, Tim S. Gray, Selina M. Stead, and Nicholas V. C. Polunin. Diverging Strategies to Planning an Ecologically Coherent Network of MPAs in the North Sea: The Roles of Advocacy, Evidence and Pragmatism in the Face of Uncertaintya. pp. 325–370.

Carlo Pipitone, Fabio Badalamenti, Tomás Vega Fernández, and Giovanni D'Anna. Spatial Management of Fisheries in the Mediterranean Sea: Problematic Issues and a Few Success Stories. pp. 371–402.

Volume 70, 2015.

Alex D. Rogers, Christopher Yesson, and Pippa Gravestock. A Biophysical and Economic Profile of South Georgia and the South Sandwich Islands as Potential Large-Scale Antarctic Protected Areas. pp. 1–286.

Volume 71, 2015.

Ricardo Calado and Miguel Costa Leal. Trophic Ecology of Benthic Marine Invertebrates with Bi-Phasic Life Cycles: What Are We Still Missing? pp. 1–70.

Jesse M.A. van der Grient and Alex D. Rogers. Body Size Versus Depth: Regional and Taxonomical Variation in Deep-Sea Meio- and Macrofaunal Organisms. pp. 71–108.

Lorena Basso, Maite Vázquez-Luis, José R. García-March, Salud Deudero, Elvira Alvarez, Nardo Vicente, Carlos M. Duarte, and Iris E. Hendriks. The Pen Shell, *Pinna nobilis*: A Review of Population Status and

Recommended Research Priorities in the Mediterranean Sea. pp. 109–160.

Volume 72, 2015.
Thomas A. Jefferson and Barbara E. Curry. Humpback Dolphins: A Brief Introduction to the Genus *Sousa*. pp. 1–16.
Sarah Piwetz, David Lundquist, and Bernd Würsig. Humpback Dolphin (Genus *Sousa*) Behavioural Responses to Human Activities. pp. 17–46.
Tim Collins. Re-assessment of the Conservation Status of the Atlantic Humpback Dolphin, *Sousa teuszii* (Kükenthal, 1892) Using the IUCN Red List Criteria. pp. 47–78.
Caroline R. Weir and Tim Collins. A Review of the Geographical Distribution and Habitat of the Atlantic Humpback Dolphin (*Sousa teuszii*). pp. 79–118.
Gill T Braulik, Ken Findlay, Salvatore Cerchio, and Robert Baldwin. Assessment of the Conservation Status of the Indian Ocean Humpback Dolphin (*Sousa plumbea*) Using the IUCN Red List Criteria. pp. 119–142.
Stephanie Plön, Victor G. Cockcroft, and William P. Froneman. The Natural History and Conservation of Indian Ocean Humpback Dolphins (*Sousa plumbea*) in South African Waters. pp. 143–162.
Salvatore Cerchio, Norbert Andrianarivelo, and Boris Andrianantenaina. Ecology and Conservation Status of Indian Ocean Humpback Dolphins (*Sousa plumbea*) in Madagascar. pp. 163–200.
Muhammad Shoaib Kiani and Koen Van Waerebeek. A Review of the Status of the Indian Ocean Humpback Dolphin *Sousa plumbea* in Pakistan. pp. 201–228.
Dipani Sutaria, Divya Panicker, Ketki Jog, Mihir Sule, Rahul Muralidharan, and Isha Bopardikar. Humpback Dolphins (Genus *Sousa*) in India: An Overview of Status and Conservation Issues. pp. 229–256.

Volume 73, 2016.
Thomas A. Jefferson and Brian D. Smith. Re-assessment of the Conservation Status of the Indo-Pacific Humpback Dolphin (*Sousa chinensis*) Using the IUCN Red List Criteria. pp. 1–26.
Leszek Karczmarski, Shiang-Lin Huang, Carmen K. M. Or, Duan Gui, Stephen C. Y. Chan, Wenzhi Lin, Lindsay Porter, Wai-Ho Wong, Ruiqiang Zheng, Yuen-Wa Ho, Scott Y. S. Chui, Angelico Jose C. Tiongson, Yaqian Mo, Wei-Lun Chang, John H. W. Kwok, Ricky W. K. Tang, Andy T. L. Lee, Sze-Wing Yiu, Mark Keith, Glenn Gailey,

and Yuping Wu. Humpback Dolphins in Hong Kong and the Pearl River Delta: Status, Threats and Conservation Challenges. pp. 27–64.
Bernd Würsig, E.C.M. Parsons, Sarah Piwetz, and Lindsay Porter. The Behavioural Ecology of Indo-Pacific Humpback Dolphins in Hong Kong. pp. 65–90.
John Y. Wang, Kimberly N. Riehl, Michelle N. Klein, Shiva Javdan, Jordan M. Hoffman, Sarah Z. Dungan, Lauren E. Dares, and Claryana Araújo-Wang. Biology and Conservation of the Taiwanese Humpback Dolphin, *Sousa chinensis taiwanensis*. pp. 91–118.
Bingyao Chen, Xinrong Xu, Thomas A. Jefferson, Paula A. Olson, Qiurong Qin, Hongke Zhang, Liwen He, and Guang Yang. Conservation Status of the Indo-Pacific Humpback Dolphin (*Sousa chinensis*) in the Northern Beibu Gulf, China. pp. 119–140.
Gianna Minton, Anna Norliza Zulkifli Poh, Cindy Peter, Lindsay Porter, and Danielle Kreb. Indo-Pacific Humpback Dolphins in Borneo: A Review of Current Knowledge with Emphasis on Sarawak. pp. 141–156.
Guido J. Parra and Daniele Cagnazzi. Conservation Status of the Australian Humpback Dolphin (*Sousa sahulensis*) Using the IUCN Red List Criteria. pp. 157–192.
Daniella M. Hanf, Tim Hunt, and Guido J. Parra. Humpback Dolphins of Western Australia: A Review of Current Knowledge and Recommendations for Future Management. pp. 193–218.
Isabel Beasley, Maria Jedensjö, Gede Mahendra Wijaya, Jim Anamiato, Benjamin Kahn, and Danielle Kreb. Observations on Australian Humpback Dolphins (*Sousa sahulensis*) in Waters of the Pacific Islands and New Guinea. pp. 219–272.
Alexander M. Brown, Lars Bejder, Guido J. Parra, Daniele Cagnazzi, Tim Hunt, Jennifer L. Smith, and Simon J. Allen. Sexual Dimorphism and Geographic Variation in Dorsal Fin Features of Australian Humpback Dolphins, *Sousa sahulensis*. pp. 273–314.

Volume 74, 2016.
J. Salinger, A.J. Hobday, R.J. Matear, T.J. O'Kane, J.S. Risbey, P. Dunstan, J.P. Eveson, E.A. Fulton, M. Feng, É.E. Plagányi, E.S. Poloczanska, A.G. Marshall, and P.A. Thompson. Decadal-Scale Forecasting of Climate Drivers for Marine Applications. pp. 1–68.
S.A. Foo and M. Byrne. Acclimatization and Adaptive Capacity of Marine Species in a Changing Ocean. pp. 69–116.

PREFACE

We thought it timely to produce a volume of *Advances in Marine Biology* that details the ecology and conservation of marine mammals in the Mediterranean Sea. However, rather than compiling a systematic account of the species in the region, we sought to provide a compendium of the novel knowledge that has been recently developed by the many scholars who have contributed to this volume.

As a hotspot of marine and coastal biodiversity, the Mediterranean region is exceptionally valuable and hosts a large proportion of endemic taxa. Although none of the marine mammal species that regularly inhabit the region are endemic, ecological mechanisms separating them from their conspecifics in the world's oceans have been at work for sufficient time to generate unique populations that must be protected and preserved.

The Mediterranean Sea is unique and valuable, but it is also a vulnerable region. It is heavily affected by the human footprint, more so than most other seas in the world. Portions of civil society have been reacting to help: scientific attention has grown exponentially in the past years, as testified by this volume, and non-governmental organizations actively advocate for environmental protection, particularly along the region's northern shores. However, very few of the governmental decision makers—whose efforts are essential to make the difference—seem to have grasped how serious the predicament is for these species and their fragile habitat. All of the necessary agreements and conventions are in place, ostensibly to protect the Mediterranean Sea and its marine mammal species, and earnest and expeditious implementation of their many resolutions could make the difference. Unfortunately, most governments do not seem to want to give to such essential conservation tools more than just lip service, and the resolved imperatives remain largely confined to the realm of unfulfilled intentions.

One day perhaps the humans who call the Mediterranean region their home will realize how fortunate they are to partake of their sea with all its natural wealth, and at the same time, how vulnerable such wealth is and how easily it is destroyed. When this realization occurs, perhaps things will start improving. From this perspective, communicating scientific results is paramount. Having this consideration in mind, we hope that this volume will provide a modest but qualified contribution to the long-term conservation of the Mediterranean Sea and its marine mammal populations.

Finally, every volume of *Advances in Marine Biology* relies on the excellence of the scientific reviewers who provide their time and expertise to evaluating contributions. This volume is no exception, and we wish to extend our most sincere appreciation to the many reviewers who participated in developing this volume. Thanks for these efforts are given to the many anonymous reviewers, and to those we are at liberty to name, including T. Agardy, P. J. Clapham, C. W. Clark, T. Genov, S. Gero, P. Mackelworth, A. Scheinin and A. Uras.

Giuseppe Notarbartolo di Sciara
Michela Podestà
Barbara E. Curry

CHAPTER ONE

Marine Mammals in the Mediterranean Sea: An Overview

G. Notarbartolo di Sciara[1]
Tethys Research Institute, Acquario Civico, Milano, Italy
[1]Corresponding author: e-mail address: disciara@gmail.com

Contents

Abstract

Despite being a small part of the world's oceans, the Mediterranean Sea hosts a diverse marine mammal fauna, with a total of 28 different species known to occur, or to have occurred, in the region. Species currently recognised as regular in the Mediterranean—the Mediterranean monk seal *(Monachus monachus)* and 11 cetaceans (fin whale, *Balaenoptera physalus*; sperm whale, *Physeter macrocephalus*; Cuvier's beaked whale, *Ziphius cavirostris*; short-beaked common dolphin, *Delphinus delphis*; long-finned pilot whale, *Globicephala melas*; Risso's dolphin, *Grampus griseus*; killer whale, *Orcinus orca*; striped dolphin, *Stenella coeruleoalba*; rough-toothed dolphin, *Steno bredanensis*; common bottlenose dolphin, *Tursiops truncatus*; harbour porpoise, *Phocoena phocoena relicta*) have adapted well to the region's environmental conditions, but their coexistence with humans is problematic. All the regular species are represented in the Mediterranean by

ISSN 0065-2881
http://dx.doi.org/10.1016/bs.amb.2016.08.005

populations genetically distinct from their North Atlantic relatives. Seventeen other species (three pinnipeds and 14 cetaceans) occur or have occurred in the Mediterranean as vagrants from adjacent regions. Impacts on the conservation status of marine mammals in the region deriving from a variety of threats include: (a) mortality caused by deliberate killing (to a large extent resulting from fisheries interactions), naval sonar, ship strikes, epizootics, fisheries bycatch, chemical pollution and ingestion of solid debris; (b) short-term redistribution caused by naval sonar, seismic surveys, vessel disturbance and vessel noise; and (c) long-term redistribution caused by fishery-induced food depletion, coastal development and possibly climate change. Accordingly, seven of the 12 marine mammals regular in the Mediterranean region are listed as Threatened on IUCN's Red List; regrettably, three are Data Deficient and two remain unassessed.

1. THE MEDITERRANEAN SEA AS A NATURAL ENVIRONMENT FOR MARINE MAMMALS

With a total surface of approximately 2.5 million km^2, the Mediterranean Sea occupies only 0.8% of the global ocean's surface. However, in spite of its small size, the region hosts a diverse marine mammal fauna, with a total of 28 different species (one of which, the harbour porpoise, *Phocoena phocoena*, is represented by two subspecies; *Phocoena p. relicta* and *P. p. phocoena*) known to occur, or to have occurred in modern times, in the Mediterranean Sea (Table 1).

The sea is contained in a deep (exceeding 5000 m) and irregular depression about 3700 km long, carved amongst the continental masses of Europe, Africa and Asia, connected to the Atlantic Ocean through the Gibraltar Strait, 14 km wide and 320 m deep, and communicating with the Black Sea through the narrow and shallow straits of the Bosphorus and Dardanelles. The Mediterranean's recent connection with the Red Sea through the Suez Canal (artificially dug in the 1860s) is oceanographically irrelevant but biologically significant, in view of the massive "Lessepsian" immigration of alien marine organisms from the Indo-Pacific realm that the Canal has permitted (Por, 1978).

The Mediterranean Sea is intersected by a series of transversal ridges arranged along a north–south direction, the most notable being a central ridge constituted by the Italian peninsula and Sicily which divides the region into a western and an eastern Basin. To facilitate a description of its diverse geomorphological and biogeographical features—inclusive of its marine mammal fauna—the Mediterranean region can be seen as subdivided into eight distinct subregions (Fig. 1): Alborán Sea/Strait of Gibraltar, Algero-Provençal Basin, Tyrrhenian Sea/eastern Ligurian Sea, Strait of

Table 1 Marine Mammal Species Occurring, or Having Occurred, in the Mediterranean Sea

	Species	English Name	Classification	Presence	Habitat	Current Status (IUCN)	Recommendations
1	*Cystophora cristata*	Hooded seal	Carnivora, Phocidae	Very rare			
2	*Monachus monachus*	Mediterranean monk seal	Carnivora, Phocidae	**Regular**	**Neritic**	Endangered	**Assess subpopulations separately**
3	*Pagophilus groenlandicus*	Harp seal	Carnivora, Phocidae	Very rare			
4	*Phoca vitulina*	Harbour seal	Carnivora, Phocidae	Very rare			
5	*Eubalaena glacialis*	North Atlantic right whale	Cetartiodactyla, Balaenidae	Very rare			
6	*Balaenoptera acutorostrata*	Common minke whale	Cetartiodactyla, Balaenopteridae	Visitor			
7	*Balaenoptera borealis*	Sei whale	Cetartiodactyla, Balaenopteridae	Very rare			
8	*Balaenoptera physalus*	Fin whale	Cetartiodactyla, Balaenopteridae	**Regular**	**Oceanic**, slope, neritic	Vulnerable	**Reassess**
9	*Megaptera novaeangliae*	Humpback whale	Cetartiodactyla, Balaenopteridae	Visitor			

Continued

Table 1 Marine Mammal Species Occurring, or Having Occurred, in the Mediterranean Sea—cont'd

	Species	English Name	Classification	Presence	Habitat	Current Status (IUCN)	Recommendations
10	*Eschrichtius robustus*	Grey whale	Cetartiodactyla, Eschrichtiidae	Very rare			
11	*Physeter macrocephalus*	Sperm whale	Cetartiodactyla, Physeteridae	**Regular**	**Slope**, oceanic	Endangered	No change
12	*Kogia sima*	Dwarf sperm whale	Cetartiodactyla, Kogiidae	Very rare			
13	*Hyperoodon ampullatus*	Northern bottlenose whale	Cetartiodactyla, Ziphiidae	Very rare			
14	*Mesoplodon bidens*	Sowerby's beaked whale	Cetartiodactyla, Ziphiidae	Very rare			
15	*Mesoplodon densirostris*	Blainville's beaked whale	Cetartiodactyla, Ziphiidae	Very rare			
16	*Mesoplodon europaeus*	Gervais' beaked whale	Cetartiodactyla, Ziphiidae	Very rare			
17	*Ziphius cavirostris*	Cuvier's beaked whale	Cetartiodactyla, Ziphiidae	**Regular**	**Slope**, oceanic	Data Deficient	**Reassess**
18	*Delphinus delphis*	Short-beaked common dolphin	Cetartiodactyla, Delphinidae	**Regular**	**Neritic**, **slope**, **oceanic**	Endangered	No change

Table 1 Marine Mammal Species Occurring, or Having Occurred, in the Mediterranean Sea—cont'd

	Species	English Name	Classification	Presence	Habitat	Current Status (IUCN)	Recommendations
19	*Globicephala macrorhynchus*	Short-finned pilot whale	Cetartiodactyla, Delphinidae	Very rare			
20	*Globicephala melas*	Long-finned pilot whale	Cetartiodactyla, Delphinidae	**Regular**	**Oceanic**, slope, neritic	Data Deficient	**Assess Mediterranean subpopulation as "Vulnerable"; assess Strait of Gibraltar subpopulation as "Critically Endangered"**
21	*Grampus griseus*	Risso's dolphin	Cetartiodactyla, Delphinidae	**Regular**	**Slope**, oceanic	Data Deficient	**Reassess**
22	*Orcinus orca*	Killer whale	Cetartiodactyla, Delphinidae	**Regular** in the Strait of Gibraltar, visitor elsewhere	**Neritic, slope, oceanic**	Not assessed	**Assess Strait of Gibraltar subpopulation as "Endangered"**
23	*Pseudorca crassidens*	False killer whale	Cetartiodactyla, Delphinidae	Visitor			
24	*Sousa plumbea*	Indo-Pacific humpback dolphin	Cetartiodactyla, Delphinidae	Very rare			
25	*Stenella coeruleoalba*	Striped dolphin	Cetartiodactyla, Delphinidae	**Regular**	**Oceanic**, slope	Vulnerable	No change

Continued

Table 1 Marine Mammal Species Occurring, or Having Occurred, in the Mediterranean Sea—cont'd

	Species	English Name	Classification	Presence	Habitat	Current Status (IUCN)	Recommendations
26	*Steno bredanensis*	Rough-toothed dolphin	Cetartiodactyla, Delphinidae	**Regular** in the Levantine Sea, visitor elsewhere	**Oceanic, slope,** neritic	Not assessed	**Assess**
27	*Tursiops truncatus*	Common bottlenose dolphin	Cetartiodactyla, Delphinidae	**Regular**	**Neritic**	Vulnerable	**No change (Mediterranean subpopulation); assess Gulf of Ambracia subpopulation as "Endangered"**
28	*Phocoena p. phocoena*	North Atlantic harbour porpoise	Cetartiodactyla, Phocoenidae	Very rare			
29	*Phocoena p. relicta*	Black Sea harbour porpoise	Cetartiodactyla, Phocoenidae	**Regular** (Aegean Sea)	**Neritic**	Endangered	No change

Habitat (preferred in *bold*) and status are indicated only for species recognised as regular.

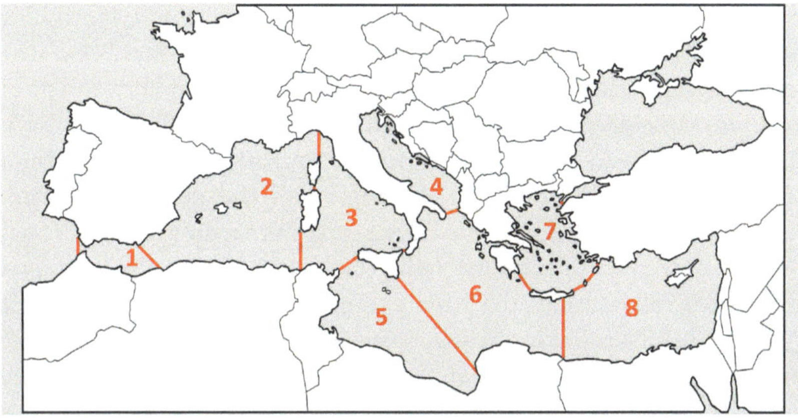

Fig. 1 Mediterranean subregions: (1) Alborán Sea/Strait of Gibraltar, (2) Algero-Provençal Basin, (3) Tyrrhenian Sea/eastern Ligurian Sea, (4) Adriatic Sea, (5) Strait of Sicily/Tunisian Plateau/Gulf of Sirte, (6) Ionian Sea/Central Mediterranean, (7) Aegean Sea, (8) Levantine Sea. *Adapted from UNEP-MAP-RAC/SPA, 2010. Overview of scientific findings and criteria relevant to identifying SPAMIs in the Mediterranean open seas, including the deep sea. By Notarbartolo di Sciara, G. and Agardy, T. Edited by RAC/SPA, Tunis (can be accessed from http://bit.ly/1W8jste).*

Sicily/Tunisian Plateau/Gulf of Sirte, Adriatic Sea, Ionian Sea/Central Mediterranean, Aegean Sea and Levantine Sea (UNEP-MAP-RAC/SPA, 2010).

The Mediterranean Sea has had a recent history of profound geological modifications which have significantly affected its biome's ecological conditions. The region's current shape, size and characteristics are what remains of the ancient Tethys Sea after it was compressed between the Eurasian and the African Plates. The two plates' converging movements severed what is today the Mediterranean Sea, first from the Indo-Pacific region around 10 million years ago (Ma), and later also from the Atlantic region, around 6 Ma, triggering what is known as Miocene's "Messinian Salinity Crisis". Isolation from the global oceans caused the marine basin to undergo almost total desiccation by virtue of the negative water balance caused by evaporation, thereby driving most of the ancient Tethys' biodiversity to extinction. This included extinct odontocetes belonging to the superfamilies Platanistoidea and Eurhinodelphinoidea, as well as extinct mysticetes from the family Cetotheriidae (Bianucci, 2015). The salinity crisis was terminated 5.3 Ma by a further movement of the Eurasian and African Plates, which reopened the connection with the Atlantic Ocean, an event marking the beginning of the Pliocene Epoch. The ensuing inflow into the Mediterranean Sea of Atlantic water (AW), known as the "Zanclean Deluge", allowed its colonisation by organisms of Atlantic origin (Bianchi and Morri, 2000), which included the true ancestors of modern cetaceans such as Pliocene balaenids, balaenopterids, delphinids, physeterids, kogiids, ziphiids and possibly phocoenids (Bianucci, 2015). A small sample of these species have survived in the Mediterranean Sea in modern times. Similarly, a lineage of pinniped Carnivora gained a foothold in the region in the same circumstances. Mediterranean monk seals, derived from early monachines such as *Pliophoca etrusca* Tavani 1942, found in Italy (Hendey, 1972), also presumably colonised (or recolonised) the Mediterranean and Black Sea regions after the end of the Messinian Salinity Crisis, from the Atlantic, where their monachine ancestors first appeared in the fossil record dating 15–17 Ma (Fyler et al., 2005).

The complex patterns of water circulation in the Mediterranean are amongst the underlying causes of the environmental conditions ultimately affecting the distribution, abundance and status of the region's marine mammals. The semi-enclosed Mediterranean Basin is characterised by a predominance of evaporation over precipitation and river runoff, causing a constant influx of Atlantic water (AW) to replace the evaporating water masses. As it progresses across the Mediterranean moving eastwards, surface AW layer is made saltier and denser by evaporation and sinks, becoming "Intermediate Levantine water", which returns westwards underneath the AW until it exits into the Atlantic Ocean through the Strait of Gibraltar. A third, deep and even denser layer, formed during winter in the northernmost portions of the Basin, occupies the remaining deeper volume of water all the way to the bottom (Millot and Taupier-Letage, 2005). The simplified trilayered model of Mediterranean circulation with predominant sinking of water masses and very reduced upwellings underpins a generalised oligotrophy of the region's pelagic waters, particularly in its eastern portion. However, localised phenomena provide notable exceptions to the rule, such as: (a) intense, large and long-lived mesoscale eddies created along the southern coasts as the AW progresses to the east; (b) the induction of mixing during winter in northern portions of the region over the entire water column bringing nutrients to the euphotic zone, caused by the sinking of cold, densified surface water (Millot and Taupier-Letage, 2004); and (c) localised upwellings generated by the dynamic effects of systems of canyons deeply indenting the continental slopes in many locations, thereby creating connections between the deep basins and the shelf (Boero, 2015).

Localised hotspots of primary productivity, thus generated in the pelagic domain, often season dependent, act as attractors for the region's more pelagic marine mammals. A mosaic of productivity locations, often of an ephemeral nature, but predictable, combined with the typical Mediterranean climatologic conditions inducing a substantive (>15°C) seasonal excursion in surface water temperatures—unsuitable during summer for Arctic and Sub-Arctic stenotherms and in winter for the tropical species—are the environmental factors that are likely responsible for the current assortment of marine mammal species that have established their home in the Mediterranean Sea.

2. SPECIES OF MARINE MAMMALS IN THE MEDITERRANEAN SEA

All the 12 marine mammal species currently recognised as regular in the Mediterranean—one seal (Order Carnivora, Superfamily Pinnipedia) and 11 cetaceans (Order Cetartiodactyla)—are represented by populations that are genetically distinct from their North Atlantic relatives. Other species have occurred or occasionally occur in the region, but are not considered to be represented by resident populations. Of these, three have been sighted on an occasional, infrequent basis, and 14 (three pinnipeds and 11 cetaceans) have been recorded only very rarely (Table 1). All the occasional and rare species are thought to enter or have entered the Mediterranean Sea from the Atlantic Ocean, with one exception (the Indian Ocean humpback dolphin, *Sousa plumbea*, a Lessepsian immigrant from the Red Sea).

2.1 Regularly Occurring Species

2.1.1 Pinnipeds

2.1.1.1 Mediterranean Monk Seal, *Monachus monachus* (Hermann, 1779)

Formerly present throughout the Mediterranean Sea, Black Sea, Marmara Sea and Northeastern Atlantic (African coast from Morocco to Mauritania and Macaronesian Archipelagos). The identification of three or four genetically distinct, isolated populations, two in the North Atlantic and possibly two in the Mediterranean, one in the Ionian Sea and the other in Aegean Sea (Karamanlidis et al., 2015), highlights the need for considering the species as consisting of different conservation units. Regular reproduction within the Mediterranean Sea is thought to persist today only in a few breeding nuclei in Greece and parts of Turkey and Cyprus (Notarbartolo di Sciara and Kotomatas, 2016). Main threats include deliberate killing, habitat degradation and loss, disturbance and to a lesser extent drowning in fishing gear. In view of slight numeric increases in specific breeding locations in recent years, the Mediterranean monk seal status at the species level (i.e. by lumping together all the distinct populations) was recently changed from Critically Endangered to Endangered in the International Union for the Conservation of Nature's (IUCN) Red List (Karamanlidis and Dendrinos, 2015).

2.1.2 Cetaceans

2.1.2.1 Fin Whale, *Balaenoptera physalus* (Linnaeus, 1758)

Present throughout the Mediterranean Sea, but observed predominantly in the western Basin. Of the two populations occurring in the region, one is resident, observed mostly from the waters north and east of the Balearic Islands to the Ionian and southern Adriatic seas; whales from the other population enter the Mediterranean seasonally across the Strait of Gibraltar from the northeastern North Atlantic Ocean. The resident population, presumed to number at most in the low thousands (and possibly in decline), is subject to several threats including ship strikes, disturbance, noise and chemical contaminants (Notarbartolo di Sciara et al., 2016a). On such basis it was assessed as Vulnerable on the IUCN Red List (Panigada and Notarbartolo di Sciara, 2012).

2.1.2.2 Sperm Whale, *Physeter macrocephalus* Linnaeus, 1758

This species is distributed over slope and deep waters throughout the Mediterranean Sea in what is believed to be a single panmictic population genetically isolated from the Atlantic, and is likely to be numbering in the low- to mid-hundreds (Rendell and Frantzis, 2016). Evidence exists of decline during the past decades, supporting the population's IUCN Red List status of Endangered (Notarbartolo di Sciara et al., 2012). Predominant threats include mortality derived from ship strikes and entanglement in driftnets, particularly before the turn of the 21st century (but still occurring occasionally). Ingestion of plastic debris, anthropogenic noise and chemical contaminants may also have significant effects on Mediterranean sperm whale conservation status (Notarbartolo di Sciara, 2014).

2.1.2.3 Cuvier's Beaked Whale, *Ziphius cavirostris* G. Cuvier, 1823

A single population in the low thousands is thought to exist in the region, genetically isolated from the Atlantic and distributed throughout the Mediterranean Sea, albeit concentrated in preferred locations along the deep continental slope, marked by specific ecological characteristics such as the presence of underwater canyons (Podestà et al., 2016). The small presumed population size, coupled with the species' known vulnerability to human pressures, most notably anthropogenic noise widely produced throughout Mediterranean waters by military and industrial activities, but also fishery bycatch and ingestion of plastics, supports the expeditious inclusion of this population in an appropriate threat category of IUCN's Red List, where it is currently listed as Data Deficient.

2.1.2.4 Short-Beaked Common Dolphin, *Delphinus delphis* Linnaeus, 1758

Once common in the Mediterranean Sea, the species experienced a major decline in the region since the mid-20th century (Bearzi et al., 2003), and was consequently assessed as Endangered on IUCN's Red List (Bearzi, 2003). Genetic studies detected significant mtDNA differentiation from Atlantic populations, as well as population structure within the Mediterranean (Natoli et al., 2008). Found in both offshore and neritic waters throughout the region, the species survives in specific locations within the Alborán Sea, the Sardinian Sea, the Sicily Strait, the eastern Ionian Sea, the Aegean Sea and the Levantine Sea off Palestine. The causes of the species' decline are unclear, but may have included culling until the 1970s and, more recently, prey depletion from overfishing and incidental mortality in fishing gear (Bearzi et al., 2016).

2.1.2.5 Long-Finned Pilot Whale, *Globicephala melas* (Traill, 1809)

Found in the western Basin only, mostly in offshore waters; its absence from waters east of Italy remains an ecological conundrum. Recent genetic and photoidentification studies have identified two populations living in the Mediterranean, both genetically differentiated from Atlantic long-finned pilot whales: (a) a population restricted to the western Basin, from the Alborán Sea to the Ligurian and Tyrrhenian seas, which has undergone significant reduction in the past decades; and (b) a very small population (<250 individuals) restricted to the Strait of Gibraltar, reduced by >26% during the past 5 years by a morbillivirus epizootic (Verborgh et al., 2016). Ongoing threats to the species are unclear, although bycatch in driftnets, ship strikes, disturbance from military sonar and contaminants all likely have significant effects. On such bases it is recommended that these populations be assessed, respectively, as Vulnerable and Critically Endangered on IUCN's Red List (Verborgh et al., 2016).

2.1.2.6 Risso's Dolphin, *Grampus griseus* (G. Cuvier, 1812)

Widespread across the Mediterranean Sea mostly over slope waters, albeit nowhere abundant, and genetically distinct from Atlantic conspecifics (Gaspari et al., 2007). Known to occur in many locations including the Alborán, Ligurian, Tyrrhenian, Adriatic, Ionian, Aegean and Levantine seas and the Strait of Sicily. Long-term monitoring in the western Ligurian Sea detected significant decreasing trends during the past decade, possibly connected to changes in the local primary production, which might be caused by environmental variability cumulated with fishery impacts (Azzellino et al., 2016). Currently assessed as Data Deficient on the IUCN Red List, although a reassessment to consider a possible decreasing trend is highly recommended.

2.1.2.7 Killer Whale, *Orcinus orca* (Linnaeus, 1758)

Found on a regular (seasonal) basis only in the Strait of Gibraltar and adjacent Atlantic waters; visitor to the rest of the Mediterranean Sea, with about 30 occurrences recorded since the late 19th century (Notarbartolo di Sciara and Birkun, 2010). A total of 39 individuals are known to currently occur in the Gibraltar Strait population (Esteban et al., 2016). The population, organised into five separate pods, congregates in the area in early summer in coincidence with the migration into the Mediterranean of its main prey, the Atlantic bluefin tuna (*Thunnus thynnus*). Killer whales in the Strait have also begun to interact with the Spanish longline (modified as a "dropline") fishery to predate tuna. Genetic evidence suggests that Gibraltar Strait killer whales may form a separate population from other North Atlantic conspecifics, in which case they should be assessed separately for the IUCN Red List (currently the IUCN regional assessment for killer whales includes all of Europe).

2.1.2.8 Striped Dolphin, *Stenella coeruleoalba* (Meyen, 1833)

The most common offshore cetacean in the Mediterranean Sea, found wherever deep waters occur, from Gibraltar to the Levantine Sea. Particularly abundant in the Alborán Sea, in the waters between the Balearic Islands and the Iberian mainland, the Gulf of Lions and the Ligurian Sea, but also frequent in the Tyrrhenian and Ionian seas and open waters of the southern Adriatic Sea. A small number of striped dolphins live in isolation in the eastern portion of the Gulf of Corinth, Greece (Bearzi et al., 2016). Known to be distinct from Atlantic conspecifics, based on morphological and genetic characters, with little or no gene flow across the Strait of Gibraltar (Aguilar and Gaspari, 2012). Subject to a wide range of threats, including die-offs caused by morbillivirus epizootics, high levels of contaminants and former massive bycatch in pelagic driftnets; on such basis, the population is listed as Vulnerable on IUCN's Red List (Aguilar and Gaspari, 2012).

2.1.2.9 Rough-Toothed Dolphin, *Steno bredanensis* (Lesson, 1828)

Formerly thought to be a visitor to the Mediterranean Sea, it is now considered regular in the eastern Mediterranean, and in particular in the Levantine Sea (albeit retaining the status of visitor elsewhere), in consideration of the frequency of sightings and strandings (including calves) of this species, in this part of the region. Preliminary investigation indicates that Mediterranean specimens of *S. bredanensis* may be genetically divergent from North Atlantic conspecifics (Kerem et al., 2016). No status assessment is possible at the moment, due to insufficient ecological information, although the species' obvious low density and restricted range should raise the highest concerns. Possible impacts include bycatch in fishing gear, anthropogenic noise and chemical contaminants.

2.1.2.10 Common Bottlenose Dolphin, *Tursiops truncatus* (Montagu, 1821)

The most common cetacean throughout the Mediterranean continental shelf, from Gibraltar to the Levantine Sea including the northern Adriatic and northern Aegean, although anthropogenic habitat degradation has likely caused population fragmentation in parts of its range. Natoli et al. (2005) detected significant population structuring between the Northeast Atlantic Ocean and the Black Sea, with boundaries coinciding with transitional features separating subregions. Further genetic studies presented evidence of fine-scale population division within the Adriatic and Levantine seas, as well as a distinction between offshore and coastal dolphins (Gaspari et al., 2015). There is no region-wide abundance estimate, with quantitative knowledge deriving from a handful of local studies such as in the Alborán Sea, Balearic Sea, Ligurian Sea, Tunisian Plateau, Northern Adriatic, Western Greece (see Gonzalvo et al., 2016) and Israel. Data from long-term studies (e.g. Fortuna, 2006 in the northern Adriatic) provide indications of population decline. Accordingly, Mediterranean common bottlenose dolphins were assessed as Vulnerable on the IUCN's Red List, based on past and continued substantial incidental mortality in fishing gear, occasional direct killings, habitat loss or degradation including coastal development, overfishing of prey and high levels of contamination (Bearzi et al., 2012).

2.1.2.11 Harbour Porpoise, Black Sea Subspecies, *Phocoena p. relicta* Abel, 1905

This subspecies' range encompasses not only the Black Sea proper and adjacent Turkish Straits System (Bosphorus, Marmara Sea, Dardanelles), but also the northern Aegean Sea (Fontaine, 2016). Sporadic strandings and sightings of harbour porpoises in the Greek northern Aegean are known since 1993, with at least nine records reported before 2008 for the Thracian Sea, Kavala Gulf, Strymonikos Gulf, Agiou Orous Gulf and Thermaikos Gulf (Birkun and Frantzis, 2008; Frantzis et al., 2001; Rosel et al., 2003). The Black Sea harbour porpoise subspecies has been assessed as Endangered on the IUCN Red List (Birkun and Frantzis, 2008), largely on the basis of pressure factors occurring in the Black Sea (past massive direct kills, ongoing bycatch in gillnets, various mass mortality events, habitat degradation).

2.2 Irregular but Repeated Occurrences: Cetaceans

2.2.1 Common Minke Whale, Balaenoptera acutorostrata *Lacépède, 1804*

A visitor from the North Atlantic Ocean, occasionally entering the Mediterranean Sea through the Strait of Gibraltar, with about 29 occurrences recorded since the late 18th century (Notarbartolo di Sciara and Birkun, 2010). Most of these occasional strandings and sightings occurred in the Algero-Provençal and Tyrrhenian subregions, although rare occurrences are known from the Levantine Sea off Israel.

2.2.2 Humpback Whale, Megaptera novaeangliae *(Borowski, 1781)*

Occasionally entering the Mediterranean Sea as a visitor from the Atlantic Ocean. Sighting and stranding events have been recorded throughout the Mediterranean—including in the Adriatic, Aegean and Levantine seas and the Gulf of Sirte—with a minimum of 23 records since 1885; notably about half of these have occurred since 2010 (Cagnolaro et al., 2015), perhaps due to a combination of greater attention for cetaceans by the public and the media in recent times, and possible increases in populations frequenting the north-east Atlantic Ocean.

2.2.3 False Killer Whale, Pseudorca crassidens *(Owen, 1846)*

A typical inhabitant of pelagic waters, but often also found over steep slope areas and continental shelf waters, occasionally strays as a visitor into the Mediterranean Sea. Sightings and stranding events have been recorded throughout the Mediterranean Sea, a minimum of 34 times since the late 18th century (Notarbartolo di Sciara and Birkun, 2010). The almost complete absence of occurrences along the African coast is likely explained by the absence of observers. Frequent occurrences of the species in the Levantine Sea, particularly off Israel, may prompt a future reconsideration of its status in the Mediterranean, from visitor to locally regular (A. Scheinin, Israel Marine Mammal Research and Assistance Center, IMMRAC, personal communication, 2016).

2.3 Very Rare Mediterranean Occurrences

2.3.1 Pinnipeds

2.3.1.1 Hooded Seal, *Cystophora cristata* (Erxleben, 1777)

Fourteen occurrences (three sightings and 11 strandings) of this Arctic/Sub-Arctic species in various locations across the Mediterranean coasts of Spain between 1996 and 2006 were reported by Bellido et al. (2008). A further Mediterranean occurrence was reported from Algeria in summer 2006 by Bouderbala et al. (2007).

2.3.1.2 Harp Seal, *Pagophilus groenlandicus* (Erxleben, 1777)

One adult male of this Arctic/Sub-Arctic species stranded near Motril, Granada, on the southern Mediterranean coast of Spain, in September 2008 (Bellido et al., 2009).

2.3.1.3 Harbour Seal, *Phoca vitulina* Linnaeus, 1758

Repeated sightings of two individuals of this cold-temperate and Arctic species were reported along the Spanish Mediterranean coast (Murcia and Valencia regions) during August 1994 by Mas et al. (1997).

2.3.2 Cetaceans

2.3.2.1 North Atlantic Right Whale, *Eubalaena glacialis* (Müller, 1776)

Once widely distributed across the North Atlantic Ocean, today a very small and endangered population survives along the North American coast, whereas the species was possibly extirpated from the Northeastern Atlantic (Reilly et al., 2012). Known to have strayed into the Mediterranean only on two historical occasions: the capture of a young female off Taranto, Italy, in 1877, and of another juvenile in the Bay of Castiglione, Algeria, in 1888 (Notarbartolo di Sciara and Birkun, 2010).

2.3.2.2 Sei Whale, *Balaenoptera borealis* Lesson, 1828

Mostly found in oceanic, productive waters where they undertake extensive, seasonal migrations, spending the summer months feeding in the subpolar higher latitudes and returning to the lower latitudes to calve in winter (Horwood, 2009). Two strandings and three sightings (considered reliable) were reported from Spain and France between 1921 and 1987 (Notarbartolo di Sciara and Birkun, 2010).

2.3.2.3 Grey Whale, *Eschrichtius robustus* (Lilljeborg, 1861)

Once found in both the North Pacific and North Atlantic Oceans, since the early 18th century the species' distribution has been limited to North Pacific waters. On 8 May 2010 an individual was sighted and photoidentified off Jaffa, Israel, and resighted off Barcelona, Spain on 6 June of the same year. How a grey whale ventured into the Mediterranean Sea from its North Pacific haunt is still a matter of conjecture (Scheinin et al., 2011).

2.3.2.4 Dwarf Sperm Whale, *Kogia sima* (Owen, 1866)

A cosmopolitan species with a clear preference for tropical and warm temperate waters, its presence in the Mediterranean Sea was recorded twice, both times in Italy: one specimen found stranded dead at the boundary between Tuscany and Latium in May 1988, and another stranded alive (and later died) near Eraclea Minoa, Sicily, in September 2002 (Notarbartolo di Sciara and Birkun, 2010).

2.3.2.5 Northern Bottlenose Whale, *Hyperoodon ampullatus* (Forster, 1770)

Found in cold-temperate and subpolar waters of the North Atlantic Ocean. Two reliable occurrences of this species were recorded in the western Mediterranean Sea in historical times: the stranding of a female with her calf in France in 1880, and a sighting in the Alborán Sea at the end of the 20th century (Notarbartolo di Sciara and Birkun, 2010).

2.3.2.6 Sowerby's Beaked Whale, *Mesoplodon bidens* (Sowerby, 1804)

Distributed in the cold-temperate waters of the North Atlantic Ocean, a specimen was sighted and photographed once off northeastern Sardinia (Central Tyrrhenian Sea) in June 2012, in the company of three Cuvier's beaked whales (Cagnolaro et al., 2015: fig. 101).

2.3.2.7 Blainville's Beaked Whale, *Mesoplodon densirostris* (Blainville, 1817)

Circumglobal in tropical and temperate waters. The only confirmed occurrence in the Mediterranean Sea was the stranding of an adult female near Castellò de la Plana, Catalonia, in February 1980 (Casinos and Filella, 1981).

2.3.2.8 Gervais' Beaked Whale, *Mesoplodon europaeus* (Gervais, 1855)

Found in temperate to tropical waters of the North and Central Atlantic Ocean, the species' occurrence in the Mediterranean Sea is known from only one specimen, an adult female stranded near Castiglioncello, Livorno, Italy, in August 2001 (Podestà et al., 2005).[1]

2.3.2.9 Short-Finned Pilot Whale, *Globicephala macrorhynchus* Gray, 1846

Found worldwide in warm temperate to tropical waters. In the Atlantic Ocean the species is common in many areas including the Gulf of Mexico, the Caribbean Sea and the waters surrounding the Macaronesian archipelagos. Vagrant in the Mediterranean, where the only known record is a documented sighting of a pod of three, off the Adriatic town of Cattolica, Italy, in May 2010 (Verborgh et al., 2016).

2.3.2.10 Indian Ocean Humpback Dolphin, *Sousa plumbea* (G. Cuvier, 1829)

Widely distributed across the coastal zone of the northwestern Indian Ocean, from the northern Red Sea to False Bay in South Africa, and east across the southern Asian coasts to the Gulf of Bengal. Rare reports exist of individuals straying through the Suez Canal into the Mediterranean as Lessepsian immigrants from the Red Sea (Notarbartolo di Sciara and Birkun, 2010). In January 2000, a single individual was sighted repeatedly in different locations along the Mediterranean coast of Israel (Kerem et al., 2001). More recently (3 February and 25 April 2016), repeated video-documented observations of humpback dolphins interacting with bottom trawling operations in the bay of Mersin, southern Turkey (Y.D. Özbilgin, Mersin University, Department of Fisheries, Turkey, personal communication, 2016) suggest the possibility of an increasing presence of the species in Eastern Mediterranean waters as Lessepsian immigrants.

2.3.2.11 Harbour Porpoise, Atlantic Subspecies, *Phocoena p. phocoena* (Linnaeus, 1758)

Regular in the North Atlantic Ocean; known to rarely stray into the westernmost reaches of the Mediterranean, Spanish coasts, where two individuals were found, stranded, in 1981 and 2006 (Notarbartolo di Sciara and Birkun, 2010). The past regular presence of *P. phocoena* in the wider

[1] A fourth species of *Mesoplodon*, the True's beaked whale *M. mirus* True, 1913, was recently reported to have occurred in southern Turkey (Amaha Öztürk et al., 2016); however, the species is not listed here pending confirmation from further genetic and morphological examination.

Mediterranean (apart from the well-known presence of the Black Sea subspecies in the Aegean Sea) is a subject of controversy (see Frantzis et al., 2001), although all the available evidence points to the absence of the species from the Mediterranean Sea in historical times (Cagnolaro et al., 2015; Fontaine, 2016).

3. DISTRIBUTIONAL PATTERNS OF MEDITERRANEAN MARINE MAMMALS

Marine mammals are unevenly distributed throughout the Mediterranean Sea, as their presence in different portions of the region is affected by the interactions amongst their specific ecological needs and geographical, ecological heterogeneities, as well as resulting from the effects of anthropogenic factors. Current understanding of marine mammal distribution and abundance in the region, albeit challenged by a persisting, unfortunate geographic unevenness of ecological knowledge evident from the following summaries (see Fig. 1 for a representation of Mediterranean subregions), points to the existence of areas hosting conspicuous diversity of species (e.g. Alborán Sea, western Ligurian Sea, Strait of Sicily, northern Aegean Sea).

3.1 Alborán Sea and Strait of Gibraltar

The Alborán Sea is a highly productive, oceanographically dynamic area, extensively surveyed in its northern and central portions, poorly known elsewhere. Of major importance for Cuvier's beaked whales, long-finned pilot whales and short-beaked common dolphins; sperm whales, striped dolphins, common bottlenose dolphins and Risso's dolphins also occur there regularly. Transient fin whales, entering from the Atlantic Ocean, are seasonally present. Monk seals, formerly present along the coastal zone in significant numbers and colonies, are today so reduced that they may be considered extirpated from the subregion. The waters of the Strait of Gibraltar have special importance for killer whales, sperm whales and long-finned pilot whales.

3.2 Algero-Provençal Basin

A wide subregion, largely consisting of deep abyssal plains and narrow shelves, and with portions of its offshore waters that are amongst the Mediterranean's most productive. Intensely surveyed in its northern portion, particularly within the boundaries of the Pelagos Sanctuary for Mediterranean Marine Mammals (Notarbartolo di Sciara et al., 2008); poorly known in its central and southern portions. The subregion contains critical habitat for many cetacean species, including Mediterranean resident fin and sperm whales. Long-finned pilot whales can be found with higher frequency in the subregion's western portion. Cuvier's beaked whales, Risso's dolphins, striped dolphins and common bottlenose dolphins are distributed throughout, wherever their preferred habitats occur. Short-beaked common dolphins are now rare in the northern portion, scattered in isolated pockets; their presence in the southern portion is uncertain. Monk seals have been extirpated from the subregion, although lone individuals may occasionally appear.

3.3 Tyrrhenian Sea and Eastern Ligurian Sea

Mostly occupied by a wide continental shelf in the north, connecting the Italian mainland to the island of Corsica, and becoming deeper to the south, strewn with a conspicuous number of seamounts. The northern portion has been intensely surveyed, with survey effort decreasing southwards, although rarer surveys were conducted also in Sicilian and Tunisian waters. This subregion contains habitat for resident fin, sperm and Cuvier's beaked whales, as well as for striped, Risso's and common bottlenose dolphins. Long-finned pilot whales have become rare and possibly extirpated from most areas in recent years. Short-beaked common dolphins are rare, only remaining in isolated pockets. Monk seals have been extirpated from this subregion in spite of the abundant suitable habitat; lone individuals are occasionally sighted in many locations.

3.4 Adriatic Sea

An elongated subregion contained between the Italian and the Balkan peninsulas, communicating with the main Mediterranean body through the narrow (75 km) Otranto Channel. The subregion was the object of recent, repeated systematic surveys (Holcer et al., 2014), which add significantly to previous, highly localised coastal studies conducted mostly in the northern portion. The Adriatic Sea is characterised by a markedly heterogeneous geomorphology, with a mean depth of few tens of metres in its northern part and >1200 m deep in the south. Human presence and use are also heterogeneous, more intense (e.g. fishing, tourism, coastal construction, offshore gas extraction) along its sandy western shore than along rocky eastern shores. Accordingly, the subregion offers a wide span of habitat conditions to its marine mammal fauna. The northern portion today is regularly populated solely by common bottlenose dolphins, whereas the south contains habitat for fin, sperm and Cuvier's beaked whales and striped, Risso's and common bottlenose dolphins. Short-beaked common dolphins occur rarely. Monk seals were formerly frequent, mostly along the eastern shores and islands (the subregion includes the species' type locality—the Kvarner Sea at Osor, see King, 1956), occurring today only as scattered, isolated individuals.

3.5 Strait of Sicily, Tunisian Plateau and Gulf of Sirte

The continental shelf predominates in this mostly shallow subregion, which constitutes the crossroads between the two deep Mediterranean basins and the submerged portion of the median land bridge connecting Europe to Africa. Intense fishing occurs here, and a ship lane crossing longitudinally the Mediterranean and passing along the south coast of Sicily is one of the world's busiest. The subregion was systematically surveyed only recently, with cetacean data having been collected before only in specific localities (e.g. Lampedusa, Malta); the Libyan shores remain practically unsurveyed. Common bottlenose dolphins are the most frequent species, and short-beaked common dolphins are still present in the area. Fin whales occur here during winter, and sperm whales are occasionally seen, perhaps during their east-west movements across the Mediterranean. Striped and Risso's dolphins also occur in the subregion, predominantly in its deeper eastern portion. Monk seals are today absent, except for occasional vagrant individuals.

3.6 Ionian Sea and Central Mediterranean

This subregion includes the Mediterranean's deepest waters, with depths exceeding 5000 m in parts of the Hellenic Trench, off south-western Greece. Marine mammals have been reasonably well surveyed in the north and the east, but with a complete lack of data along Cyrenaica (Libya). Fin whales seasonally occur off eastern Sicily and western Greece. The Hellenic Trench provides one of the Mediterranean's most suitable habitats for sperm whales. The subregion is also important for Cuvier's beaked whales, both along the Sicilian and Greek coasts; a system of canyons north of Cyrenaica could also be suitable for the species; however, no information on its presence there is available. Common bottlenose, Risso's and striped dolphins are frequent wherever their habitats occur, including in the narrow Gulf of Corinth (Greece) (Bearzi et al., 2016). Common bottlenose dolphins also inhabit the inner Gulf of Ambracia, where their density is the Mediterranean's highest (Gonzalvo et al., 2016). Short-beaked common dolphins are still found in the subregion, mostly along the west coast of Greece. Breeding nuclei of monk seals survive off Greece, and perhaps along the Cyrenaican coast as well.

3.7 Aegean Sea

This subregion is shallower than the Mediterranean average, and interspersed with a large number of islands, islets and rocks. This is one of the least surveyed portion of the Mediterranean in terms of cetacean presence. The Aegean Sea is the main remaining stronghold of monk seals in the Mediterranean, with several scattered breeding nuclei in Greece and Turkey; it is also the only Mediterranean location where the Black Sea subspecies of harbour porpoises regularly occurs. Other species regularly found in the subregion, wherever suitable habitat occurs, include common bottlenose, Risso's, striped, and short-beaked common dolphins. Fin, sperm and Cuvier's beaked whales also occur, but rarely.

3.8 Levantine Sea

A highly modified marine environment due to maverick geoengineering initiatives (i.e. the opening of Suez Canal and river Nile water sequestration in the Aswan Dam in Egypt) having major ecological consequences over the entire subregion and beyond. This is the only subregion where rough-tooted dolphins occur with some regularity. Cuvier's beaked whales find suitable habitat along the deep waters in the north, connecting Crete to the southwest portion of the Anatolian peninsula. Scattered breeding nuclei of monk seals occur along the south coast of Turkey and Cyprus. Common bottlenose dolphins are frequent throughout the shelf area, and striped dolphins can be found more offshore. Large herds of short-beaked common dolphins have been repeatedly sighted off southern Israel, Gaza and possibly northeastern Egypt. Ecological knowledge of marine mammals along the extensive Egyptian EEZ is inexistent.

4. STATUS OF MEDITERRANEAN MARINE MAMMALS AND THREATS

The Mediterranean region, an outstanding hotspot of marine and coastal biodiversity (Bianchi and Morri, 2000), is under significant threat due to the many human activities occurring along its coasts and in its waters; the region is home to almost half a billion human inhabitants, and in addition it receives half as many visitors every summer (Cuttelod et al., 2008). Impacting activities include coastal development and sprawl, overfishing, destructive fishing, contamination of sediments and biota caused by pollution, nutrient over-enrichment, disturbance and noise pollution caused by shipping, marine industries and navies, spread of invasive species in many cases facilitated by climate change and degradation of transitional and estuarine areas (UNEP-MAP, 2012). Accordingly, all these pressure factors exert a significant impact on the Mediterranean biota and taxa, leading to an increasing number of species facing a high risk of extinction.

Marine mammals are affected by anthropogenic pressure due to their ecological and life history traits (e.g. long lifespan, low reproductive potential, small population sizes, late maturity) that make them especially vulnerable. A range of different anthropogenic threats and their impacts on marine mammal populations regular in the Mediterranean is listed in Table 2. Impacts can be defined as inducing: (a) direct mortality (including deliberate killing, naval sonar, ship strikes, epizootics, fishery bycatch, chemical pollution, ingestion of solid debris), (b) redistribution caused by short-term habitat degradation (naval sonar, seismic surveys, vessel disturbance, vessel noise), and (c) redistribution caused by long-term habitat degradation (fishery-induced food depletion, coastal development, climate change).

Accordingly, of the 12 regular marine mammal species in the Mediterranean, six are considered to be Mediterranean "subpopulations" (*sensu* IUCN), and are listed in a Threatened category on IUCN's Red List. Three of these—Mediterranean monk seals, sperm whales, short-beaked common dolphins—are listed as Endangered, as is the Black Sea subspecies of harbour porpoise, and three others—fin whales, striped dolphins and common bottlenose dolphins—are listed as Vulnerable. For three other regular species, the Mediterranean subpopulations are considered Data Deficient (Cuvier's beaked whales, long-finned pilot whales and Risso's dolphins). Killer whales and rough-toothed dolphins remain unassessed, and have not been formally designated as IUCN subpopulations in the Mediterranean region.

Table 2 Threats Impacting on Marine Mammals in the Mediterranean Sea

Threat	Root Cause	Types of Impact	Metrics to Quantify Trends	Species Most Affected	Selected References
Naval sonar	Military activities	Direct mortality	Population size	Cuvier's beaked whale	Fernàndez et al. (2004), Frantzis (1998), Jepson et al. (2003), and Podestà et al. (2016)
	Military activities	Short-term habitat degradation causing medium-scale redistribution	Distribution	Cuvier's beaked whale, long-finned pilot whale	Aguilar Soto et al. (2006), DeRuiter et al. (2013), Dolman et al. (2011), Podestà et al. (2016), Rendell and Gordon (1999), Tyack et al. (2011), and Verborgh et al. (2016)
Seismic exploration	Offshore oil and gas industry	Short-term habitat degradation causing medium-scale redistribution	Distribution	All	Castellote et al. (2012), Kerem et al. (2016), and Podestà et al. (2016)
Disturbance from whale watching, motorised vessels	Unregulated whale watching, irresponsible tourism	Short-term habitat degradation causing small-scale redistribution	Distribution	All	Esteban et al. (2016), Jahoda et al. (2003), Rendell and Frantzis (2016), and Verborgh et al. (2016)
	Irresponsible tourists entering breeding caves during breeding season	Short-term habitat degradation disrupting reproduction	Population size, distribution	Monk seal	Notarbartolo di Sciara and Kotomatas (2016)

Continued

Table 2 Threats Impacting on Marine Mammals in the Mediterranean Sea—cont'd

Threat	Root Cause	Types of Impact	Metrics to Quantify Trends	Species Most Affected	Selected References
Vessel noise, vessel traffic	Powered vessel traffic	Short-term habitat degradation causing small-scale redistribution	Distribution	All	Campana et al. (2015), Esteban et al. (2016), Rako et al. (2013), Rendell and Frantzis (2016), Verborgh et al. (2016), and Williams et al. (2015)
Ship strikes	Powered vessel traffic	Direct mortality	Population size	Fin whale, sperm whale, long-finned pilot whale	Jensen et al. (2004), Notarbartolo di Sciara et al. (2016a), Panigada et al. (2006), Rendell and Frantzis (2016), Verborgh et al. (2016), and Weinrich et al. (2006)
Epizootics	Contagious disease (mostly morbillivirus), possibly enhanced by contaminants	Direct mortality	Population size	Fin whale, long-finned pilot whale, common bottlenose dolphin, striped dolphin, Black Sea harbour porpoise (known to have been affected in the Black Sea only)	Aguilar and Raga (1993), Di Guardo et al. (2011), Mazzariol et al. (2016), Müller et al. (2000), and Verborgh et al. (2016)

Table 2 Threats Impacting on Marine Mammals in the Mediterranean Sea—cont'd

Threat	Root Cause	Types of Impact	Metrics to Quantify Trends	Species Most Affected	Selected References
Fisheries	Unregulated or inadequately regulated fishery	Direct mortality (through bycatch)	Population size	Monk seal, striped dolphin, sperm whale, long-finned pilot whale, Risso's dolphin, Cuvier's beaked whale, common bottlenose dolphin, short-beaked common dolphin, rough-toothed dolphin	Azzellino et al. (2016), Kerem et al. (2016), Notarbartolo di Sciara (1990), Notarbartolo di Sciara and Kotomatas (2016), Öztürk (2015), Podestà et al. (2016), and Rendell and Frantzis (2016)
	Various including overfishing and animal behaviour	Direct mortality (from competition with fisheries for shared prey)	Population size	Monk seal, common bottlenose dolphin, killer whale	Bearzi et al. (2004), Esteban et al. (2016), Notarbartolo di Sciara and Birkun (2010), and Notarbartolo di Sciara and Kotomatas (2016)
	Various including poorly managed fishery, possibly combined with climate change	Long-term habitat degradation (depletion of prey) causing medium-scale redistribution	Population size, distribution	Killer whale, Risso's dolphin, common bottlenose dolphin, short-beaked common dolphin	Azzellino et al. (2016), Bearzi et al. (2006), Esteban et al. (2016), and Tudela (2004)

Continued

Table 2 Threats Impacting on Marine Mammals in the Mediterranean Sea—cont'd

Threat	Root Cause	Types of Impact	Metrics to Quantify Trends	Species Most Affected	Selected References
Pollution	Urban and agricultural runoff, industrial effluent, navigation accidents	Mortality/health impairment deriving from contact/ingestion of oil and chemical spills, ingestion of noxious substances through food	Population size	All	Esteban et al. (2016), Fossi et al. (2002, 2012, 2014), Gonzalvo et al. (2016), Jepson and Law (2016), Jepson et al. (2016), Kerem et al. (2016), Pinzone et al. (2015), Rendell and Frantzis (2016), and Squadrone et al. (2015)
	Poor solid waste management, illegal dumping of fishing gear	Direct mortality by entanglement in/ingestion of solid debris	Population size	Sperm whale, Cuvier's beaked whale	de Stephanis et al. (2013), Podestà et al. (2016), and Rendell and Frantzis (2016)
Coastal development	Unregulated or inadequately regulated coastal planning	Long-term habitat degradation causing medium-scale redistribution	Distribution	Monk seal, common bottlenose dolphin	Bearzi et al. (2012), Coll et al. (2012), Halpern et al. (2008), and Notarbartolo di Sciara and Kotomatas (2016)
Climate change	Atmospheric carbon loading	Long-term habitat degradation causing large-scale redistribution	Population size, distribution	All	Lejeusne et al. (2010) and Simmonds et al. (2012)

5. CONCLUSIONS

Coexistence between top marine predators, such as marine mammals, and humans cannot be expected to be easy in a marine region like the Mediterranean Sea, which is characterised by heavy human presence. The existence of widespread and diverse conservation problems for marine mammals in the Mediterranean is, in fact, evident from the IUCN Red List assessments of the region's taxa, none of which can be ascribed to a nonthreatened category. Furthermore, the persisting condition of three taxa as "Data Deficient" and two as "Not Assessed" does not speak highly for regional and national conservation efforts. There is no evident justification for such continuing uncertainty, because scientific knowledge of marine mammal ecology in the region has recently grown substantively, and has made of the Mediterranean, in such respect, one of the best known of the world's regions, as testified by the contributions contained in this volume. Having gotten off to a great start in terms of marine mammal zoology 25 centuries ago with the pioneering works of Aristotle, since the beginning of the Christian era the region remained in woeful conditions in terms of marine mammal ecological knowledge until the end of the 20th century. It is only during the last three decades that newly developed research techniques and technologies started being applied to marine mammal investigations promoting substantive ecological understanding (Notarbartolo di Sciara, 2012). Unfortunately, such knowledge is still quite heterogeneous in terms of geographical coverage, with much greater effort deployed in the northern and western portions of the Mediterranean (Fig. 2), and, even there, mostly related to the summer season.

Specifically, it seems appropriate that the following IUCN Red List assessment effort be considered expeditiously (Table 1): (a) the recognised Mediterranean monk seal "subpopulations" (Karamanlidis et al., 2015) should be assessed separately, based upon the hypothesis that some may still qualify as Critically Endangered; (b) Cuvier's beaked whales, long-finned pilot whales and Risso's dolphins are almost certainly no longer Data Deficient (Azzellino et al., 2016; Podestà et al., 2016; Verborgh et al., 2016), and the same applies to killer whales that are Not Assessed (Esteban et al., 2016); (c) efforts to collect further data on rough-toothed dolphins in the Levantine Sea should be intensified in order to enable the assessment of the population in the Mediterranean Sea (Kerem et al., 2016); (d) the resident fin whale "subpopulation", currently assessed as Vulnerable, should be reassessed as it may qualify for a higher category of threat (Notarbartolo di Sciara et al., 2016a); finally, (e) clearly separate

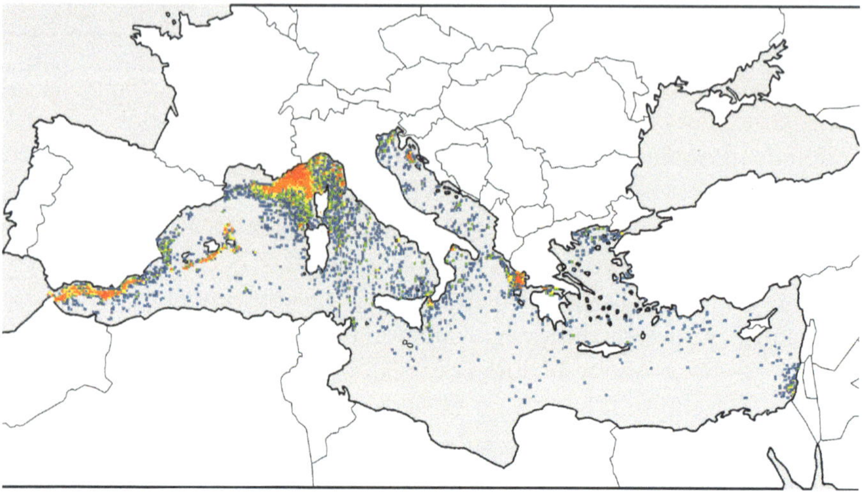

Fig. 2 Marine mammal sightings derived from the OBIS SEAMAP Database (Halpin et al., 2009) illustrate the geographic heterogeneity of search effort conducted in the Mediterranean Sea (http://seamap.env.duke.edu). Records are by 1 degree cells (*red*: >100 records per cell; *orange*: 51–100; *yellow*: 11–50; *green*: 6–10; *blue*: 1–5).

"units to conserve", exemplified by common bottlenose dolphins in the Gulf of Ambracia, Greece (Gonzalvo et al., 2016) and striped dolphins in the Gulf of Corinth, Greece (Bearzi et al., 2016) are likely to benefit from a possible future assessment on IUCN's Red List.

Red List assessments, however, are not a solution *per se*. Although the authoritative, science-based arguments that they contain can attract the attention of policy- and decision-makers on the need to act, Red List assessments do not substitute conservation or management action at sea. Threats to marine mammals in the Mediterranean are today reasonably well known (Table 2), and they impact on populations because they are not being effectively addressed, such as when countries fail to: (a) stem illegal, highly destructive fishing practices like the use of explosives and the long-banned pelagic driftnets; (b) enforce the rule of law addressing the deliberate killing of monk seals, recognised as the single most important factor of the species' decline in the region (Notarbartolo di Sciara and Kotomatas, 2016); or (c) address the increasingly pervasive introduction in the marine environment of noxious anthropogenic noise, such as from naval exercises, seismic surveys, shipping and coastal construction, impacting in particular acoustically sensible species such as Cuvier's beaked whales.

Conserving marine biodiversity—and marine mammals in particular—in the Mediterranean too often clashes with economic interests, and when a compromise is sought, economic concerns always get the upper hand; in most cases, however, compromise is not even considered and conservation remains a hollow term. The obsession for extracting resources and services from the sea to the exclusive human benefit, recently dubbed with the captivating term of "blue growth" (European Commission, 2012), is reducing and often obliterating living space to everything nonhuman, under the mistaken assumption that nature's material benefits to people is all that people need (Notarbartolo di Sciara, 2015). Parties to the Convention on Migratory Species Agreement on the Conservation of Cetaceans in the Black Sea, Mediterranean Sea and Contiguous Atlantic Area (ACCOBAMS) (see Scovazzi, 2016) have repeatedly resolved to designate Marine Protected Areas (MPAs) for cetaceans under the recommendation of the Agreement's Scientific Committee; however, not a single MPA dedicated to cetacean conservation has been established in the Mediterranean since the Agreement came into force in 2002. For instance, impacts from ongoing military activity in the Mediterranean Sea underscore the need for the identification of "Areas of Special Concern" for beaked whales, incorporating buffer zones to limit high risk activities around known areas where beaked whales occur at high densities (Podestà et al., 2016: Fig. 6). Hopefully, a recent effort of identifying Important Marine Mammal Areas (IMMAs) in the world's oceans, seas and rivers, the Mediterranean Sea included, is set to provide a peer-reviewed tool to inform decision-makers and planners about the presence and extent of areas containing important habitat for marine mammals, thereby improving harmonisation between human activities at sea and marine mammal conservation, even in the absence of more effective dedicated MPAs (Notarbartolo di Sciara et al., 2016b).

Conserving Mediterranean marine mammals, ostensibly at the forefront of Mediterranean coastal nations' concern, must also become so in reality. Mediterranean marine mammal populations have persisted until now, with variable success, despite the weakness and ineffectiveness of the conservation efforts in the region. It would seem foolish to assume that such luck will last for much longer in the future.

ACKNOWLEDGEMENTS

I am grateful to Barbara E. Curry and Thomas A. Jefferson for their helpful comments and suggestions.

REFERENCES

Aguilar Soto, N., Johnson, M., Madsen, P.T., Tyack, P.L., Bocconcelli, A., Borsani, F.J., 2006. Does intense ship noise disrupt foraging in deep-diving Cuvier's beaked whales (*Ziphius cavirostris*)? Mar. Mamm. Sci. 22 (3), 690–699.

Aguilar, A., Gaspari, S., 2012. *Stenella coeruleoalba* (Mediterranean subpopulation). In: The IUCN Red List of Threatened Species 2012. e.T16674437A16674052, http://dx.doi.org/10.2305/IUCN.UK.2012-1.RLTS.T16674437A16674052.en.

Aguilar, A., Raga, J.A., 1993. The striped dolphin epizootic in the Mediterranean Sea. Ambio 22 (8), 524–528.

Amaha Öztürk, A., Dede, A., Tonay, A.M., Danyer, E., Danyer, I.A., Özbek, E.O., Ercan, D., Öztürk, B., 2016. The first record of True's beaked whale, *Mesoplodon mirus*, from the Mediterranean coast of Turkey during multiple strandings in June 2016. J. Black Sea/Mediterr. Environ. 22 (2), 194–199.

Azzellino, A., Airoldi, S., Gaspari, S., Lanfredi, C., Moulins, A., Podestà, M., Rosso, M., Tepsich, P., 2016. Risso's dolphin, *Grampus griseus*, in the Western Ligurian Sea: trends in population size and habitat use. In: Notarbartolo di Sciara, G., Podestà, M., Curry, B.E. (Eds.), Mediterranean Marine Mammal Ecology and Conservation. Advances in Marine Biology, vol. 75. Elsevier, Amsterdam, pp. 205–232.

Bearzi, G., 2003. *Delphinus delphis*: Mediterranean population. In: The IUCN Red List of Threatened Species 2003. e.T41762A10557372, http://dx.doi.org/10.2305/IUCN.UK.2003.RLTS.T41762A10557372.en.

Bearzi, G., Reeves, R.R., Notarbartolo di Sciara, G., Politi, E., Cañadas, A., Frantzis, A., Mussi, B., 2003. Ecology, status and conservation of short-beaked common dolphins *Delphinus delphis* in the Mediterranean Sea. Mammal Rev. 33 (3), 224–252.

Bearzi, G., Holcer, D., Notarbartolo di Sciara, G., 2004. The role of historical dolphin takes and habitat degradation in shaping the present status of northern Adriatic cetaceans. Aquat. Conserv. 14, 363–379.

Bearzi, G., Fortuna, C.M., Reeves, R.R., 2012. *Tursiops truncatus* (Mediterranean subpopulation). In: The IUCN Red List of Threatened Species 2012. e.T16369383A16369386, http://dx.doi.org/10.2305/IUCN.UK.2012-1.RLTS.T16369383A16369386.en.

Bearzi, G., Bonizzoni, S., Santostasi, N.L., Furey, N.B., Eddy, L., Valavanis, V.D., Gimenez, O., 2016. Dolphins in a scaled-down Mediterranean: the Gulf of Corinth's odontocetes. In: Notarbartolo di Sciara, G., Podestà, M., Curry, B.E. (Eds.), Mediterranean Marine Mammal Ecology and Conservation. Advances in Marine Biology, vol. 75. Elsevier, Amsterdam, pp. 297–331.

Bellido, J.J., Castillo, J.J., Farfan, M.A., Martin, J.J., Mons, J.L., Real, R., 2008. First records of hooded seals (*Cystophora cristata*) in the Mediterranean Sea. Mar. Biodiv. Rec. 1, e74. http://dx.doi.org/10.1017/S1755267207007804.

Bellido, J.J., Cabot, J., Castillo, J.J., Baez, J.C., Martin, J.J., Mons, J.L., Larios, J., Rubia, J., Real, R., 2009. First record of the harp seal (*Pagophilus groenlandicus*) extralimital presence in the Mediterranean Sea. Mar. Biodiv. Rec. 2, e169. http://dx.doi.org/10.1017/S1755267209990959.

Bianchi, C.N., Morri, C., 2000. Marine biodiversity of the Mediterranean Sea: situation, problems and prospects for future research. Mar. Pollut. Bull. 40 (5), 367–376.

Bianucci, G., 2015. Paleontologia. In: Cagnolaro, L., Cozzi, B., Notarbartolo di Sciara, G., Podestà, M. (Eds.), Fauna d'Italia. Mammalia IV. Cetacea. Calderini Edagricole, Milano, pp. 8–32.

Birkun Jr., A.A., Frantzis, A., 2008. *Phocoena phocoena* ssp. *relicta*. In: The IUCN Red List of Threatened Species 2008. e.T17030A6737111, http://dx.doi.org/10.2305/IUCN.UK.2008.RLTS.T17030A6737111.en.

Boero, F., 2015. The future of the Mediterranean Sea ecosystem: towards a different tomorrow. Rend. Fis. Acc. Lincei 26, 3. http://dx.doi.org/10.1007/s12210-014-0340-y.

Bouderbala, M., Bouras, D., Bekrattou, D., Doukara, K., Abdelghani, M.F., Boutiba, Z., 2007. First recorded instance of a hooded seal (*Cystophora cristata*) in Algeria. The Monachus Guardian 10 (1). http://www.monachus-guardian.org/mguard19/1931scien.htm.

Cagnolaro, L., Cozzi, B., Notarbartolo di Sciara, G., Podestà, M., 2015. Fauna d'Italia—Mammalia IV. Cetacea. Calderini—Edagricole, Bologna.

Campana, I., Crosti, R., Angeletti, D., Carosso, L., David, L., Di-Meglio, N., Moulins, A., Rosso, M., Tepsich, P., Arcangeli, A., 2015. Cetacean response to summer maritime traffic in the Western Mediterranean Sea. Mar. Environ. Res. 109, 1–8. http://dx.doi.org/10.1016/j.marenvres.2015.05.009.

Casinos, A., Filella, S., 1981. Notes on cetaceans of the Iberian coasts: IV. A specimen of *Mesoplodon densirostris* (Cetacea, Hyperoodontidae) stranded on the Spanish littoral. Säugetierk. Mitt. 29 (4), 61–67.

Castellote, M., Clark, C.W., Lammers, M.O., 2012. Acoustic and behavioural changes by fin whales (*Balaenoptera physalus*) in response to shipping and airgun noise. Biol. Conserv. 147 (1), 115–122. http://dx.doi.org/10.1016/j.biocon.2011.12.021.

Coll, M., Piroddi, C., Albouy, C., Ben Rais Lasram, F., Cheung, W.W., Christensen, V., Karpouzi, V.S., Guilhaumon, F., Mouillot, D., Paleczny, M., Palomares, M.L., 2012. The Mediterranean Sea under siege: spatial overlap between marine biodiversity, cumulative threats and marine reserves. Global Ecol. Biogeogr. 21 (4), 465–480.

Cuttelod, A., García, N., Abdul Malak, D., Temple, H., Katariya, V., 2008. The Mediterranean: a biodiversity hotspot under threat. In: Vié, J.-C., Hilton-Taylor, C., Stuart, S.N. (Eds.), The 2008 Review of the IUCN Red List of Threatened Species. IUCN Gland, Switzerland, pp. 1–13.

de Stephanis, R., Gimenez, J., Carpinelli, E., Gutierrez-Exposito, C., Cañadas, A., 2013. As main meal for sperm whales: plastics debris. Mar. Pollut. Bull. 69 (1–2), 206–214. http://dx.doi.org/10.1016/j.marpolbul.2013.01.033.

DeRuiter, S.L., Southall, B.L., Calambokidis, J., Zimmer, W.M., Sadykova, D., Falcone, E.A., Friedlaender, A.S., Joseph, J.E., Moretti, D., Schorr, G.S., Thomas, L., 2013. First direct measurements of behavioural responses by Cuvier's beaked whales to mid-frequency active sonar. Biol. Lett. 9 (4), 20130223.

Di Guardo, G., Mazzariol, S., Fernández, A., 2011. Biologically threatened dolphins and whales. Environ. Microbiol. 13 (11), 2833–2834. http://dx.doi.org/10.1111/j.1462-2920.2011.02565.x.

Dolman, S.J., Evans, P.G., Notarbartolo-di-Sciara, G., Frisch, H., 2011. Active sonar, beaked whales and European regional policy. Mar. Pollut. Bull. 63 (1), 27–34.

Esteban, R., Verborgh, P., Gauffier, P., Alarcón, A., Salazar-Sierra, J.M., Giménez, J., Foote, A.D., de Stephanis, R., 2016. Conservation status of killer whales, *Orcinus orca*, in the Strait of Gibraltar. In: Notarbartolo di Sciara, G., Podestà, M., Curry, B.E. (Eds.), Mediterranean Marine Mammal Ecology and Conservation. Advances in Marine Biology, vol. 75. Elsevier, Amsterdam, pp. 141–172.

European Commission, 2012. Blue Growth: opportunities for marine and maritime sustainable growth. Communication from the Commission to the European Parliament, the Council, the European Economic and Social Committee and the Committee of the Regions. Brussels, 13.9.2012. COM(2012) 494 final.

Fernàndez, A., Arbelo, M., Deaville, R., Paterson, I.A.P., Castro, P., Baker, J.R., Degollada, E., Ross, H.M., Herraez, P., Pocknell, A.M., Rodriguez, F., Howie, F.E., Espinosa, A., Reid, R.J., Jaber, J.R., Martin, V., Cunningham, A.A., Jepson, P.D., 2004. Whales, sonar and decompression sickness. Nature 428 (6984), U1–U2. http://dx.doi.org/10.1038/nature02528.

Fontaine, M.C., 2016. Harbour porpoises, *Phocoena phocoena*, in the Mediterranean Sea and adjacent regions: biogeographic relicts of the last glacial period. In: Notarbartolo di Sciara, G., Podestà, M., Curry, B.E. (Eds.), Mediterranean Marine Mammal Ecology and Conservation. Advances in Marine Biology, vol. 75. Elsevier, Amsterdam, pp. 333–358.

Fortuna, C.M., 2006. Ecology and conservation of bottlenose dolphins (*Tursiops truncatus*) in the north-eastern Adriatic Sea. University of St. Andrews. A thesis submitted in partial fulfilment of the requirements for the degree of Doctor of Philosophy.

Fossi, M.C., Casini, S., Marsili, L., Ancora, S., Mori, G., Neri, G., Ausili, A., Romeo, T., Moscatelli, A., Notarbartolo di Sciara, G., 2002. Biomarkers of exposure and effects for assessing toxicological risk of endocrine disrupters in top predators of the Mediterranean Sea. Mar. Ecol. 23 (1), 184–189.

Fossi, M.C., Panti, C., Guerranti, C., Coppola, D., Giannetti, M., Marsili, L., Minutoli, R., 2012. Are baleen whales exposed to the threat of microplastics? A case study of the Mediterranean fin whale (*Balaenoptera physalus*). Mar. Pollut. Bull. 64 (11), 2374–2379.

Fossi, M.C., Coppola, D., Baini, M., Giannetti, M., Guerranti, C., Marsili, L., Panti, C., de Sabata, E., Clò, S., 2014. Large filter feeding marine organisms as indicators of microplastic in the pelagic environment: the case studies of the Mediterranean basking shark (*Cetorhinus maximus*) and fin whale (*Balaenoptera physalus*). Mar. Environ. Res. 100, 17–24.

Frantzis, A., 1998. Does acoustic testing strand whales? Nature 392, 29.

Frantzis, A., Gordon, J., Hassidis, G., Komnenou, A., 2001. The enigma of harbor porpoise presence in the Mediterranean Sea. Mar. Mamm. Sci. 17 (4), 937–944.

Fyler, C.A., Reeder, T.W., Berta, A., Antonelis, G., Aguilar, A., Androukaki, E., 2005. Historical biogeography and phylogeny of monachine seals (Pinnipedia: Phocidae) based on mitochondrial and nuclear DNA data. J. Biogeogr. 32, 1267–1279.

Gaspari, S., Airoldi, S., Hoelzel, R., 2007. Risso's dolphins (*Grampus griseus*) in UK waters are differentiated from a population in the Mediterranean Sea and genetically less diverse. Conserv. Genet. 8, 727–732.

Gaspari, S., Scheinin, A., Holcer, D., Fortuna, C., Natali, C., Genov, T., Frantzis, A., Chelazzi, G., Moura, A.E., 2015. Drivers of population structure of the bottlenose dolphin (*Tursiops truncatus*) in the Eastern Mediterranean Sea. Evol. Biol. 42, 177. http://dx.doi.org/10.1007/s11692-015-9309-8.

Gonzalvo, J., Lauriano, G., Hammond, P.S., Viaud-Martinez, K.A., Fossi, M.C., Natoli, A., Marsili, L., 2016. The Gulf of Ambracia's common bottlenose dolphins, *Tursiops truncatus*: a highly dense and yet threatened population. In: Notarbartolo di Sciara, G., Podestà, M., Curry, B.E. (Eds.), Mediterranean Marine Mammal Ecology and Conservation. Advances in Marine Biology, vol. 75. Elsevier, Amsterdam, pp. 259–296.

Halpern, B.S., Walbridge, S., Selkoe, K.A., Kappel, C.V., Micheli, F., D'Agrosa, C., Bruno, J.F., Casey, K.S., Ebert, C., Fox, H.E., Fujita, R., Heinemann, D., Lenihan, H.S., Madin, E.M.P., Perry, M.T., Selig, E.R., Spalding, M., Steneck, R., Watson, R., 2008. A global map of human impact on marine ecosystems. Science 319, 948–952.

Halpin, P.N., Read, A.J., Fujioka, E., Best, B.D., Donnelly, B., Hazen, L.J., Kot, K., Urian, K., LaBrecque, E., Dimatteo, A., Cleary, J., Good, C., Crowder, L.B., Hyrenbach, K.D., 2009. OBIS-SEAMAP: the world data center for marine mammal, sea bird, and sea turtle distributions. Oceanography 22 (2), 104–115.

Hendey, Q.B., 1972. The evolution and dispersal of the Monachinae (Mammalia: Pinnipedia). Ann. South African Mus. 1972, 99–113.

Holcer, D., Fortuna, C.M., Mackelworth, P.C., 2014. Important areas for conservation of cetaceans, sea turtles and giant devil rays in the Adriatic Sea: summary of existing knowledge: UNEP/MAP-RAC/SPA Internal report. Contract No. 08/RAC/SPA_2013 MedOpenSeas.

Horwood, J., 2009. Sei whale, *Balaenoptera borealis*. In: Würsig, B., Thewissen, J.G.M., Perrin, W.F. (Eds.), Encyclopedia of Marine Mammals. Academic Press, San Diego, San Francisco, New York, Boston, London, Sydney, Tokyo, pp. 1001–1003.

Jahoda, M., Lafortuna, C.L., Biassoni, N., Almirante, C., Azzellino, A., Panigada, S., Zanardelli, M., Notarbartolo di Sciara, G., 2003. Mediterranean fin whale's (*Balaenoptera physalus*) response to small vessels and biopsy sampling assessed through passive tracking and timing of respiration. Mar. Mamm. Sci. 19 (1), 96–110.

Jensen, A.S., Silber, G.K., Calambokidis, J., 2004. Large whale ship strike database. US Department of Commerce, National Oceanic and Atmospheric Administration, Washington, DC. Technical Memorandum NMFS-OPR, 37 p.

Jepson, P.D., Law, R.J., 2016. Persistent pollutants, persistent threats. Science 352 (6292), 1388–1389.

Jepson, P.D., Arbelo, M., Deaville, R., Patterson, I.A.P., Castro, P., Baker, J.R., Degollada, E., Ross, H.M., Herraez, P., Pocknell, A.M., Rodriguez, F., Howie, F.E., Espinosa, A., Reid, R.J., Jaber, J.R., Martin, V., Cunningham, A.A., Fernàndez, A., 2003. Gas-bubble lesions in stranded cetaceans: was sonar responsible for a spate of whale deaths after an Atlantic military exercise? Nature 425, 575–576.

Jepson, P.D., Deaville, R., Barber, J.L., Aguilar, A., Borrell, A., Murphy, S., Barry, J., Brownlow, A., Barnett, J., Berrow, S., Cunningham, A.A., Davison, N.J., ten Doeschate, M., Esteban, R., Ferreira, M., Foote, A.D., Genov, T., Giménez, J., Loveridge, J., Llavona, A., Martin, V., Maxwell, D.L., Papachlimitzou, A., Penrose, R., Perkins, M.W., Smith, B., de Stephanis, R., Tregenza, N., Verborgh, P., Fernandez, A., Law, R.J., 2016. PCB pollution continues to impact populations of orcas and other dolphins in European waters. Scientific Reports 6, 18573. http://dx.doi.org/10.1038/srep18573.

Karamanlidis, A., Dendrinos, P., 2015. *Monachus monachus*. In: The IUCN Red List of Threatened Species 2015. e.T13653A45227543, http://dx.doi.org/10.2305/IUCN.UK.2015-4.RLTS.T13653A45227543.en.

Karamanlidis, A.A., Dendrinos, P., Fernández de Larrinoa, P., Gücü, A.C., Johnson, W.M., Kiraç, C.O., Pires, R., 2015. The Mediterranean monk seal *Monachus monachus*: status, biology, threats, and conservation priorities. Mammal Rev. 46 (2), 92–105. http://dx.doi.org/10.1111/mam.12053.

Kerem, D., Goffman, O., Spanier, E., 2001. Sighting of a single humpback dolphin (*Sousa* sp.) along the Mediterranean coast of Israel. Mar. Mamm. Sci. 17 (1), 170–171.

Kerem, D., Goffman, O., Elasar, M., Hadar, N., Scheinin, A., Lewis, T., 2016. The rough-toothed dolphin, *Steno bredanensis*, in the Eastern Mediterranean Sea: a relict population? In: Notarbartolo di Sciara, G., Podestà, M., Curry, B.E. (Eds.), Mediterranean Marine Mammal Ecology and Conservation. Advances in Marine Biology, vol. 75. Elsevier, Amsterdam, pp. 233–258.

King, J.E., 1956. The monk seals (genus *Monachus*). Bull. Brit. Mus. Nat. Hist. (Zool.) 3, 203–256.

Lejeusne, C., Chevaldonné, P., Pergent-Martini, C., Boudouresque, C., Pérez, T., 2010. Climate change effects on a miniature ocean: the highly diverse, highly impacted Mediterranean Sea. Trends Ecol. Evol. 25, 250–260. http://dx.doi.org/10.1016/j.tree.2009.10.009.

Mas, J., Jimenez, J., Raga, J.A., 1997. First record of harbour seals in the western Mediterranean Sea. Eur. Res. Cetaceans 10, 328.

Mazzariol, S., Centelleghe, C., Beffagna, G., Povinelli, M., Terracciano, G., Cocumelli, C., Pintore, A., Denurra, D., Casalone, C., Pautasso, A., Di Francesco, C.E., Di Guardo, G.,

2016. Mediterranean fin whales (Balaenoptera physalus) threatened by dolphin morbillivirus. Emerg. Infect. Dis. 22 (2), 302–305. http://dx.doi.org/10.3201/eid2202.150882.

Millot, C., Taupier-Letage, I., 2004. The cetacean world as seen by physical and biological oceanographers. In: CIESM, Investigating the Role of Cetaceans in Marine Ecosystems. CIESM Workshop Monograph n. 25, Monaco, pp. 107–116.

Millot, C., Taupier-Letage, I., 2005. Circulation in the Mediterranean Sea. In: The Handbook of Environmental Chemistry. The Natural Environment and the Biological Cycles, vol. 5. Part K, Springer-Verlag, Berlin, pp. 29–66. http://dx.doi.org/10.1007/b107143.

Müller, G., Wunschmann, A., Baumgartner, W., Birkun, A., Komakhidze, A., Stanev, T., Joiris, C.R., 2000. Immunohistological and serological investigation of morbillivirus infection in harbour porpoises (*Phocoena phocoena*) from the Black Sea. Vet. Microbiol. 87 (2), 183–190.

Natoli, A., Birkun, A., Aguilar, A., Lopez, A., Hoelzel, A.R., 2005. Habitat structure and the dispersal of male and female bottlenose dolphins (*Tursiops truncatus*). Proc. R. Soc. B Biol. Sci. 272 (1569), 1217–1226.

Natoli, A., Cañadas, A., Vaquero, C., Politi, E., Fernandez-Navarro, P., Hoelzel, A.R., 2008. Conservation genetics of the short-beaked common dolphin (*Delphinus delphis*) in the Mediterranean Sea and in the eastern North Atlantic Ocean. Conserv. Genet. 9, 1479–1487. http://dx.doi.org/10.1007/s10592-007-9481-1.

Notarbartolo di Sciara, G., 1990. A note on the cetaceans incidental catch in the Italian driftnet swordfish fishery, 1986–1988: Reports of the International Whaling Commission. 40, pp, 459–460.

Notarbartolo di Sciara, G., 2012. Ancient waves, recent concerns: the budding of marine mammal conservation science in Italy. Aquat. Mamm. 38 (4), 441–455. http://dx.doi.org/10.1578/AM.38.4.2012.441.

Notarbartolo di Sciara, G., 2014. Sperm whales, *Physeter macrocephalus*, in the Mediterranean Sea: a summary of status, threats, and conservation recommendations. Aquat. Conserv. 24 (Suppl. 1), 4–10. http://dx.doi.org/10.1002/aqc.2457.

Notarbartolo di Sciara, G., 2015. Marine conservation. In: Smith, H.D., Suarez de Vivero, J.L., Agardy, T.S. (Eds.), Routledge Handbook of Ocean Resources and Management. Routledge, London and New York, pp. 145–156.

Notarbartolo di Sciara, G., Birkun, A., Jr., 2010. Conserving whales, dolphins and porpoises in the Mediterranean and Black Seas: an ACCOBAMS status report, 2010. ACCOBAMS and Department of External Relations, Principality of Monaco (can be accessed from http://bit.ly/23RYpdH).

Notarbartolo di Sciara, G., Kotomatas, S., 2016. Are Mediterranean monk seals, *Monachus monachus*, being left to save themselves from extinction? In: Notarbartolo di Sciara, G., Podestà, M., Curry, B.E. (Eds.), Mediterranean Marine Mammal Ecology and Conservation. Advances in Marine Biology, vol. 75. Elsevier, Amsterdam, pp. 359–386.

Notarbartolo di Sciara, G., Agardy, T., Hyrenbach, D., Scovazzi, T., Van Klaveren, P., 2008. The Pelagos Sanctuary for Mediterranean marine mammals. Aquat. Conserv. 18, 367–391. http://dx.doi.org/10.1002/aqc.855.

Notarbartolo di Sciara, G., Frantzis, A., Bearzi, G., Reeves, R., 2012. *Physeter macrocephalus* (Mediterranean subpopulation). In: The IUCN Red List of Threatened Species 2012. e.T16370739A16370477, http://dx.doi.org/10.2305/IUCN.UK.2012-1.RLTS. T16370739A16370477.en.

Notarbartolo di Sciara, G., Castellote, M., Druon, J.-N., Panigada, S., 2016a. Fin whales, *Balaenoptera physalus*: at home in a changing Mediterranean Sea? In: Notarbartolo di Sciara, G., Podestà, M., Curry, B.E. (Eds.), Mediterranean Marine Mammal Ecology and Conservation. Advances in Marine Biology, vol. 75. Elsevier, Amsterdam, pp. 75–101.

Notarbartolo di Sciara, G., Hoyt, E., Reeves, R.R., Ardron, J., Marsh, H., Vongraven, D., Barr, B., 2016b. Place-based approaches to marine mammal conservation. Aquat. Conserv. 26 (Suppl. 2), 85–100. http://dx.doi.org/10.1002/aqc.2642.

Öztürk, B., 2015. Nature and extent of the illegal, unreported and unregulated (IUU) fishing in the Mediterranean Sea. J. Black Sea/Mediterr. Environ. 21 (1), 67–91.

Panigada, S., Notarbartolo di Sciara, G., 2012. *Balaenoptera physalus*—Mediterranean subpopulation. In: The IUCN Red List of Threatened Species 2012. e.T16208224A17549588, http://dx.doi.org/10.2305/IUCN.UK.2012.RLTS.T16208224A17549588.en.

Panigada, S., Pesante, G., Zanardelli, M., Capoulade, F., Gannier, A., Weinrich, M.T., 2006. Mediterranean fin whales at risk from fatal ship strikes. Mar. Pollut. Bull. 52, 1287–1298.

Pinzone, M., Budzinski, H., Tasciotti, A., Ody, D., Lepoint, G., Schnitzler, J., Scholl, G., Thomé, J.P., Tapie, N., Eppe, G., Das, K., 2015. POPs in free-ranging pilot whales, sperm whales and fin whales from the Mediterranean Sea: influence of biological and ecological factors. Environ. Res. 142, 185–196.

Podestà, M., Cagnolaro, L., Cozzi, B., 2005. First record of a stranded Gervais' beaked whale, *Mesoplodon europaeus* (Gervais, 1855), in the Mediterranean waters. Atti Soc. Ital. Sci. Nat. Museo Civ. Storia Nat. Milano 146 (1), 109–116.

Podestà, M., Azzellino, A., Cañadas, A., Frantzis, A., Moulins, A., Rosso, M., Tepsich, P., Lanfredi, C., 2016. Cuvier's beaked whale, *Ziphius cavirostris*, distribution and occurrence in the Mediterranean Sea: high-use areas and conservation threats. In: Notarbartolo di Sciara, G., Podestà, M., Curry, B.E. (Eds.), Mediterranean Marine Mammal Ecology and Conservation. Advances in Marine Biology, vol. 75. Elsevier, Amsterdam, pp. 103–140.

Por, F., 1978. Lessepsian migration: the influx of Red Sea biota into the Mediterranean by way of the Suez Canal. In: Billings, W.D., Golley, F., Lange, O.L., Olson, J.S. (Eds.), Ecological Studies. Springer-Verlag, Berlin, p. 228.

Rako, N., Fortuna, C.M., Holcer, D., Mackelworth, P., Nimak-Wood, M., Pleslic, G., Sebastianutto, L., Vilibic, I., Wiemann, A., Picciulin, M., 2013. Leisure boating noise as a trigger for the displacement of the bottlenose dolphins of the Cres-Losinj Archipelago (northern Adriatic Sea, Croatia). Mar. Pollut. Bull. 68 (1–2), 77–84. http://dx.doi.org/10.1016/j.marpolbul.2012.12.019.

Reilly, S.B., Bannister, J.L., Best, P.B., Brown, M., Brownell Jr., R.L., Butterworth, D.S., Clapham, P.J., Cooke, J., Donovan, G., Urbán, J., Zerbini, A.N., 2012. *Eubalaena glacialis*. In: The IUCN Red List of Threatened Species 2012. e.T41712A17084065, http://dx.doi.org/10.2305/IUCN.UK.2012.RLTS.T41712A17084065.en.

Rendell, L., Frantzis, A., 2016. Mediterranean sperm whales, *Physeter macrocephalus*: the precarious state of a lost tribe. In: Notarbartolo di Sciara, G., Podestà, M., Curry, B.E. (Eds.), Mediterranean Marine Mammal Ecology and Conservation. Advances in Marine Biology, vol. 75. Elsevier, Amsterdam, pp. 37–74.

Rendell, L.E., Gordon, J.C.D., 1999. Vocal response of long-finned pilot whales to military sonar in the Ligurian Sea. Mar. Mamm. Sci. 15 (1), 198–204.

Rosel, P.E., Frantzis, A., Lockyer, C., Komnenou, A., 2003. Source of Aegean Sea harbour porpoises. Mar. Ecol. Progr. Ser. 247, 257–261.

Scheinin, A.P., Kerem, D., MacLeod, C.D., Gazo, M., Chicote, C.A., Castellote, M., 2011. Gray whale (*Eschrichtius robustus*) in the Mediterranean Sea: anomalous event or early sign of climate-driven distribution change? Mar. Biodiv. Rec. 4, e28. http://dx.doi.org/10.1017/S1755267211000042.

Scovazzi, T., 2016. The international legal framework for marine mammal conservation in the Mediterranean Sea. In: Notarbartolo di Sciara, G., Podestà, M., Curry, B.E. (Eds.), Mediterranean Marine Mammal Ecology and Conservation. Advances in Marine Biology, vol. 75. Elsevier, Amsterdam, pp. 387–416.

Simmonds, M.P., Gambaiani, D., Notarbartolo di Sciara, G., 2012. Climate change effects on Mediterranean cetaceans: time for action. In: Stambler, N. (Ed.), Life in the Mediterranean Sea: A Look at Habitat Changes. Nova Science Publishers, Inc, Hauppauge, New York, pp. 685–701.

Squadrone, S., Brizio, P., Chiaravalle, E., Abete, M.C., 2015. Sperm whales (*Physeter macrocephalus*), found stranded along the Adriatic coast (Southern Italy, Mediterranean Sea), as bioindicators of essential and non-essential trace elements in the environment. Ecol. Indic. 58, 418–425.

Tudela, S., 2004. Ecosystem effects of fishing in the Mediterranean: an analysis of the major threats of fishing gear and practices to biodiversity and marine habitats. Gen. Fish. Counc. Mediterr. Stud. Rev. 74, 1–44.

Tyack, P.L., Zimmer, W.M., Moretti, D., Southall, B.L., Claridge, D.E., Durban, J.W., Clark, C.W., D'Amico, A., Di Marzio, N., Jarvis, S., McCarthy, E., 2011. Beaked whales respond to simulated and actual navy sonar. PLoS One 6 (3), e17009.

UNEP-MAP, 2012. State of the Mediterranean marine and coastal environment. UNEP/MAP, Barcelona Convention, Athens.

UNEP-MAP-RAC/SPA, 2010. Overview of scientific findings and criteria relevant to identifying SPAMIs in the Mediterranean open seas, including the deep sea. By Notarbartolo di Sciara, G. and Agardy, T. Edited by RAC/SPA, Tunis (can be accessed from http://bit.ly/1W8jste).

Verborgh, P., Gauffier, P., Esteban, R., Giménez, J., Cañadas, A., Salazar-Sierra, J.M., de Stephanis, R., 2016. Conservation status of long-finned pilot whales, *Globicephala melas*, in the Mediterranean Sea. In: Notarbartolo di Sciara, G., Podestà, M., Curry, B.E. (Eds.), Mediterranean Marine Mammal Ecology and Conservation. Advances in Marine Biology, vol. 75. Elsevier, Amsterdam, pp. 173–203.

Weinrich, M.T., Panigada, S., Guinet, C., 2006. ACCOBAMS Workshop on Large Whale Ship Strikes in the Mediterranean Sea: Report to the Secretariat of ACCOBAMS. p. 43. Monaco, 14–15 November 2005.

Williams, R., Wright, A.J., Ashe, E., Blight, L.K., Bruintjes, R., Canessa, R., Clark, C.W., Cullis-Suzuki, S., Dakin, D.T., Erbe, C., Hammond, P.S., Merchant, N.D., O'Hara, P.D., Purser, J., Radford, A.N., Simpson, S.D., Thomas, L., Wale, M.A., 2015. Impacts of anthropogenic noise on marine life: publication patterns, new discoveries, and future directions in research and management. Ocean Coastal Manage. 115, 17–24. http://dx.doi.org/10.1016/j.ocecoaman.2015.05.021.

CHAPTER TWO

Mediterranean Sperm Whales, *Physeter macrocephalus*: The Precarious State of a Lost Tribe

L. Rendell*[,1], A. Frantzis[†]
*Sea Mammal Research Unit, University of St Andrews, St Andrews, Fife, United Kingdom
[†]Pelagos Cetacean Research Institute, Vouliagmeni, Greece
[1]Corresponding author: e-mail address: ler4@st-andrews.ac.uk

Contents

Abstract

First observed in the classical era, a population of sperm whales (*Physeter macrocephalus*) persists to this day in the deep waters of the Mediterranean Sea. Genetic and observational evidence support the notion that this is an isolated population, separated from its Atlantic neighbours. These whales depend on mesopelagic squid for food, and appear to occupy a very similar ecological niche to sperm whales in the open oceans. Recent evidence proving that individuals can pass between the eastern and western deep water basins confirms that this is a single population, not isolated into western and eastern stocks. We lack robust information on their population status, but they could number in the hundreds rather than thousands, and current densities appear to be much lower than those reported in the 1950s, suggesting that we should be very concerned about the conservation status of this population. This makes it vitally important to address the serious threats posed by ship strikes and entanglement in fishing nets, especially driftnets, and to carefully monitor other potential sources of anthropogenic impact. A step change in funding to collect better data and a clear shift in policy priorities are needed if we are to be serious about conserving this population.

Advances in Marine Biology, Volume 75
ISSN 0065-2881
http://dx.doi.org/10.1016/bs.amb.2016.08.001

1. INTRODUCTION

The Mediterranean Sea is bounded on nearly all sides by crowded, largely industrialised coastlines, criss-crossed by hundreds of thousands of vessels, subject to exploration for and extraction of fossil fuels, and heavily harvested for food. Yet it still has its depths, and dramatic seafloor relief, and, perhaps astonishingly, is home to its own population of the largest toothed whale, the dominant mesopelagic predator, the sperm whale (*Physeter macrocephalus*, Linnaeus, 1758; Fig. 1).

Worldwide, sperm whales live in pelagic waters, typically, although not exclusively, foraging on squid or fish at depths of several hundreds of metres. The current global population is estimated to be around 360,000 (Whitehead, 2002), reduced from a prewhaling abundance of over 1 million, although these estimates were obtained by inference and extrapolation rather than direct measurement, so are inherently uncertain. Nonetheless, there is little doubt that this is an ecologically successful species—sperm whales are found in the deep waters of every ocean on Earth, and their range extends from the tropics where female social groups are congregated to latitudes at least as high as 70°, beyond which only large males are found (Rice, 1989). Although there are surely many factors underlying their ecological success, two stand out in particular. The first is the sperm whale's highly evolved sound production system, the most intense animal sonar on Earth, which is capable of producing highly directional and extremely powerful

Fig. 1 A sperm whale (*Physeter macrocephalus*) breaches in the waters of the Hellenic Trench, Greece. *Photograph: Copyright A. Frantzis, Pelagos Cetacean Research Institute.*

pulses of sound (Møhl et al., 2000; Zimmer et al., 2005b) that the whales use to sense their surrounding and detect and pursue prey (Miller et al., 2004). The second is their social behaviour, incorporating long-term bonds and multiple, hierarchically organised layers of structure. The evolution of this social structure is generally thought to have been driven by the dilemma that dependent calves, unable to dive to depth, impose upon their mothers who need to leave them vulnerable at the surface while they dive to forage (Whitehead, 2003).

Sperm whale society is sharply divided between males over about 15 years of age, and females and immatures of both sexes. The most basic structure of sperm whale society is the long-term social *unit*, comprising one or more matrilines of female whales (Christal et al., 1998; Mesnick, 2001; Ortega-Ortiz et al., 2012; Whitehead and Weilgart, 2000; Whitehead et al., 1991). All the available evidence suggests that the basic function of these units is to provide care for calves (Gero et al., 2009, 2013; Mesnick et al., 2003; Whitehead, 1996), although they also likely provide protection for adults (Pitman et al., 2001). Calves are apparently unable to make prolonged foraging dives for the first few years of their lives, and remain dependent upon their mother's milk during that time. Mothers however need to forage continuously to sustain lactation and so have no choice but to leave their calves at the surface, during which time they are exposed to predation risk in the refuge-less pelagic surface waters. Calves left at the surface are regularly escorted by other adults in their natal social unit, and the suggestion is that sociality in female sperm whales is largely driven by these interactions (Gero et al., 2013; Whitehead, 2003; Whitehead and Weilgart, 2000). Wherever they have been studied, female sperm whales live in these social units, although they may vary in size between populations (Whitehead et al., 2012). In some areas, these units also form short-term associations with each other, lasting on average around 10 days, and hence the social structure has, in the Pacific Ocean at least, a fission–fusion quality at the level of unit, rather than individual (Whitehead, 2003). In other regions however the picture is different; in the Atlantic, for example, associations between units are typically much shorter, such that units are typically encountered alone, and unit sizes can also vary between regions (Gero et al., 2015; Whitehead et al., 2012). The reasons behind this variation are still not understood.

Our knowledge of sperm whales in the Mediterranean dates back to classical antiquity. Aristotle, who lived and worked on a Mediterranean shore, provided the first descriptions of the morphology and natural history of the sperm whale. In his work 'Των περί τα ζώα Ιστοριών', known as *Historia*

Animalium, written almost 2370 years ago, Aristotle (2004) describes how 'the air-passage of the dolphin goes through its back, but that of the whale is found in its forehead'. There are two reasons to think that, by 'whale', he is almost certainly referring to the sperm whale, as opposed to baleen whales. First, his description of the blowhole position cannot correspond to anything but the sperm whale's unique characteristic among cetaceans, and second, he refers separately elsewhere to the 'μυστakokήτος' (the root of the modern term 'mysticete' for baleen whales), which he notes 'instead of teeth has hairs in its mouth resembling pigs' bristles'. Although Aristotle made no reference of the exact location of the observations that he reported, his knowledge was likely to have originated mainly from the north Aegean Sea, since he was born on the Chalkidiki Peninsula, off which sperm whales can still be found today. Ironically, despite this auspicious beginning, research on sperm whales has been driven since then by studies in other regions, notably the eastern tropical Pacific Ocean (Whitehead, 2003). Our understanding of the lives of these animals in the Mediterranean Sea lagged behind until the start of dedicated long-term studies in several areas in the past two decades that are, as we review later, now beginning to bear fruit.

The absence of any significant whaling targeting sperm whales in Mediterranean waters was probably a crucial factor in their survival in the region to the present day. Although ninteenth century whalers knew about and exploited the Gibraltar Straits ground, this effort was, according to the logbooks of the time, concentrated on the Atlantic side of the Strait. An analysis of 317 logbooks from the era showed only two recorded expeditions into the Mediterranean itself, and that these led to a minimum removal estimate of 237 animals in the period from 1862 to 1899 (Aguilar and Borrell, 2007). It has also been reported that sperm whales were hunted more recently with explosives around the Straits of Messina in the years immediately following World War II (e.g. Bolognari, 1949), but accurate records were not kept so we do not know how many animals were killed during that time (Frantzis et al., 2011).

Little scientific progress was made on understanding sperm whales in the Mediterranean Sea between Aristotle's times and the late twentieth century, except for studies performed on stranded animals by anatomists and zoologists. The postwar observations of Arturo Bolognari, based at the University of Messina in Italy, are a notable exception. He recorded frequent sightings of large sperm whale herds off the coast of north-eastern Sicily, and published his observations in a series of papers that appeared between 1949 and 1957 (e.g. Bolognari, 1950, 1951; summarised in Frantzis et al., 2011).

Sightings of groups containing up to 30 individuals were not unusual in Bolognari's reports, but are unheard of today. Indeed, in the personal experience of both authors of this chapter, many people in the Mediterranean region are not aware that sperm whales swim there today, and are astonished when they learn it.

Today the scientific study of Mediterranean sperm whales is being driven forward by a number of small research groups that generally receive very little governmental support, often (although not always) operating outside the 'traditional' context of academic university-based research, and relying on dedicated efforts by volunteers from all walks of life (Table 1). Specific contributions have also been made by the United Kingdom (UK) based, Marine Conservation Research (www.marineconservationresearch.co.uk) group, which has conducted a number of large spatial scale surveys of Mediterranean waters. This is by no means an exhaustive list, but illustrates the range of efforts that are ongoing. Most of the broader insights we have gained in recent years are the result of *ad hoc* efforts to coordinate and pool genetic and/or photo-identification data among these studies.

The International Union for the Conservation of Nature (IUCN) currently considers the Mediterranean sperm whale to be the only subpopulation of sperm whales with a distinct conservation status, Endangered (Notarbartolo di Sciara et al., 2012a). This assessment is not based on hard data regarding population trends—as we shall discuss, no such data exist—but rather on the justifiable concern that observed mortality levels could not be considered sustainable under any reasonable extrapolation of population estimates from specific subregions to the entire Mediterranean. The assessment is based on IUCN Red List criterion C2a(ii) which refers to a 'population size estimated to number fewer than 2500 mature individuals and either: a continuing decline, observed, projected, or inferred, in numbers of mature individuals and at least 95% of mature individuals in one subpopulation'. There is great concern that scientific efforts to understand the ecology and population dynamics of the Mediterranean sperm whale population lag far behind the needs of conservation managers and governmental decision makers (Pace et al., 2014b). In this chapter, we review current scientific knowledge, organising information around a number of key questions that pertain to the critical issues for conserving this population. First we consider whether the Mediterranean sperm whale should be considered 'special', both in terms of the extent to which it is an isolated population, and whether it has unique behavioural and ecological features. The question of isolation is vital, since it determines how we should respond to declining numbers in

Table 1 Organisations That Have Conducted Major Long-Term Research on Sperm Whales (*Physeter macrocephalus*) in the Mediterranean Sea

Organisation	Base	Study Started	Study Area(s)
CIRCE (Conservación, Información y Estudio sobre Cetáceos; www.circe.info)	Cádiz, Spain	2001	Strait of Gibraltar
GREC (Groupe de Recherche sur les Cétacés; www.cetaces.org)	Antibes, France	1991	Entire Mediterranean Sea
Tethys Research Institute (www.tethys.org)	Milan, Italy	1986	Corso-Ligurian-Provencal basin, the Adriatic and Ionian seas
Pelagos Cetacean Research Institute (www.pelagosinstitute.gr)	Athens, Greece	1998	Greek waters, Hellenic Trench
Oceanomare-Delphis Onlus[a] (ODO; www.oceanomaredelphis.org)	Ischia, Italy	2002	Deep waters around island of Ischia in Bay of Naples
SEAME Sardinia (Scientific Education and Activities in the Marine Environment; www.seame.it)	Sardinia, Italy	2010	Waters off Northeastern Sardinia
Balearic Sperm Whale Project (synergy.st-andrews.ac.uk/balearicspermwhales)	Sea Mammal Research Unit, University of St Andrews and Asociación Tursiops (www.asociaciontursiops.org), Palma de Mallorca, Spain	2003	Waters around the Balearic archipelago

[a]Organised the first international workshop on the ecology and conservation of Mediterranean sperm whales in 2011 (Pace et al., 2014b).
This is by no means an exhaustive list, but illustrates the range of efforts that are ongoing.

this population. We then review the evidence regarding how sperm whales make their living in the Mediterranean, both from direct studies of diet, and from studies of habitat use and preference that provide clues to the ecology that we cannot directly study in the meso- and bathypelagic waters where the whales forage. Understanding the ecology of this population properly is crucial to appropriately designing and implementing measures to conserve it. We then consider the evidence pertinent to our understanding of the structure of this population, since conservation measures cannot be effective if they are not underpinned by proper understanding of the structure of the population or populations one is seeking to conserve. In the penultimate section of the chapter, we collate the available evidence on the size of the Mediterranean sperm whale population, and finally, we focus on conservation directly by attempting to prioritise the manifold threats to its survival.

2. IS THE MEDITERRANEAN SPERM WHALE SPECIAL?

The Mediterranean Sea is closed off from the world's oceans except for the narrow Strait of Gibraltar. Its geography raises one critical question about the sperm whale population there—to what degree is it isolated from the larger populations in the Atlantic? The answer has major implications for how we think about sperm whales in the Mediterranean, and can also tell us about how large marine mammals can adapt to life in an 'oceanic island'. Do these animals represent a 'lost tribe', descendants from a single colonisation event that has since remained isolated on its own trajectories, or are they an extended 'lobe' of the huge North Atlantic population? Clearly the answer to that question will dramatically impact the conservation perspective we should take with respect to the Mediterranean sperm whale, because if it is isolated, then the population is highly vulnerable since it is uncertain if it can ever be replenished from neighbouring stocks. Views on this issue have varied dramatically. As recently as the 1980s, some researchers expressed an assumption that sperm whales in the Mediterranean were temporary residents from the large Atlantic stock, even going so far as to suggest that the Mediterranean might be used as a special 'nursery' area for that population (Di Natale and Mangano, 1985). However, none of the data collected since then have provided any support for this idea. On the contrary, although logically it is impossible to prove that animals never enter or leave the Mediterranean Sea, all the evidence available for us to review below is consistent with the notion that the Mediterranean sperm whale population is separate and isolated from that of the Atlantic.

In geological terms, the history of sperm whales' presence in the Mediterranean is relatively recent; this is because the Mediterranean itself, as it appears today, is a recent sea. Between six and five million years ago the landmasses of Africa and Europe, in their converging motion, had squeezed the ancient Tethys Sea to the point of cutting it off completely from the world's oceans, thereby causing most of the water to evaporate and transforming the region into a desert interspersed by series of shallow, hypersaline lagoons (Krijgsman et al., 1999). This upheaval, the Messinian Salinity Crisis, obliterated most of the Tethys' ancient marine life forms. Subsequent tectonic readjustments eventually recreated an opening between Europe and Africa, in the site known today as the Strait of Gibraltar, and our exploration of the isolation question starts at that location.

Sperm whales are regularly seen in the Strait of Gibraltar, even on the Atlantic side where the Strait opens to the ocean (De Stephanis et al., 2008), so it seems clear that there is no physical barrier that rules out *a priori* any exchange of animals with the Atlantic. Photo-identification has revealed that animals sighted in the Strait are regularly observed all across the Western Mediterranean Sea, right up to the northeast corner of the Ligurian Sea. Of 47 animals identified in the Strait from 1999 to 2011, 15% were identified in other parts of the western Basin during the time period from 1994 to 2011 (Carpinelli et al., 2014). However, there has been no photo-identification effort on the Atlantic side of the Strait that we are aware of, and the closest records are from the Azores, some 1800 km distant (Matthews et al., 2001). No matches between Mediterranean and Azores catalogues have resulted from comparisons conducted up to the time of writing (Lisa Steiner, Whale Watch Azores, Portugal, personal communication, 2 August 2016), but a lack of matches cannot rule out the possibility that animals move in and out of the Mediterranean, especially as there is no *a priori* reason to think they should head to or come from the Azores after exiting or before entering the Strait of Gibraltar. Boisseau et al. (2010) conducted visual and acoustic surveys in the adjacent Atlantic Ocean and reported single sightings of sperm whales in Moroccan waters to the south, and Spanish waters to the north, of the Strait, and none directly west, but survey effort was too low to draw robust conclusions. Thus, the survey data from both line-transect and photo-identification studies are currently insufficient to definitely answer the isolation question with confidence.

The powerful tools of population genetics have also been applied to this question. One study collected skin samples from 13 live individuals in the western Basin and the Ionian Sea in the eastern Basin, and compared the first 200 bp of the mitochondrial DNA (mtDNA) control region from these samples to sequences obtained from animals stranded on the coasts of the British Isles and northern Europe, but not adjacent Atlantic waters (Drouot et al., 2004a). All 13 Mediterranean individuals shared an identical sequence, which was also found in 28 (49%) of the 57 samples from outside; 23 of the remaining non-Mediterranean animals had a second sequence, and a third was found in a single individual. However, 200 bp is not a very long sequence in the context of current sequencing technology, so more variation may be revealed with mitogenomic analyses. More recently, Engelhaupt et al. (2009) analysed samples from 293 individuals, of which 44 were collected from the Mediterranean, and assessed variation in both mtDNA (up to 399 bp) and 16 nuclear microsatellite loci. Like Drouot et al. (2004a), Engelhaupt et al. (2009) found that all the Mediterranean samples shared an identical mitochondrial sequence, which they identified as haplotype 'C'. This haplotype is not however unique to the Mediterranean population, as globally it is one of the three most common sequences in sperm whales, found as far afield as the eastern tropical Pacific (Rendell et al., 2012). More recently, Alexander et al. (2016) also found just a single myDNA haplotype among 40 individuals sampled from the Mediterranean. This tells us that only a single matriline has colonised the Mediterranean, but does not rule out the possibility of individuals belonging to that matriline entering and leaving. The picture becomes clearer however when nuclear DNA analyses are also considered. Engelhaupt et al.'s (2009) analysis of microsatellite variation across the entire North Atlantic showed unambiguously that the Mediterranean population, and only that population, was significantly differentiated, and analyses that attempted to segregate the samples into putative breeding populations indicated that there were only two such populations in the sample—the Mediterranean and everywhere else. Thus the data from population genetics strongly suggest that the Strait of Gibraltar is a significant or plausibly even isolating barrier to gene flow and are certainly inconsistent with any degree of regular movements of animals into and out of the Mediterranean Sea.

There is a non-genetic aspect of population structure in sperm whales, the variation in vocal dialects of stereotyped patterns of broadband clicks called 'codas' (Rendell et al., 2012; Whitehead et al., 1998). These dialects can divide populations in sympatry (Rendell and Whitehead, 2003), and there are some types of codas that appear to mark or delineate discrete subpopulations (Gero et al., 2016). Like sperm whales everywhere, Mediterranean animals primarily make codas while socialising, and at the beginning and end of dives (Teloni, 2005; but see Frantzis and Alexiadou, 2008). Every study published to date agrees that vocal dialects in the Mediterranean are distinctive.

Mediterranean coda dialects are dominated by one type of coda—the '3 + 1' type, comprising three regularly spaced clicks followed by a longer pause before the last click. Pavan et al. (2000) analysed 138 codas from 15 encounters over a 12-year period ending 1997, in the Ligurian and Tyrrhenian Seas (i.e. the western Basin) and found that 134 (97%) were of the 3 + 1 pattern. Similarly, Teloni (2005) analysed 131 codas from 27 different whales in the years 2001 to 2003, of which 128 (98%) were of the 3 + 1 pattern, again from the western Basin. These studies give the impression that the coda repertoire of Mediterranean sperm whales is almost completely dominated by a single type. This is very unusual compared to every other study published from recordings made in other regions that report a much higher diversity (Gero et al., 2016; Rendell and Whitehead, 2003; Weilgart and Whitehead, 1993). However, the codas analysed by Pavan et al. (2000) and Teloni (2005) did not come from identified social units but from singletons, presumably males, while the larger sample of 751 codas recorded from 13 different encounters analysed by Drouot et al. (2004c) were produced by units. This latter study, which included data from the eastern Basin, used extended sampling of social units to reveal a wider coda repertoire and a broader diversity of coda structures, albeit still dominated by the 3 + 1 pattern, which accounted for around 67% of all the codas recorded. In addition to reporting the presence of the 'standard' 3 + 1 coda in the eastern Basin population, as well as a diversity similar to that obtained by Drouot et al. (2004c), Frantzis and Alexiadou (2008) report the only evidence we are aware of where specific types of codas have been linked to behavioural contexts. This latter study described codas of a type termed 'root' that were characterised as being relatively short in absolute durations compared to other types. 'Root' codas were identified by the presence of a series of very rapid clicks at the start, were produced at the surface, and strongly associated with contexts where regular foraging dive cycles had been interrupted by some kind of disturbance, such as the proximity of the research vessel, or a swimmer entering the water. This led Frantzis and Alexiadou (2008) to suggest that these root-type codas might function as alarm calls, a common feature in terrestrial vocal communication (Macedonia and Evans, 1993). Studies from other regions have reported coda repertoires that have one or two major types (for example, Gero et al. (2016) showed that a single 1 + 1 + 3 type accounted for 39% of the codas recorded off Dominica) but they do not dominate the repertoires in the same way as the 3 + 1 does in the Mediterranean. Since coda repertoires are almost certainly acquired by cultural transmission (Rendell et al., 2012), this suggests that cultural evolutionary processes may have affected coda dialects differently in the Mediterranean compared to other regions. Some bird species also acquire vocal dialects by cultural transmission, and studies of populations on islands have highlighted a loss of syllable diversity relative to mainland populations as a consequence of cultural evolution in this isolated context (Catchpole and Slater, 2008). The apparent dominance of the 3+1 type in Mediterranean sperm whales may therefore be another piece of evidence suggesting that the population is isolated.

We can also compare other aspects of sperm whale behaviour within and outwith the Mediterranean Sea. One such aspect is segregation between age and sex classes. Globally, sperm whale populations show strong divisions among age/sex classes, with tropical and subtropical social units of females plus calves and juveniles segregating from mid-latitude groups of maturing males (often labelled 'bachelor' schools) and large high-latitude and typically singleton mature males (Whitehead, 2003; Whitehead and Weilgart, 2000). In contrast, the restricted latitudinal range available in the Mediterranean does not allow for such segregation, and it appears to be the norm there for these age/sex classes to inhabit the same waters, even if they do not closely associate. Along the Hellenic Trench, social units contain an average of around eight individuals, but appear to have a much more fluid structure in comparison to open ocean populations, with repeated observed instances of animals apparently changing units (Frantzis et al., 2014). In this area, solitary males, loose male aggregations, social units and small bachelor groups have been observed to coexist. Similarly, during nine years of surveys conducted off the Bay of Naples, Italy, Pace et al. (2014a) observed all the major types of groupings that have been identified elsewhere (female social groups, singleton males and bachelor groups), occurring in the same general study site. The same coexistence of age/sex classes is seen in the waters around the Balearic Islands (Pirotta et al., 2011). These studies suggest that the Mediterranean sperm whale population has a relatively compressed spatial social structure compared to their ocean counterparts, and may therefore experience more intraspecific competition for prey resources. Drouot et al. (2004b) did however observe a difference in the group sizes encountered in the northern and southern halves of the western Basin, where over 70% of encounters in the northern half were singletons, presumably males, and 80% of sightings in the southern half were of groups that contained calves. This variation led Drouot et al. (2004b) to suggest that even though the latitudinal range of the Mediterranean Sea is much smaller than for any ocean inhabited by sperm whales, at least in the western Basin they still maintain a scaled down version of the social division between males at higher latitudes and female groups at lower ones seen in the wider oceans (Whitehead, 2003;

Whitehead and Weilgart, 2000). Drouot et al.'s (2004b) survey did not reach much further south than the Balearics; however, since that study, long-term research around the Balearic archipelago has shown that the area is equally if not more heavily used by singleton males than female groups (Pirotta et al., 2011). It therefore appears that sperm whale age/sex classes which are typically strongly allopatric elsewhere are found in sympatry in the Mediterranean. It is unclear whether this reflects a fundamentally different behavioural response or simply a latitudinal compression of the 'normal' pattern of age/sex segregation.

Nonetheless, repeated observations in the Mediterranean reveal a sympatry of solitary males and female groups, with no obvious reproductive purpose (Frantzis et al., 2014; Pace et al., 2014a; Pirotta et al., 2011), that is not typically seen in other study sites (Whitehead, 2003). Such overlapping ranges could alter the way that different age/sex classes of sperm whales divide up ecological niches, and could in theory lead to increased competition for resources between female groups and solitary or very loosely associated juvenile males. Our understanding of sperm whale feeding ecology is currently too poor to raise this above the level of speculation, although comparisons based on the modelling work of Pirotta et al. (2011) suggest that, within a given region, there might be fine-scale partitioning evident in the differing habitat preferences of singletons and groups, offering a potential way forward (Jones et al., 2016). Detailed study of diving and feeding behaviour using passive acoustics and/or archival suction cup tags (e.g. Gannier et al., 2012; Watwood et al., 2006) could help us understand whether there are likely to be competition effects as a result of this age/sex class sympatry that is not seen in other oceans. This is important because if, for example, lactating female sperm whales are facing competition for resources from subadult males that they do not face in other populations, this could lead to constraints on population growth rate that are not predicted by studies outside the Mediterranean Sea.

Female sperm whales the world over live in long-term social units, but patterns of group formation appear to vary among populations (Whitehead et al., 2012). Since cetacean group sizes result from a balance of evolutionary and ecological forces that can change in both space and time (Connor, 2000), examining unit sizes in the Mediterranean Sea can give us clues as to any special conditions that may pertain there. As ever, we do not have as much data available as we would like, and that which we do have is not completely consistent. In 51 encounters over five years in the western Basin, Drouot et al. (2004b) did not encounter any groups of more than seven animals. Moulins and Würtz (2005) reported a 'herd' of 10 females and calves sighted in the Ligurian Sea off Monaco in 2001, but there have been no such reports since, suggesting that this was a rare and exceptional observation for the northern part of the western Basin. In their surveys along the Hellenic Trench from 1998 to 2009, Frantzis et al. (2014) identified 16 social units, with 13 resighted across multiple years, containing four to 13 individuals, with an average of eight. The authors also observed several instances of apparently temporary aggregations containing multiple social units with up to 15 individuals present, and on four occasions what they termed 'gatherings' containing 17 to 20 individuals within a 10 km radius, comprising either two social units or a social unit with a loose aggregation of males. We do not yet have definitive evidence, but the available data hint that female social units in the Mediterranean may be smaller than those encountered in the Pacific and the North Atlantic, which typically contain around 11 to 12 females and immatures (Whitehead et al., 2012). There are however two populations in which comparable unit sizes have been reported, the Gulf of Mexico and Dominica, where social units contain five to six individuals on average (Whitehead et al., 2012). It is possible then that similar evolutionary and ecological forces are acting on social unit size both in these latter populations and in the Mediterranean, although the nature of these forces remains a matter of speculation.

Variation in ecological conditions can result in varying growth patterns in different populations of the same species. There is in fact good evidence to suggest that Mediterranean sperm whales may show such variation compared to other populations. During the entire twentieth century, and until the time of writing, there have been no reliable records of stranded sperm whales that exceeded 15 m in total length. Bearzi et al.'s (2011) exhaustive survey of strandings in the Adriatic Sea revealed only one instance of a whale greater than 15 m length, a 16.5 m measurement reported in the 1920s from Croatia, but the author of the original report (Hirtz, 1921) did not measure the whale himself. A recent mass stranding in the Adriatic consisted entirely of males less than 12.5 m in length (Mazzariol et al., 2011). The ages of two whales stranded in Greece were estimated by counting growth layer groups in tooth sections, showing that a 25-year-old female had attained a length of only 10 m and, even more striking, a 44-year-old male had attained only 12.8 m in length (Frantzis et al., 2003). However, while measurements of stranded animals are generally accurate and can give us important information, they are not a random sample of a population—there might be a correlation between being relatively small and risk of stranding if growth has been retarded by malnutrition, for example.

Although it is very difficult to measure the length of sperm whales at sea, their unique sound production anatomy results in an acoustic index of the size of the vocalising animal. Each click contains multiple pulses resulting from echoes of the original pulse reflected of air-sacs at each end of the spermaceti organ, and measurement of this interpulse interval (IPI) can therefore inform us of the size of vocalising animals (Gordon, 1991; Growcott et al., 2011; Zimmer et al., 2005a). Drouot et al. (2004b) used this technique to estimate the sizes of 31 animals sampled across the entire Mediterranean region, with a maximum size estimate of 13.5 m. Along the Hellenic Trench, similar acoustic measurements suggested a mean length of 11.4 m and a range of 8.9–14.6 m for 19 males, and a mean of 9.1 m with a range of 8.6–9.5 m for 9 females (Frantzis and Alexiadou, 2008; Frantzis et al., 2014). Caruso et al. (2015) took advantage of the deployment of two deep-sea hydrophones as part of a project to test technologies for the construction of a neutrino telescope to analyse 156 different acoustic encounters, resulting in 183 IPI estimates from an unknown number of sperm whales that contained no estimate greater than 14 m. This contrasts with measurements from other regions, where male sperm whales consistently grow longer than 15 m and regularly attain lengths in excess of 17 m, although the sizes of females do not appear quite so anomalous (Whitehead, 2003). It is impossible to say whether this size difference is due to an evolved reduction in growth in this population or a phenotypic response resulting from a lack of access to the highly productive high-latitude waters in which male sperm whales outside the Mediterranean Sea grow to their mature lengths (Whitehead and Weilgart, 2000), but it does suggest that Mediterranean sperm whales may be subject to different ecological and evolutionary forces than those experienced by their open ocean counterparts.

3. MAKING A LIVING IN A SMALL AND OLIGOTROPHIC OCEAN

Any effort to conserve Mediterranean sperm whales will be seriously impaired if we do not understand their foraging ecology. Unfortunately, it is very challenging to study the foraging of a mesopelagic predator directly. Instead, we typically rely on indirect techniques such as examining the stomach contents of stranded animals, measuring stable isotope profiles, making inferences from echolocation behaviour and, perhaps most commonly, trying to understand their ecology from studies of habitat use.

According to all the evidence we are aware of, the Mediterranean sperm whale's diet is almost exclusively squid, with rare instances of fish and octopus being eaten. De Stephanis et al. (2008) identified squid beaks in the stomachs of a whale stranded on the Spanish coast, and Mazzariol et al. (2011) identified 'well-digested' squid beaks in the otherwise empty stomachs of seven animals stranded in the Adriatic region. Roberts (2003) reported that the stomach of a male sperm whale found floating dead near Crete, Greece, contained nearly 3000 squid beaks from seven squid species, all known to occur in the Mediterranean. Garibaldi and Podestà (2014) similarly report the stomach contents of a single whale stranded in NW Italy as being dominated by histioteuthid beaks, having identified 233 upper and 291 lower beaks. Finally, the largest sample comes from the stomach contents of eight sperm whales recently stranded along the Greek coast, which comprised remains belonging almost exclusively to mesopelagic squids, represented by over 30,000 lower squid beaks from 12 squid species, all known to occur in the Mediterranean. Among them were one common octopus beak and remains of 12 ray-finned fishes of total length less than 11 cm (I. Foskolos/Pelagos Cetacean Research Institute, unpublished data). These results illustrate the overwhelming importance of squid in the diet of the Mediterranean sperm whales, as is the case for most sperm whales around the world (Whitehead, 2003). This dependence on deep-sea squids is surely a key factor in the persistence of such a large predator in the oligotrophic Mediterranean Sea, since it confers a degree of independence from the main food chain that passes from phytoplankton to zooplankton to fish in the surface waters, although just how independent they can be remains to be seen. Nonetheless, as elsewhere, in the Mediterranean the sperm whale is a key top predator of the meso- and bathypelagic habitat. Praca and Gannier (2008) compared the ecological niches of sperm whales and two other teuthophagous cetaceans, long-finned pilot whales (*Globicephala melas*) and Risso's dolphins (*Grampus griseus*), and reported that while sperm whales appeared to have the broadest habitat use, in terms of using habitat across a range of depths, they also had the most specialised diet, and a distribution that seemed to be influenced by both properties of the water mass and topographic features, such as the shelf edge.

Since sperm whales use echolocation to forage (Miller et al., 2004), it is possible to learn something about their activities in deep water using acoustic signatures of prey capture attempts. When attempting to capture an individual prey item, sperm whale click rates increase rapidly to produce a 'buzz', and these buzzes are associated with rapid changes in direction indicating strong manoeuvring in the water column (Miller et al., 2004). Teloni (2005) analysed towed hydrophone recordings of 279 dives from 204 different encounters of singleton whales and found a mean foraging period per dive, as characterised by the production of both regular echolocation clicks and buzzes, of 35 min. These figures are very similar to the values reported from the Ligurian Sea using recordings from acoustic recording suction cup dTags placed on 12 animals (28 min), and also very similar to data obtained from whales in the northeast Atlantic and the Gulf of Mexico (Watwood et al., 2006). This study also found that whales in the Ligurian Sea produced very similar average numbers of buzzes per foraging dive (18.5) to whales sampled in the other areas. Using surface hydrophones rather than on-animal tags, Gannier et al. (2012) measured an average of 25 buzzes per dive from a larger sample of 156 dives during 52 sperm whale sightings. It is not clear whether the difference between the two studies is simply sampling noise or some systematic difference between the remote and on-animal recording methods, but in either case the broader point still stands—sperm whale foraging behaviour in the Mediterranean appears to be very typical of the species as a whole, suggesting, along with the stomach contents analyses, that this population occupies a very similar ecological niche to those in the oceans.

The relatively novel technique of tracking isotope signatures across growth layer groups in the teeth of stranded animals offers another way to understand the feeding ecology of these animals. To the best of our knowledge, this technique has only been performed on teeth from two Mediterranean sperm whales, a male and a female, both stranded in Greece, as part of a study that also analysed teeth from locations in Iceland, Scotland and the Azores (Mendes et al., 2007). Notably, both animals from the Mediterranean had lower $\delta^{15}N$ levels than did any of the other samples, which the authors interpreted as reflecting the oligotrophic nature of the Mediterranean Sea, since $\delta^{15}N$ depletion is characteristic of reduced levels of nitrate assimilation by phytoplankton. By contrast, the $\delta^{15}C$ levels of both whales were not unusual within the samples analysed, indicating that the animals were not utilising food sources that were closely associated with coastal waters, which is consistent with all the evidence above that Mediterranean sperm whales have a similar squid-based diet to sperm whales all over the world.

Since the distribution of marine predators is strongly affected by that of their prey, studying how sperm whales distribute themselves across their habitat can provide insights into their foraging ecology. This is a relatively well-studied aspect of Mediterranean sperm whale ecology. A series of studies have shown how sperm whale distribution in the Mediterranean is not uniform, but instead concentrated in specific regions and associated with specific topographic and oceanographic features (Frantzis et al., 2014; Gannier and Praca, 2007; Gannier et al., 2002; Pirotta et al., 2011; Praca and Gannier, 2008; Praca et al., 2009). Across the broadest spatial scales, Gannier et al. (2002) analysed four years of combined visual and acoustic survey effort from a single vessel, spanning the entire region from Gibraltar to the Ionian Sea and reported high relative abundance in two areas. They highlighted the Gulf of Lion, the southwestern Basin and the Ionian Sea as areas of relatively high density, and the Ligurian Sea as an area of intermediate density. The geographic scale of this study was impressive, being one of the few we are aware of that surveyed both east and west basins in the same effort, but it is also necessarily the case that using a single vessel across such a large spatial scale resulted in low coverage of any particular area, so we cannot expect these estimates to be highly precise. The same is true for the surveys reported by Boisseau et al. (2010), which did not completely overlap the study areas of Gannier et al. (2002) but still highlighted the western Basin (in this case, the southern portion) and the waters around Crete, as areas of high encounter rates. Praca and Gannier (2008) used combined visual and acoustic surveys to show that sperm whales have a strong preference for shelf waters both along the French and Italian coasts, as well as those along the eastern Balearics; this also appears to be true of the shelf waters off southern Spain (Cañadas et al., 2002). Praca et al. (2009) used multiple statistical techniques to produce habitat suitability models based on data collected in the northern part of the western Basin, and found that all approaches highlighted continental shelf edges, where the depth transitions from shallow coastal waters to abyssal plains, as particularly important areas for sperm whales. Some of Praca et al.'s (2009) models also suggested that the deeper waters over the abyssal plains were somewhat important, although an ongoing paucity of data from these deep water areas make this suggestion difficult to verify.

There can be very strong associations between sperm whales and certain bathymetric features, for reasons we have yet to fully understand, but that almost certainly reflect variations in prey density. One such feature is the Hellenic Trench running from the west of the Ionian Islands to the west and south of Crete and south and east of Rhodes Island. In this area, sperm whales have a strong and clear density peak around the 1000-m depth contour which drops off rapidly as the water gets either shallower or deeper either side of the contour (Frantzis et al., 2014). Other such zones of high density include the waters of a similar depth found to the south of Mallorca and Ibiza in the Balearic Islands (Pirotta et al., 2011; Fig. 2), and a further example occurs in Turkish waters, where the majority of sperm whale sightings over an 18-year period were concentrated in the Fethiye Canyon, one of the deepest parts of the Mediterranean Sea at 4500 m (Öztürk et al., 2013). Similarly, Tepsich et al. (2014) found a strong association between sperm whale distribution and submarine canyons in the northern Ligurian Sea. In all these cases it seems likely that physical oceanographic factors, such as current interactions with bathymetry, have profound effects on the spatial ecology of squid, but we have yet to understand such processes in any meaningful way. Non-uniform distributions indicating links between sperm whales and bathymetric features have also been identified on relatively small

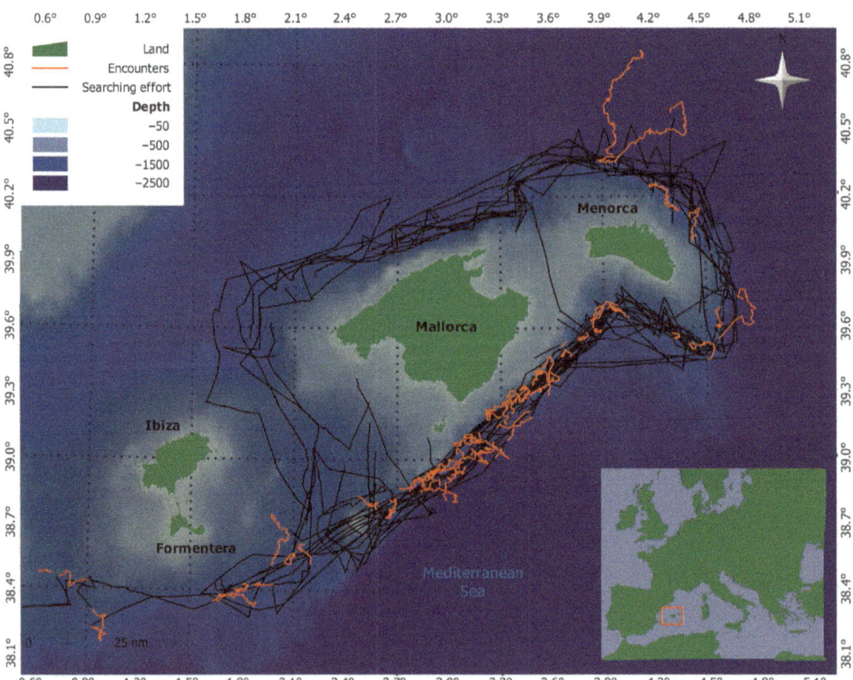

Fig. 2 Search effort and sperm whale (*Physeter macrocephalus*) encounters around the Balearic Islands, 2003–08. *After Pirotta, E., Matthiopoulos, J., MacKenzie, M., Scott-Hayward, L., Rendell, L., 2011. Modelling sperm whale habitat preference: a novel approach combining transect and follow data. Mar. Ecol. Prog. Ser. 436, 257–272. doi: 10.3354/meps09236.*

spatial scales. For example, Mussi et al. (2014) analysed visual survey data collected over nine years in the months between June and October and described a strong association between sperm whales and a bathymetric feature called Cuma Canyon, which is less than 20 km wide and located to the northwest of the island of Ischia just outside the Bay of Naples in Italy.

Aside from fixed topographical features, physical oceanographic factors can produce measurable but mobile and sometimes ephemeral features associated with water movements and fronts between water masses. Praca and Gannier's (2008) study of sperm whale distribution in the western Basin suggested that the whales preferred waters with lower surface temperatures (perhaps indicating relatively recent upwelling) and higher chlorophyll (consistent with productivity boosts resulting from the upwelling of nutrient-rich waters). This latter observation suggests that temporal and spatial lags between primary production and the availability of prey may not be as pronounced for Mediterranean sperm whales as they are in other parts of the globe, such as the tropical Pacific (Jaquet, 1996). Another study from the same researchers highlighted what appears to be a strong link between sperm whale distribution and fronts separating water masses in the deep pelagic waters of the north-western basin, especially the North Balearic Front to the north and west of the Balearic archipelago that separates the remnants of Atlantic surface water inflows from the colder waters of the Ligurian Basin to the north (Gannier and Praca, 2007). This relationship is not unusual for sperm whales, as studies from other regions have illustrated associations between sperm whales and sea surface temperature features such as warm-core eddies from the Gulf Stream in the North Atlantic (e.g. Griffin, 1999). These results again speak to an underlying similarity in the niches occupied by sperm whales in the Mediterranean to those of other populations.

The overall picture then, is that Mediterranean sperm whales make their living in broadly the same way as conspecifics in all the world's oceans—from mesopelagic squid. The limited accessibility of this ecological resource from human fisheries has almost certainly been a huge advantage for the survival of Mediterranean sperm whales, since it has insulated them from the sustained degradation that has been inflicted on other biota—currently, 85% of assessed fish stocks in the Mediterranean Sea are harvested unsustainably (Colloca et al., 2013; FAO, 2016). Despite these insights, important gaps in our understanding remain. Our knowledge is strongly restricted to particular months of the year, for example. Data are currently very limited in the winter months because of the more challenging weather conditions, but hopefully new passive acoustic technologies will improve this picture in the coming years. Surveys using moored devices have been able to confirm that sperm whale presence in the Ligurian Sea is not limited to summer months, with animals recorded in the northeast zone in December (Giorli et al., 2016), although surveys using towed hydrophones on a wider spatial scale still suggest some seasonality in sperm whale presence (Laran and Drouot-Dulau, 2007). Giorli et al. (2016) also reported an apparent shift towards concentrating foraging at night outside of the summer months in their study area, which could indicate seasonality in either prey behaviour or main prey species. Nonetheless, more information on seasonal patterns in habitat use should be an important ongoing research goal, not least because economically costly conservation actions are more robustly defensible when based on knowledge of where key habitat is found at different times of the year.

4. POPULATION STRUCTURE WITHIN THE MEDITERRANEAN SEA: IS THERE MORE THAN ONE POPULATION?

We have already considered population structure with respect to the isolation of the Mediterranean sperm whale population from the neighbouring Atlantic stock, but another critical area of knowledge we need to review before discussing conservation directly concerns population structure within the Mediterranean itself. The relatively shallow and narrow channels that separate the western and eastern basins are at least potentially barriers to movement for a species that has a strong preference for deep water. Consequently, it is not obvious whether we are considering a single well-mixed population or whether there is significant population structure within the Mediterranean itself. If there were, this would make the conservation status of the two populations—western and eastern—even more precarious.

All photo-identification evidence currently available suggests that the population in the western Basin is well mixed. Drouot-Dulau and Gannier (2007) documented movements between the Balearic Islands, and the Ligurian Sea and Gulf of Lion, both journeys in excess of 400 km in less than a month, and also documented one animal making two complete round trips between the Ligurian Sea and the Balearics between August 1999 and July 2004. Carpinelli et al. (2014) compared multiple photo-identification catalogues, including those from the North Atlantic and the

Hellenic Trench, with that of the sperm whales sighted in the Strait of Gibraltar. No matches were found with either the North Atlantic or Hellenic Trench catalogues, but 15 of the 47 (32%) whales identified in the Strait were sighted in other parts of the western Basin, with the longest distance between resightings being 1600 km. Rendell et al. (2014) analysed movements of 180 individuals from photo-identification records originating from three long-term studies in the northern part of the western Basin, and estimated a typical home range to be 1000 km across, which was very similar to estimates from studies in the Pacific (Whitehead, 2001), with typical displacements of around 400–600 km across an one year timescale. Rendell et al. (2014) also documented multiple movements between areas of concentration around the Balearic Islands and the Ligurian Sea (Fig. 3). Together, the results of these studies clearly show that whales move freely around the entire western Basin, although it is important to acknowledge a critical data gap concerning the waters off North Africa.

Until recently, the question of whether the eastern and western basins housed mutually isolated populations was a matter of speculation, but recent

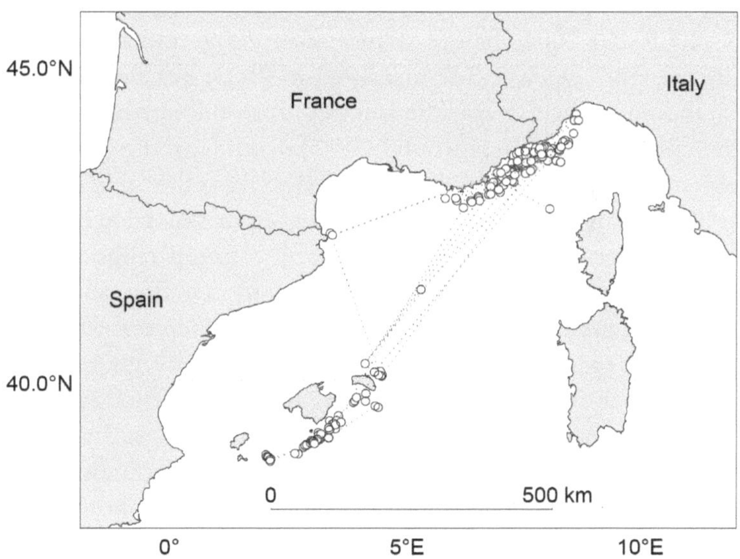

Fig. 3 Movements of sperm whales (*Physeter macrocephalus*) sighted in more than one year in the northwest Mediterranean Basin. Sighting locations are marked with ○, and *dashed lines* link identifications of the same individual. *After Rendell, L., Simião, S., Brotons, J.M., Airoldi, S., Fasano, D., Gannier, A., 2014. Abundance and movements of sperm whales in the western Mediterranean basin. Aquat. Conserv. Mar. Freshw. Ecosyst. 24, 31–40. doi: 10.1002/aqc.2426.*

fortuitous observations have answered this question definitively. Part of the evidence comes from the stranding of seven sperm whales on the southern Adriatic coast of Italy in December 2009. The whales, all male, stranded alive, but all died during the following two days (Mazzariol et al., 2011). Comparisons of the flukes and pigmentation patterns of the stranded animals with catalogues from a number of field projects revealed that three of the seven had been previously identified, two in the western Basin (i.e. not the one they stranded in) and one in the eastern Basin (Frantzis et al., 2011). The male that did not change basins had been seen several times along the Hellenic Trench in Greece, first in 2000 as a juvenile member of a social unit, likely his natal unit, and seven more times in 2002 and 2005, always with the other members of this social unit. When stranded in 2009, this animal measured 10.5 m and tooth sectioning resulted in an age estimate of 15 years. The straight-line distance between the stranding location in the Adriatic Sea and the last recorded field observation of this individual was 630 km. Two other individuals from the stranded group had previously been observed and photo-identified in the western Mediterranean, specifically the north-western Ligurian Sea. One individual (12.1 m, 20–21 years of age) was first photo-identified in 2002, and observed five more times in 2003, 2005 and 2007, before stranding in 2009, and the other (12.2 m, 19–20 years of age) was first identified in 2003, but not seen again until the stranding event. By sea, the journey from the northwest Ligurian Sea to the stranding location is around 2000 km, and, most importantly, would necessitate traversing either the Strait of Messina or the Strait of Sicily. These observations show that male sperm whales originating from both the eastern and western basins came together to form a group (although we do not know how long they had been together prior to stranding) and provides hard evidence that animal do pass between the two deep water basins. Finally, another male sperm whale has been documented by photo-identification data to have moved from the western to the eastern Basin. This whale was first photographed in the western Basin in 1991 and was reidentified along the Hellenic Trench in the eastern Basin in 2004 (Frantzis et al., 2011).

Another piece of evidence consistent with the photo-identification results described above is the sharp shift in growth layer group $\delta^{15}N$ and $\delta^{13}C$ isotope levels observed by Mendes et al. (2007) in a tooth obtained from a stranded male. This shift occurred as the animal attained 20 years of age, which is around the age male sperm whales in other oceans typically make large movements from feeding to breeding grounds (Whitehead, 2003; Whitehead and Weilgart, 2000), and indicates a significant change in dietary sources of these stable isotopes. It is known that there are significant variations in isotope levels between the eastern and western basins, and this led Mendes et al. (2007) to suggest that the observed shift in isotope levels could indicate that the male had moved from the western to the eastern Basin. This suggestion was made prior to the 2009 Adriatic stranding, showing that it was indeed possible.

Although the data are few and the numbers involved are low, it is clear that sperm whales do move between the eastern and western Mediterranean, and therefore these populations are not isolated from each other. Although the evidence we have all comes from males, it is unclear at this stage whether such movements are restricted to one sex. Irrespective of this, the likelihood of significant gene flow, sufficient to stop the populations from diverging, is very high. In conservation terms, this could be seen as a good thing since such exchanges raise the effective population size by maintaining gene flow and mixing throughout the entire range. Without those links, we would be considering two smaller and consequently even more vulnerable populations, rather than a single vulnerable one. Keeping the east–west population links open is therefore likely to be crucial to the long-term viability of the population, and the potential impact of any human activities in the Straits of Messina and Sicily that could inhibit this process needs to be carefully considered.

5. HOW MANY SPERM WHALES ARE THERE IN THE MEDITERRANEAN SEA?

Risibly, we currently do not even know with any certainty how many sperm whales there are in the Mediterranean. Photo-identification and line-transect studies have been used to assess populations in specific areas, but it is very difficult to extrapolate these to a basin scale when so much of the region remains unsurveyed, an astonishing fact given the highly developed status of most of the nations bordering it.

There is a relatively recent population estimate of just 62 individuals for the entire Ionian Sea (Lewis et al., 2007). Photo-identification research conducted between 1998 and 2009 along the Hellenic Trench identified 181 individuals, 17 of which did not survive until the end of this period (Frantzis et al., 2014). Considering the high rate of photographic recapture and that some of the individuals observed in the Aegean Sea were also photo-identified in the Hellenic Trench, any estimate for the entire subpopulation of sperm whales living in the Greek Seas is unlikely to exceed

250 animals. Since large-scale surveys conducted over the entire eastern Mediterranean Basin indicate that the majority of the sperm whales concentrate along the Hellenic Trench (Boisseau et al., 2010; Lewis et al., 2007), the total number of sperm whales for the eastern Basin is very likely to be in the low hundreds. Nine years of photo-identification effort around the island of Ischia just outside the Bay of Naples in Italy has resulted in a catalogue of 60 individuals from a survey area approximately 8800 km^2 (Pace et al., 2014a), although the discovery curve showed no sign of a decreasing rate of new identifications, so this site may be part of a much larger home range for the western Basin population. Rendell et al. (2014) analysed 507 photo-identifications from across the northern part of the western Basin. They identified 180 individuals between 1990 and 2008 and estimated the population size using a variety of analytical approaches, with the result that none of the upper confidence bounds on the estimates exceeded 900 individuals once sampling bias was taken into account, and lower bounds were less than 200. These studies present the best data available on sperm whale abundance, and on face value they give an alarming picture of a very small total population size. None of the datasets are perfect however, being restricted in time, spatial coverage, unevenness of sampling effort, and combinations thereof, so we are far from achieving a robust abundance estimate. When we lack such basic knowledge as a baseline of how many sperm whales there are in the Mediterranean Sea, effective conservation will be challenging at best. At the very least however, all the studies carried out thus far indicate that Notarbartolo di Sciara et al. (2012a) were accurate in suggesting that 'overall sperm whale numbers in the Mediterranean are likely to be only in the low to mid hundreds' when they assessed the population as Endangered for the IUCN Red List.

Estimating cetacean population trends is notoriously difficult even for the best studied populations (Wilson et al., 1999), and if estimating the population size is difficult, then information on population trends in Mediterranean sperm whales is, not surprisingly, extremely sparse. Accounts from as recently as the 1950s suggest historically higher densities than we observe today, but such information is impossible to assess quantitatively (Notarbartolo di Sciara et al., 2012a). Bearzi et al. (2011) noted a decline in the frequency of stranding events on the Adriatic coast in the latter half of the twentieth century, contrary to an expectation of increased reporting efficiency in modern times. While this could indicate a potential population decline, assuming a consistent per capita stranding rate, the decline in stranding reports is equally consistent with distributional shifts, potentially driven by a changing ecology (Bearzi et al., 2011; see also Notarbartolo di Sciara et al., 2012a).

Recruitment of young is a crucial factor in the population dynamics of any mammal, but here again we have only the most superficial observations for Mediterranean sperm whales. There is some evidence that sperm whales in the eastern Basin have a reasonable calving success, as Frantzis et al. (2014) reported that 15 of the 16 social units they observed between 1998 and 2009 had a calf with them at least once in that period, and that 79% of encounters with social groups featured calves. It is difficult to draw strong conclusions however, since calf presence does not necessarily lead to recruitment into the adult population, and the same study also reported that calf and juvenile mortality was likely to be high (>40% and >27%, respectively). Thus, to date, we remain woefully lacking in hard data on what is happening to Mediterranean sperm whale populations, even though there are ominous indicators from ongoing mortality that all is not well.

6. WHAT IS THREATENING SPERM WHALES IN THE MEDITERRANEAN?

The Mediterranean sperm whale population is a fragile thing. In a geographically restricted habitat, an 'oceanic island' that has been more heavily impacted by human activity than perhaps any other marine habitat, and apparently cut-off from the much larger pool of conspecifics in the neighbouring Atlantic Ocean, the population is threatened. The threats to their sustainable presence in the region are manifold. In a recent comprehensive review of conservation issues affecting sperm whales in the Mediterranean, Notarbartolo di Sciara (2014) identified six important problems: fishery bycatch (of which so-called 'ghost-fishing' by abandoned drift nets is considered highly significant), ship strikes, ingestion of marine debris (principally plastics), chemical pollutants, anthropogenic noise and disturbance from poorly managed whale watching operations. The general concerns of climate change and prey depletion were also highlighted as important issues, but ones that are much more difficult to assess (Notarbartolo di Sciara, 2014). In our view, the two most critical and immediate sources of anthropogenic mortality in this population are being hit by large ships and entanglement in fishing gear. Habitat degradation due to chronic noise pollution is likely to also be a significant factor, but it is much harder to assess of the degree of impact at the population level in this case (Notarbartolo di Sciara, 2014).

Fig. 4 A stranded sperm whale (*Physeter macrocephalus*) observed 15 February 2014, 25 km north of Pylos in southwest Peloponnese, Greece. Three big cuts from a large propeller were apparent on this very freshly dead whale. *Photograph: Courtesy of Filiatranet/Archive of the Pelagos Cetacean Research Institute.*

Large cetaceans in the Mediterranean Sea are particularly susceptible to being hit by ships due to the high density of shipping routes over sensitive deep-sea ecosystems. The consequences are often fatal (Fig. 4). Over the past half century, shipping has greatly expanded in the Mediterranean Sea (e.g. Anonymous, 2008).[1] About 220,000 vessels having a displacement greater than 100 tonnes cross the Mediterranean each year. This sea, although with a surface of only 0.8% of the world's oceans, was estimated 10 years ago to carry 30% of the world's total merchant shipping and 20% of its oil shipping. The total number of large cargo vessels that are cruising the Mediterranean Sea at any moment is >2000. Furthermore, a total of over 9000 other commercial vessels, including ferries, fast ferries and hydrofoils, as well as countless military, fishing, tourism and leisure craft, daily navigate the waters of the western Basin alone. The relatively high densities of sperm whales along the 1000 m contour beside the Hellenic Trench coincide almost exactly with major foci of maritime traffic, especially west of Zakynthos Island (Fig. 5; Frantzis et al., 2014), such that the risk of whales being affected by ship strikes has to be regarded as non-trivial. Propeller marks and/or cut flukes have been observed on photo-identified sperm whales from the catalogues of both authors of this chapter, so in both the western and eastern Mediterranean basins (L. Rendell, A. Frantzis, unpublished data). The known numbers of sperm whale ship strikes in Greece alone illustrate the scale of the problem. Of 23 sperm whales stranded in Greece that were examined since 1992, 15 (65%) had definite (12) or possible (3) collision marks on their bodies, indicating a ship strike as the likely cause of death (Frantzis et al., 2015). These figures necessarily represent a subset of the true mortality, since many

[1] See also http://www.unepmap.org/index.php?module=content2&catid=001003002, accessed 4/8/2016.

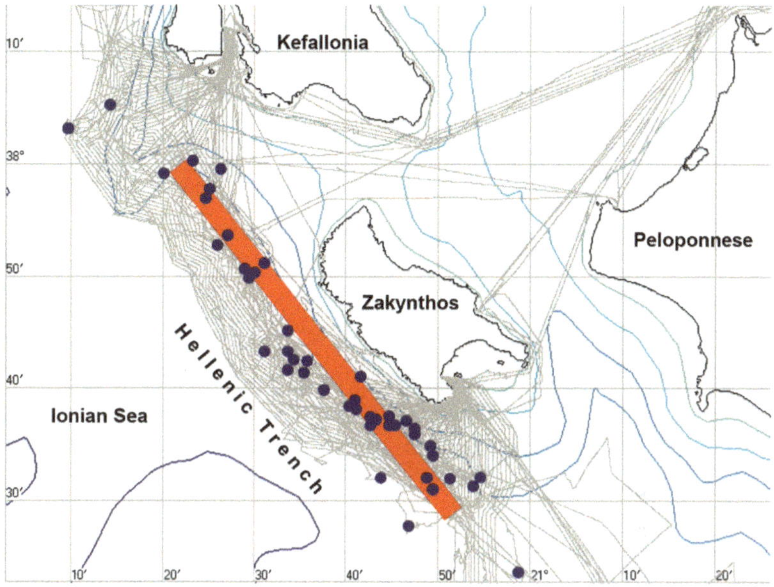

Fig. 5 Survey tracks and sperm whale (*Physeter macrocephalus*) sightings adjacent to the Hellenic Trench. *Thick red line* indicates 'centre line' of shipping route between 37°30′N and 38°N. *Modified after Frantzis, A., Alexiadou, P., Gkikopoulou, K.C., 2014. Sperm whale occurrence, site fidelity and population structure along the Hellenic Trench (Greece, Mediterranean Sea). Aquat. Conserv. Mar. Freshw. Ecosyst. 24, 83–102. doi: 10.1002/aqc.2435.*

victims of ship strike never reach the shore. Such a rate of collisions is considered to be unsustainable for the Endangered Mediterranean population and is a major reason for its current IUCN status (Notarbartolo di Sciara et al., 2012a). Furthermore, even if animals escape being hit, it appears that sperm whales, like many cetacean species, avoid areas of high traffic density (Campana et al., 2015), which can cause problems if that traffic occurs in regions of critical habitat.

In theory, the shipping problem can be mitigated using sensible restrictions on the speed and routing of maritime traffic, but we are aware of only one case in which measures have been put in place—the Spanish Government has imposed speed restriction in the Strait of Gibraltar, where high-speed ferries are a major collision risk (Abdulla and Linden, 2008). Since every restriction has an economic impact, the International Maritime Organisation, although willing to engage in conservation efforts, understandably has policies that require a strong scientific case for mandating changes of routing and/or speed, and the paucity of funding for comprehensive year-round habitat studies means that these cases can be difficult to develop. In our view, it is a vital priority for the ongoing conservation of Mediterranean sperm whales that there is a step change in the level of investment dedicated to obtaining a robust picture of the areas of critical habitat for these animals so that maritime traffic can be appropriately managed in sensitive areas.

Entanglement in nets is the second major threat, and by far the most devastating impact has been from driftnets, the so-called 'walls of death', averaging 20 km in length (Notarbartolo di Sciara, 1990). There is little doubt that entanglement with driftnets was a major cause of mortality for sperm whales in the latter decades of the last century—Notarbartolo di Sciara et al. (2004) analysed 229 sperm whale strandings that occurred in the north-western Mediterranean between 1971 and 2003 and found that the vast majority were overwhelmingly the result of entanglement with nets (in many cases, the nets were still present). The authors noted two large peaks in strandings over the study period, one in the mid-1980s associated with the establishment of a national stranding network in Italy, and one in the mid-1990s associated with a westward expansion of the Italian driftnet fleet, but they also saw a subsequent and steep decline after that last peak. This decline in the face of an ever-increasing probability of strandings being reported, and with no discernible decrease in fishing effort, was interpreted as indicative of a population decline (Notarbartolo di Sciara et al., 2004). There have been numerous reports of sperm whale carcasses found either with nets themselves or with the characteristic marks the nets leave on the animals (Notarbartolo di Sciara, 2014; Notarbartolo di Sciara et al., 2004). The scale of the problem is vividly illustrated by the account offered by Pace et al. (2008) of an incident in which five whales, possibly an entire social group, became entangled in a single net approximately 74 km from the Italian coast. Extraordinary efforts by the Italian coastguard over two days at sea resulted in the eventual release of these animals, but it was through luck alone that they were discovered before it was too late. The estimates of mortality from entanglement do not include animals killed by nets whose carcasses never reached the coasts. These nets are not just a problem for sperm whales, but kill marine megafauna indiscriminately (Tudela et al., 2005). In 2002, a number of international regulatory bodies banned the use of driftnets in the Mediterranean, including the European Commission, the International Commission for the Conservation of Atlantic Tuna, the General Fisheries Commission for the Mediterranean, and the Convention on Migratory Species Agreement on the Conservation of Cetaceans of the

Black Sea, Mediterranean Sea and Contiguous Atlantic Area (ACCOBAMS). Enforcement has been painfully slow however, and it is likely that vessels from a number of nations are still engaged in the illegal use of driftnets (Notarbartolo di Sciara, 2014). In our view, it is absolutely critical to the continued survival of Mediterranean sperm whales that such illegal fishing is stamped out entirely. Even if all such fishing ceased immediately, abandoned nets will likely continue to be a threat to sperm whales for many years to come.

Marine debris, primarily plastics, is regularly found in the stomachs of stranded sperm whales. The ingestion of plastic has been implicated in the death of at least two sperm whales in the Mediterranean Sea. One, stranded on the southeast coast of Spain, was a 10 m male with over 17 kg of plastics in its stomach (de Stephanis et al., 2013). The plastic had impacted internally and ultimately caused the gastric rupture that was the proximate cause of the death. The animal was also highly emaciated, so it is unclear whether hunger led to an increased ingestion of plastic debris or whether ingestion of plastics led to starvation (de Stephanis et al., 2013). Similarly a 5.35 m juvenile male that was found floating in Greek waters off Mykonos Island in 2006, had ingested nearly 100 plastic items including heavy duty plastic bags (Notarbartolo di Sciara et al., 2012b). Finding an animal with plastic in its stomach does not necessarily mean that this was the cause of death, because starving animals can become less selective in what they ingest, but it may well be a contributing factor in many strandings (Mazzariol et al., 2011). Thus it is difficult to assess to what extent such debris is a principal cause of mortality, but it is uncontroversial to suggest that it can make a bad situation worse. Society as a whole has an important role to play in reducing the entry of plastics into the marine environment.

Anthropogenic noise has become increasingly important on the list of conservation priorities in recent years, not least because of events such as the mortality of Cuvier's beaked whales (*Ziphius cavirostris*) associated with military sonar exercises in the Ionian Sea (Frantzis, 1998; see also Podestà et al., 2016). Geopolitical instability and consequent military activity, high levels of maritime traffic and increasing exploration for deep-sea hydrocarbons (Notarbartolo di Sciara, 2014) are all contributing to elevated noise levels in the Mediterranean, with the potential to cause chronic effects, such as elevated stress levels (Rolland et al., 2012), as well as the possibility of causing acute effects such as hearing damage and stranding (Frantzis, 1998). It is however notoriously difficult to quantify such effects and to extrapolate them to population level consequences, even in the best studied populations (King et al., 2015). Thus, we are not in a position to say very much scientifically about how noise is affecting Mediterranean sperm whales except to note that it must necessarily contribute to a general degradation of habitat associated with high levels of human impact from tourism, industrial and military activities.

Chemical pollution is always a concern for marine mammals, whose large lipid stores can act as bioaccumulators of various compounds that are released into the marine environment. At present, we have little information on how serious a threat this is to Mediterranean sperm whales. A recent stranding of seven whales in the Adriatic Sea provided an opportunity for extensive toxicological analysis (Marsili et al., 2014; Mazzariol et al., 2011). These studies found evidence of elevated levels of persistent organic pollutants, especially for three classes of compounds: hexachlorobenzene (HCB), dichlorodiphenyltrichloroethane (DDT) and polychlorinated biphenyls (PCBs). Furthermore, Marsili et al. (2014) suggested a mechanism by which these pollutants could exacerbate problems of nutritional stress (the stomachs of the stranded animals were empty save a few 'well-digested' squid beaks), whereby increased metabolism of lipid deposits could lead to the relatively rapid liberation into physiologically active tissues of pollutants accumulated in lipid storage over longer time periods. These high levels could, in theory at least, impair neurological and immune function at a time when the animal is already suffering nutritional stress. Such a condition would likely have increased the chances of mortality. Squadrone et al.'s (2015) analysis of heavy metal concentrations in the bodies of three female sperm whales from a second mass stranding in the Adriatic in 2014 indicated that of the 20 trace elements analysed, most did not appear at levels high enough to cause concern, but did note high concentrations of aluminium in some tissues. It may be that the unusual mesopelagic foraging habits of sperm whales distance them somewhat from the sources of chemical pollutants (Notarbartolo di Sciara, 2014), but continued vigilance will be required until anthropogenic chemical inputs are significantly reduced.

7. CONCLUSIONS

The answers we have outlined in response to the questions posed in this review can be succinctly stated. Mediterranean sperm whales are in all likelihood an isolated population, separated from their Atlantic neighbours by geography and also by divergent population genetics and cultural traits. Despite this divergence, like sperm whales the world over, the Mediterranean whales still depend on mesopelagic squid for food. Recent evidence confirms that it is a single population, not isolated into western and eastern stocks. We lack robust information on their population status, but they could well number in the hundreds rather than thousands, which makes it vitally important to address the serious threats posed by ship strikes and entanglement in fishing nets, especially driftnets, and carefully monitor other potential sources on anthropogenic impact.

Perhaps our most significant conclusion however is that our knowledge of the Mediterranean sperm whale population is severely limited, shockingly so given the presence of so many relatively wealthy nations on the shores of the Mediterranean. Despite the important environmental legislation and directives of the European Union (EU) which affect almost all the northern Mediterranean countries, the research and conservation effort on these animals are minimal, because EU policy considers the IUCN-designated Endangered Mediterranean sperm whale population of minor conservation importance in comparison to, for example, the IUCN-designated Vulnerable Mediterranean bottlenose dolphin population (see Habitat Directive,[2] Annexes II and IV). In our view this inexplicable mistake, which has persisted for more than two decades, is almost certainly one of the reasons why we are still lacking appropriate tools for more research and improved conservation of this fragile population today. Without more data to provide scientific support to conservation action, there is no guarantee that sperm whales will persist in the Mediterranean Sea, and once they are significantly reduced or gone, all the evidence we have suggests we cannot take recovery or recolonisation for granted.

ACKNOWLEDGEMENTS

L.R. was supported by the MASTS pooling initiative (The Marine Alliance for Science and Technology for Scotland) and their support is gratefully acknowledged. MASTS is funded by the Scottish Funding Council (grant reference HR09011) and contributing institutions. A.F. was supported by OceanCare (Switzerland) who are also gratefully acknowledged for yearly supporting sperm whale research and conservation in Greece, since 2008. We thank Shane Gero and an anonymous reviewer, as well as the editors Giuseppe Notarbartolo di Sciara, Michela Podestà, and Barbara Curry, for comments that improved the manuscript.

[2] http://ec.europa.eu/environment/nature/legislation/habitatsdirective/index_en.htm

REFERENCES

Abdulla, A., Linden, O. (Eds.), 2008. Maritime Traffic Effects on Biodiversity in the Mediterranean Sea: Review of Impacts, Priority Areas and Mitigation Measures. IUCN Centre for Mediterranean Cooperation, Malaga, Spain. Available at: https://cmsdata.iucn.org/downloads/maritime_v1_lr.pdf.

Aguilar, A., Borrell, A., 2007. Open-boat whaling on the straits of Gibraltar ground and adjacent waters. Mar. Mammal Sci. 23, 322–342. http://dx.doi.org/10.1111/j.1748-7692.2007.00111.x.

Alexander, A., Steel, D., Hoekzema, K., Mesnick, S., Engelhaupt, D., Kerr, I., Payne, R., Baker, C.S., 2016. What influences the worldwide genetic structure of sperm whales (*Physeter macrocephalus*)? Mol. Ecol. 25, 2754–2772. http://dx.doi.org/10.1111/mec.13638.

Anonymous, 2008. Study of Maritime Traffic Flows in the Mediterranean Sea. A report prepared for the Regional Marine Pollution Emergency Response Centre for the Mediterranean Sea (REMPEC) by Lloyd's Marine Intelligence Unit.

Aristotle, 2004. Αριστοτέλης Άπαντα 15. Των περί τα ζώα Ιστοριών. Cactus Editions, Athens, Greece. (in Greek).

Bearzi, G., Pierantonio, N., Affronte, M., Holcer, D., Maio, N., Notarbartolo Di Sciara, G., 2011. Overview of sperm whale *Physeter macrocephalus* mortality events in the Adriatic Sea, 1555–2009. Mammal Rev. 41, 276–293. http://dx.doi.org/10.1111/j.1365-2907.2010.00171.x.

Boisseau, O., Lacey, C., Lewis, T., Moscrop, A., Danbolt, M., McLanaghan, R., 2010. Encounter rates of cetaceans in the Mediterranean Sea and contiguous Atlantic area. J. Mar. Biol. Assoc. UK 90, 1589–1599. http://dx.doi.org/10.1017/S0025315410000342.

Bolognari, A., 1949. A proposito della recente cattura di alcuni esemplari di capodoglio (*Physeter macrocephalus* L.) nel Mediterraneo. letin l'Institut Océanographique (Fondation Albert 1er. Prince Monaco) 949, 1–43.

Bolognari, A., 1950. Ancora sulla comparsa del Capodoglio: (*Physeter macrocephalus* L.) nel Mediterraneo. Ital. J. Zool. 17, 29–37.

Bolognari, A., 1951. La migrazione del capodoglio nel Mediterraneo. Ital. J. Zool. 18, 253–256.

Campana, I., Crosti, R., Angeletti, D., Carosso, L., David, L., Di-Méglio, N., Moulins, A., Rosso, M., Tepsich, P., Arcangeli, A., 2015. Cetacean response to summer maritime traffic in the Western Mediterranean Sea. Mar. Environ. Res. 109, 1–8. http://dx.doi.org/10.1016/j.marenvres.2015.05.009.

Cañadas, A., Sagarminaga, R., García-Tiscar, S., 2002. Cetacean distribution related with depth and slope in the Mediterranean waters off southern Spain. Deep. Res. 49, 2053–2073. http://dx.doi.org/10.1016/S0967-0637(02)00123-1.

Carpinelli, E., Gauffier, P., Verborgh, P., Airoldi, S., David, L., Di-Méglio, N., Cañadas, A., Frantzis, A., Rendell, L., Lewis, T., Mussi, B., Pace, D.S., De Stephanis, R., 2014. Assessing sperm whale (*Physeter macrocephalus*) movements within the western Mediterranean Sea through photo-identification. Aquat. Conserv. Mar. Freshwat. Ecosyst. 24, 23–30. http://dx.doi.org/10.1002/aqc.2446.

Caruso, F., Sciacca, V., Bellia, G., De Domenico, E., Larosa, G., Papale, E., Pellegrino, C., Pulvirenti, S., Riccobene, G., Simeone, F., Speziale, F., Viola, S., Pavan, G., 2015. Size distribution of sperm whales acoustically identified during long term deep-sea monitoring in the Ionian Sea. PLoS One 10, 1–16. http://dx.doi.org/10.1371/journal.pone.0144503.

Catchpole, C.K., Slater, P.J.B., 2008. Bird song: Biological Themes and Variations, second ed. Cambridge University Press, Cambridge.

Christal, J., Whitehead, H., Lettevall, E., 1998. Sperm whale social units: variation and change. Can. J. Zool. 76, 1431–1440. http://dx.doi.org/10.1139/z98-087.

Colloca, F., Cardinale, M., Maynou, F., Giannoulaki, M., Scarcella, G., Jenko, K., Bellido, J.M., Fiorentino, F., 2013. Rebuilding Mediterranean fisheries: a new paradigm for ecological sustainability. Fish Fish. 14, 89–109. http://dx.doi.org/10.1111/j.1467-2979.2011.00453.x.

Connor, R.C., 2000. Group living in whales and dolphins. In: Mann, J., Connor, R.C., Tyack, P.L., Whitehead, H. (Eds.), Cetacean Societies: Field Studies of Dolphins and Whales. University of Chicago Press, Chicago, pp. 199–218.

De Stephanis, R., Cornulier, T., Verborgh, P., Sierra, J.S., Gimeno, N.P., Guinet, C., 2008. Summer spatial distribution of cetaceans in the Strait of Gibraltar in relation to the oceanographic context. Mar. Ecol. Prog. Ser. 353, 275–288. http://dx.doi.org/10.3354/meps07164.

de Stephanis, R., Giménez, J., Carpinelli, E., Gutierrez-Exposito, C., Cañadas, A., 2013. As main meal for sperm whales: plastics debris. Mar. Pollut. Bull. 69, 206–214. http://dx.doi.org/10.1016/j.marpolbul.2013.01.033.

Di Natale, A., Mangano, A., 1985. Mating and calving of the Sperm Whale in the central Mediterranean Sea. Aquat. Mamm. 1, 7–9.

Drouot, V., Berube, M., Gannier, A., Goold, J.C., Reid, R.J., Palsboll, P.J., 2004a. A note on genetic isolation of Mediterranean sperm whales (*Physeter macrocephalus*) suggested by mitochondrial DNA. J. Cetacean Res. Manag. 6, 29–32.

Drouot, V., Gannier, A., Goold, J.C., 2004b. Summer social distribution of sperm whales (*Physeter macrocephalus*) in the Mediterranean Sea. J. Mar. Biol. Assoc. UK 84, 675–680. http://dx.doi.org/10.1017/S0025315404009749h.

Drouot, V., Goold, J.C., Gannier, A., 2004c. Regional diversity in the social vocalizations of sperm whale in the Mediterranean Sea. Rev. Ecol. Terre Vie 59, 545–558.

Drouot-Dulau, V., Gannier, A., 2007. Movements of sperm whale in the western Mediterranean Sea: preliminary photo-identification results. J. Mar. Biol. Assoc. UK 87, 195–200. http://dx.doi.org/10.1017/S0025315407054860.

Engelhaupt, D., Rus Hoelzel, A., Nicholson, C., Frantzis, A., Mesnick, S., Gero, S., Whitehead, H., Rendell, L., Miller, P., De Stefanis, R., Cañadas, A., Airoldi, S., Mignucci-Giannoni, A.A., 2009. Female philopatry in coastal basins and male dispersion across the North Atlantic in a highly mobile marine species, the sperm whale (*Physeter macrocephalus*). Mol. Ecol. 18, 4193–4205. http://dx.doi.org/10.1111/j.1365-294X.2009.04355.x.

FAO, 2016. The State of Mediterranean and Black Sea Fisheries. General Fisheries Commission for the Mediterranean, Rome, Italy. http://www.fao.org/3/a-i5496e.pdf.

Frantzis, A., 1998. Does acoustic testing strand whales? Nature 392, 29. http://dx.doi.org/10.1038/32068.

Frantzis, A., Alexiadou, P., 2008. Male sperm whale (*Physeter macrocephalus*) coda production and coda-type usage depend on the presence of conspecifics and the behavioural context. Can. J. Zool. 86, 62–75. http://dx.doi.org/10.1139/Z07-114.

Frantzis, A., Alexiadou, P., Paximadis, G., Politi, E., Gannier, A., Corsina-Foka, M., 2003. Current knowledge of the cetacean fauna of the Greek Seas. J. Cetacean Res. Manag. 5, 219–232.

Frantzis, A., Airoldi, S., Notarbartolo-di-Sciara, G., Johnson, C., Mazzariol, S., 2011. Inter-basin movements of Mediterranean sperm whales provide insight into their population structure and conservation. Deep Sea Res. Part I Oceanogr. Res. Pap. 58, 454–459.

Frantzis, A., Alexiadou, P., Gkikopoulou, K.C., 2014. Sperm whale occurrence, site fidelity and population structure along the Hellenic Trench (Greece, Mediterranean Sea). Aquat. Conserv. Mar. Freshwat. Ecosyst. 24, 83–102. http://dx.doi.org/10.1002/aqc.2435.

Frantzis, A., Leaper, R., Paraskevi, A., Lekkas, D., 2015. Update on sperm whale ship strike risk in the Hellenic Trench, Greece. In: Paper Presented to IWC Scientific Committee, San Diego, CA, USA, 22 May–3 June 2015, SC/66a/HIM06.

Gannier, A., Praca, E., 2007. SST fronts and the summer sperm whale distribution in the north-west Mediterranean Sea. J. Mar. Biol. Assoc. UK 87, 187–193. http://dx.doi.org/10.1017/S0025315407054689.

Gannier, A., Drouot, V., Goold, J.C., 2002. Distribution and relative abundance of sperm whales in the Mediterranean Sea. Mar. Ecol. Ser. 243, 281–293. http://dx.doi.org/10.3354/meps243281.

Gannier, A., Petiau, E., Dulau, V., Rendell, L., 2012. Foraging dives of sperm whales in the north-western Mediterranean Sea. J. Mar. Biol. Assoc. UK 92, 1799–1808. http://dx.doi.org/10.1017/S0025315412001087.

Garibaldi, F., Podestà, M., 2014. Stomach contents of a sperm whale (*Physeter macrocephalus*) stranded in Italy (Ligurian Sea, north-western Mediterranean). J. Mar. Biol. Assoc. UK 94, 1087–1091. http://dx.doi.org/10.1017/S0025315413000428.

Gero, S., Engelhaupt, D., Rendell, L., Whitehead, H., 2009. Who cares? Between-group variation in alloparental care-giving in sperm whales. Behav. Ecol. 20, 838–843. http://dx.doi.org/10.1093/beheco/arp068.

Gero, S., Gordon, J., Whitehead, H., 2013. Calves as social hubs: dynamics of the social network within sperm whale units. Proc. R. Soc. B 280, 1763–1772. http://dx.doi.org/10.1098/rspb.2013.1113.

Gero, S., Gordon, J., Whitehead, H., 2015. Individualized social preferences and long-term social fidelity between social units of sperm whales. Anim. Behav. 102, 15–23. http://dx.doi.org/10.1016/j.anbehav.2015.01.008.

Gero, S., Whitehead, H., Rendell, L., 2016. Individual, unit and vocal clan level identity cues in sperm whale codas. R. Soc. Open Sci. 3http://dx.doi.org/10.1098/rsos.150372.

Giorli, G., Au, W.W.L., Neuheimer, A., 2016. Differences in foraging activity of deep sea diving odontocetes in the Ligurian Sea as determined by passive acoustic recorders. Deep. Res. Part I Oceanogr. Res. Pap. 107, 1–8. http://dx.doi.org/10.1016/j.dsr.2015.10.002.

Gordon, J.C.D., 1991. Evaluation of a method for determining the length of sperm whales (*Physeter macrocephalus*) from their vocalizations. J. Zool. 224, 301–314. http://dx.doi.org/10.1111/j.1469-7998.1991.tb04807.x.

Griffin, R.B., 1999. Sperm whale distributions and community ecology associated with a warm-core ring off Georges Bank. Mar. Mammal Sci. 15, 33–51. http://dx.doi.org/10.1111/j.1748-7692.1999.tb00780.x.

Growcott, A., Miller, B., Sirguey, P., Slooten, E., Dawson, S., 2011. Measuring body length of male sperm whales from their clicks: the relationship between inter-pulse intervals and photogrammetrically measured lengths. J. Acoust. Soc. Am. 130, 568–573. http://dx.doi.org/10.1121/1.3578455.

Hirtz, M., 1921. Kitovi u našim vodama. Lovačko-Ribarski Vjesn 30, 34–35. (in Croatian).

Jaquet, N., 1996. How spatial and temporal scales influence understanding of sperm whale distribution: a review. Mammal Rev. 26, 51–65. http://dx.doi.org/10.1111/j.1365-2907.1996.tb00146.x.

Jones, E.L., Rendell, L., Pirotta, E., Long, J.A., 2016. Novel application of a quantitative spatial comparison tool to species distribution data. Ecol. Indic. 70, 67–76. http://dx.doi.org/10.1016/j.ecolind.2016.05.051.

King, S.L., Schick, R.S., Donovan, C., Booth, C.G., Burgman, M., Thomas, L., Harwood, J., 2015. An interim framework for assessing the population consequences of disturbance. Methods Ecol. Evol. 6, 1150–1158. http://dx.doi.org/10.1111/2041-210X.12411.

Krijgsman, W., Hilgen, F.J., Raffi, I., Sierro, F.J., Wilson, D.S., 1999. Chronology, causes and progression of the Messinian salinity crisis. Nature 400, 652–655. http://dx.doi.org/10.1038/2323.

Laran, S., Drouot-Dulau, V., 2007. Seasonal variation of striped dolphins, fin- and sperm whales' abundance in the Ligurian Sea (Mediterranean Sea). J. Mar. Biol. Assoc. UK 87, 345. http://dx.doi.org/10.1017/S0025315407054719.

Lewis, T., Gillespie, D., Lacey, C., Matthews, J., Danbolt, M., Leaper, R., McLanaghan, R., Moscrop, A., 2007. Sperm whale abundance estimates from acoustic surveys of the Ionian Sea and Straits of Sicily in 2003. J. Mar. Biol. Assoc. UK 87, 353–357. http://dx.doi.org/10.1017/S0025315407054896.

Macedonia, J.M., Evans, C.S., 1993. Essay on contemporary issues in ethology: variation among mammalian alarm call systems and the problem of meaning in animal signals. Ethology 93, 177–197. http://dx.doi.org/10.1111/j.1439-0310.1993.tb00988.x.

Marsili, L., Maltese, S., Coppola, D., Carletti, L., Mazzariol, S., Fossi, M.C., 2014. Ecotoxicological status of seven sperm whales (*Physeter macrocephalus*) stranded along the Adriatic coast of Southern Italy. Aquat. Conserv. Mar. Freshwat. Ecosyst. 24, 103–118. http://dx.doi.org/10.1002/aqc.2447.

Matthews, J.N., Steiner, L., Gordon, J., 2001. Mark-recapture analysis of sperm whale (*Physeter macrocephalus*) photo-id data from the Azores (1987–1995). J. Cetacean Res. Manag. 3, 219–226.

Mazzariol, S., Di Guardo, G., Petrella, A., Marsili, L., Fossi, C.M., Leonzio, C., Zizzo, N., Vizzini, S., Gaspari, S., Pavan, G., Podestà, M., Garibaldi, F., Ferrante, M., Copat, C., Traversa, D., Marcer, F., Airoldi, S., Frantzis, A., de Beraldo Quirós, Y., Cozzi, B., Fernández, A., 2011. Sometimes sperm whales (*Physeter macrocephalus*) cannot find their way back to the high seas: a multidisciplinary study on a mass stranding. PLoS One 6 http://dx.doi.org/10.1371/journal.pone.0019417, e19417.

Mendes, S., Newton, J., Reid, R.J., Frantzis, A., Pierce, G.J., 2007. Stable isotope profiles in sperm whale teeth: variations between areas and sexes. J. Mar. Biol. Assoc. UK 87, 621–627. http://dx.doi.org/10.1017/S0025315407056019.

Mesnick, S.L., 2001. Genetic relatedness in sperm whales: evidence and cultural implications. Behav. Brain Sci. 24, 346–347. http://dx.doi.org/10.1017/S0140525X01463965.

Mesnick, S.L., Evans, K., Taylor, B.L., Hyde, J., Escorza-Treviño, S., Dizon, A.E., 2003. Sperm whale social structure: why it takes a village to raise a child. In: de Waal, F.B.M., Tyack, P. (Eds.), Animal Social Complexity: Intelligence, Culture and Individualized Societies. Harvard University Press, Cambridge, MA, pp. 170–174.

Miller, P.J.O., Johnson, M.P., Tyack, P.L., 2004. Sperm whale behaviour indicates the use of echolocation click buzzes "creaks" in prey capture. Proc. R. Soc. B 271, 2239–2247. http://dx.doi.org/10.1098/rspb.2004.2863.

Møhl, B., Wahlberg, M., Madsen, P.T., Miller, L.A., Surlykke, A., 2000. Sperm whale clicks: directionality and source level revisited. J. Acoust. Soc. Am. 107, 638–648. http://dx.doi.org/10.1121/1.428329.

Moulins, A., Würtz, M., 2005. Occurrence of a herd of female sperm whales and their calves (*Physeter catodon*), off Monaco, in the Ligurian Sea. J. Mar. Biol. Assoc. UK 85, 213–214. http://dx.doi.org/10.1017/S0025315405011094h.

Mussi, B., Miragliuolo, A., Zucchini, A., Pace, D.S., 2014. Occurrence and spatio-temporal distribution of sperm whale (*Physeter macrocephalus*) in the submarine canyon of Cuma (Tyrrhenian Sea, Italy). Aquat. Conserv. Mar. Freshwat. Ecosyst. 24, 59–70. http://dx.doi.org/10.1002/aqc.2460.

Notarbartolo di Sciara, G., 1990. A note on the cetaceans incidental catch in the Italian driftnet swordfish fishery, 1986–1988. Rep. Int. Whal. Comm. 40, 459–460.

Notarbartolo di Sciara, G., 2014. Sperm whales, *Physeter macrocephalus*, in the Mediterranean Sea: a summary of status, threats, and conservation recommendations. Aquat. Conserv. Mar. Freshwat. Ecosyst. 24, 4–10. http://dx.doi.org/10.1002/aqc.2409.

Notarbartolo di Sciara, G., Bearzi, G., Cañadas, A., Frantzis, A., 2004. High mortality of sperm whales in the north-western Mediterranean, 1971–2003. In: Paper SC/56/BC10 Presented to the Scientific Committee of the International Whaling Commission, Sorrento, 29 June–10 July 2004.

Notarbartolo di Sciara, G., Frantzis, A., Bearzi, G., Reeves, R., 2012a. *Physeter macrocephalus* (Mediterranean subpopulation), in: IUCN Red List of Threatened Species. Version 2013.1. Downloaded on 04 February 2013 from Www.iucnredlist.org., In: IUCN 2010. IUCN Red List of Threatened Species. Version 2010.4. Downloaded on 04 February 2011 from Www.iucnredlist.org.

Notarbartolo di Sciara, G., Frantzis, A., Rendell, L., 2012b. Sperm whales in the Mediterranean: the difficult art of coexisting with humans in a crowded sea. J. Am. Cetacean Soc. 41, 30–37.

Ortega-Ortiz, J.G., Engelhaupt, D., Winsor, M., Mate, B.R., Hoelzel, A.R., 2012. Kinship of long-term associates in the highly social sperm whale. Mol. Ecol. 21, 732–744. http://dx.doi.org/10.1111/j.1365-294X.2011.05274.x.

Öztürk, A.A., Tonay, A.M., Dede, A., 2013. Sperm whale (*Physeter macrocephalus*) sightings in the Aegean and Mediterranean part of Turkish waters. J. Black Sea/Mediterranean Environ. 19, 169–177.

Pace, D.S., Miragliuolo, A., Mussi, B., 2008. Behaviour of a social unit of sperm whales (*Physeter macrocephalus*) entangled in a driftnet off Capo Palinuro (Southern Tyrrhenian Sea, Italy). J. Cetacean Res. Manag. 10, 131–135.

Pace, D.S., Miragliuolo, A., Mariani, M., Vivaldi, C., Mussi, B., 2014a. Sociality of sperm whale off Ischia Island (Tyrrhenian Sea, Italy). Aquat. Conserv. Mar. Freshwat. Ecosyst. 24, 71–82. http://dx.doi.org/10.1002/aqc.2459.

Pace, D.S., Mussi, B., Gordon, J.C.D., Würtz, M., 2014b. Foreword. Aquat. Conserv. Mar. Freshwat. Ecosyst. 24, 1–3. http://dx.doi.org/10.1002/aqc.2457.

Pavan, G., Hayward, T.J., Borsani, J.F., Priano, M., Manghi, M., Fossati, C., Gordon, J., 2000. Time patterns of sperm whale codas recorded in the Mediterranean Sea 1985–1996. J. Acoust. Soc. Am. 107, 3487–3495. http://dx.doi.org/10.1121/1.429419.

Pirotta, E., Matthiopoulos, J., MacKenzie, M., Scott-Hayward, L., Rendell, L., 2011. Modelling sperm whale habitat preference: a novel approach combining transect and follow data. Mar. Ecol. Prog. Ser. 436, 257–272. http://dx.doi.org/10.3354/meps09236.

Pitman, R.L., Ballance, L.T., Mesnick, S.L., Chivers, S.J., 2001. Killer whale predation on sperm whales: observations and implications. Mar. Mammal Sci. 17, 494–507. http://dx.doi.org/10.1111/j.1748-7692.2001.tb01000.x.

Podestà, M., Azzellino, A., Cañadas, A., Frantzis, A., Moulins, A., Rosso, M., Tepsich, P., Lanfredi, C., 2016. Cuvier's beaked whale *Ziphius cavirostris*, distribution and occurrence in the Mediterranean Sea: High-Use areas and conservation threats. In: Notarbartolo di Sciara, G., Podestà, M., Curry, B.E. (Eds.), Mediterranean Marine Mammals Ecology and Conservation. Advances in Marine Biology. vol. 75. Elsevier, Amsterdam, pp. 103–140.

Praca, E., Gannier, A., 2008. Ecological niches of three teuthophageous odontocetes in the northwestern Mediterranean Sea. Ocean Sci. 4, 49–59. http://dx.doi.org/10.5194/os-4-49-2008.

Praca, E., Gannier, A., Das, K., Laran, S., 2009. Modelling the habitat suitability of cetaceans: example of the sperm whale in the northwestern Mediterranean Sea. Deep. Res. Part I Oceanogr. Res. Pap. 56, 648–657. http://dx.doi.org/10.1016/j.dsr.2008.11.001.

Rendell, L.E., Whitehead, H., 2003. Vocal clans in sperm whales (*Physeter macrocephalus*). Proc. R. Soc. B 270, 225–231. http://dx.doi.org/10.1098/rspb.2002.2239.

Rendell, L., Mesnick, S.L., Dalebout, M.L., Burtenshaw, J., Whitehead, H., 2012. Can genetic differences explain vocal dialect variation in sperm whales, Physeter macrocephalus? Behav. Genet. 42, 332–343. http://dx.doi.org/10.1007/s10519-011-9513-y.

Rendell, L., Simião, S., Brotons, J.M., Airoldi, S., Fasano, D., Gannier, A., 2014. Abundance and movements of sperm whales in the western Mediterranean basin. Aquat. Conserv. Mar. Freshwat. Ecosyst. 24, 31–40. http://dx.doi.org/10.1002/aqc.2426.

Rice, D.W., 1989. Sperm whale, *Physeter macrocephalus*, Linnaeus, 1758. In: Ridgway, S.H., Harrison, R. (Eds.), Handbook of Marine Mammals. Academic Press, London, pp. 177–233.

Roberts, S.M., 2003. Examination of the stomach contents from a Mediterranean sperm whale found south of Crete, Greece. J. Mar. Biol. Assoc. UK 83, 667–670. http://dx.doi.org/10.1017/S0025315403007628h.

Rolland, R.M., Parks, S.E., Hunt, K.E., Castellote, M., Corkeron, P.J., Nowacek, D.P., Wasser, S.K., Kraus, S.D., 2012. Evidence that ship noise increases stress in right whales. Proc. R. Soc. B 279, 2363–2368. http://dx.doi.org/10.1098/rspb.2011.2429.

Squadrone, S., Brizio, P., Chiaravalle, E., Abete, M.C., 2015. Sperm whales (*Physeter macrocephalus*), found stranded along the Adriatic coast (Southern Italy, Mediterranean Sea), as bioindicators of essential and non-essential trace elements in the environment. Ecol. Indic. 58, 418–425. http://dx.doi.org/10.1016/j.ecolind.2015.06.016.

Teloni, V., 2005. Patterns of sound production in diving sperm whales in the northwestern Mediterranean. Mar. Mammal Sci. 21, 446–457. http://dx.doi.org/10.1111/j.1748-7692.2005.tb01243.x.

Tepsich, P., Rosso, M., Halpin, P.N., Moulins, A., 2014. Habitat preferences of two deep-diving cetacean species in the northern Ligurian Sea. Mar. Ecol. Prog. Ser. 508, 247–260. http://dx.doi.org/10.3354/meps10851.

Tudela, S., Kai Kai, A., Maynou, F., El Andalossi, M., Guglielmi, P., 2005. Driftnet fishing and biodiversity conservation: the case study of the large-scale Moroccan driftnet fleet operating in the Alboran Sea (SW Mediterranean). Biol. Conserv. 121, 65–78. http://dx.doi.org/10.1016/j.biocon.2004.04.010.

Watwood, S.L., Miller, P.J.O., Johnson, M., Madsen, P.T., Tyack, P.L., 2006. Deep-diving foraging behaviour of sperm whales (*Physeter macrocephalus*). J. Anim. Ecol. 75, 814–825. http://dx.doi.org/10.1111/j.1365-2656.2006.01101.x.

Weilgart, L., Whitehead, H., 1993. Coda vocalizations in sperm whales (*Physeter macrocephalus*) off the Galapagos Islands. Can. J. Zool. 71, 744–752. http://dx.doi.org/10.1139/z93-098.

Whitehead, H., 1996. Babysitting, dive synchrony, and indications of alloparental care in sperm whales. Behav. Ecol. Sociobiol. 38, 237–244. http://dx.doi.org/10.1007/s002650050238.

Whitehead, H., 2001. Analysis of animal movement using opportunistic individual identifications: application to sperm whales. Ecology 82, 1417–1432. http://dx.doi.org/10.1890/0012-9658(2001)082[1417:AOAMUO]2.0.CO;2.

Whitehead, H., 2002. Estimates of the current global population size and historical trajectory for sperm whales. Mar. Ecol. Prog. Ser. 242, 295–304. http://dx.doi.org/10.3354/meps242295.

Whitehead, H., 2003. Sperm Whales: Social Evolution in the Ocean. University of Chicago Press, Chicago.

Whitehead, H., Weilgart, L., 2000. The sperm whale: social females and roving males. In: Mann, J., Connor, R.C., Tyack, P.L., Whitehead, H. (Eds.), Cetacean Societies: Field Studies of Dolphins and Whales. University of Chicago Press, Chicago, pp. 154–172.

Whitehead, H., Waters, S., Lyrholm, T., 1991. Social organization in female sperm whales and their offspring: constant companions and casual acquaintances. Behav. Ecol. Sociobiol. 29, 385–389. http://dx.doi.org/10.1007/BF00165964DO.

Whitehead, H., Dillon, M., Dufault, S., Weilgart, L., Wright, J., 1998. Non-geographically based population structure of South Pacific sperm whales: dialects, fluke-markings and genetics. J. Anim. Ecol. 67, 253–262. http://dx.doi.org/10.1046/j.1365-2656.1998.00187.x.

Whitehead, H., Antunes, R., Gero, S., Wong, S.N.P., Engelhaupt, D., Rendell, L., 2012. Multilevel societies of female sperm whales (*Physeter macrocephalus*) in the Atlantic and Pacific: why are they so different? Int. J. Primatol. 33, 1142–1164. http://dx.doi.org/10.1007/s10764-012-9598-z.

Wilson, B., Hammond, P.S., Thompson, P.M., 1999. Estimating size and assessing trends in a coastal bottlenose dolphin population. Ecol. Appl. 91, 288–300. http://dx.doi.org/10.1890/1051-0761(1999)009[0288:ESAATI]2.0.CO;2.

Zimmer, W., Madsen, P.T., Teloni, V., Johnson, M.P., Tyack, P.L., 2005a. Off-axis effects on the multipulse structure of sperm whale usual clicks with implications for sound production. J. Acoust. Soc. Am. 118, 3337–3345. http://dx.doi.org/10.1121/1.2082707.

Zimmer, W., Tyack, P.L., Johnson, M.P., Madsen, P.T., 2005b. Three-dimensional beam pattern of regular sperm whale clicks confirms bent-horn hypothesis. J. Acoust. Soc. Am. 117, 1473–1485. http://dx.doi.org/10.1121/1.1828501.

CHAPTER THREE

Fin Whales, *Balaenoptera physalus*: At Home in a Changing Mediterranean Sea?

G. Notarbartolo di Sciara*[,1], M. Castellote[†], J.-N. Druon[‡], S. Panigada*

*Tethys Research Institute, Acquario Civico, Milano, Italy
[†]National Marine Mammal Laboratory, Alaska Fisheries Science Center/NOAA, Seattle, WA, United States
[‡]European Commission, DG Joint Research Centre, Directorate D—Sustainable Resources, Unit D.02 Water and Marine Resources, Ispra, Italy
[1]Corresponding author: e-mail address: disciara@gmail.com

Contents

Abstract

1. The relationship of Mediterranean fin whales (*Balaenoptera physalus*) to their Atlantic conspecifics has puzzled zoologists for centuries. Recent data indicate the occurrence of two distinct populations, one resident in the Mediterranean Sea and the other a seasonal visitor to the western Mediterranean from the northeastern North Atlantic Ocean.
2. Resident Mediterranean fin whales are nomadic opportunists that have adapted to exploit localised mesoscale hotspots of productivity that are highly variable in space and time. These appear to be fairly widespread across the region during winter, whereas in summer favourable feeding habitat is dramatically reduced, concentrating mostly in the western Ligurian Sea and Gulf of Lion. This prompts a reinterpretation of the movement pattern of resident fin whales, based on a contraction/dispersion hypothesis caused by seasonal variability in available feeding habitat, as opposed to a pattern of migrations occurring along defined directions as is common in other Mysticetes.
3. Calving peaks in autumn but has been observed year-round throughout the Mediterranean, suggesting that resident fin whales engage in breeding activities whenever favourable physiological conditions occur. It can be assumed that the Mediterranean environment, which is relatively forgiving in comparison to oceanic

Advances in Marine Biology, Volume 75
ISSN 0065-2881
http://dx.doi.org/10.1016/bs.amb.2016.08.002

habitats, combined with negligible predation pressure and high potential for sound-mediated socialisation due to the region's relatively small size, might have provided year-round resident fin whales an extended and more flexible calendar of breeding and feeding opportunities.

4. Considering the Mediterranean fin whales' small and possibly decreasing population size, low survival rate and the high pressure from many threats deriving from human activities such as vessel traffic, noise, chemical pollution and likely climate change, their status raises considerable concern and conservation measures should be urgently implemented.

1. INTRODUCTION

Fin whales (Fig. 1) are circumglobal cetaceans, found in all the world's major oceans. They occur mainly, although not exclusively, in offshore temperate and polar waters, and rarely in the tropics (Edwards et al., 2015). Most of the world's fin whales are thought to be migratory; however, their movements do not seem to follow a simple pattern (Jefferson et al., 2015). Fin whale seasonal migration has been traditionally considered, mostly on the basis of research from whaling vessels, to be regularly occurring between supposed temperate winter breeding grounds and higher-latitude summer feeding grounds (Gambell, 1985). However, it is now recognised that unlike other mysticetes such as humpbacks and grey whales, fin whale movements cannot be shoehorned into such simple schemes, and that they are subject to

Fig. 1 A fin whale (*Balaenoptera physalus*) displays the asymmetrical colouration of its head region as it dives into the clear waters of the Pelagos Sanctuary for Mediterranean Marine Mammals. *Photograph courtesy of Danny Kessler © and Tethys Research Institute.*

a much more complex set of factors affecting habitat use. This was already understood long ago, when Kellogg (1929) stated that: 'The migration routes of finbacks are not so well known as those of humpbacks, and the observed facts of gestation indicate that their journeys do not have any especial connection with their breeding habits [...] Climate seemingly has little influence in curtailing their wanderings, for finbacks appear to be indifferent alike to Tropic and Arctic temperatures, and travel where they will'.

While fin whale presence during summer in high-latitude feeding grounds is commonplace in both hemispheres, their winter migrations towards lower latitudes show a much less clear pattern. There have been an increasing number of observations of fin whales overwintering in their polar and subpolar feeding grounds, and no certain existence of specific winter breeding grounds anywhere (for a summary of these observations, see Geijer et al., 2016). An increasing body of knowledge, partly deriving from still-unpublished satellite and acoustic tracking efforts of oceanic fin whales, points to a continuum of migratory strategies, ranging between a more 'traditional' latitudinal round-trip displacement model and more opportunistic nomadism (defined by Jonzen et al., 2011, as: 'irregular movements' at seasonal timescales 'in response to environmental fluctuations, and typically also characterised by between-year variability in the geographic location of reproductive events'). Such nomadic habits cause the whales to move between locations characterised by the presence, often temporary, of favourable feeding conditions (Geijer et al., 2016). Furthermore, fin whales can also be nonmigrating, permanent residents in specific low- or mid-latitude locations containing persistently favourable habitat, e.g. in the East China Sea and northern Sea of Japan (Mizroch et al., 2009), in the Gulf of California (Tershy et al., 1993) where genetic evidence attests to their isolation (Bérubé et al., 2002), and in the Mediterranean Sea, as explained in this chapter.

Fin whales are a common mysticete in the North Atlantic Ocean, where a total of roughly 53,000 individuals was estimated to exist around the year 2000 (Reilly et al., 2013). These were classified by the International Whaling Commission (IWC) into seven management units, based largely on catch and marking data: Nova Scotia, Newfoundland-Labrador, West Greenland, East Greenland-Iceland, North Norway, West Norway-Faroe Islands and British Isles-Spain-Portugal (Donovan, 1991). Unsurprisingly, evidence exists that some movement occurs across the boundaries of these management units, indicating that the units are not discrete. The current IWC model proposes seven hypotheses for stock structure within these management units. The model assumes a central group of stocks that feed in the area between East Greenland and the Faroe Islands; a Spanish stock; and, under most hypotheses, an eastern and western group of stocks (IWC, 2010). Mediterranean fin whales are currently defined as a distinct population from those in the North Atlantic, with a range that perhaps extends out to southern Portugal (IWC, 2009).

2. POPULATIONS OF FIN WHALES IN THE MEDITERRANEAN SEA

The question of whether or not fin whales in the Mediterranean are isolated from fin whales in the Atlantic Ocean, has been debated for a long time. The first to suggest that fin whales in the Mediterranean may have been an isolated, nonmigrating population, was the Norwegian marine zoologist G.O. Sars in 1881, noting simultaneous fin whale sightings off Norway and in the Mediterranean (Jonsgård, 1966). Two contrasting theories—one involving resident isolation and the other seasonal immigration from the North Atlantic Ocean—have occupied zoologists for the better part of two centuries, and are described in detail in Notarbartolo di Sciara et al. (2003). Discussions, however, were always based upon speculation and indirect inference until significant levels of divergence and heterogeneity in both mitochondrial and nuclear DNA were found between Mediterranean and Eastern North Atlantic fin whales; this was based on genetic analyses performed on skin tissue remotely collected from free-ranging individuals in the Ligurian Sea (Bérubé et al., 1998).

The notion of a Mediterranean genetically distinct breeding and feeding population, isolated from the Atlantic Ocean, was also supported by contaminant analyses (Aguilar et al., 2002) and satellite tracking studies (Bentaleb et al., 2011; Cotté et al., 2011; Panigada et al., 2015). However, this was inconsistent with the concept of fin whales moving in and out of the Mediterranean from the Atlantic Ocean through the Strait of Gibraltar. This concept is corroborated by: (a) the historical presence of the species near the Gibraltar Strait in North Atlantic waters, where thousands of fin whales were caught by short-lived whaling activities that occurred there between 1921 and 1954 (Clapham et al., 2008; Sanpera and Aguilar, 1992), and (b) recent observations of fin whales in the Gibraltar area, which were seen crossing the Strait primarily westward in summer and eastward in winter (Gauffier et al., 2009, 2012).

An explanation of this apparent contradiction was provided by acoustic monitoring performed through the deployment of archival bottom-mounted audio recorders in nine different locations of the Western Mediterranean and northeastern North Atlantic Ocean between 2006 and 2009 (Castellote et al., 2012a). The use of song characteristics to describe population affiliation over a broad geographical range had been already successfully applied to mysticete species including fin whales (Hatch and Clark, 2004). Recordings by Castellote et al. (2012a) revealed long sequences of typical fin whale songs—i.e. reproductive displays (Croll et al., 2002) occurring during most of the year (Clark and Gagnon, 2002)—which fell into two patterns that were consistently and significantly distinct on the basis of inter-note interval and note bandwidth. One pattern was recorded in the northeastern North Atlantic Ocean from the Azores to the Gibraltar Strait, and across the Strait into the Mediterranean Sea all the way to the Balearic Basin. The other, identical to songs previously recorded in the Ligurian Sea (Clark et al., 2002), was recorded east of the Balearic Basin into the Provençal Basin. These recordings are indicative of the simultaneous presence in the Mediterranean Sea, of two different fin whale populations: a genetically and culturally distinct population of resident Mediterranean fin whales (hereafter the "MED whales") found between the Provençal Basin and the Balearic Basin, and members of a northeastern North Atlantic population ("NENA whales") travelling into the westernmost portion of the Mediterranean. The latter animals very likely cross the Strait of Gibraltar in winter and remain in the Mediterranean Sea until summer (Fig. 2).

Many questions still remain about the use and partitioning of the Mediterranean region by members of these two populations. These include the extent of the interactions between MED and NENA whales, the importance of the Mediterranean habitat for the NENA whales, and the extent of the dispersal of the MED whales westward, possibly all the way to the Atlantic Ocean. Acoustic studies conducted in March 2011 (Castellote et al., 2012a) confirmed the coexistence of song types from both NENA and MED whales in the Balearic Basin, but not in the Provençal Basin, thus indicating (a) the existence of seasonal sympatry between males of the two populations, (b) that the Balearic Basin seemed to mark the easternmost range limit of the NENA males, and (c) the presence of NENA singers in the Mediterranean during both winter and summer (Castellote et al., 2011, 2012a). The presence of NENA male singers in the Western Mediterranean is indicative of breeding-related behaviour in the area, however no information is yet available about the migratory behaviour of NENA females.

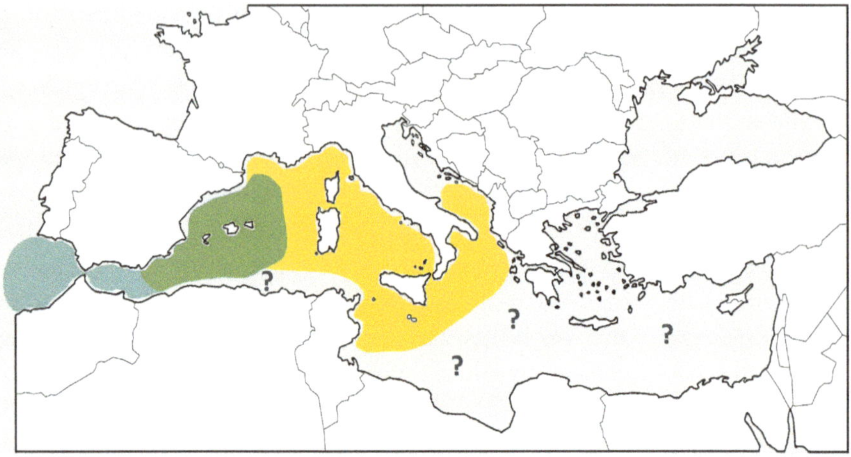

Fig. 2 Presumed distribution of fin whale (*Balaenoptera physalus*) populations in the Mediterranean Sea. *Blue*: north-east North Atlantic population (NENA whales). *Yellow*: Mediterranean population (MED whales). In *green* the presumed overlap between the two populations. NENA whales' distribution in the wider Atlantic Ocean is not shown.

Analyses of stable isotopes conducted on Mediterranean fin whale baleen plates add further elements to the picture. Bentaleb et al. (2011) compared fin whale baleen plate stable isotopes from a sample of nine stranded individuals in the Western Mediterranean with isotopes found in the whales' main prey, the euphausiid *Meganyctiphanes norvegica*, collected both in the Atlantic Ocean and in the Mediterranean Sea. In their analysis, most of the fin whale isotopes in their sample were consistent with the Mediterranean *M. norvegica* isotopic signature, indicating that feeding by the sampled whales had occurred only in the Mediterranean Sea. Those authors, however, also discovered two outliers from plates collected near Malaga, Spain, in the Alborán Sea, with δ^{13}C values intermediate between those of Atlantic and Mediterranean *M. norvegica*, further confirmed by Ryan et al. (2013) as occurring within the isotopic niches of both Biscayan and Irish/United Kingdom fin whales. These two outliers were interpreted by Castellote et al. (2013) as being NENA whales stranded during their Mediterranean visit, whereas Gimenez et al. (2013) suggested that they might have been MED whales that had previously foraged in the North Atlantic Ocean. A similar question refers to a single fin whale (#10842) tagged in the Provençal Basin in summer 2003. This animal travelled into the North Atlantic, unlike seven other whales tagged in the same experiment and who remained in the tagging area throughout the subsequent winter (Bentaleb et al., 2011). Was whale #10842 a NENA visitor returning to the North Atlantic, as suggested by Castellote et al. (2012a, 2013), or was it a MED whale venturing into the North Atlantic, as suggested by Gimenez et al. (2013)? Considering that passive acoustic data indicate regular movements by the NENA whales between the North Atlantic and Mediterranean basins, and considering the absence of the MED song type in the Alborán Sea, the Strait of Gibraltar and the Azores, the first hypothesis seems more likely. Genetic studies supporting a limited but recurrent, male-mediated gene flow in MED whales between the Ligurian Sea and the North Atlantic Ocean (Palsbøll et al., 2004) do not preclude the second hypothesis. However, such gene flow could also be consistent with just male NENA whales occasionally mating with female MED whales inside the Mediterranean.

In conclusion, and in spite of the remaining uncertainties, the current scientific knowledge indicates that two fin whale populations coexist within the Mediterranean Basin, with low but recurrent gene flow between them: a visiting NENA population and a permanent MED population. The remainder of this chapter will examine in greater detail the ecology of this latter population.

3. THE MEDITERRANEAN SEA AS FIN WHALE HABITAT

As mentioned earlier, satellite telemetry experiments point to the prolonged permanence of fin whales in the Mediterranean. Of eight fin whales tracked by satellite in the northwestern Mediterranean during summer 2003, all except one remained in the tagging area through autumn and winter (Bentaleb et al., 2011; Cotté et al., 2011). Further tagging studies performed in September 2012 in the Ligurian Sea (Fig. 3) resulted in the tagging of another eight fin whales, some of which retained tags for extended periods (up to 142 days). This clearly indicates the propensity of the tagged whales to remain in Mediterranean waters (Panigada et al., 2015).

The MED whales occur throughout the Mediterranean, from the Balearic Islands to the Levantine Sea, although they are in large part found in a subregion between the Gulf of Lion in France and southern Italy, as well as farther to the south into the Strait of Sicily and the wide Tunisian shelf (Notarbartolo di Sciara et al., 2003). An area of particular importance for fin whales, comprising the Ligurian, Corsican, Sardinian and Tyrrhenian seas, was designated as a protected area named the 'Pelagos Sanctuary for Mediterranean Marine Mammals'; this was established in 1999 by a treaty

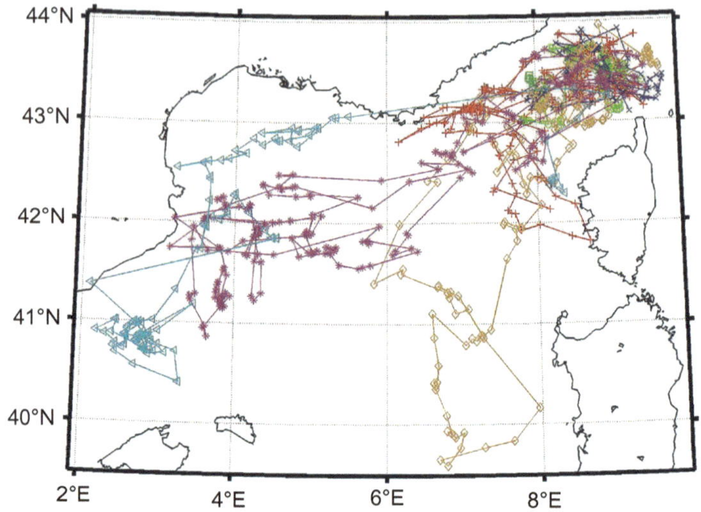

Fig. 3 Tracks of eight fin whales (*Balaenoptera physalus*) which were fitted with satellite tags in the Ligurian Sea in September 2012.

among France, Italy and the Monaco Principality (Notarbartolo di Sciara et al., 2008). East of Italy, in the Ionian Sea and southern Adriatic Sea, fin whales are also found with some regularity, although apparently in smaller numbers than in the Western Mediterranean. By contrast, records of the species' occurrence in the Eastern Mediterranean and along its southern shores (e.g. in the Aegean and Levantine seas) are much rarer, likely resulting in part from low density, and in part from lack of systematic observations (Notarbartolo di Sciara et al., 2003).

Year-round residency by fin whales in the Mediterranean Sea has required ecological and behavioural adaptations to regional specificities, notably regarding the whales' feeding and breeding needs. Fin whales have been observed engaging in feeding in the Mediterranean throughout the year. Most of the observations concern summer, e.g. in the Ligurian Sea (Notarbartolo di Sciara et al., 2003; Orsi Relini and Giordano, 1992), off eastern Sicily (Puzzolo and Tringali, 2001) and in the southern Tyrrhenian Sea off the island of Ischia (Mussi et al., 1999). In the latter area, feeding was inferred from swimming behaviour and frequently observed defecation episodes. Feeding was also seen in spring off eastern Sicily (Catalano et al., 2001) and in winter in the central Tyrrhenian Sea off northeastern Sardinia (Magnone et al., 2011), as well as in the Strait of Sicily near Lampedusa Island (Canese et al., 2006) where whales were frequently observed foraging at the surface by swimming in formation.

The observations of fin whales moving across a marine region such as the Mediterranean Sea, known to be largely oligotrophic (e.g. Huertas et al., 2012), to feed in specific and predictable locations and times of the year, imply that the following two conditions are met: (a) the existence of a mosaic of mesoscale productive features, highly variable in space and time, favouring zooplankton growth, and (b) the whales' ability to locate and exploit such features in a timely fashion. Based on these assumptions and on experience gathered from the Atlantic bluefin tuna, *Thunnus thynnus* (another predator of *M. norvegica*, Druon et al., 2011), by relating the proximity of over 10 years of fin whale sighting locations ($n=1451$) to concurrent remotely sensed oceanic fronts of chlorophyll a (Chl a), Druon et al. (2012) developed a model to detect and map on a daily timescale fin whale potential feeding habitat. Although not accounting for a portion of primary productivity which occurs at depth, which is undetected by satellite remote sensing (Macias et al., 2014), Druon et al.'s (2012) model proved to be very accurate when ground truthed with actual fin whale movements monitored by satellite tracking. Subsurface primary productivity is indeed substantially lower than the surface productivity hot spots linked to mesoscale features (chlorophyll fronts), notably due to the exponential decrease of light with depth and the decrease of the chlorophyll/carbon ratio.

Proof of the accuracy of the Druon et al. (2012) model is provided by the tracks of two whales which were satellite tagged in March 2015 off Lampedusa Island in the Strait of Sicily, and tracked for the subsequent 29 (whale n. 87776) and 44 (whale n. 87780) days (Panigada et al., 2015). During the first 15 days of tracking the whales remained in the waters of the Strait of Sicily, which were at that time highly productive; the whales clearly engaged in feeding behaviour as was evident from direct observation and defecation episodes as well as from inference of the tracked swim pattern (Fig. 4A). During the two subsequent fortnights both whales had moved to the north, abandoning the Sicily Strait where productivity was waning; the whales took advantage along the way of short spells of productivity in the Tyrrhenian Sea (whale n. 87776: Fig. 4B), and ultimately ended in the Ligurian Seas, where by that time the productivity conditions had sharply improved (whale n. 87780: Fig. 4C).

Based on the Druon et al.'s (2012) model, potential fin whale feeding habitat in the Western Mediterranean Sea undergoes considerable seasonal variation, ranging from a highly diffused condition in winter and spring to extreme summer concentration in the Ligurian Sea and Gulf of Lion area and, to a minor extent, along the southeastern shores of Italy (Fig. 5).

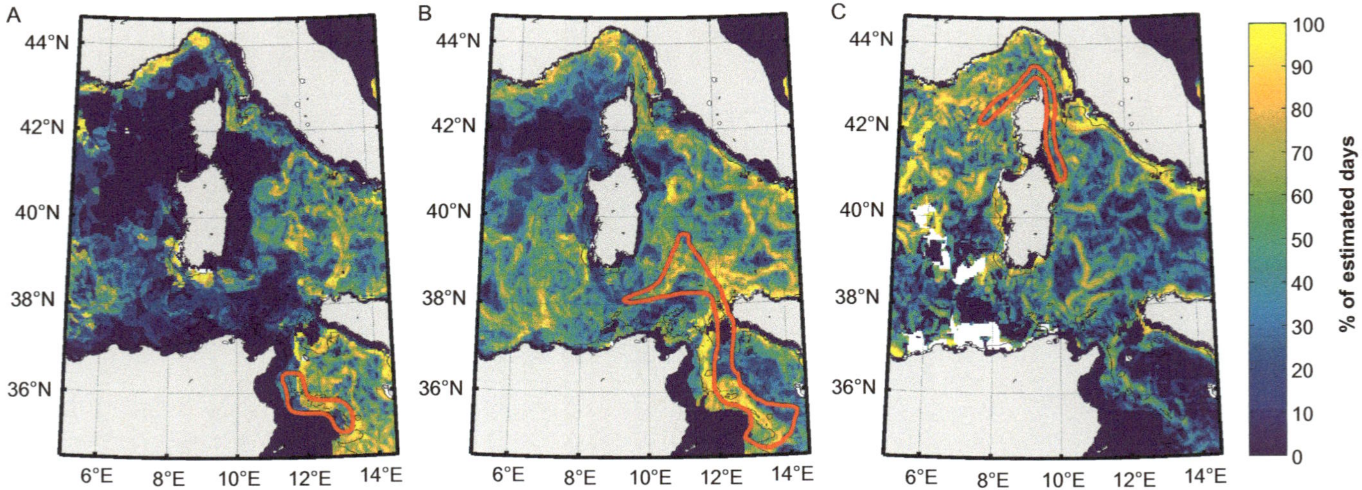

Fig. 4 Dynamic representation of fin whale (*Balaenoptera physalus*) feeding habitat (% of occurrence) from 16 March to 30 April 2015. The *encircled areas* indicate the positions of satellite-tagged fin whales during the three successive periods. (A) 16–31 March (e-tags #87776 and 87780); (B) 1–15 April (e-tags #87776 and 87780) and (C) 16–30 April (e-tag #87780). Depth contour is 200 m; blank is habitat coverage <5% or CHL coverage <1%.

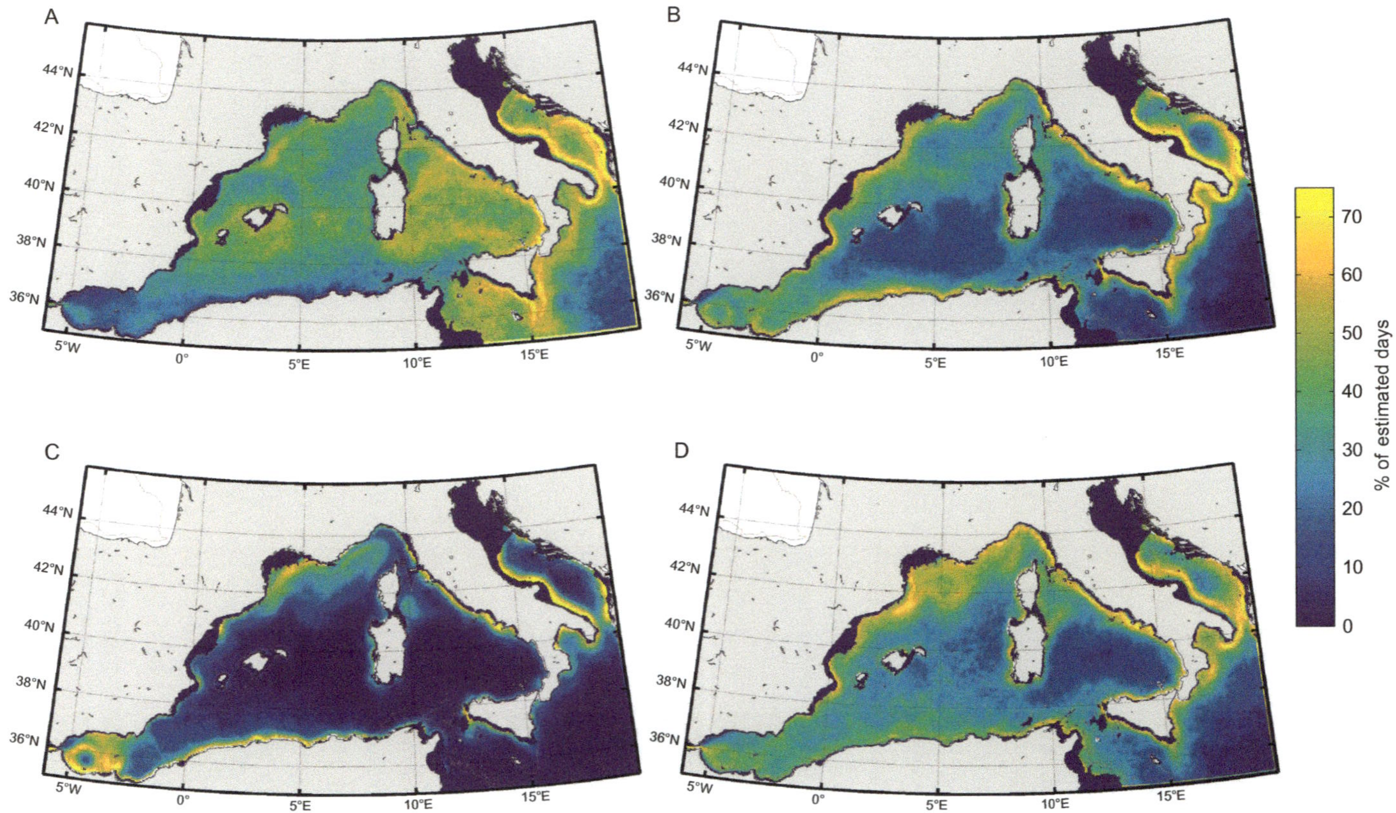

Fig. 5 Occurrence of potential fin whale (*Balaenoptera physalus*) feeding habitat (in % of occurrence) in the western Mediterranean, averaged over 13 years (2003–15) subdivided between winter (A: January–March); spring (B, April–June); summer (C, July–September) and autumn (D, October–December) (200 m depth contour).

These observations require a reinterpretation of the movement patterns of MED whales in the Western Mediterranean. These are often depicted as fixed migratory movement patterns from unknown winter breeding grounds to the Ligurian-Provençal Basin summer feeding grounds along well-defined corridors (e.g. Marini et al., 1996). We suggest that: (a) the whales' gathering in the northwestern Mediterranean Sea in summer is caused by the contraction of feeding opportunities elsewhere (although such gathering may appear to be occurring by recurrent travelling along a defined route where a passage is confined by coastlines, e.g. between Corsica and the Italian mainland, or between the Balearic Islands and the Iberian Peninsula), and (b) that the whales' dispersal from the Ligurian Sea after summer is a consequence of the diffused reappearance of favourable feeding conditions over a much wider Mediterranean surface. Such reinterpretation of fin whale movements in the Western Mediterranean Sea is based upon a contraction/dispersal hypothesis caused by the seasonal variability of available feeding habitat, as opposed to a migration hypothesis occurring along defined routes. The former is corroborated by the wide movements irradiating in many directions out of the Ligurian Sea at the end of summer, demonstrated in eight fin whales tagged in the Ligurian Sea at the end of summer 2012 (Fig. 3). It also explains the observed continued presence of whales in the Ligurian-Provençal Basin during winter, albeit at much lower densities, as evidenced by surface observations (Gannier and Gannier, 1993), winter aerial surveys (Panigada et al., 2011a) and passive acoustic monitoring (C.W. Clark, Cornell University, personal communication, 2016).

Druon et al.'s (2012) niche model further reveals substantial interannual differences in the distribution of potential fin whale feeding habitat, e.g. between 2003 and 2005 (Fig. 6). This was strikingly reflected by corresponding differences in fin whale encounter rates during extensive summer vessel-based surveys in the Pelagos Sanctuary, which increased by a factor of four from 2003 to 2005 (A. Azzellino, Politecnico di Milano, personal communication, 2016). Considerable interannual changes in patterns of fin whale occurrence in the adjacent central Tyrrhenian Sea over a 20-year period were also reported by Arcangeli et al. (2012, 2014).

Much uncertainty still remains concerning the presence, movements and habitat use of fin whales in the Eastern Mediterranean Basin and along the entire length of the African coast, due to the dearth of direct observations in those areas. Seasonal changes in the mean yearly distribution of potential fin whale feeding habitat in the Eastern Mediterranean Sea derived from the Druon et al. (2012) model shows that feeding habitat in the Eastern Basin

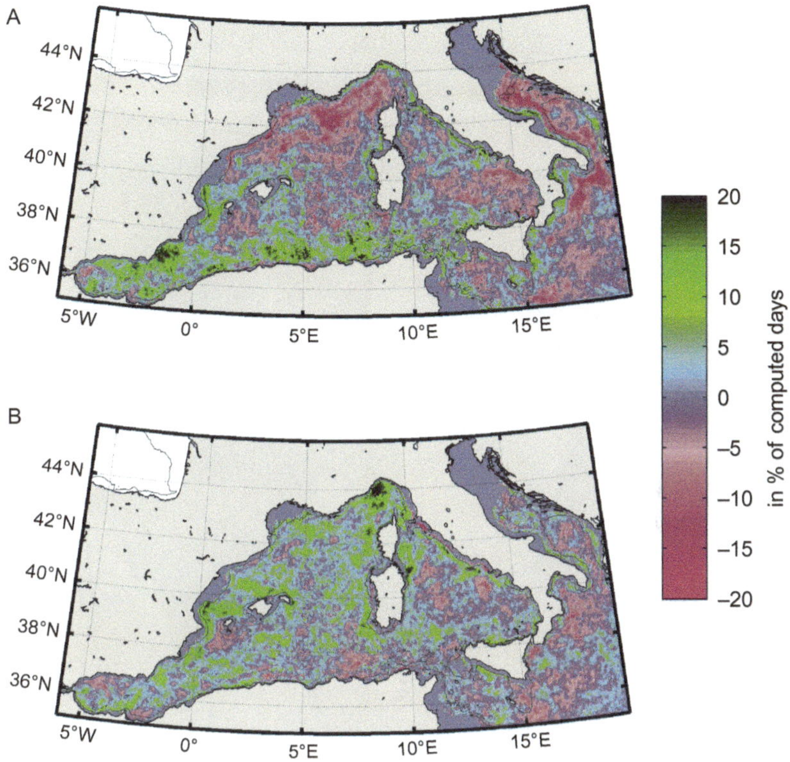

Fig. 6 Occurrence of potential fin whale (*Balaenoptera physalus*) feeding habitat (in % of occurrence) in the Western Mediterranean. (A) 2003 average and (B) 2005 average (200 m depth contour).

during summer almost completely disappears (Fig. 7C). By contrast, several locations seem to be apt to host temporary aggregations of feeding fin whales in the other seasons, most notably in winter; these include the western Ionian Sea from the east coast of Sicily to the south towards Libya, the southern Adriatic Sea (see also Fig. 4), parts of the Aegean Sea, the waters comprised between the eastern tip of Crete and Cyprus including the conspicuous Rhodes Gyre, and the coastal waters of Syria, Lebanon and Israel. Confirmation of the actual presence of fin whales in these locations unfortunately only comes from a handful of direct observations, limited to the central and southern Adriatic (Lipej et al., 2004; Pierantonio and Bearzi, 2012), the Sicilian Ionian coastal waters (Aissi et al., 2008; Puzzolo and Tringali, 2001; Sciacca et al., 2015; Tringali et al., 1999, 2006) and Israel (Kerem et al., 2012). Building on the contraction/dispersal

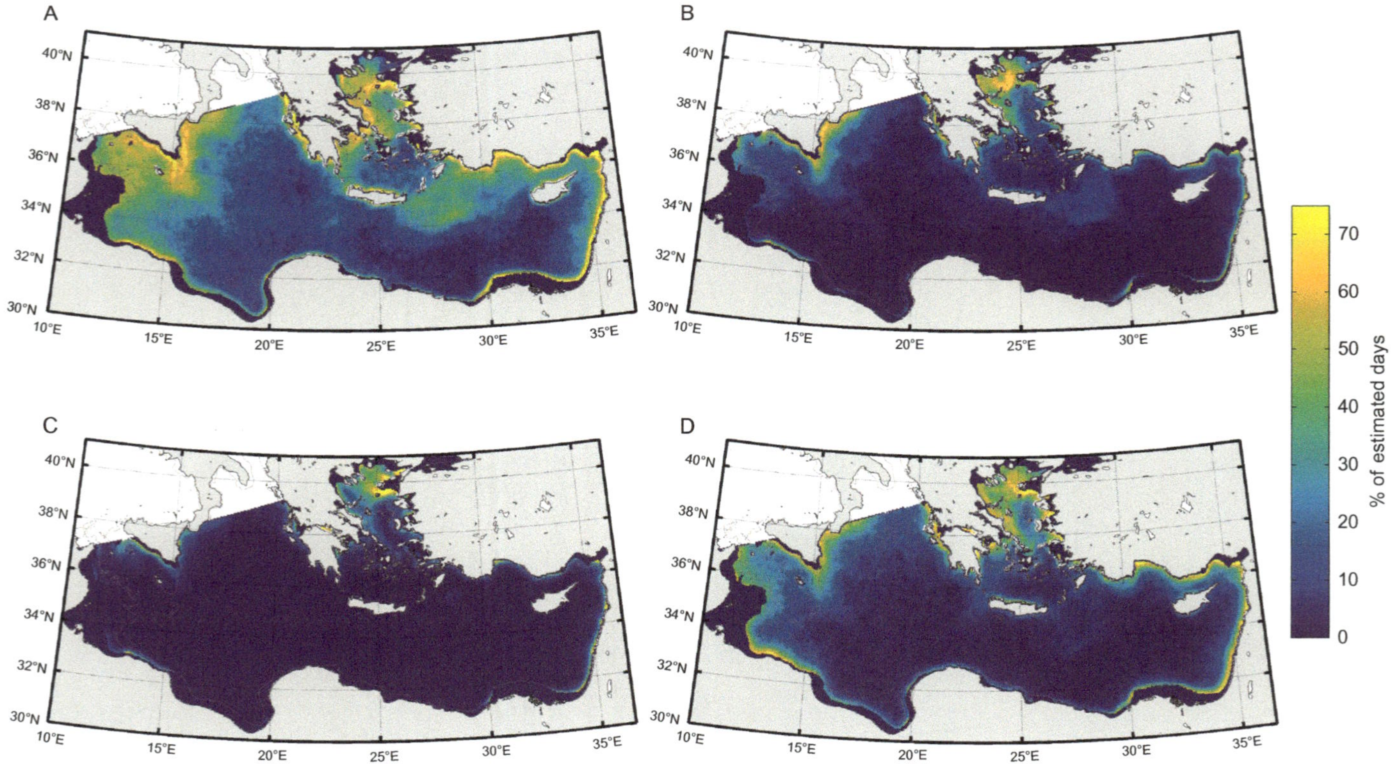

Fig. 7 Occurrence of potential fin whale (*Balaenoptera physalus*) feeding habitat (in % of occurrence) in the Eastern Mediterranean, averaged over 13 years (2003–15) subdivided between winter (A, January–March); spring (B, April–June); summer (C, July–September) and autumn (D, October–December) (200 m depth contour).

hypothesis, it would be sensible to presume that fin whales occur in feeding gatherings between autumn and spring in many of the Eastern Mediterranean locations, where substantive potential feeding habitat is predicted (i.e. above ca. 40% of occurrence), like that observed around Lampedusa Island in winter.

Similar to the absence of a specific feeding season for MED whales, breeding in this population occurs all year long; however, a peak of births during autumn is evident on the basis of a record of 56 newborn fin whales reported from the Mediterranean in Notarbartolo di Sciara et al. (2003). As expected, records of newborn whales originated mostly from the western portion of the region where whale density is higher. However, newborns also occurred in the Eastern Mediterranean, supporting the hypothesis that MED whales, rather than gathering in specific breeding grounds, engage in breeding activities wherever favourable physiological conditions occur. Unlike for other mysticetes such as humpback and grey whales, major uncertainties concerning the locations of specific fin whale breeding grounds still exist also with regard to oceanic populations (Kellogg, 1929; Mizroch et al., 2009).

Piecing together the past knowledge with recent information from telemetry and habitat modelling, the available evidence supports the suggestion by Geijer et al. (2016) that MED whales, instead of regular migrants, are nomadic opportunists moving between specific sites where oceanographic conditions cause the recurring appearance of localised prey abundance. As proposed in Notarbartolo di Sciara et al. (2003), it can be assumed that the Mediterranean environment (which is meteorologically and climatically relatively forgiving in comparison to oceanic habitats), combined with negligible predation pressure and high potential for sound-mediated socialisation due to the region's small size, might have provided year-round resident fin whales an extended calendar of breeding and feeding opportunities. At the same time, unlike their oceanic conspecifics needing to migrate to calve out of cold water for energetic reasons, this condition would release MED whales from the need to migrate to distant locations for calving purposes.

4. STATUS AND THREATS

The Mediterranean population of fin whales was assessed as Vulnerable on the International Union for the Conservation of Nature's (IUCN) Red List (Panigada and Notarbartolo di Sciara, 2012). The listing was justified on the basis of the following statements:

1. The Mediterranean subpopulation, which is genetically distinct from fin whales in the Atlantic, contains fewer than 10,000 mature individuals.
2. The subpopulation experiences an inferred continuing decline in number of mature individuals.
3. All mature individuals are in one subpopulation.

Some of the above statements now need to be reexamined, based upon more recent information on Mediterranean fin whale ecology (Castellote et al., 2012a), which has been gathered since the population was assessed in the IUCN Red List. Statement (1) is still valid in terms of the Mediterranean population being distinct from North Atlantic fin whales, however the population estimate of >3500 individuals by Forcada et al. (1996), generated by line transect surveys performed between the Strait of Gibraltar and the Ligurian Sea, likely sampled NENA whales as well as MED whales, thereby contradicting statement (3). By consequence, a population estimate limited to MED whales could be significantly smaller, and perhaps consistent with a higher threat classification.

Conversely, statement (2) likely remains valid. Threats to fin whale survival in the Mediterranean Sea include direct mortality caused by collisions with vessels and, to a lesser extent, bycatch and disease. Vessel disturbance, chronic negative effects and habitat exclusion caused by anthropogenic noise, pollution and climate change are also of concern.

Ship strikes are likely the main source of nonnatural, human-induced mortality in MED whales (Panigada et al., 2006), particularly in areas of heavy vessel traffic. A study of potential risk of ship strike using habitat estimates and density of maritime traffic showed that the collision risk is concentrated along the main maritime routes and substantially grows in the Pelagos Sanctuary area during summer—the whales' core habitat in that season—due to both the seasonal increase of maritime traffic (passenger transport) and the shrinking of favourable fin whale feeding habitat (Vaes and Druon, 2013). Ship traffic in the area already grew fourfold between 1992 and 2012, and is expected to further grow by 23% over the next decade (Panigada et al., 2008). In a scenario involving future steep increase of maritime traffic in the Mediterranean Sea, fin whale mortality caused by vessel collisions is a mounting concern. A further concern is that an increase in shipping entails an increase in noise in the fin whale communication band. This might, through habituation to the chronic noise production by distant shipping, impair the whales' ability to respond with effective avoidance to approaching vessels, as hypothesised in the case of the endangered North Atlantic right whale, *Eubalaena glacialis* (Terhune and Verboom, 1999).

Even in the absence of direct mortality caused by ship strikes, the disturbance caused by intense traffic through whale habitat is a reason for concern. Vaes and Druon (2013) noted the near absence of whale sightings along the shipping route connecting the Strait of Gibraltar to the Strait of Sicily, one of the world's busiest, in spite of the presence of favourable feeding habitat predicted along that route by Druon et al.'s (2012) model. Vessel disturbance can also be of concern if caused by unregulated whale watching, particularly in the crammed Ligurian Sea feeding grounds, where disturbed whales have been shown to interrupt their feeding activities for unknown durations (Jahoda et al., 2003).

Anthropogenic underwater noise is now recognised as a worldwide cetacean conservation problem, and recent studies have shown a broad range of negative effects in a variety of taxa (Williams et al., 2015). Background noise levels in the Mediterranean Sea are higher than in any other ocean basin (Ross, 2005), with ship traffic and seismic surveys being among the primary sources of noise (Maglio et al., 2015). Low-frequency specialists, in particular mysticetes, may well be particularly susceptible to the effects of anthropogenic noise on their communication (Hildebrand, 2005). Therefore, it can be assumed that Mediterranean fin whales are disproportionately affected by chronically exposed high levels of noise. Acoustic and behavioural changes by both MED and NENA whales have been documented in response to two different types of anthropogenic noise typically occurring in the Mediterranean Sea: shipping and seismic airgun noise. Fin whales modified their song characteristics under increased background noise conditions and under seismic airgun activity conditions; and they abandoned the area for a time period well beyond the 10-day duration of seismic airgun activity (Castellote et al., 2012b). Acoustic changes in fin whale songs might compensate the masking effects of increased background noise, but the metabolic costs to the animals could be higher. Furthermore, song modifications might affect song effectiveness, in particular changes in note interval and bandwidth, since these seem to be related to population identity (Castellote et al., 2012a) and might play an important role in reproduction (Croll et al., 2002).

Interactions with fishery activities are considered to have a minor effect on Mediterranean fin whales (Notarbartolo di Sciara et al., 2003). Even during the 1980s and 1990s, a period in which driftnet fishing for large pelagic fish was at its peak in the northwestern Mediterranean Sea and caused elevated incidental mortality in many cetacean species, documented bycatch of fin whales was rare (Notarbartolo di Sciara, 1990; Podestà and Magnaghi, 1989). Fin whale prey depletion by fishing is not a concern, given the predominance in the whales' diet of euphausiids, which are not targeted by fisheries in the region.

Like all marine top predators in the Mediterranean Sea, and in spite of their position at a lower trophic level than odontocetes such as striped and bottlenose dolphins, fin whales are exposed to significant levels of chemical pollution. This includes organochlorines, trace elements, dichlorodiphenyltrichloroethane (DDT) metabolites, and endocrine-disrupting chemicals. Recent attention is being focused on emergent contaminants, such as bisphenol A, brominated flame retardants and phthalates, which have been found to occur in high concentrations in the blubber of fin whales; the latter are related to the presence in the Mediterranean environment of large amounts of marine litter, and in particular plastic debris and microplastics (Cózar et al., 2015; Fossi et al., 2012, 2014, 2016). Contaminant load can might negatively affect fin whale reproductive functions due to their reported estrogenic and antiandrogenic effects (Fossi et al., 2003, 2007), and a potential weakening of the whales' immune response. Such consideration raises particular concern in view of the recent evidence, deriving from direct (immunohistochemical and biomolecular) and indirect (serologic) investigations, of dolphin morbillivirus (DMV) infection or exposure in five out of nine fin whales stranded in Italy between 2011 and 2013. This suggests the potential of an epidemic cluster of fatal DMV in MED whales, and stresses that DMV should be regarded as one of the major biological threats to fin whales in the Mediterranean Sea (Mazzariol et al., 2016).

Finally, the effects of climate change and ocean acidification on the Mediterranean marine environment have been flagged as a potentially increasing threat to MED whales; in particular such changes might affect the viability, abundance and distribution of *M. norvegica*, the whales' main food supply in the region (Gambaiani et al., 2009). Climate change-induced alteration of water mass circulation in the Mediterranean, possibly detected already decades ago (Bethoux and Gentili, 1996), is likely to affect the spatiotemporal pattern of the region's marine productivity; however, the complexity of the involved ecological phenomena still presents a challenge to unambiguous cause–effect interpretation (Wyatt, 2010). Alterations in Mediterranean marine productivity patterns may present a serious impediment to the ability of fin whales to locate transient food sources, although the whales' nomadic opportunist attitudes might help them to overcome such challenge.

Considering all of the threats to MED whales, actual and potential, listed earlier, the observed population decrease raises considerable concern. A comparison between line transect population estimates conducted within the boundaries of the Pelagos Sanctuary 18 years apart indicates a >60%

decline since the 1992 estimate of 901 whales (CV=0.217; 95% CI: 591–1374) (Forcada et al., 1995; Panigada et al., 2011b). The 1992 estimate was corroborated by comparable results obtained during independent survey efforts by Gannier (1997). Furthermore, photo-identification data collected over 18 consecutive summers (1990–2007) and analysed by Zanardelli et al. (2011) with a Jolly–Seber open population model yielded a population size in 1990 of 980 whales (CV: 0.20; 95% CI: 670–1437), a rate of population change of 0.99 (95% CI: 0.92–1.07), and an unexpectedly low apparent survival rate (0.88, 95% CI: 0.76–0.94). The observed decrease of fin whales can be explained in several different ways, and likely reflect the high observed interannual variability of feeding habitat (Fig. 6); for instance, a higher contraction of summer feeding habitat in the early 1990s might have resulted in higher whale concentrations in the Pelagos Sanctuary in those years, with density levels there that may have no longer occurred in subsequent years. Thus, successive relocation of MED whales to different areas within the boundaries of their known range is not unlikely.

However, the alternative explanation of population decline caused by a decrease in survival rate and/or reproductive success, reinforced by the low measured level of apparent survival, cannot be discounted until synoptic surveys encompassing the entire population range (as advocated by Agreement on the Conservation of Cetaceans in the Black Sea Mediterranean Sea and Contiguous Atlantic Area, ACCOBAMS, for more than a decade) have been conducted.

5. RECOMMENDATIONS

Considering the conditions of the MED whale population described earlier, the uncertainties that still surround them, and the cumulative effects in a resident population of the many impacting pressures in a semienclosed region heavily affected by human activities, the highest precaution is recommended. The MED whales should be treated as a high regional conservation priority, and their IUCN Red List status should be reassessed.

Well-identified threats such as vessel collisions and anthropogenic noise should be addressed through the enactment of timely and effective measures. Priority should be given to define appropriate ways to minimise vessel collisions and reduce acoustic factors that could contribute to exclusion from or loss of fin whale habitat.

Concerning ship strikes, emphasis should be placed on the implementation of routing schemes and speed reduction zones in areas and periods where distribution modelling exercises identify potential for high fin whale densities at appropriate temporal and geographical scales. As is well known, the most effective and until now the only demonstrated method to address lethal strikes consists in reducing vessel speed. This has been shown in a number of studies, in large part concerning North Atlantic right whales (Conn and Silber, 2013; Laist et al., 2014); these results are valid also for fin whales (Laist et al., 2001; Panigada et al., 2006). Furthermore, reducing speed also reduces ship noise, therefore this should be considered a double mitigation action.

Other sources of anthropogenic noise are also a concern, particularly considering the diffused proliferation of commercial and scientific seismic surveys in the Mediterranean Sea (Maglio et al., 2015). Control, regulation and permit application procedures for these activities in the Mediterranean Sea should be addressed in accordance to the European Union Marine Strategy Framework Directive (European Union, 2008), as well as the Habitat Directive (European Union, 1992) requirements, and environmental impact assessments should consider the mitigation of fin whale noise impact a priority. To effectively address noise pressure on this highly sensitive Mediterranean species, mitigation should consider both direct close-range physiological effects (e.g. through the identification of exclusion zones, the involvement of independent observers and the adoption of manoeuvres such as power downs, shut downs, ramp ups), and long-range behavioural effects (e.g. through spatial and temporal limitations to avoid ensonifying known or predicted high-density areas or times, or by establishing buffer zones around sensitive areas).

Mindful that the threat of shipping noise to marine life has been recognised, among others, by the IMO (International Maritime Organization, 2014) and by the European Union (2008), noise fields within fin whale important habitat should be regularly monitored and mapped, thereby providing guidance and insight on the need for and ways of mitigating negative effects. Commercial whale watching operations targeting fin whales in the species' important habitat should be regulated and carefully monitored.

Enhancing place-based protection of MED whale habitat is another important consideration. In the light of the seasonal dynamics of MED whales' feeding habitat and their consequences on the whales' distribution and movements, a reassessment of the Pelagos Sanctuary boundaries and seasonal importance is strongly urged. For example, seasonally and/or dynamically managed protected zones could be implemented to address the danger of ship strikes, as occurs in North Atlantic right whale habitat in US waters (Asaro, 2012). An extension to the west of the boundaries of the Pelagos Sanctuary should also be considered in order to encompass the entirety of fin whale summer feeding habitat, notably given the dramatic reduction of such habitat during summer at the Mediterranean scale. Conversely, habitat protection in recurrent seasonal movement areas, as well as during the colder seasons in other Mediterranean areas known to host important fin whale concentration should be given appropriate consideration through the designation of a network of seasonally managed marine protected areas, thereby increasing the percentage of fin whale important habitat falling under conservative management regimes (Notarbartolo di Sciara et al., in press).

Finally, existing gaps in fin whale ecological knowledge should be addressed through directed research studies to properly inform and strengthen conservation and management actions. These studies should include aerial- and ship-based targeted surveys, population structure studies including behavioural and reproductive interactions between the NENA and MED populations, as well as satellite tracking experiments and passive acoustic monitoring to gain a detailed understanding of the species' presence in the southern and eastern portions of the Mediterranean Sea.

ACKNOWLEDGEMENTS

We wish to express our heartfelt thanks to colleagues Phil J. Clapham, Chris W. Clark and Barbara E. Curry for their advice, edits and suggestions which greatly improved this manuscript. We also thank Danny Kessler for permitting the use of his picture of a fin whale (Fig. 1).

REFERENCES

Aguilar, A., Borrell, A., Reijnders, P.J.H., 2002. Geographical and temporal variation in levels of organochlorine contaminants in marine mammals. Mar. Environ. Res. 53, 425–452.

Aissi, M., Celona, A., Comparetto, G., Mangano, R., Würtz, M., Moulins, A., 2008. Large-scale seasonal distribution of fin whales (*Balaenoptera physalus*) in the central Mediterranean Sea. J. Mar. Biol. Assoc. UK 88 (6), 1253–1261. http://dx.doi.org/10.1017/S0025315408000891.

Arcangeli, A., Marini, L., Crosti, R., 2012. Changes in cetacean presence, relative abundance and distribution over 20 years along a trans-regional fixed line transect in the Central Tyrrhenian Sea. Mar. Ecol. 2012, 1–10. http://dx.doi.org/10.1111/maec.12006.

Arcangeli, A., Orasi, A., Carcassi, S.P., Crosti, R., 2014. Exploring thermal and trophic preference of *Balaenoptera physalus* in the central Tyrrhenian Sea: a new summer feeding ground? Mar. Biol. 161, 427–436. http://dx.doi.org/10.1007/s00227-013-2348-8.

Asaro, M.J., 2012. Geospatial analysis of management areas implemented for protection of the North Atlantic right whale along the northern Atlantic coast of the United States. Mar. Policy 36 (4), 915–921. http://dx.doi.org/10.1016/j.marpol.2012.01.004.

Bentaleb, I., Martin, C., Vrac, M., Mate, B., Mayzaud, P., Siret, D., de Stephanis, R., Guinet, C., 2011. Foraging ecology of Mediterranean fin whales in a changing environment elucidated by satellite tracking and baleen plate stable isotopes. Mar. Ecol. Prog. Ser. 438, 285–302. http://dx.doi.org/10.3354/meps09269.

Bérubé, M., Aguilar, A., Dendanto, D., Larsen, F., Notarbartolo di Sciara, G., Sears, R., Sigurjonsson, J., Urban, J., Palsbøll, P.J., 1998. Population genetic structure of North Atlantic, Mediterranean Sea and Sea of Cortez fin whales, *Balaenoptera physalus* (Linnaeus 1758): analysis of mitochondrial and nuclear loci. Mol. Ecol. 7 (5), 585–599.

Bérubé, M., Urban, J., Dizon, A.E., Brownell, R.L., Palsbøll, P.J., 2002. Genetic identification of a small and highly isolated population of a fin whales (*Balaenoptera physalus*) in the Sea of Cortez, Mexico. Conserv. Ecol. 3, 183–190.

Bethoux, J.P., Gentili, B., 1996. The Mediterranean Sea, coastal and deep-sea signatures of climatic and environmental changes. J. Mar. Syst. 7, 383–394.

Canese, S., Cardinali, A., Fortuna, C.M., Giusti, M., Lauriano, G., Salvati, E., Greco, S., 2006. The first identified winter feeding ground of fin whales (*Balaenoptera physalus*) in the Mediterranean Sea. J. Mar. Biol. Assoc. UK 86 (5119), 1–5.

Castellote, M., Alvarez, C., Laran, S., La Manna, G., Gazo, M., Manghi, M., Lammers, M.O., 2011. Further acoustic insights into the Mediterranean fin whale migration and population identity. In: Gauffier, P., Verborgh, P. (Eds.), Abstracts, 25th Conference of the European Cetacean Society, Cádiz, Spain, 21–23 March 2011, p. 70.

Castellote, M., Clark, C.W., Lammers, M.O., 2012a. Fin whale (*Balaenoptera physalus*) population identity in the western Mediterranean Sea. Mar. Mamm. Sci. 28 (2), 325–344. http://dx.doi.org/10.1111/j.1748-7692.2011.00491.x.

Castellote, M., Clark, C.W., Lammers, M.O., 2012b. Acoustic and behavioural changes by fin whales (*Balaenoptera physalus*) in response to shipping and airgun noise. Biol. Conserv. 147 (1), 115–122. http://dx.doi.org/10.1016/j.biocon.2011.12.021.

Castellote, M., Clark, C.W., Lammers, M.O., 2013. Dispersal of North Atlantic fin whales (*Balaenoptera physalus*) into the Mediterranean Sea and exchange between populations: response to Giménez et al., Rapid Communications in Mass Spectrometry 2013, 27, 1801–1806. Rapid Commun. Mass Spectrom. 28, 665–667. http://dx.doi.org/10.1002/rcm.6819.

Catalano, D., Tringali, L.M., Petralia, R., Caltavuturo, G., 2001. Mediterranean fin whales, *Balaenoptera physalus*, foraging off the eastern coast of Sicily, Ionian Sea. Eur. Res. Cetaceans 15, 56–58.

Clapham, P.T., Aguilar, A., Hatch, L.T., 2008. Determining spatial and temporal scales for management: lessons from whaling. Mar. Mamm. Sci. 24 (1), 183–201.

Clark, C.W., Gagnon, G.C., 2002. Low-frequency vocal behaviors of baleen whales in the North Atlantic: insights from IUSS detections, locations and tracking from 1992 to 1996. J. Underw. Acoust. (USN) 52 (3), 609–640.

Clark, C.W., Borsani, J.F., Notarbartolo di Sciara, G., 2002. Vocal activity of fin whales, *Balaenoptera physalus*, in the Ligurian Sea. Mar. Mamm. Sci. 18 (1), 286–295.

Conn, P.B., Silber, G.K., 2013. Vessel speed restrictions reduce risk of collision-related mortality for North Atlantic right whales. Ecosphere 4 (4), 43. http://dx.doi.org/10.1890/ES13-00004.1.

Cotté, C., d'Ovidio, F., Chaigneau, A., Lévy, M., Taupier-Letage, I., Mate, B., Guinet, C., 2011. Scale-dependent interactions of Mediterranean whales with marine dynamics. Limnol. Oceanogr. 56 (1), 219–232. http://dx.doi.org/10.4319/lo.2011.56.1.0219.

Cózar, A., Sanz-Martín, M., Martí, E., González-Gordillo, J.I., Ubeda, B., Gálvez, J.Á., Irigoien, X., Duarte, C.M., 2015. Plastic accumulation in the Mediterranean Sea. PLoS One 10, e0121762. http://dx.doi.org/10.1371/journal.pone.0121762.

Croll, D.A., Clark, C.W., Acevedo, A., Tershy, B., Flores, S., Gedamke, J., Urban, J., 2002. Only male fin whales sing loud songs. Nature 417, 809–811.

Donovan, G.P., 1991. A review of IWC stock boundaries. Rep. Int. Whaling Comm. 13, 39–68.

Druon, J.N., Fromentin, J.M., Aulanier, F., Heikkonen, J., 2011. Potential feeding and spawning habitats of Atlantic bluefin tuna in the Mediterranean Sea. Mar. Ecol. Prog. Ser. 439, 223–240.

Druon, J.-N., Panigada, S., David, L., Gannier, A., Mayol, P., Arcangeli, A., Cañadas, A., Laran, S., Di Méglio, N., Gauffier, P., 2012. Potential feeding habitat of fin whales in the western Mediterranean Sea: an environmental niche model. Mar. Ecol. Prog. Ser. 464, 289–306. http://dx.doi.org/10.3354/meps09810.

Edwards, E.F., Hall, C., Moore, T.J., Sheredy, C., Redfern, J.V., 2015. Global distribution of fin whales *Balaenoptera physalus* in the post-whaling era (1980–2012). Mammal Rev. 45 (4), 197–214. http://dx.doi.org/10.1111/mam.12048.

European Union, 1992. Council Directive 92/43/EEC of 21 May 1992 on the conservation of natural habitats and of wild fauna and flora. Off. J. Eur. Communities L 206 (35), 7–50.

European Union, 2008. Directive 2008/56/EC of the European Parliament and of the Council of 17 June 2008 establishing a framework for community action in the field of marine environmental policy (Marine Strategy Framework Directive). Off. J. Eur. Union L 164, 19–40.

Forcada, J., Notarbartolo di Sciara, G., Fabbri, F., 1995. Abundance of fin whales and striped dolphins summering in the Corso-Ligurian Basin. Mammalia 59 (1), 127–140.

Forcada, J., Aguilar, A., Hammond, P., Pastor, X., Aguilar, R., 1996. Distribution and abundance of fin whales (*Balaenoptera physalus*) in the western Mediterranean Sea during the summer. J. Zool. 238, 23–34.

Fossi, M.C., Marsili, L., Neri, G., Natoli, A., Politi, E., Panigada, S., 2003. The use of a non-lethal tool for evaluating toxicological hazard of organochlorine contaminants in Mediterranean cetaceans: new data 10 years after the first paper published in MPB. Mar. Pollut. Bull. 46 (8), 972–982.

Fossi, M.C., Casini, S., Marsili, L., 2007. Potential toxicological hazard due to endocrine-disrupting chemicals on Mediterranean top predators: state of art, gender differences and methodological tools. Environ. Res. 104, 174–182.

Fossi, M.C., Panti, C., Guerranti, C., Coppola, D., Giannetti, M., Marsili, L., Minutoli, R., 2012. Are baleen whales exposed to the threat of microplastics? A case study of the Mediterranean fin whale (*Balaenoptera physalus*). Mar. Pollut. Bull. 64, 2374–2379.

Fossi, M.C., Coppola, D., Baini, M., Giannetti, M., Guerranti, C., Marsili, L., Panti, C., de Sabata, E., Clò, S., 2014. Large filter feeding marine organisms as indicators of microplastic in the pelagic environment: the case studies of the Mediterranean basking shark (*Cetorhinus maximus*) and fin whale (*Balaenoptera physalus*). Mar. Environ. Res. 100, 17–24.

Fossi, M.C., Marsili, L., Baini, M., Giannetti, M., Guerranti, C., Caliani, I., Minutoli, R., Lauriano, G., Finoia, M.G., Rubegni, F., Panigada, S., Bérubé, M., Urban, J., Panti, C., 2016. Fin whales and microplastics: the Mediterranean Sea and the Sea of Cortez scenarios. Environ. Pollut. 209, 68–78. http://dx.doi.org/10.1016/j.envpol.2015.11.022.

Gambaiani, D.D., Mayol, P., Isaac, S.J., Simmonds, M.P., 2009. Potential impacts of climate change and greenhouse gas emissions on Mediterranean marine ecosystems and cetaceans. J. Mar. Biol. Assoc. UK 89 (1), 179–201.

Gambell, R., 1985. Fin whale—*Balaenoptera physalus*. In: Ridgway, S.H., Harrison, R. (Eds.), The Sirenians and Baleen Whales. In: Handbook of Marine Mammals, vol. 3. Academic Press, New York, pp. 171–192.

Gannier, A., 1997. Estimation de l'abondance estivale du rorqual commun *Balaenoptera physalus* (Linné, 1758) dans le bassin liguro-provençal (Méditerranée occidentale). Rev. Ecol. (Terre Vie) 52, 69–86.

Gannier, A., Gannier, O., 1993. The winter presence of the fin whale in the Liguro-Provencal Basin: preliminary study. Eur. Res. Cetaceans 7, 131–134.

Gauffier, P., Verborgh, P., Andreu, E., Esteban, R., Medina, B., Gallego, P., de Stephanis, R., 2009. An update on fin whales (*Balaenoptera physalus*) migration through intense maritime traffic in the Strait of Gibraltar. In: Paper SC/61/BC6 presented to the International Whaling Commission Scientific Committee, May 2009, Madeira, Portugal.

Gauffier, P., Moral Cendon, M., Blasi, A., Carpinelli, E., Jimenez-Torres, C., Gimenez Verdugo, J., Esteban, R., de Stephanis, R., Verborgh, P., 2012. Winter presence of large cetaceans in the Strait of Gibraltar. In: 26th Annual Conference of the European Cetacean Society, Galway, Ireland, 2012, p. 209.

Geijer, C.K.A., Notarbartolo di Sciara, G., Panigada, S., 2016. Mysticete migration revisited: are Mediterranean fin whales an anomaly? Mammal Rev. http://dx.doi.org/10.1111/mam.12069.

Gimenez, J., Gomez-Campos, E., Borrell, A., Cardona, L., Aguilar, A., 2013. Isotopic evidence of limited exchange between Mediterranean and eastern North Atlantic fin whales. Rapid Commun. Mass Spectrom. 27 (15), 1801–1806. http://dx.doi.org/10.1002/rcm.6633.

Hatch, L.T., Clark, C.W., 2004. Acoustic differentiation between fin whales in both the North Atlantic and North Pacific Oceans, and integration with genetic estimates of divergence. International Whaling Commission Document SC/56/SD6, Sorrento, Italy.

Hildebrand, J., 2005. Impacts of anthropogenic sound. In: Reynolds, J.E., Perrin, W.F., Reeves, R.R., Montgomery, S., Ragen, T.J. (Eds.), Marine Mammal Research: Conservation Beyond Crisis. The Johns Hopkins University Press, Baltimore, pp. 101–124.

Huertas, I.E., Rios, A.F., Garcia-Lafuente, J., Navarro, G., Makaoui, A., Sanchez-Roman, A., Rodriguez-Galvez, S., Orbi, A., Ruiz, J., Perez, F.F., 2012. Atlantic forcing of the Mediterranean oligotrophy. Global Biogeochem. Cycles 26, GB2022. http://dx.doi.org/10.1029/2011GB004167.

International Maritime Organization, 2014. Guidelines for the reduction of underwater noise from commercial shipping to address adverse impacts on marine life. MEPC.1/Circ.833, 7 April 2014. London.

IWC, 2009. Report of the first intersessional RMP workshop on North Atlantic fin whales. Copenhagen, March 31st–April 4th 2008. J. Cetacean Res. Manag. 11 (Suppl. 2), 425–452.

IWC, 2010. Report of the 2nd intersessional workshop on the North Atlantic fin whale implementation. J. Cetacean Res. Manag. 11 (Suppl. 2), 587–627.

Jahoda, M., Lafortuna, C.L., Biassoni, N., Almirante, C., Azzellino, A., Panigada, S., Zanardelli, M., Notarbartolo di Sciara, G., 2003. Mediterranean fin whale's (*Balaenoptera physalus*) response to small vessels and biopsy sampling assessed through passive tracking and timing of respiration. Mar. Mamm. Sci. 19 (1), 96–110.

Jefferson, T.A., Webber, M.A., Pitman, R.L., Gorter, U., 2015. Marine Mammals of the World: A Comprehensive Guide to Their Identification, second ed. Elsevier, San Diego, CA.

Jonsgård, Å., 1966. Biology of the North Atlantic fin whale *Balaenoptera physalus* (L.): taxonomy, distribution, migration and food. Hvalradets Skrifter 49, 1–62.

Jonzen, N., Knudsen, E., Holt, R.D., Saeter, B.E., 2011. Uncertainty and predictability: the niches of migrants and nomads. In: Milner-Gulland, E.J., Fryxell, J.M., Sinclair, A.R.E. (Eds.), Animal Migration: A Synthesis. Oxford University Press, Oxford, pp. 91–109.

Kellogg, R., 1929. What is known on the migrations of some of the whalebone whales? Smithsonian Institution Annual Report, Washington 1928, 467–494.

Kerem, D., Hadar, N., Goffman, O., Scheinin, A., Kent, R., Boisseau, O., Schattner, U., 2012. Update on the cetacean fauna of the Mediterranean Levantine Basin. Open Mar. Biol. J. 6, 6–27.

Laist, D.W., Knowlton, A.R., Mead, J.G., Collet, A.S., Podestà, M., 2001. Collisions between ships and whales. Mar. Mamm. Sci. 17 (1), 35–75.

Laist, D.W., Knowlton, A.R., Pendleton, D., 2014. Effectiveness of mandatory vessel speed limits for protecting North Atlantic right whales. Endanger. Species Res. 23, 133–147. http://dx.doi.org/10.3354/esr00586.

Lipej, L., Dulcic, J., Krystufek, B., 2004. On the occurrence of the fin whale (*Balaenoptera physalus*) in the northern Adriatic. J. Mar. Biol. Assoc. UK 84 (4), 861–862.

Macias, D., Stips, A., Garcia-Gorriz, E., 2014. The relevance of deep chlorophyll maximum in the open Mediterranean Sea evaluated through 3D hydrodynamic-biogeochemical coupled simulations. Ecol. Modell. 281, 26–37. http://dx.doi.org/10.1016/j.ecolmodel.2014.03.002.

Maglio, A., Pavan, G., Castellote, M., Frey, S., 2015. Overview of the noise hotspots in the ACCOBAMS area. Final Report to the ACCOBAMS Secretariat. 45 p.

Magnone, F., Trainito, E., Picollo, V., De Lazzari, A., Fozzi, A., Napolitano, E., Vitale, S., 2011. Osservazione del comportamento alimentare di balenottera comune (*Balaenoptera physalus*) lungo la costa nord-orientale della Sardegna. In: Atti del 42° Congresso della Società Italiana di Biologia Marina, Olbia, 23–28 maggio 2011, pp. 162–163.

Marini, L., Consiglio, C., Angradi, A.M., Catalano, B., Sanna, A., Valentini, T., Finoia, M.G., Villetti, G., 1996. Distribution, abundance and seasonality of cetaceans sighted during scheduled ferry crossings in the central Tyrrhenian Sea: 1989–1992. Ital. J. Zool. 63 (4), 381–388.

Mazzariol, S., Centelleghe, C., Beffagna, G., Povinelli, M., Terracciano, G., Cocumelli, C., Pintore, A., Denurra, D., Casalone, C., Pautasso, A., Di Francesco, C.E., Di Guardo, G., 2016. Mediterranean fin whales (*Balaenoptera physalus*) threatened by dolphin morbillivirus. Emerg. Infect. Dis. 22 (2), 302–305. http://dx.doi.org/10.3201/eid2202.150882.

Mizroch, S.A., Rice, D., Zwiefelhofer, D., Waite, J., Perryman, W., 2009. Distribution and movements of fin whales in the North Pacific Ocean. Mammal Rev. 39, 193–227.

Mussi, B., Miragliuolo, A., Monzini, E., Diaz Lopez, B., Battaglia, M., 1999. Fin whale (*Balaenoptera physalus*) feeding ground in the coastal waters of Ischia (Archipelago Campano). Eur. Res. Cetaceans 13, 330–335.

Notarbartolo di Sciara, G., 1990. A note on the cetaceans incidental catch in the Italian driftnet swordfish fishery, 1986–1988. Rep. Int. Whaling Comm. 40, 459–460.

Notarbartolo di Sciara, G., Zanardelli, M., Jahoda, M., Panigada, S., Airoldi, S., 2003. The fin whale, *Balaenoptera physalus* (L. 1758), in the Mediterranean Sea. Mammal Rev. 33 (2), 105–150. http://dx.doi.org/10.1046/j.1365-2907.2003.00005.x.

Notarbartolo di Sciara, G., Agardy, T., Hyrenbach, D., Scovazzi, T., Van Klaveren, P., 2008. The Pelagos Sanctuary for Mediterranean marine mammals. Aquat. Conserv. 18, 367–391. http://dx.doi.org/10.1002/aqc.855.

Notarbartolo di Sciara, G., Hoyt, E., Reeves, R.R., Ardron, J., Marsh, H., Vongraven, D., Barr, B., in press. Place-based approaches to marine mammal conservation. Aquat. Conserv. doi: http://dx.doi.org/10.1002/aqc.2642.

Orsi Relini, L., Giordano, A., 1992. Summer feeding of the fin whale, *Balaenoptera physalus*, in the Liguro-Provencal Basin. Eur. Res. Cetaceans 6, 138–141.

Palsbøll, P.J., Berubé, M., Aguilar, A., Notarbartolo di Sciara, G., Nielsen, R., 2004. Discerning between recurrent gene flow and recent divergence under a finite-site mutation model applied to North Atlantic and Mediterranean Sea fin whale (*Balaenoptera physalus*) populations. Evolution 57 (3), 670–675.

Panigada, S., Notarbartolo di Sciara, G., 2012. *Balaenoptera physalus* (Mediterranean subpopulation). In: The IUCN Red List of Threatened Species. http://dx.doi.org/10.2305/IUCN.UK.2012.RLTS.T16208224A17549588.en.

Panigada, S., Pesante, G., Zanardelli, M., Capoulade, F., Gannier, A., Weinrich, M.T., 2006. Mediterranean fin whales at risk from fatal ship strikes. Mar. Pollut. Bull. 52, 1287–1298.

Panigada, S., Pavan, G., Borg, J.A., Galil, B.S., Vallini, C., 2008. Biodiversity impacts of ship movement, noise, grounding and anchoring. In: Abdulla, A., Linden, O. (Eds.),

Maritime Traffic Effects on Biodiversity in the Mediterranean Sea: Review of Impacts, Priority Areas and Mitigation Measures. IUCN Centre for Mediterranean Cooperation, Malaga, Spain, pp. 9–56.

Panigada, S., Lauriano, G., Burt, L., Pierantonio, N., Donovan, G., 2011a. Monitoring winter and summer abundance of cetaceans in the Pelagos Sanctuary (Northwestern Mediterranean Sea) through aerial surveys. PLoS One 6 (7), e22878. http://dx.doi.org/10.1371/journal.pone.0022878.

Panigada, S., Lauriano, G., Pierantonio, N., Donovan, G., 2011b. Monitoring cetaceans populations through aerial surveys in the Central Mediterranean Sea. In: Gauffier, P., Verborgh, P. (Eds.), Abstracts, 25th Conference of the European Cetacean Society, Cádiz, Spain, 21–23 March 2011, p. 104.

Panigada, S., Donovan, G.P., Druon, J.N., Lauriano, G., Pierantonio, N., Pirotta, E., Zanardelli, M., Zerbini, A., 2015. Satellite telemetry on Mediterranean fin whales to identify critical habitats and mitigate ship strikes. In: 21st Biennial Society for Marine Mammalogy Conference, San Francisco, USA, 13–18 December 2015.

Pierantonio, N., Bearzi, G., 2012. Review of fin whale mortality events in the Adriatic Sea (1728–2012), with a description of a previously unreported killing. Mar. Biodiv. Rec. 5, e109. http://dx.doi.org/10.1017/S1755267212000930. 6 p.

Podestà, M., Magnaghi, L., 1989. Unusual number of cetacean bycatches in the Ligurian Sea. Eur. Res. Cetaceans 3, 67–70.

Puzzolo, V., Tringali, L.M., 2001. Monitoring the seasonal Mediterranean fin whales, *Balaenoptera physalus*, passage through the Strait of Messina (Ionian Sea) using GIS techniques. Eur. Res. Cetaceans 15, 458–460.

Reilly, S.B., Bannister, J.L., Best, P.B., Brown, M., Brownell Jr., R.L., Butterworth, D.S., Clapham, P.J., Cooke, J., Donovan, G.P., Urbán, J., Zerbini, A.N., 2013. *Balaenoptera physalus*. In: The IUCN Red List of Threatened Species. http://dx.doi.org/10.2305/IUCN.UK.2013-1.RLTS.T2478A44210520.en.

Ross, D., 2005. Ship sources of ambient noise. IEEE J. Ocean Eng. 30, 257–261.

Ryan, C., McHugh, B., Trueman, C.N., Sabin, R., Deaville, R., Harrod, C., Berrow, S.D., O'Connor, I., 2013. Stable isotope analysis of baleen reveals resource partitioning among sympatric rorquals and population structure in fin whales. Mar. Ecol. Prog. Ser. 479, 251–261. http://dx.doi.org/10.3354/meps10231.

Sanpera, C., Aguilar, A., 1992. Modern whaling off the Iberian Peninsula during the 20th century. Rep. Int. Whaling Comm. 42, 723–730.

Sciacca, V., Caruso, F., Beranzoli, L., Chierici, F., Domenico, E., Embriaco, D., Favali, P., Giovanetti, G., Larosa, G., Marinaro, G., Papale, E., Pavan, G., Pellegrino, C., Pulvirenti, S., Simeone, F., Viola, S., Riccobene, G., 2015. Annual acoustic presence of fin whale (*Balaenoptera physalus*) offshore eastern Sicily, Central Mediterranean Sea. PLoS One 10 (11), e0141838. http://dx.doi.org/10.1371/journal.pone.0141838.

Terhune, J.M., Verboom, W.C., 1999. Right whales and ship noises. Mar. Mamm. Sci. 15 (1), 256–258.

Tershy, B.R., Urban, J., Breese, D., Rojas-Bracho, L., Findley, L.Y., 1993. Are fin whales resident to the Gulf of California? Rev. Investig. Cient., UABCS 1 (No. Esp. SOMMEMA 1), 69–72.

Tringali, L.M., Caltavuturo, G., Gurrieri, G., Di Martino, V., 1999. Osservazioni preliminari sulla presenza estiva di balenottera comune (*Balaenoptera physalus*) e tursiope (*Tursiops truncatus*) nelle acque dello Stretto di Messina. Biol. Mar. Mediterr. 6 (1), 619–622.

Tringali, L.M., La Manna, G., Internullo, E., Summa, A., Floridia, S., 2006. Occurrence of cetacean in the Southern Ionian Sea and Strait of Sicily. In: Abstract Book, 20th Conference of the European Cetacean Society, Gdynia, Poland, 2–7 April 2006, p. 198.

Vaes, T., Druon, J.N., 2013. Mapping of potential risk of ship strike with fin whales in the Western Mediterranean Sea. JRC Scientific and Policy Reports, Ispra. 28 pp.

Williams, R., Wright, A.J., Ashe, E., Blight, L.K., Bruintjes, R., Canessa, R., Clark, C.W., Cullis-Suzuki, S., Dakin, D.T., Erbe, C., Hammond, P.S., Merchant, N.D., O'Hara, P.D., Purser, J., Radford, A.N., Simpson, S.D., Thomas, L., Wale, M.A., 2015. Impacts of anthropogenic noise on marine life: publication patterns, new discoveries, and future directions in research and management. Ocean Coast. Manag. 115, 17–24. http://dx.doi.org/10.1016/j.ocecoaman.2015.05.021.

Wyatt, T., 2010. Can we detect meaningful changes in Mediterranean phytoplankton? In: Briand, F. (Ed.), Phytoplankton Responses to Mediterranean Environmental Changes. CIESM Workshop Monographs, CIESM, Monaco, pp. 15–18.

Zanardelli, M., Airoldi, S., Beaubrun, P., Bérubé, M., Borsani, J.F., Gannier, A., Guinet, C., Hammond, P., Jahoda, M., Lauriano, G., Notarbartolo di Sciara, G., Panigada, S., 2011. Long-term photo-identification study of fin whales in the Pelagos Sanctuary (NW Mediterranean). In: Gauffier, P., Verborgh, P. (Eds.), Abstracts, 25th Conference of the European Cetacean Society, Cádiz, Spain, 21–23 March 2011, p. 103.

CHAPTER FOUR

Cuvier's Beaked Whale, *Ziphius cavirostris*, Distribution and Occurrence in the Mediterranean Sea: High-Use Areas and Conservation Threats

M. Podestà[*,1], A. Azzellino[†,‡], A. Cañadas[§], A. Frantzis[¶], A. Moulins[||], M. Rosso[||], P. Tepsich[||,#], C. Lanfredi[†,‡]

[*]Museum of Natural History of Milan, Milano, Italy
[†]Politecnico di Milano, University of Technology, Milano, Italy
[‡]Tethys Research Institute, Milano, Italy
[§]ALNILAM Research and Conservation, Navacerrada, Madrid, Spain
[¶]Pelagos Cetacean Research Institute, Vouliagmeni, Greece
[||]CIMA Research Foundation, Savona, Italy
[#]University of Genoa, Genoa, Italy
[1]Corresponding author: e-mail address: michela_podesta@hotmail.com

Contents

Advances in Marine Biology, Volume 75
ISSN 0065-2881
http://dx.doi.org/10.1016/bs.amb.2016.07.007

Abstract

Cuvier's beaked whale (*Ziphius cavirostris* G. Cuvier, 1823) is the only beaked whale species commonly found in the Mediterranean Sea. Until recently, species presence in this area was only inferred from stranding events. Dedicated cetacean surveys have increased our knowledge of the distribution of Cuvier's beaked whales, even though many areas still remain unexplored. Here, we present an updated analysis of available sighting and stranding data, focusing on the atypical mass strandings that have occurred in the Mediterranean Sea since 1963. We describe in detail the five more recent events (2006–14), highlighting their relationship with naval exercises that used mid-frequency active sonar. The distribution of the species is apparently characterized by areas of high density where animals seem to be relatively abundant, including the Alborán Sea, Ligurian Sea, Central Tyrrhenian Sea, southern Adriatic Sea and the Hellenic Trench, but other such areas may exist where little or no survey work has been conducted. Population size has been estimated for the Alborán and Ligurian seas. Habitat modelling studies for those areas, confirmed the species preference for the continental slope and its particular association with submarine canyons, as has also been found to be the case in other areas of the world. The application of results from habitat modelling to areas different from their calibration sites is proposed as a management tool for minimizing the potential impacts of human activities at sea. Military sonar is known worldwide as a threat for this species and is suggested to be a major threat for Cuvier's beaked whale in the Mediterranean Sea.

1. INTRODUCTION

Cuvier's beaked whale (*Ziphius cavirostris* G. Cuvier, 1823) is one of the best known species of the family Ziphiidae, and has a cosmopolitan distribution in all oceans, with the exception of very high-latitude polar regions of both hemispheres (MacLeod et al., 2006). Cuvier's beaked whales occur in deep waters (>200 m) and are often found over the continental slope, apparently frequenting slope areas with a steep seafloor (Taylor et al., 2008). The elusive behaviour of this medium-sized odontocete, with short surfacing durations and inconspicuous blows, as well as its offshore distribution, have made the species very difficult to study at sea (Heyning, 1989). During the last decade, field research conducted with data logging tags (Johnson and Tyack, 2003) characterized these animals as extreme divers, since they can routinely dive deeper than 1 km for an hour or more (Baird et al., 2006; Tyack et al., 2006). The mammalian dive record has recently been reported by Schorr et al. (2014) for a Cuvier's beaked whale tagged off the Southern California coast that reached a depth of 2992 m with a dive duration of 137.5 min (both new mammalian dive records).

Fig. 1 An adult male Cuvier's beaked whale (*Ziphius cavirostris*) breaching in the Ligurian Sea (ID 31308071 CIMA RF database). Note the presence of erupted teeth extending beyond the rostrum and dense scarring along the anterior part of the animal. *Photograph: M. Rosso, CIMA Research Foundation.*

Cuvier's beaked whale is the only beaked whale species commonly found in the Mediterranean Sea (Fig. 1). A few other species (*Mesoplodon* spp.) are reported occasionally as rare sightings or strandings (Notarbartolo di Sciara and Birkun, 2010; Podestà et al., 2005). In the past, Cuvier's beaked whale presence in the region has mainly been inferred via stranding data (Podestà et al., 2006). Over the last 30 years, however, dedicated cetacean surveys have greatly increased our knowledge on the distribution of this species, despite the fact that survey effort has not covered the entire Mediterranean Basin and many areas remain unexplored (Cañadas, 2012). Occurrence of Cuvier's beaked whale has been confirmed for the entire Mediterranean Basin, from the Western Mediterranean (Alborán Sea) to the far eastern part of the Levantine Sea. Species distribution is apparently characterized by areas of high density, where individuals seem to be relatively abundant, such as in the Alborán Sea, Ligurian Sea, Central Tyrrhenian Sea, South Adriatic Sea and the Hellenic Trench (Cañadas et al., 2013).

In the Ligurian Sea, Cuvier's beaked whale diving and foraging behaviour has been investigated using archival DTag technology (Aguilar de Soto et al., 2006; Johnson et al., 2004). Dive profiles were characterized by a series of shallow dives, lasting no more than 22 min and reaching a maximum depth of 425 m, followed by deep dives to a maximum depth of 1888 m with a maximum duration of 85 min (Tyack et al., 2006; Zimmer et al., 2005). A recent study in the Ligurian Sea used photo-identification techniques based on natural markings and found that notches, large scars, scrapes and other markings (some the result of inter/intraspecific aggression) are long-lasting marks that may be used for future studies, including mark-recapture studies to provide estimates of abundance for this species (Rosso et al., 2011).

Cuvier's beaked whales are suction feeders, frequently preying upon deep-sea cephalopods, but also occasionally feed on fish and crustaceans (MacLeod et al., 2003). Feeding habits in the Mediterranean Sea have been described based on stomach contents of specimens from Spanish waters (Blanco and Raga, 2000), the Ligurian Sea (Orsi Relini and Garibaldi, 2005), the Tyrrhenian Sea (Carlini et al., 1992; Pedà et al., 2015; Podestà and Meotti, 1991), the southern Adriatic Sea (Kovačić et al., 2010) and the Ionian Sea (Garibaldi et al., 2015; Lefkaditou and Poulopoulos, 1998). These results confirm that in the Mediterranean, deep-sea squids (primarily histioteuthids) represent the main prey for this species, although in a single case mesopelagic fish was found to be a significant part of the stomach contents (Garibaldi et al., 2015). Woodside et al. (2006) suggested that seafloor gouge marks on mud volcanoes observed during geological surveys in the eastern Mediterranean Sea (at depths of 1700–2100 m) could have been made by Cuvier's beaked whales during foraging dives.

Habitat preference of beaked whales has been investigated worldwide (Baumgartner et al., 2001; Davis et al., 1998, 2002; Ferguson et al., 2006; MacLeod, 2000; MacLeod and D'Amico, 2006; MacLeod and Mitchell, 2006; Mannocci et al., 2011; Waring et al., 2001). Most studies reported a clear relationship with the topographic features of the sea bottom (Baumgartner et al., 2001; Davis et al., 1998, 2002; Ferguson et al., 2006; Mannocci et al., 2011; Waring et al., 2001), with beaked whales regularly observed over the continental slope in waters up to 2000 m of depth (Hamazaky, 2002; Hooker et al., 2002; MacLeod and Zuur, 2005; MacLeod et al., 2006; Waring et al., 2001) and near submarine canyons (Wimmer and Whitehead, 2004). This is consistent with findings for Cuvier's beaked whales in the Mediterranean Sea where the species has also been found to be associated with the continental slope, and in particular, with submarine canyon areas (Azzellino et al., 2008, 2011, 2012; Cañadas and Vázquez, 2014; Cañadas et al., 2002; Coomber, 2016; D'Amico et al., 2003; Moulins et al., 2007; Tepsich et al., 2014). Indeed, the species' habitat preferences are mainly driven by its diet. The main high-density areas, where habitat preferences of this species have been studied in the Mediterranean Sea, are the Genoa Canyon in the Ligurian Sea, located within the Pelagos Sanctuary for Mediterranean Marine Mammals (the largest Mediterranean Marine Protected Area, MPA, encompassing an area between southeastern France, Monaco, northwestern Italy, Northern Sardinia and surrounding Corsica and the Tuscan Archipelago), the Alborán Sea and the Hellenic Trench. MacLeod and Mitchell (2006) classified these three areas as Cuvier's beaked whale 'key areas' in the Mediterranean Sea. However, the records of species occurrence and the geographical complexity of some additional areas, such as the Tyrrhenian Sea (Arcangeli et al., 2015; Gannier, 2011) and the southern Adriatic Sea (Holcer et al., 2007), suggest that potential areas of key habitat and high Cuvier's beaked whales density may exist in other regions of the Mediterranean Sea.

In a worldwide molecular genetic analysis of the species, a high degree of differentiation was observed between Atlantic Cuvier's beaked whales and the Mediterranean population (Dalebout et al., 2005). Moreover, haplotype diversity was lower in the Mediterranean Sea than in other regions investigated, suggesting that this population may be isolated and relatively small (Dalebout et al., 2005).

Globally, the species is assessed as being of Least Concern on the International Union for the Conservation of Nature (IUCN) Red List (Taylor et al., 2008), while the Mediterranean sub-population is classified as Data Deficient (Cañadas, 2012), and a proposal to change the current listing to Vulnerable is currently under review. This proposal is prompted by the multiple mass strandings of Cuvier's beaked whales that occurred in the basin during the past five decades, causing the death of at least 100 animals, and demonstrated to have been related to naval exercises using mid-frequency active sonar (ACCOBAMS, 2016; Frantzis, 1998, 2015; Podestà et al., 2006) (see Section 2.2).

Here we review and synthesize information on distribution, abundance and habitat of Cuvier's beaked whale in the Mediterranean Sea. We begin with a review of recent Cuvier's beaked whale strandings, including records of single strandings and a description of recent atypical mass stranding events. Next, we outline what is known about the distribution of Cuvier's beaked whales in the Mediterranean, describing the known areas of high density, and including information on apparently preferred habitat in the Mediterranean Sea—the Alborán Sea, Ligurian Sea, Central Tyrrhenian Sea, southern Adriatic Sea and the Hellenic Trench. We include a discussion of the results of a recent collaborative effort implemented by the Agreement on the Conservation of Cetaceans in the Black Sea, Mediterranean Sea and Contiguous Atlantic Area (ACCOBAMS) to map areas of high use by beaked whales in the Mediterranean region. Major anthropogenic threats affecting the species in the Mediterranean are highlighted, demonstrating that urgent conservation measures are needed to ensure the protection of the Cuvier's beaked whales in the Mediterranean Sea.

2. STRANDINGS OF CUVIER'S BEAKED WHALES IN THE MEDITERRANEAN REGION

2.1 The Importance of Strandings for Understanding Cuvier's Beaked Whale Ecology

Comparisons of stranding data with sighting data have demonstrated that stranding records can be a good indicator of species composition at sea and can sometimes be used to infer species distribution (Maldini et al., 2005; Peltier and Ridoux, 2015; Peltier et al., 2012; Pyenson, 2010, 2011).

Effort for the collection of stranded marine mammals along Mediterranean coasts is extremely variable, but very few stranding networks are organized at a national level (these include France, Israel, Italy, Croatia, and to a lesser extent, Greece and Spain), and many areas are managed by local institutions that have limited geographical coverage. Coastal topography presents another difficulty, including areas that are inaccessible, and the expanse of small islands cannot be carefully monitored for stranded animals (e.g. Greece).

Podestà et al. (2006) reviewed Cuvier's beaked whale strandings in the Mediterranean Sea from the first stranding record in 1803 to 2003. A total of 316 stranded individuals were reported from Albania, Algeria, Croatia, Egypt, France, Greece, Israel, Italy, Malta, Spain and Turkey. Countries with a higher number of strandings included Italy (118), Greece (86), Spain (38) and France (34) (atypical mass stranding events were also reported for these countries).

To update stranding records for the Mediterranean coasts, we reviewed the literature and searched for unpublished records (2004–15). Strandings included single individuals and mass strandings of multiple individuals, including some for which there was no exact number. Therefore, an approximate total of 100 Cuvier's beaked whales were found stranded in Algeria, Croatia, France, Greece, Israel, Italy, Lebanon, Malta, Spain, Syria, Turkey (Arbelo et al., 2008; Banca Dati Italiana Spiaggiamenti, 2015; Dalton, 2006; Dhermain, 2012; Dhermain et al., 2011, 2015; Gomerčić et al., 2006; Gozalbes-Aparicio and Raga, 2015; Holcer et al., 2007; Kerem et al., 2012; Medaces, 2015; Öztürk et al., 2011; S. Muscat, Marine Rescue Team, Malta, personal communication, 23 June 2011). Strandings had previously been reported from each of these countries (see Podestà et al.,

2006), with the exceptions of Lebanon and Syria. In both cases, a single stranding was reported, one in 2008 for Lebanon and one in 2005 for Syria (Akkaya Bas et al., 2016). We also found two previously unreported strandings in Cyprus, dating 2001 and 2002 (S. Michaelidis, Department of Fishery and Marine Research, Cyprus, personal communication, 11 February 2016), adding this country to the range of known strandings for this species in the Mediterranean region.

Geographic distribution of Cuvier's beaked whale strandings from 1803 to 2015 indicates the presence of the species in the entire Mediterranean Basin (Fig. 2), including in the more eastern areas where recent at sea research supports the stranding data (Akkaya Bas et al., 2016; Kerem et al., 2012; Öztürk et al., 2011). The majority of the strandings are reported along the coasts of Spain, France, Italy (excluding the northern and central Adriatic Sea) and Greece, where monitoring effort is generally higher. Stranding rate for single animal strandings in these areas is generally only a few cases per year.

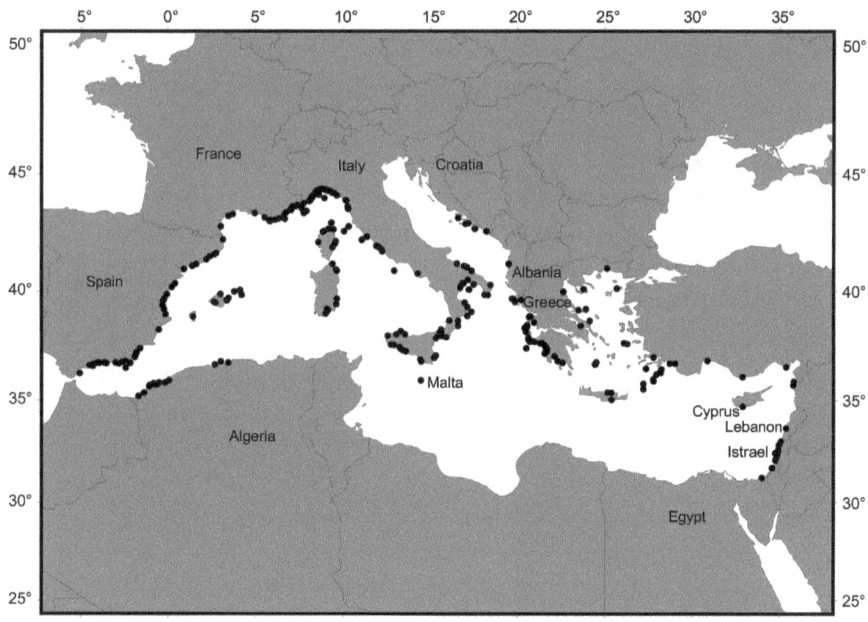

Fig. 2 Distribution of Cuvier's beaked whale (*Ziphius cavirostris*) strandings recorded in the Mediterranean Sea from 1803 to 2015.

Cuvier's beaked whales in the Mediterranean Sea, as well as in other areas of the world, have also been involved in atypical mass strandings, as described in detail with updated information in the following section.

2.2 Atypical Mass Strandings

Frantzis (1998) defined an atypical mass stranding as one that involves two, or usually more, individuals and refers to an unusual temporal and spatial distribution of stranded cetaceans, as opposed to typical mass strandings known, for example, from pilot whales (*Globicephala* spp.) and false killer whales (*Pseudorca crassidens*) (Geraci and Lounsbury, 1993). The cetaceans involved in an atypical mass stranding may belong to one or more species that strand during a relatively short period of time (one or few days) at nearby, but separate locations, sometimes along many kilometres of coastline. Beaked whales have been reported to be particularly prone to atypical mass strandings, which have been reported for ziphiids in different areas of the world and have almost always been shown to have occurred in coincidence with naval exercises, when military data were available (e.g. Greece in 1996, 1997, 2011 and 2014; Bahamas in 2000; Canary Islands in 1985, 1986, 1988, 1989, 2002 and 2004: Frantzis, 1998, 2004, 2015; Balcomb and Claridge, 2001; Martín, 2002; Martín et al., 2004).

Podestà et al. (2006) reported that at least 12 atypical mass stranding events, totalling a minimum of 80 Cuvier's beaked whales, occurred in Italy (Liguria and Calabria), France (Corsica) and Greece, between 1963 and 1999. The first reported atypical mass stranding of Cuvier's beaked whales in the Mediterranean region occurred in 1963 along the Ligurian coasts of Northern Italy. In that year, three mass strandings were reported in three different time periods: the end of January to the beginning of February, May and November (involving 5, 15 and 15 specimens, respectively). The May 1963 stranding occurred over more than 70 km of Ligurian coastline and was originally described as a single stranding event (Tortonese, 1963). Littardi et al. (2004) reviewed the available information on military activity in the area during the early 1960s and found a temporal link between the strandings and the presence of military ships in the area. Some stranded animals in that May 1963 event were dead of gunshot wounds and others were still alive, and were described as having clear signs of 'sickness' (Littardi et al., 2004; Tortonese, 1963).

Between 2006 and 2015, five atypical mass strandings of Cuvier's beaked whales occurred along the Mediterranean coasts of Spain, Italy and Greece, totalling a minimum of 28 individuals (Table 1; Fig. 3). Here, we briefly summarize these five events.

Table 1 Atypical Mass Strandings of Cuvier's Beaked Whales (*Ziphius cavirostris*) Recorded Between 2006 and 2014 Along the Coasts of the Mediterranean Sea

ID	Date	Location(s)	Number of Animals	Sex (Age Class)	Length (m)	Naval Activity in the Region
A	26 January 2006	Almeria, Spain	4	2 juvenile females, 2 adult males	Unknown	NATO Active Sonar Training, 25–26 January 2006 (DON, 2008; see also ACCOBAMS, 2016)
B	11 April 2006	Messina, Italy	4–5	2 females	4.8; 5.5	Unknown
C	8 February 2011	Siracusa, Italy	2–3	1 female	5.05	NATO Exercise *Proud Manta*, 4–17 February 2011
D	30 November to 19 December 2011	Corfu, Greece, Crotone, Italy	12	1 female	5.38	Military exercise *Mare Aperto/Amphex* 2011 (Italy), 28 November to 5 December 2011
E	1–6 April 2014	Crete, Greece	6–10	1 female (pregnant)	Unknown	Military exercise *Noble Dina* 2014 (Greece, Israel, USA), 26 March to 10 April 2014

ID, identifier shown in Fig. 3. Sex, age class and length data of individuals are included where known.

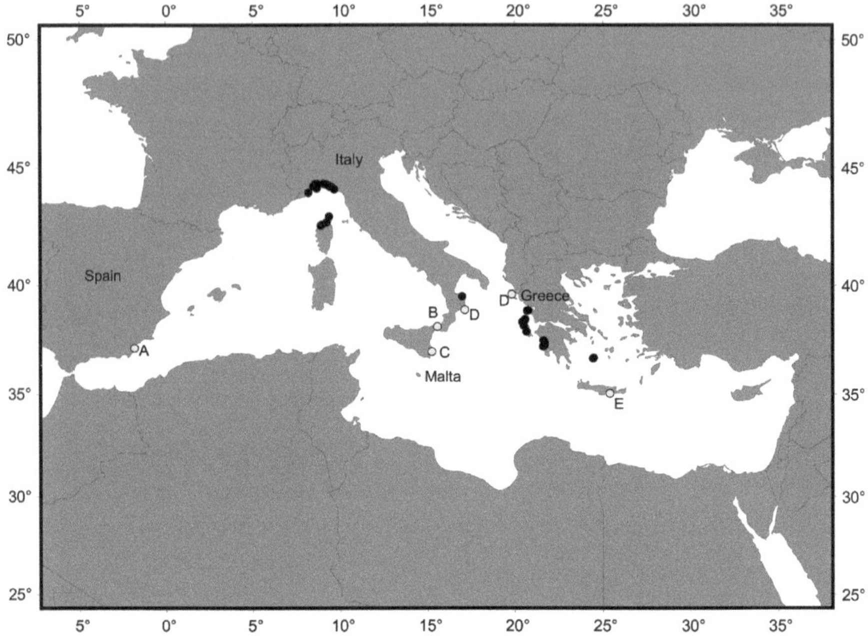

Fig. 3 Distribution of Cuvier's beaked whale (*Ziphius cavirostris*) atypical mass stranding events (three or more animals) in the Mediterranean Sea (*black dots*: 1963–2004 events; *yellow dots*: 2006–14 events).

On 26 January 2006, an atypical mass stranding of four Cuvier's beaked whales (two juvenile females and two adult males) occurred on the southern coast of Almeria, Spain (Arbelo et al., 2008). Two individuals were already dead, and two were still alive but died soon after being found. The two live animals were reported to show signs of illness. All animals appeared to be in good body condition. Necropsy documented gas bubble-associated lesions and fat emboli in the vessels and parenchyma of organs, and these were similar to previous findings from mass strandings associated with naval military exercises (Arbelo et al., 2008; Dalton, 2006). Military activity occurred in the region from 25 to 26 January 2006, when seven North Atlantic Treaty Organization (NATO) surface ships and a Spanish submarine conducted active sonar training within 93 km of the stranding site (DON, 2008; see ACCOBAMS, 2016).

Another atypical mass stranding was reported close to Messina, Sicily, Italy, on 11 April 2006 (Cozzi et al., 2011). Three Cuvier's beaked whales stranded alive 10 km south of Messina and subsequently floated back to sea. A fourth female individual stranded alive, 1 km south of Messina, and died soon after. Fat emboli were detected in alveolar vessels of this animal (Cozzi et al., 2011), suggesting gas and fat embolic syndrome, the pathological condition in beaked whales associated to the exposure to mid-frequency military sonar (Fernández et al., 2004, 2005). On 20 April 2006, a 5.5 m long female Cuvier's beaked whale was found dead (in an advanced state of decomposition) 20 km south of Messina, in Alì Terme. No data were recorded for the three beaked whales live stranded on 11 April; thus it is uncertain whether this animal was a part of that mass stranding event.

On 8 February 2011, two Cuvier's beaked whales stranded alive in Fontane Bianche, 13 km south of Siracusa, Sicily, on the Ionian coast, and were towed to open waters by the Coast Guard (Cozzi et al., 2011). On 9 February, one of these two animals (recognizable by the numerous scars and marks on the body) stranded alive in the same area. The animal (5.05 m long, probably a female, as no protruding teeth were present in the lower jaw) was constantly trying to head toward shore, and there were signs of repeated scrapes (loss of the superficial layers of the skin) from the rocky shoreline. Copepods, *Pennella* spp., were observed all over the body. The animal was towed (once more) to open waters and was released approximately 6 km off Capo Murro di Porco. Once released from the tow gears, the animal spontaneously swam away and was no longer sighted (Cozzi et al., 2011). One month later, on 15 March 2011, a dead Cuvier's beaked whale was found stranded in Eraclea Minoa, Agrigento, in an advanced state of decomposition. From 4 to 17 February 2011, the *Proud Manta* exercise, which consisted of 'intense Anti-Submarine warfare activity', was conducted by NATO in the Ionian Sea region and, at the time, was reported to be the largest annual event of this type ever conducted by the Alliance, including 10 NATO nations with six submarines, 19 aircraft (including ship-borne helicopters) and eight surface ships.[1]

From 30 November to 19 December 2011, a total of 12 Cuvier's beaked whales stranded on the coasts of Greece and southern Italy. Ten Cuvier's beaked whales stranded along 23 km of coast of the western island of Corfu, Greece, and two stranded 240 km away on the southern Italian coast of Calabria, in Irto, Crotone. Necropsies on two of the specimens stranded in Greece indicated a gas and fat embolic pathology (A. Fernández, University of Las Palmas de Gran Canaria, personal communication, 4 June 2012; see also Bernaldo de Quirós et al., 2012; Fernández et al., 2005). Analysis of gas amount and gas composition in one whale revealed a condition

[1] See www.dvidshub.net/news/printable/65852.

compatible with decompression sickness (DCS) (A. Fernández, University of Las Palmas de Gran Canaria, personal communication, 4 June 2012; see also Bernaldo de Quirós et al., 2012). No inflammatory or neoplastic processes were noted, and no pathogens were identified as responsible for the pathology. An animal that stranded and died in Italy (on 1 December 2011) had fat emboli of the inner ear and peri-auricular vascular plexus (S. Mazzariol, Cetacean Emergency Response Team, University of Padua, personal communication, 7 March 2016). The Italian naval exercise *Mare Aperto/Amphex* 2011 was being conducted in the Ionian Sea and southern Adriatic Sea (Gulf of Taranto, from the Ionian coast of Calabria to approximately 130 km west of Corfu) from 28 November to 5 December, and involved anti-submarine warfare including 13 ships and two submarines.[2]

On 1 April 2014 another atypical mass stranding of Cuvier's beaked whales occurred along the southern coast of the island of Crete, Greece (Frantzis, 2015). Several strandings of one, two and three live animals were reported along nearly 70 km of coast (Fig. 4). Two of these whales died, and the remaining whales were refloated to the open sea (but stranded again). In the following days (2, 5 and 6 April), three whales single stranded dead in the

Fig. 4 Two live Cuvier's beaked whales (*Ziphius cavirostris*) that stranded on the southern coast of Crete, Greece, in an atypical mass stranding event during April 2014. This stranding, involving from six to 10 Cuvier's beaked whales, occurred during the large-scale anti-submarine military exercise, *Noble Dina*, which was conducted in the Eastern Mediterranean Sea from 26 March to 10 April. See Frantzis (2015). *Photograph Copyright: L. Aggelopoulos, Pelagos Cetacean Research Institute.*

[2] See www.marina.difesa.it/conosciamoci/notizie/Pagine/20111205_amphex.aspx.

same area. Subsequent analysis of photographs taken during the different strandings suggested that a minimum of six and a maximum of 10 animals were involved in the mass stranding (Frantzis, 2015). A trilateral military exercise, *Noble Dina* 2014, involving surface and air defence (including two guided-missile destroyer ships, a replenishment oiler and P-3 aircraft), with anti-submarine and mine-swept channel exercises, was conducted by Greek, Israeli and United States navies from 26 March to 10 April in the Eastern Mediterranean Sea.[3]

3. DISTRIBUTION, ABUNDANCE AND HABITAT PREFERENCES OF CUVIER'S BEAKED WHALES IN THE MEDITERRANEAN SEA

A collaborative initiative, supported by ACCOBAMS, used Cuvier's beaked whale sighting data to describe the species' habitat distribution in the Mediterranean Sea. Data were obtained from a total of 420,050 km of survey effort, in good to moderate searching conditions (Beaufort 3 or less), yielding 456 sightings of Cuvier's beaked whale including 1036 individuals, and covering a time span of 21 years, from 1990 to 2010 (Cañadas et al., 2013). Habitat modelling incorporated grid of cells with a resolution of 0.2 degree (22.2 × 22.2 km; 494 km^2), and a number of geographical and environmental covariates were associated with each grid cell. Data were modelled by using the Generalized Additive Model (see Cañadas et al., 2013).

Results of habitat modelling highlighted three areas with the highest relative density (more than 40%) of Cuvier's beaked whale: (1) the Alborán Sea, (2) the northern Ligurian Sea and (3) the Hellenic Trench (and the area north of Crete). In addition, the northwestern Tyrrhenian Sea, the southern Adriatic Sea and some areas to the north of the Balearic Islands and south of Sicily showed relatively high predicted densities (around 40%) compared to the rest of the Mediterranean Sea. Model results highlighted an interesting area in the far east of the Mediterranean Sea, off the coast of Syria, where there has not been survey coverage, but where there was a relatively high prediction (>40%) of Cuvier's beaked whale preferred habitat. This area should be surveyed to confirm the presence of Cuvier's beaked whale. The model did not identify any habitat along the African coast, but this could be due to the use of latitude as a covariate in the model, together with the lack of survey effort (and observations) in the area (Cañadas et al., 2013).

[3] See www.navy.mil/submit/display.asp?story_id=80259.

Here, we describe information on Cuvier's beaked whale distribution, abundance and habitat preference based on the above-mentioned modelling effort and the available literature. In particular, we focus on known Cuvier's beaked whale high-density areas including the Alborán Sea, the Ligurian Sea, the central Tyrrhenian Sea, the southern Adriatic Sea and the Hellenic Trench, while noting other areas of possible importance.

3.1 Distribution

Despite increased knowledge resulting from dedicated sighting surveys conducted in the Mediterranean Sea, knowledge gaps regarding the distribution of Cuvier's beaked whale still exist, especially for the southern part of the Basin. Thus, strandings of this species can provide useful distribution data. Even though stranding sites may be strongly influenced by multiple factors, including surface current movements and coastal topography, these sites can provide useful information for the inference areas of Cuvier's beaked whale occurrence beyond known areas for the species. The distribution of stranding and sighting data is shown in Fig. 5.

Data from strandings and sightings seem to confirm that Cuvier's beaked whales occur at high densities in some regions (e.g. Alborán Sea, Genoa Canyon area, Central Tyrrhenian Sea, southern Adriatic Sea, Hellenic Trench), while in other areas only occasional strandings are reported. These occasional strandings may be indicative of additional habitat for the species. While Cuvier's beaked whale distribution appears to be characterized by high-use areas (Fig. 6) where animals seem to be relatively abundant, we note that these areas are those in which more research effort has been undertaken.

3.1.1 Alborán Sea

In the Alborán Sea, Cuvier's beaked whales have been regularly observed in the last 20 years during dedicated cetacean surveys conducted since the early 1990s (Cañadas et al., 2002, 2005) that highlighted the presence of the species in an area with borders that coincide with the 1000 m isobath. The data suggest that the Alborán Sea has one of the world's highest densities for this species, and Cañadas and Vázquez (2014) have recently proposed to increase protection in the region. There are no stranding or sighting records from the Strait of Gibraltar at the western most region of the Alborán Sea (Cañadas et al., 2005).

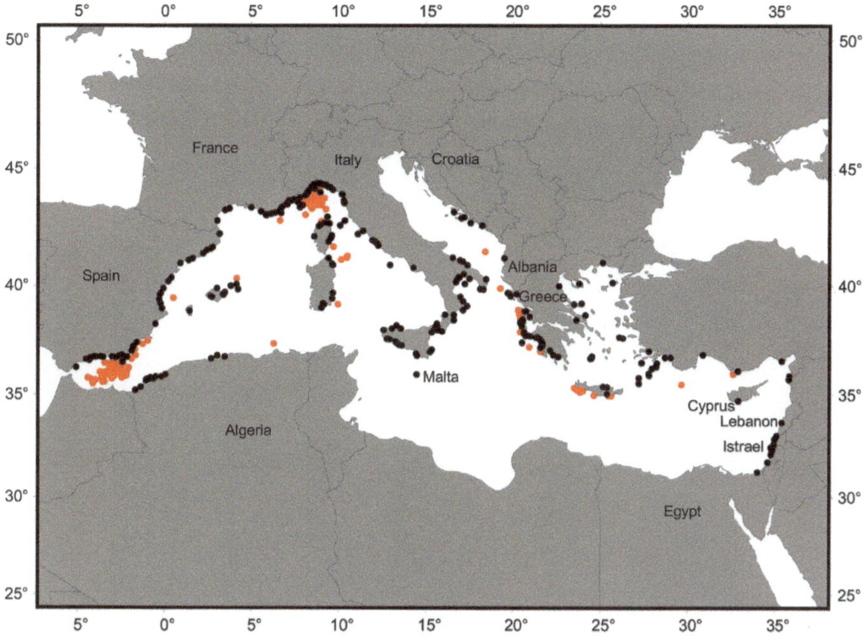

Fig. 5 Distribution of strandings and sightings of Cuvier's beaked whale (*Ziphius cavirostris*) in the Mediterranean Sea (*black dots*: strandings; *red dots*: sightings). *Sightings data from Cañadas, A., B-Nagy, A., Bearzi, G., Cotte, C., Fortuna, C., Frantzis, A., Gannier, A., Laran, S., Lauriano, G., Lewis, T., Moulins, A., Mussi, B., Pastor, X., Politi, E., Pulcini, M., Raga, J.A., Rendell, L., 2013. ACCOBAMS collaborative effort to map high-use areas by beaked whales in the Mediterranean, Monaco, p. 24.*

3.1.2 Ligurian Sea

Another Mediterranean area where cetacean surveys have been conducted since the late 1980s is the Pelagos Sanctuary, in which Cuvier's beaked whales have been sighted especially in waters over and around canyons (Azzellino et al. 2008, 2011, 2012; D'Amico et al., 2003). In particular, the Genoa Canyon area has been identified as a high-density area for Cuvier's beaked whales (MacLeod and Mitchell, 2006; Moulins et al., 2007; Tepsich et al., 2014).

3.1.3 Central Tyrrhenian Sea

Systematic cetacean surveys conducted from ferries in the Central Tyrrhenian Sea, south of the Pelagos Sanctuary, between the Italian Peninsula and the islands of Corsica and Sardinia during two different time periods (1990–92; 2007–11), documented the occurrence of Cuvier's beaked whales and indicated site fidelity in the area (Arcangeli et al., 2015;

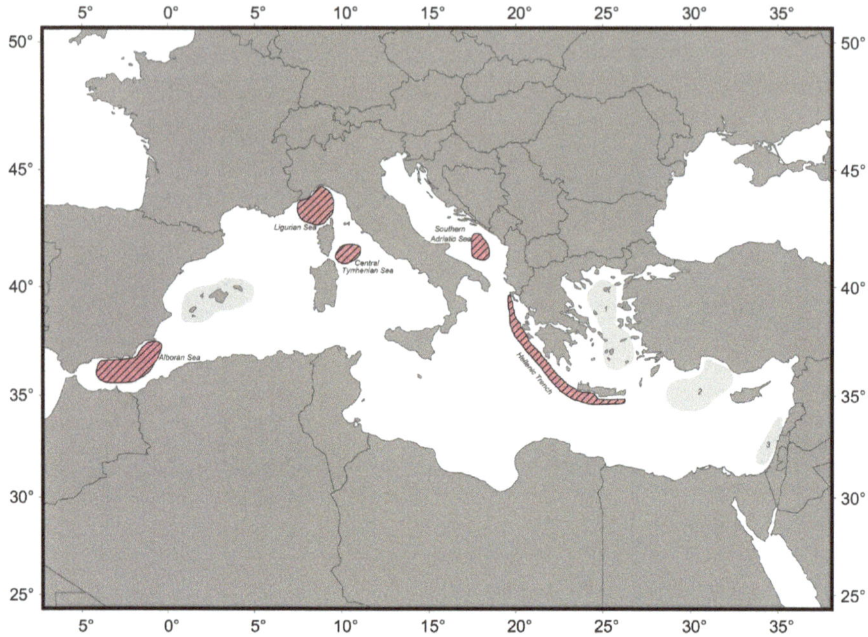

Fig. 6 High-density areas of Cuvier's beaked whale (*Ziphius cavirostris*) occurrence in the Mediterranean Sea (*purple/stripes*). Other possible high-use areas are shown in *grey* (1. Greek waters; 2. Greek–Turkish waters; 3. Levantine Sea; 4. Balearic area).

Marini et al., 1992). Sightings in the same region were also reported by Gannier (2011, 2015), based on small vessel surveys.

3.1.4 Southern Adriatic Sea

Cuvier's beaked whale presence in the southern Adriatic Sea is supported by recent stranding and sighting data reported from the Italian and Croatian coasts (Gomerčić et al., 2006; Holcer et al., 2007). Aerial surveys in the area confirmed the presence of the species in correspondence to a deep, wide depression (>1000 m depth), along the northeastern part of the South Adriatic Basin (Holcer et al., 2014).

Five sightings were recently reported by Bräger et al. (2014) in Albanian and northern Greek waters (north of the island of Corfu). Groups of two and three individuals were observed during the encounters, with a cow–calf pair present in two of the five encounters.

These observations suggest the possibility that the southern Adriatic high-density area might represent a continuum with the Hellenic Trench area.

3.1.5 Hellenic Trench

Cuvier's beaked whales have been observed in Greek seas all along the Hellenic Trench, from northwestern Corfu to east Rodos Island. The areas with the highest number of sightings are south of Crete and west to Lefkada (Frantzis et al., 2003; Pelagos Cetacean Research Institute, unpublished data).

3.1.6 Other Areas of the Mediterranean Sea

Cuvier's beaked whale observations are available for Greek waters (Fig. 6) over all the steep depressions of the Aegean Plateau (northern Aegean Trench from northern Sporades to north of Limnos Island, Ikarion Sea, South Milos Island, west and northwest of Karpathos Island and in the North Cretan Sea) (Frantzis et al., 2003; Pelagos Cetacean Research Institute, unpublished data).

Recent sighting data from the Levantine Sea (Fig. 6) suggest a regular presence of the species in the area, where the Anaximander seamounts, Antalya Canyon and Adana Trough could be areas of particular importance for the species (Frantzis, 2009; Frantzis et al., 2003). More recently, a survey carried out in Turkish waters of the Levantine Sea confirmed the presence of Cuvier's beaked whales in the Antalya Bay (Akkaya Bas et al., 2016).

Results from dedicated surveys conducted in the Levantine Sea in the past decades reported occasional Cuvier's beaked whale sightings off the Israeli and Turkish coasts (Fig. 6) (Kerem et al., 2012).

Sightings have also been reported for the Balearic Sea region (Fig. 6) (Gannier and Epinat, 2008), suggesting the occurrence of this species in the area, although recent dedicated surveys did not confirm this (Cañadas et al. 2013).

3.2 Abundance

There are no estimates of total abundance for Cuvier's beaked whale in the Mediterranean Sea. There are subregional estimates for only two areas, the Alborán Sea and the Ligurian Sea.

3.2.1 Alborán Sea

Research in the Alborán Sea, including spatial modelling based on a detection function with data from 1992 to 2009, yielded an abundance estimate of 429 individuals (CV=0.22; 95% CI=334–557) (Cañadas and Vázquez, 2014). Density estimates from line transect surveys are usually subject to 'availability bias' due to animals not always being available for detection while within detectable range (Buckland et al., 2004), and to 'perception bias' due to observers failing to detect animals even though they are available (Buckland et al., 1993), both causing a negative bias. Deep diving species such as beaked whales are even more subject to such negative bias. To account for this, a correction for availability bias was included in the estimate of Cuvier's beaked whale abundance in the Alborán Sea (Cañadas and Vázquez, 2014).

Based on the modelling of these data, the Alborán Sea presents one of the highest densities of Cuvier's beaked whales in the world, together with Hawaii and the California Current (Barlow et al., 2006; Cañadas and Vázquez, 2014).

3.2.2 Ligurian Sea

A long-term photo-identification study was carried out in the Ligurian Sea to ascertain reliable identification features such as colour patterns and natural marking of Cuvier's beaked whales (Coomber et al., 2016; Rosso, 2010; Rosso et al., 2011). The population analysed in the Genoa Canyon region was found to be generally well and reliably marked along the visible flank (right side and left side), and although 96% of individuals were marked, only 71% of the population was considered reliably marked over multiple years.

Estimates of population size and size trends were calculated separately for left and right side identifications using different models available in the POPAN module of SOCPROG 2.4 (Whitehead, 2009). The '*mortality*' model was the best fit (Table 2). This model estimates that the population of reliably marked individuals was on average composed of 71 (for right side identifications) and 68 individuals (for left side identifications). Results indicated that a small number of Cuvier's beaked whales inhabited the pelagic waters of the Genoa Canyon between 2002 and 2008. The estimated total population size (N_{total}) of Cuvier's beaked whales in the Genoa Canyon was around 100 individuals: 98 individuals for the right side dataset (CV = 0.10; 95% CI ranged = 81–116) and 95 individuals for the left side dataset (CV = 0.09; 95% CI = 79–112).

Because there were no previous estimates of abundance, it is impossible to assess whether the population size has been stable, increasing or decreasing. However, the *mortality + trend* model showed a negative, but not significant, trend (Table 2). It is possible that the population might still be recovering from atypical mass strandings that have occurred since the early 1960s wherein at least 35 individuals (around one-third of the current population size) have stranded in the area (Podestà et al., 2006; Tortonese, 1963; see Section 2.2).

Table 2 Abundance Estimates of Population Size for Cuvier's Beaked Whales (*Ziphius cavirostris*) in the Ligurian Sea

Right Side Sample	θ	**SE**	N_{marked}	**SE**	N_{total}	**SE**	**95% CI**	**Trend (%)**	**95% CI**	**AICc**
Population ($n = 82$)	*0.73*	*0.16*								
Closed			87	4	119	7	103–136			380
Mortality			71	7	98	10	81–116			375
Mortality + trend			73	9	100	13	75–136	−3	−15–7	377
Reimmigration			71	9	98	13	66–114			377
Reimmigration + mortality			69	7	95	10	77–115			379
Left Side Sample	θ	**SE**	N_{marked}	**SE**	N_{total}	**SE**	**95% CI**	**Trend (%)**	**95% CI**	**AICc**
Population ($n = 82$)	*0.72*	*0.18*								
Closed			89	4	113	6	96–133			395
Mortality			68	6	95	9	79–112			383
Mortality + trend			71	8	99	12	82–132	−3	−15–6	385
Reimmigration			67	8	93	12	64–113			385
Reimmigration + mortality			65	7	90	10	74–113			387

Results include right and left side identifications. θ, proportion of identifiable animals; N_{marked}, estimate of reliably marked individuals; N_{total}, estimate of total population size; *AICc*, Akaike Information Criterion value. Models are (i) closed: this model assumes a closed population, whose size is estimated by maximum likelihood; (ii) mortality: this assumes a population of constant size, where mortality (which may include permanent emigration) is balanced by birth (which may include immigration); (iii) mortality + trend: this assumes a population growing or declining at a constant rate; (iv) reimmigration: this is the model in which members of a closed population move from (emigration rate) and into (reimmigration rate) a study area; (v) reimmigration + mortality: this is model 'reimmigration' with the exception that mortality (which may include permanent emigration from the total population) is balanced by birth (which may include immigration).

Modified from Rosso, M., 2010. Population size, residency patterns and energy demand of Cuvier's beaked whales (Ziphius cavirostris) in the north western Mediterranean sea. PhD thesis, University of Basilicata, Potenza, Italy.

Considering that no past or present abundance estimate is available for the entire range of the Mediterranean Cuvier's beaked whale population, and that abundance data are available only for limited areas within the region, inference regarding the total number of Cuvier's beaked whales in the Mediterranean Sea is currently impossible. Considering the fundamental conservation importance of such knowledge, it seems recommendable that all efforts be invested in obtaining it.

3.3 Habitat

3.3.1 Alborán Sea

Habitat modelling was performed for the Alborán Sea, to identify the important habitats for Cuvier's beaked whales in the area (Cañadas and Vázquez, 2014). Data used for these analysis came from two sources: (a) data collected during summers 2008–09 on board the NATO research vessel *NRV Alliance* during the research surveys Sirena08 and Med09, and (b) data collected during surveys carried out by the Non-Governmental Organization, Alnitak, on board three vessels: *Toftevaag* (1992–2010), *Thomas Donagh* (2009) and the Fisheries Patrol boat of the General Secretariat of Maritime Fisheries (2003–09) (see Cañadas and Vázquez, 2014). Highest Cuvier's beaked whale density was associated with the area over 500 m depth, and especially around 1000 m or deeper waters, which is the most suitable habitat for this species that is rarely found in shallower waters. In comparison with the available information, the Alborán Sea is clearly a very important area for Cuvier's beaked whale within the Atlantic and Mediterranean system, with one of the highest densities recorded (mean density of 0.0054 animals/km^2, CV = 22%). Habitat modelling results have been used as key tool to design a proposal for the designation of a 'Critical Area' (Fig. 7), or MPA, by the Spanish government, and as the basis for a Cuvier's beaked whale Management Plan (Cañadas and Vázquez, 2014).

3.3.2 Ligurian Sea

Cuvier's beaked whale habitat preference has been widely investigated in the northern part of the Pelagos Sanctuary, based on a sighting dataset deriving from dedicated ship-based surveys (Azzellino et al., 2008, 2011, 2012; Coomber, 2016; D'Amico et al., 2003; Gannier and Epinat, 2008; Moulins et al., 2007, 2008; Tepsich et al., 2014).

Dedicated surveys conducted between 2000 and 2006 in the northern part of the Pelagos Sanctuary reported higher Cuvier's beaked whale

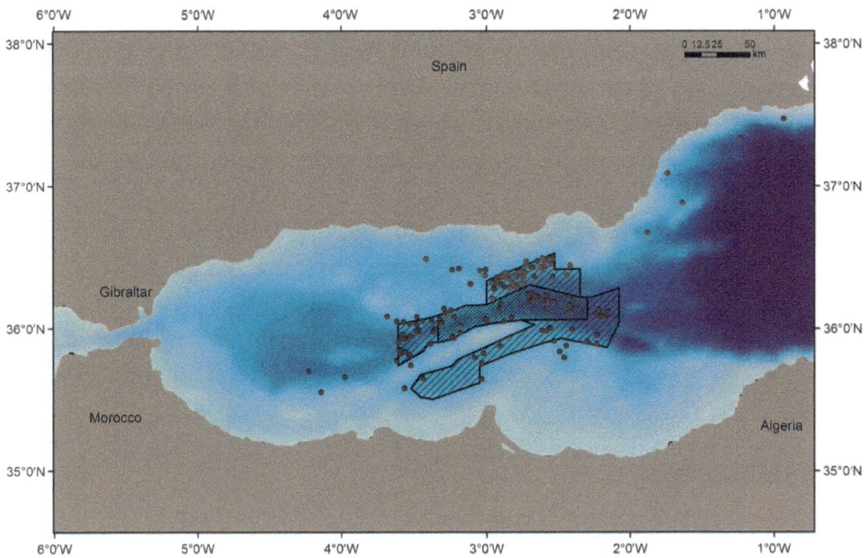

Fig. 7 'Critical Area' proposed by Cañadas and Vázquez (2014) in the Alborán Sea to be used for management measures for the protection of Cuvier's beaked whales (*Ziphius cavirostris*).

sighting frequency in deep waters (between 1000 and 2000 m) within a mid-closed basin with its boundary limited by the 1000 m isobath (Moulins et al., 2007). The majority of sightings were located between a depth of 756 and 1389 m, but the encounter rate was highest between 1389 and 2021 m (Moulins et al., 2007).

Cuvier's beaked whale habitat preferences have been modelled primarily through the use of topographic descriptors, and studies highlighted a strong habitat preference for the upper and lower slopes along the Ligurian–Provençal coast, at depths ranging from 1000 to 2500 m (Azzellino et al., 2008; Moulins et al., 2007). A strong preference was also found for areas with complex topography such as canyons and seamounts (Azzellino et al., 2012; Gannier and Epinat, 2008; Moulins et al., 2007, 2008). Core habitat area was identified in the Genoa Valley, within the Genoa Canyon axis, as well as in the small half-basin with gentle slope located off western coast of Liguria (D'Amico et al., 2003; Moulins et al., 2007, 2008). The Genoa Canyon is the largest and northern-most canyon of the Western Mediterranean Sea. It is 20–30 km across and 60 km long at its 1000 m isobath. This canyon starts about 6 km from the Port of Genoa, which is one of the main commercial harbours in the Mediterranean Basin. As evidenced by Moulins et al. (2008), species presence is also regular around a seamount located at the mouth of the Genoa Canyon. Recently, Tepsich et al. (2014) noted the importance of the deeper portion of the Ligurian Basin, confirming the existence of additional habitat for this species in deeper waters not closely related to peculiar topographic structures. In addition, dynamic predictors, such as remotely sensed physical oceanographic parameters and modelled biogeochemical and physical parameters, have also been used to investigate Cuvier's beaked whale habitat preference in the Gulf of Genoa (Azzellino et al., 2011; Coomber, 2016; Lanfredi, 2014). Results suggest that dynamic predictors may act as proxy for macroscale features (i.e. upwelling/downwelling motion) that indirectly delineate beaked whale habitat in the area.

3.3.3 Hellenic Trench

Information regarding range and habitat of Cuvier's beaked whale in Greek waters (Ionian and Aegean seas and Sea of Crete), including in the Hellenic Trench area, comes from dedicated surveys conducted from 1991 to 2015 (79 Cuvier's beaked whale sightings) and secondarily from the 147 strandings in the region (see Frantzis, 2009). Seventy of the sightings were recorded along the Hellenic Trench (especially along southwest Crete and west of Lefkada Island), where search effort has been much greater than that in the Aegean Sea (Frantzis, 2009). In Greek waters, Cuvier's beaked whales occurred mainly over the continental slope. They were less frequently observed in the pelagic waters of the region. The mean water depth for 63 Cuvier's beaked whale sightings made along the Hellenic Trench was 1066 m (range 491–2279 m; sd = 343), and mean distance from the coast was 8.6 km (range 2.1–26.5 km; sd = 6.1). These values should only be considered representative for Cuvier's beaked whales along the slope or above the Aegean plateau, since there are no available sightings over the abyssal plains.

3.4 Cuvier's Beaked Whale Habitat Model Transferability and Potential for Improvement

Cuvier's beaked whale habitat models may provide predictions about the range of the species in areas of the Mediterranean Sea where its distribution is not well known, and this could be useful to support species conservation in the region.

Although many models with different statistical approaches have been used to predict the presence or absence of sensitive species, in very few instances have such models been evaluated for their transferability to areas different from their calibration sites. Azzellino et al. (2011) evaluated the transferability of habitat predictions for Cuvier's beaked whales deriving from a model developed for the Ligurian Sea area. Data used for this modelling exercise came from the Ligurian Sea Sirena 02 survey, and model evaluation benefited from data collected in Alborán Sea Sirena 08 survey (see Section 3.3.1). The Ligurian Sea dataset was used for the model calibration, and then the Alborán Sea dataset was used to evaluate the Ligurian Sea models. The presence/absence classification performances of the models developed in the Ligurian Sea were evaluated using both dynamic (i.e. remote-sensed chlorophyll *a*) and static (i.e. sea bottom topographic features) predictors. Accuracy was slightly lower when using dynamic predictors with respect to static predictors (i.e. 73% vs 87%). However, despite differences in accuracy, the two models showed good agreement (Fig. 8). A prediction map of presence and absence cells based on the Ligurian Sea models was produced for the '*a priori*' evaluation of the Alborán Sea area. Model accuracy was evaluated by overlaying the Cuvier's beaked whale observations (both visual and acoustic) collected in the field. Model predictions, based on either chlorophyll or bathymetry features, were surprisingly comparable in both the study areas (Azzellino et al., 2011). Results indicated that *a priori* predictions were significantly correlated with Cuvier's beaked whale sightings in the Alborán Sea (evaluation area), although, as expected, the model overall accuracy was much lower than the accuracy estimated for the Ligurian Sea (calibration area). Moreover, based on the fact that model high-risk predictions (i.e. higher presence probability of a species sensitive to anthropogenic impact) were found no more than 7–8 km from the closest beaked whale sighting and significantly closer to visual sightings or acoustic detections than the cells predicted as low risk, this distance was proposed as a spatial uncertainty factor to be attributed to the *a priori* predictions. Considering such a spatial uncertainty factor, the *a priori* predictions can be considered robust enough to support knowledge-based decisions for determining the ranking of priority areas that may be sensitive to anthropogenic impact within a region (Azzellino et al., 2011).

Azzellino et al. (2011) demonstrated that human activity impact risk maps can be drawn based on *a priori* predictions of this kind (Fig. 9) and used as knowledge-based support for minimizing the potential impacts induced by human activities at sea.

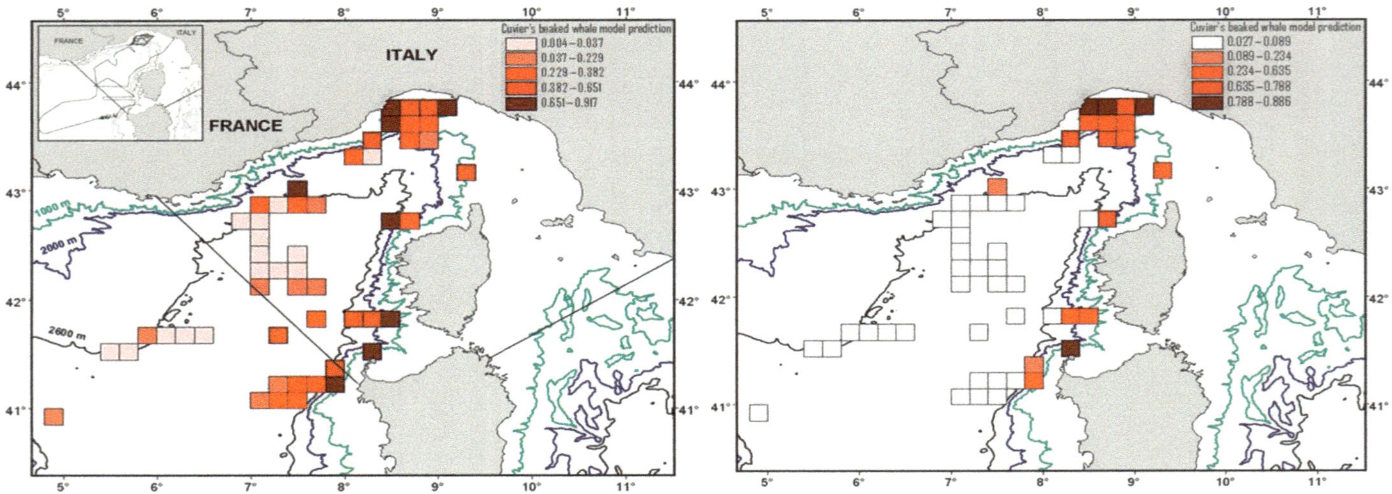

Fig. 8 Cuvier's beaked whale (*Ziphius cavirostris*) presence probability predictions for Ligurian Sea Basin according to the bathymetry model (*right*) and chlorophyll model (*left*). The Pelagos Sanctuary boundaries are also shown (Azzellino et al., 2011).

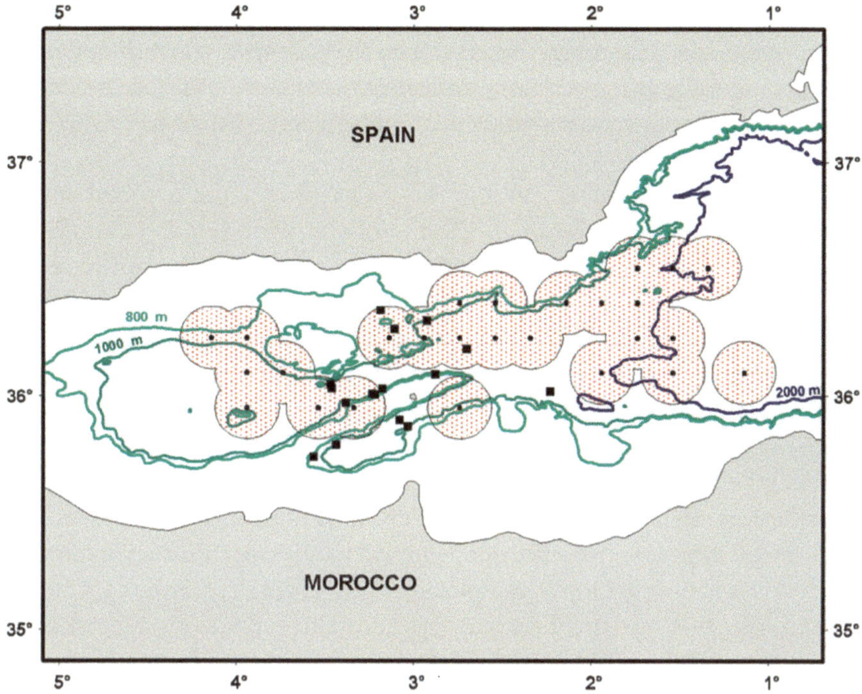

Fig. 9 Risk prediction map for the Alborán Sea area (Azzellino et al., 2011). Cuvier's beaked whale (*Ziphius cavirostris*) observations are also shown as *dark full squares.*

4. THREATS

Cuvier's beaked whales in the Mediterranean Sea are likely impacted by many threats linked to human presence in the semi-enclosed basin, but few studies have quantified the problems affecting this population. Conclusive necropsies (e.g. those described in Section 2.2) are rare for the species in this region, largely due to the rarity of strandings and the difficulty of expeditiously recovering newly stranded specimens (more often found in an advanced state of decomposition). Known threats include bycatch in fishing activities, ingestion of plastics, possible chemical contamination and anthropogenic noise.

Only scattered information concerning Cuvier's beaked whale fisheries interactions is available in the scientific literature. In the past (prior to the driftnet ban of 2002), occasional bycatch in pelagic driftnets was reported from Spanish, French and Italian waters (Banca Dati Italiana Spiaggiamenti, 2015; Cañadas, 2012; Podestà and Magnaghi, 1989). Rosso et al. (2011)

reported one specimen photographed in the Ligurian Sea with a linear amputation of the dorsal fin, suggesting that it could have been the result of entanglement in fishing gear, such as monofilament lines used in the local sword fish fishery. Intentional captures in French and Spanish waters have also been reported (Northridge, 1994).

Plastic debris has been found in the stomachs of some stranded animals (Cañadas, 2012; Gomerčić et al., 2006; Podestà and Meotti, 1991), and in some cases was considered a possible cause of death (Cagnolaro et al., 1986; Frantzis, 2009; Holcer et al., 2003).

No information exists on the actual impact of chemical contaminants on Mediterranean Cuvier's beaked whale survival. High concentrations of mercury, selenium and cadmium have been detected for Cuvier's beaked whales from the Ligurian Sea (Capelli et al., 2008). Ecotoxicological status of Cuvier's beaked whales was recently investigated by Baini et al. (2016). Cytochrome P450 (CYP1A1 and CYP2B isoforms) were used as biomarkers of exposure to anthropogenic contaminants. Protein expression was evaluated using biopsy samples of tissue from free ranging Cuvier's beaked whales from the Ligurian Sea. Protein expression seemed to be linked to age and sex. This method may provide a useful means of assessing ecotoxicological status for this species in the future.

4.1 Anthropogenic Noise

4.1.1 Military Sonar Activity

One of the main threats affecting the Cuvier's beaked whale population in the Mediterranean Sea, is anthropogenic noise resulting from military activities as is highlighted by the atypical mass strandings that have occurred since 1963 (see Section 2.2). Gas bubble-associated lesions and fat embolism in the vessels and parenchyma of vital organs were described by Jepson et al. (2003) and by Fernández et al. (2004, 2005) in beaked whales found stranded in the Canary Islands, evidencing a DCS similar to that in human divers. The syndrome was suggested to have been induced by exposure to mid-frequency sonar signals, and the strandings were temporally and spatially coincident with naval exercises employing this acoustic source (Cox et al., 2006; D'Amico et al., 2009; Martín et al., 2004). Recent analyses of sperm whale (*Physeter macrocephalus*) and beaked whale diving physiology (Fahlman et al., 2014; Hooker et al., 2012; Tyack et al., 2006; Zimmer and Tyack, 2007) suggest altered behaviour of these species in response to naval sonar could increase the risk of gas bubble embolism. Changes in dive time and consequent variation in physiological parameters could explain the DCS symptoms found in beaked whales stranded in atypical mass events. Cuvier's beaked whale atypical mass strandings that occurred in the Mediterranean Sea region prior to 2006 have been correlated with naval activity (Filadelfo et al., 2009; Frantzis, 1998, 2015; Podestà et al., 2006). Based on pathology from the five recent mass stranding events of 2006–14, the most likely cause of death was naval activities in the areas (ACCOBAMS, 2016; Arbelo et al., 2008; Bernaldo de Quirós et al., 2012; Cozzi et al., 2011; Dalton, 2006; DON, 2008; A. Frantzis, unpublished data; A. Fernández, University of Las Palmas de Gran Canarias, S. Mazzariol, University of Padua, Italy, personal communication, 7 March 2016) (see Section 2.2). During these five recent events a total of at least 28 animals were found stranded, mainly concentrated in the South of Italy and Greece, where similar atypical mass strandings had already occurred for this species in the past (Podestà et al., 2006). Furthermore, the total number of animals that died without reaching the coasts may have been much higher than the number of reported strandings. Large portions of both the Western and Eastern Mediterranean, including some areas that are known Cuvier's beaked whale high-use areas, have been affected by military activities, including naval exercises using low- and mid-frequency active sonar and underwater and surface detonations (ACCOBAMS, 2016; DON, 2008) (see Section 2.2). Mortality is the most significant impact to these whales, but response to nonlethal acoustic exposure can also be expressed by avoidance, and can cause disruption of foraging behaviour (New et al., 2013; Tyack et al., 2011), and alterations in dive profiles, that could affect metabolism and reduce individual fitness (DeRuiter et al., 2013).

As a result of these concerns regarding cetacean exposure to anthropogenic noise resulting from military activity, ACCOBAMS (2013) strongly recommended that, during naval exercises using sonar or underwater explosions, there should be absolute avoidance within an approximate 90 km buffer zone around all areas that have been designated as 'Areas of Special Concern for Beaked Whales' in the Mediterranean Sea (see Fig. 7) (ACCOBAMS, 2013). This did not prevent military activities from occurring in one such area only one year later, when *Noble Dina* was temporally and spatially connected to the atypical mass stranding event in the Hellenic Trench off Crete in 2014 (Frantzis, 2015; see Section 2.2).

While abundance estimates and trends remain unknown for the global Mediterranean Sea population, the impact of mortalities occurring during atypical mass strandings could be significant. For example, for Cuvier's beaked whales in the Hellenic Trench area, a decrease in sightings and strandings (with no reduction in search effort) during the last two decades has been reported (Frantzis, 2009).

4.1.2 Seismic Activity

Another source of anthropogenic noise that could be a threat to Cuvier's beaked whales in the Mediterranean Sea is oil and gas exploration through seismic surveys. A recent overview of high impact areas for noise in the Mediterranean demonstrated that the entire basin is heavily impacted by seismic surveys, with 830 active areas surveyed in the past 10 years (ACCOBAMS, 2016). Main areas of exploration were concentrated in the Gulf of Valencia, Alborán Sea, Strait of Sicily, Ionian Sea, Levantine Sea and Adriatic Sea. The Adriatic Sea has more than 130 different gas and oil extraction installations that represent an additional possible threat to the marine fauna of this semi-enclosed basin (Holcer et al., 2014). Two Cuvier's beaked whale stranding events (four whales in the Galapagos islands in 2000, and two whales in the Gulf of California in 2002) have been cautiously linked to seismic pulses (Gentry, 2002; Gordon et al., 2003; Malakoff, 2002). Considering the apparently heightened sensitivity of Cuvier's beaked whale to acoustic noise such as the military sonar, the extensive use of air guns could have a cumulative impact and in some cases may be exacerbated by the overlapping of Cuvier's beaked whale high-use areas (e.g. Alborán Sea, Ligurian Sea, Central Tyrrhenian Sea, southern Adriatic Sea, and the Hellenic Trench) with the main areas of seismic exploration.

4.1.3 Maritime Traffic

Little is known about the effects of maritime traffic and ship noise on Cuvier's beaked whales. Changes in Cuvier's beaked whale dive and foraging behaviour in response to ship noise have been reported in the Ligurian Sea (Aguilar de Soto et al., 2006). Behavioural reactions to vessel noise were also observed for Blainville's beaked whales (*Mesoplodon densirostris*) in the Tongue of the Ocean, Bahamas (Pirotta et al., 2012). The main commercial harbours of the Ligurian Sea are located within the Pelagos Sanctuary, which, as a consequence, is a crossing for all main commercial routes in the area. Marine traffic and noise pollution in this region have some of the highest levels within the entire Mediterranean Basin (Coomber, 2016; LMIU, 2008). Accordingly, the potential impact of maritime traffic (including frequent naval traffic) on the species habitat should be considered as a possible threat affecting the Cuvier's beaked whale population.

No evidence of ship collision or injury due to propellers in the Mediterranean Sea has been reported for Cuvier's beaked whales. Recently, the Secretariat of the Pelagos Sanctuary Agreement funded a project dedicated to determining the potential impact of shipping and its associated noise on Cuvier's beaked whale distribution in the Ligurian Sea (Azzellino et al., 2016). The potential species response to the naval traffic density, inferred from Automatic Identification System (AIS) data, and the related generated noise (inferred from a simulation model) was analysed (Azzellino et al., 2016). The study highlighted the close association of Cuvier's beaked whale preferred habitat with areas of intense naval traffic. Moreover, Cuvier's beaked whales avoided zones with a higher than average density of naval traffic. Another analysis focused on the Genoa Canyon area (considered to be optimal beaked whale habitat) and found that Cuvier's beaked whales preferentially avoided zones with a higher density of naval traffic than the zonal average (Azzellino et al., 2016). Similarly, Coomber (2016) used Generalized Additive Mixed Models to examine the effects of maritime shipping traffic in the Genoa Canyon and found that the level of shipping had a negative linear correlation with sighting rates of Cuvier's beaked whales.

5. CONCLUSION AND RECOMMENDATIONS

Genetic analysis has indicated a high degree of differentiation from the Atlantic population (Dalebout et al., 2005) and suggests that Cuvier's beaked whales in the Mediterranean Sea should be considered as a separate Evolutionarily Significant Unit, distinct from other populations. Moreover, we suggest that Cuvier's beaked whales in the Mediterranean Sea should be designated as Vulnerable or Threatened on the IUCN Red List.

Military activities, seismic exploration and ship noise all represent major threats to Cuvier's beaked whales in the Mediterranean Sea, especially in areas where the species occurs in high densities, and maritime traffic is particularly intense, such as the Genoa Canyon area in the Pelagos Sanctuary. Research in the Genoa Canyon has shown that the species attempts to avoid areas with high shipping density (Azzellino et al., 2016), and there is some evidence of a possible association of Cuvier's beaked whale strandings with seismic explorations (Gordon et al., 2003). Mitigation measures should be applied in the Pelagos Sanctuary, and in other Cuvier's beaked whale high-density areas, to alleviate the impact of maritime traffic and anthropogenic noise. Furthermore, application of mitigation measures to restrict military exercises can provide a successful tool for reducing the impact of noise on this species, in particular by restricting such operations from occurring in its preferred habitat. Indeed, the effectiveness of such spatial mitigation has apparently been successful for the Canary Islands (an area where atypical mass strandings had previously occurred in relation to military exercises), where no atypical mass stranding of beaked whales has been recorded since 2004, after a moratorium of military activities using sonar was declared there (Fernández et al., 2012).

Urgent conservation measures need to be applied to ensure the protection of Cuvier's beaked whale in the Mediterranean Sea. Important Cuvier's beaked whale habitats, including the Alborán Sea, Ligurian Sea, Tyrrhenian Sea, southern Adriatic Sea, Hellenic Trench and likely other less studied areas (e.g. the Levantine Sea), have a high conservation value for beaked whales in the Mediterranean and, therefore, are worthy of conservation actions. Major efforts should be undertaken to fill the current knowledge gaps regarding distribution, population size and trends and anthropogenic disturbances of Cuvier's beaked whale in the Mediterranean region. Results from predictive habitat models may be used as a basis for better designing survey effort in unsurveyed areas. Similarly, research on stranded specimens should be better organized with a standardized protocol and collection of samples implemented all along the Mediterranean coasts. Research to gain a better understanding of the effects of other anthropogenic pressures on Cuvier's beaked whales should also be pursued. Detailed analyses investigating the causes for mortality in stranded Cuvier's beaked whales will be useful to properly assess the impacts of the threats potentially affecting the population in the Mediterranean Basin, and to implement the appropriate mitigation measures. Moreover, management measures are urgently needed to restrict military activities within and around Cuvier's beaked whale high-density areas in the Mediterranean Sea.

ACKNOWLEDGEMENTS

This review includes research findings obtained by different projects: (1) 'Noise impact on sperm whale (*P. macrocephalus*) and Cuvier's beaked whale (*Z. cavirostris*), estimated from the marine traffic' Convention No. 01/2014 financed by the Permanent Secretariat of the Pelagos Agreement establishing the Sanctuary for the marine mammals in the Mediterranean Sea; (2) 'Analysis of the distribution and absolute/relative abundance of sperm whale (*P. macrocephalus*), Risso's dolphin (*Grampus griseus*) and Cuvier's beaked whale (*Z. cavirostris*) in the Pelagos Sanctuary in function of environmental changes and anthropogenic pressures' financed by the Italian Ministry for the Environment

(N.0003302/PNM 19/02/2014). We are also deeply grateful to A. Akkaya Bas, P. Alexiadou, M. Arbelo, M. Baini, C. Cesarini, F.G. Coomber, A. Fernández, C. Fossi, D. Kerem, A. Maglio, S. Mazzariol, S. Michaelidis, S. Muscat, G. Pavan, T. Raga, V. Ridoux and A. Vella for providing information and data. Last but not least we would like to thank the two anonymous reviewers and the Editors Barbara E. Curry and Giuseppe Notarbartolo di Sciara for greatly improving the quality of this manuscript.

REFERENCES

ACCOBAMS, 2013. In: Report of the Fifth Meeting of the Parties to Accobams. Tangier, 5–8 November 2013, p. 152.

ACCOBAMS, 2016. Overview of the noise hotspots in the ACCOBAMS area part I—Mediterranean Sea: Final report by Maglio, A., Pavan, G., Castellote, M., Frey, S. 45 pp. http://dx.doi.org/10.13140/RG.2.1.2574.8560.

Aguilar de Soto, N., Johnson, M., Madsen, P.T., Tyack, P.L., Bocconcelli, A., Borsani, J.F., 2006. Does intense ship noise disrupt foraging in deep-diving Cuvier's beaked whales *Ziphius cavirostris*? Mar. Mamm. Sci. 22, 690–699.

Akkaya Bas, A., Lagoa, G.C., Atchoi, E., 2016. New records of Cuvier's beaked whales (*Ziphius cavirostris*) from the Turkish Levantine Sea. Turk. J. Zool. 40, 454–460. http://dx.doi.org/10.3906/zoo-1509-19.

Arbelo, M., Bernaldo de Quirós, Y., Sierra, E., Méndez, M., Godinho, A., Ramírez, G., Caballero, M.J., Fernandez, A., 2008. Atypical beaked whale mass stranding in Almeria's coast: pathological study. Int. J. Anim. Sound Rec. 17 (10), 295–297.

Arcangeli, A., Campana, I., Marini, L., MacLeod, C.D., 2015. Long-term presence and habitat use of Cuvier's beaked whale (*Ziphius cavirostris*) in the Central Tyrrhenian Sea. Mar. Ecol. 37, 269–282. http://dx.doi.org/10.1111/maec.12272.

Azzellino, A., Airoldi, S., Gaspari, S., Nani, B., 2008. Habitat use of cetaceans along the Continental Slope and adjacent waters in the Western Ligurian Sea. Deep Sea Res. I 55, 296–323.

Azzellino, A., Lanfredi, C., D'Amico, A., Pavan, G., Podestà, M., Haun, J., 2011. Risk mapping for sensitive species to underwater anthropogenic sound emissions: model development and validation in two Mediterranean areas. Mar. Poll. Bull. 63, 56–70.

Azzellino, A., Panigada, S., Lanfredi, C., Zanardelli, M., Airoldi, S., Notarbartolo di Sciara, G., 2012. Predictive habitat models for managing marine areas: spatial and temporal distribution of marine mammals within the Pelagos Sanctuary (Northwestern Mediterranean Sea). Ocean Coast. Manag. 67, 63–74.

Azzellino, A., David, L., Di Meglio, N., Labach, H., Origné, L., Amar, D., Agliati, P.O., Coomber, F., Rosso, M., Tepisich, P., Moulins, A., 2016. Noise impact on sperm whale (*P. macrocephalus*) and Cuvier's beaked whale (*Z. cavirostris*), estimated from the marine traffic: Final report. p. 140. Convention PELAGOS Sanctuary No. 2014-01.

Baini, M., Fossi, M.C., Panti, C., Marsili, L., Coomber, F., Tepsich, P., Moulins, A., Rosso, M., 2016. Cytochrome P450 1A1 and 2B protein expression as biomarker for the first assessment of the ecotoxicological status of Cuvier's beaked (*Ziphius cavirostris*) in the NW Mediterranean sea. In: 30th Annual Conference of the European Cetacean Society, 14–16 March 2016, Madeira.

Baird, R.W., Webster, D.L., McSweeney, D.J., Ligon, A.D., Schorr, G.S., Barlow, J., 2006. Diving behavior and ecology of Cuvier's (*Ziphius cavirostris*) and Blainville's (*Mesoplodon densirostris*) beaked whales in Hawaii. Can. J. Zool. 84, 1120–1128.

Balcomb III, K.C., Claridge, D.E., 2001. A mass stranding of cetaceans caused by naval sonar in the Bahamas. Bahamas J. Sci. 2, 2–12.

Banca Dati Italiana Spiaggiamenti, 2015. Available at: http://mammiferimarini.unipv.it (accessed 30 January 2016).

Barlow, J., Ferguson, M.C., Perrin, W.F., Ballance, L., Gerrodette, T., Joyce, G., Macleod, C.D., Mullin, K., Palka, D.L., Waring, G., 2006. Abundance and densities of beaked and bottlenose whales (family Ziphiidae). J. Cetacean Res. Manag. 7 (3), 263–270.

Baumgartner, M.F., Mullin, K.D., May, L.N., Leming, T.D., 2001. Cetaceans habitats in the northern Gulf of Mexico. Fish. Bull. 99, 219–239.

Bernaldo de Quirós, Y., Gonzales-Diaz, O., Arbelo, M., Sierra, E., Sacchini, S., Fernández, A., 2012. Decompression versus decomposition: distribution, quantity and gas composition of bubbles in stranded marine mammals. Front. Aquat. Physiol. 3, 177. http://dx.doi.org/10.3389/fphys.2012.00177.

Blanco, C., Raga, J.A., 2000. Cephalopod prey of two *Ziphius cavirostris* (Cetacea) stranded on the western Mediterranean coast. J. Mar. Biol. Assoc. U. K. 80, 381–382.

Bräger, S., Kopcsányi, T., Bräger, Z., 2014. First sightings of living Cuvier's beaked whales (*Ziphius cavirostris*) in Albanian waters. Mar. Biodiv. 44, 553–557.

Buckland, S.T., Anderson, R., Burnham, K.P., Laake, J.L., 1993. Distance Sampling: Estimating Abundance of Biological Populations. Chapman and Hall, London. Available from: http://www.ruwpa.st-and.ac.uk/distance.

Buckland, S.T., Anderson, R., Burnham, K.P., Laake, J.L., Borchers, D.L., Thomas, L., 2004. Advance Distance Sampling. Examining Abundance of Biological Populations. Oxford University Press, Oxford, UK.

Cagnolaro, L., Cozzi, B., Magnaghi, L., Podestà, M., Poggi, R., Tangerini, P., 1986. Su 18 cetacei spiaggiati sulle coste italiane dal 1981 al 1985: rilevamento biometrico ed osservazioni necroscopiche (Mammalia Cetacea). Atti Soc. Ital. Sci. Nat. Museo Civ. Stor. Nat. Milano 127, 79–106 (in Italian).

Cañadas, A., 2012. *Ziphius cavirostris* (Mediterranean subpopulation). In: The IUCN Red List of Threatened Species 2012. e.T16381144A16382769, http://dx.doi.org/10.2305/IUCN.UK.2012-1.RLTS.T16381144A16382769.en.

Cañadas, A., Vázquez, J.A., 2014. Conserving Cuvier's beaked whales in the Alborán Sea (SW Mediterranean): identification of high density areas to be avoided by intense man-made sound. Biol. Conserv. 178, 155–162.

Cañadas, A., Sagarminaga, R., García-Tiscar, S., 2002. Cetacean distribution related with depth and slope in the Mediterranean waters off southern Spain. Deep Sea Res. I 49 (11), 2053–2073.

Cañadas, A., Sagarminaga, R., de Stephanis, R., Urquiola, E., Hammond, P.S., 2005. Habitat preference modelling as a conservation tool: proposal of marine protected areas for cetaceans in Southern Spain. Aquat. Conserv. 15, 495–521.

Cañadas, A., B-Nagy, A., Bearzi, G., Cotte, C., Fortuna, C., Frantzis, A., Gannier, A., Laran, S., Lauriano, G., Lewis, T., Moulins, A., Mussi, B., Pastor, X., Politi, E., Pulcini, M., Raga, J.A., Rendell, L., 2013. ACCOBAMS collaborative effort to map high-use areas by beaked whales in the Mediterranean, Monaco. p. 24.

Capelli, R., Das, K., De Pellegrini, R., Drava, G., Lepoint, G., Miglio, C., Minganti, V., Poggi, R., 2008. Distribution of trace elements in organs of six species of cetaceans from the Ligurian Sea (Mediterranean), and the relationship with stable carbon and nitrogen ratios. Sci. Total Environ. 390, 569–578.

Carlini, R., Pulcini, M., Wurtz, M., 1992. Cephalopods from the stomachs of Cuvier's beaked whale (*Ziphius cavirostris* Cuvier, 1823) stranded at Fiumicino, Central Tyrrhenian Sea. Eur. Res. Cetaceans 6, 190–191.

Coomber, F., 2016. Ecological impact assessment of maritime traffic and its associated noise pollution on Cuvier's beaked whales (*Ziphius cavirostris* Cuvier, 1823) in the Pelagos Sanctuary. PhD thesis, University of Genoa, Italy.

Coomber, F., Moulins, A., Tepsich, P., Rosso, M., 2016. Sexing free-ranging adult Cuvier's beaked whales (Ziphius cavirostris) using natural marking thresholds and pigmentation patterns. J. Mammal. 97, 879–890. http://dx.doi.org/10.1093/jmammal/gyw033.

Cox, T.M., Ragen, T.J., Read, A.J., Vos, E., Baird, R.W., Balcomb, K., Barlow, J., Caldwell, J., Cranford, T., Crum, L., D'Amico, A., D'Spain, G., Fernández, A., Finneran, J., Gentry, R., Gerth, W., Gulland, F., Hildebrand, J., Houser, D., Hullar, T., Jepson, P.D., Ketten, D., MacLeod, C.D., Miller, P., Moore, S., Mountain, D., Palka, D., Ponganis, P., Rommel, S., Rowles, T., Taylor, B., Tyack, P., Wartzok, D., Gisiner, R., Mead, J., Benner, L., 2006. Understanding the impacts of anthropogenic sound on beaked whales. J. Cetacean Res. Manag. 7 (3), 177–187.

Cozzi, B., Podestà, M., Mazzariol, S., 2011. Strandings of beaked whales in the Italian waters: a perspective of 25 years. Document presented to the Scientific Committee of the International Whaling Commission. Paper SC/63/SM5.

D'Amico, A., Bergamasco, A., Zanasca, P., Carniel, E., Portunato, N., Teloni, V., Mori, C., Barbanti, R., 2003. Qualitative correlation of marine mammals with physical and biological parameters in the Ligurian Sea. IEEE J. Ocean. Eng. 28 (1), 29–43.

D'Amico, A., Gisiner, R.C., Ketten, D.R., Hammock, J.A., Johnson, C., Tyack, P.L., Mead, J., 2009. Beaked whale strandings and naval exercises. Aquat. Mamm. 35 (4), 452–472.

Dalebout, M.L., Robertson, K.M., Frantzis, A., Engelhaupt, D., Mignucci, A.A., Rosario-Delestre, R.J., Baker, C.S., 2005. Worldwide structure of mtDNA diversity among Cuvier's beaked whales (*Ziphius cavirostris*): implications for threatened populations. Mol. Ecol. 14, 3353–3371.

Dalton, R., 2006. More whale strandings are linked to sonar. Nature 440, 593. http://dx.doi.org/10.1038/440593.

Davis, R.W., Fargion, G.S., May, N., Leming, T.D., Baumgartner, M.F., 1998. Physical habitat of cetaceans along the continental slope in the north-central and western Gulf of Mexico. Mar. Mamm. Sci. 14 (3), 490–507.

Davis, R.W., Ortega-Ortiz, J.G., Ribic, C.A., Evans, W.E., Biggs, D.C., Ressler, P.H., Cady, R.B., Leben, R.R., Mullin, K.D., Wursig, B., 2002. Cetacean habitat in the northern oceanic Gulf of Mexico. Deep Sea Res. I 49, 121–142.

DeRuiter, S.L., Southall, B.L., Calambokidis, J., Zimmer, W.M.X., Sadykova, D., Falcone, E.A., Friedlaender, A.S., Joseph, J.E., Moretti, D., Schorr, G.S., Thomas, L., Tyack, P.L., 2013. First direct measurements of behavioural responses by Cuvier's beaked whales to mid-frequency active sonar. Biol. Lett. 9, 20130223. http://dx.doi.org/10.1098/rsbl.2013.0223.

Dhermain, F., 2012. Réseau Echouage Méditerranéen Suivi des échouages sur les côtes méditerranéennes françaises. Années 2009–2012: Rapport GECEM/GIS3M, contract 10-058-83400 pour le Parc national de Port-Cros. GECEM Publ., Marseille, pp. 1–140.

Dhermain, F., Dupraz, F., Dupont, L., Keck, N., Godenir, J., Cesarini, C., Wafo, E., 2011. Recensement des échouages de cétacés sur les côtes françaises de Méditerranée. Années 2005–2009. Sci. Rep. Port-Cros Natl. Park Fr. 25, 121–141.

Dhermain, F., Astruc, G., Cesarini, C., Dupont, L., Dupraz, F., Godenir, J., Keck, N., Labach, H., Wafo, E., 2015. Recensement des échouages de cétacés sur les côtes françaises de Méditerranée, entre 2010 et 2012. Sci. Rep. Port-Cros Natl. Park 29, 103–126.

DON, 2008. Final Atlantic Fleet Active Sonar Training Environmental Impact Statement/ Overseas Environmental Impact Statement. United States Department of the Navy, Norfolk, VA.

Fahlman, A., Tyack, P.L., Miller, P.J.O., Kvadsheim, P.H., 2014. How man-made interference might cause gas bubble emboli in deep diving whales. Front. Physiol. 5 (13), 1–6.

Ferguson, M.C., Barlow, J., Reilly, S.B., Gerrodette, T., 2006. Predicting Cuvier's (*Ziphius cavirostris*) and Mesoplodon beaked whale population density from habitat characteristics in the eastern tropical Pacific Ocean. J. Cetacean Res. Manag. 7 (3), 287–299.

Fernández, A., Arbelo, M., Deaville, R., Patterson, I.A.P., Castro, P., Baker, J.R., Degollada, E., Ross, H.M., Herraez, P., Pocknell, A.M., Rodriguez, F., Howie, F.E., Espinosa, A., Reid, R.J., Jaber, J.R., Martin, V., Cunninghan, A.A., Jepson, P.D., 2004. Beaked whales, sonar and decompression sickness. Nature 10, 1038.
Fernández, A., Edwards, J.F., Rodríguea, F., Espinosa de los Monteros, A., Herráez, P., Castro, P., Jaber, J.R., Martin, V., Arbelo, M., 2005. "Gas and Fat Embolic Syndrome" involving a mass stranding of beaked whales (Family *Ziphiidae*) exposed to anthropogenic sonar signals. Vet. Pathol. 42, 446–457.
Fernández, A., Sierra, E., Martín, E., Méndez, M., Sacchinni, S., Bernaldo de Quirós, Y., Andrada, M., Rivero, M., Quesada, O., Tejedor, M., Arbelo, M., 2012. Last "atypical" beaked whales mass stranding in the Canary Islands (July, 2004). J. Marine Sci. Res. Dev. 2, 107. http://dx.doi.org/10.4172/2155-9910.1000107.
Filadelfo, R., Mintz, J., Michlovich, E., D'Amico, A., Tyack, P.L., Ketten, D.R., 2009. Correlating military sonar use with beaked whale mass strandings: what do the historical data show? Aquat. Mamm. 35 (4), 435–444.
Frantzis, A., 1998. Does acoustic testing strand whales? Nature 392, 29.
Frantzis, A., 2004. The first mass stranding that was associated with the use of active sonar (Kyparissiakos Gulf, Greece, 1996). In: Proceedings of the Workshop: "Active Sonar and Cetaceans", pp. 14–20. 8 March 2003, Las Palmas, Gran Canaria. ECS Newsletter 42 (special issue).
Frantzis, A., 2009. Cetaceans in Greece: present status of knowledge. Initiative for the Conservation of Cetaceans in Greece, Athens, Greece. 94 pp.
Frantzis, A., 2015. Short report on the mass stranding of Cuvier's beaked whales that occurred on the 1st of April 2014 in South Crete, Greece, during naval exercises. FINS 6 (1), 10–11.
Frantzis, A., Alexiadou, P., Paximadis, G., Politi, E., Gannier, A., Corsini-Foka, M., 2003. Current knowledge of the cetacean fauna of the Greek Seas. J. Cetacean Res. Manag. 5, 219–232.
Gannier, A., 2011. Using existing data and focused surveys to highlight Cuvier's beaked whales favourable areas: a case study in the central Tyrrhenian Sea. Mar. Poll. Bull. 63 (1–4), 10–17.
Gannier, A., 2015. Cuvier's beaked whale (*Ziphius cavirostris*) diving behavior as obtained by visual observation methods and consequences in terms of visual detection during surveys. Sci. Rep. Port-Cros Natl. Park 29, 127–134.
Gannier, A., Epinat, J., 2008. Cuvier's beaked whale distribution in the Mediterranean Sea: results from small boat surveys 1996–2007. J. Mar. Biol. Assoc. U. K. 88 (6), 1245–1251.
Garibaldi, F., Cataldini, G., Insacco, G., Podestà, M., Garibaldi, F., Cataldini, G., Insacco, G., Podestà, M., 2015. Stomach contents of Cuvier's beaked whales, *Ziphius cavirostris*, stranded in Southern Italy. In: 29th Annual Conference of the European Cetacean Society, March, Malta.
Gentry, R.L., 2002. Mass stranding of beaked whales in the Galapagos islands, April 2000. Available at: http://www.nmfs.noaa.gov/pr/pdfs/health/galapagos_stranding.pdf.
Geraci, J.R., Lounsbury, V., 1993. Marine Mammals Ashore. A Field Guide for Strandings. Texas A.M. Sea Grant Publications, Galveston, TX.
Gomerčić, H., Đuras Gomerčić, M., Gomerčić, T., Lucić, H., Dalebout, M., Galov, A., Škrtić, D., Ćurković, S., Vuković, S., Huber, Đ., 2006. Biological aspects of Cuvier's beaked whale (*Ziphius cavirostris*) recorded in the Croatian part of the Adriatic Sea. Eur. J. Wildl. Res. 52, 182–187.
Gordon, J., Gillespie, D., Potter, J., Frantzis, A., Simmonds, M.P., Swift, R., Thompson, D., 2003. A review of the effects of seismic surveys on marine mammals. Mar. Tech. Soc. J. 37 (4), 16–34.
Gozalbes-Aparicio, P., Raga, J.A., 2015. Progress report 2014 on the Mediterranean database of cetacean strandings. p. 44. Contract RAC/SPA, No. 24/2013/RAC/SPA.

Hamazaky, T., 2002. Spatiotemporal prediction models of North Atlantic Ocean (from Cape Hatteras, North Carolina, U.S.A. to Nova Scotia, Canada) cetacean habitats in the mid-western. Mar. Mamm. Sci. 18 (4), 920–939.

Heyning, J.E., 1989. Cuvier's beaked whale, *Ziphius cavirostris* G. Cuvier, 1823. In: - Ridgway, S.H., Harrison, R. (Eds.), Handbook of Marine Mammals, vol. 4. Academic Press, London, pp. 289–308.

Holcer, D., Notarbartolo di Sciara, G., Fortuna, C.M., Onofri, V., Lazar, B., Tvrtkovic, N., 2003. The occurrence of Cuvier's beaked whale (*Ziphius cavirostris*) in Croatian Adriatic waters: historical and recent findings. In: Besendorfer, V., Kopjar, N. (Eds.), Proceedings of the 8th Croatian Biological Congress, Zagreb, pp. 255–256.

Holcer, D., Notarbartolo di Sciara, G., Fortuna, C.M., Lazar, B., Onofri, V., 2007. Occurrence of Cuvier's beaked whale in the Southern Adriatic Sea evidence of an important Mediterranean habitat. J. Mar. Biol. Assoc. U. K. 87, 359–362.

Holcer, D., Fortuna, C.M., Mackelworth, P., 2014. Important areas for conservation of cetaceans, sea turtles and giant devil rays in the Adriatic Sea: summary of existing knowledge: UNEP/MAP-RAC/SPA Internal report. Contract No. 08/RAC/SPA_2013 MedOpenSeas, 67 pp.

Hooker, S.K., Whitehead, H., Gowans, S., Baird, R.W., 2002. Fluctuations in distribution and patterns of individual range use of northern bottlenose whales. Mar. Ecol. Prog. Ser. 225, 287–297.

Hooker, S.K., Fahlman, A., Moore, M.J., Aguilar De Soto, N., Bernaldo De Quiros, Y., Brubakk, A.O., Costa, D.P., Costidis, A.M., Dennison, S., Falke, K.J., Fernandez, A., Ferrigno, M., Fitz-Clarke, J.R., Garner, M.M., Houser, D.S., Jepson, P.D., Ketten, D.R., Kvadsheim, P.H., Madsen, P.T., Pollock, N.W., Rotstein, D.S., Rowles, T.K., Simmons, S.E., Van Bonn, W., Weathersby, P.K., Weise, M.J., Williams, T.M., Tyack, P.L., 2012. Deadly diving? Physiological and behavioural management of decompression stress in diving mammals. Proc. R. Soc. Lond. B Biol. Sci. 279, 1041–1050. http://dx.doi.org/10.1098/rspb.2011.2088.

Jepson, P.D., Arbelo, M., Deaville, R., Patterson, I.A.P., Castro, P., Baker, J.R., Degollada, E., Ross, H.M., Herráez, P., Pocknell, A.M., Rodríguez, F., Howiell, F.E., Espinosa, A., Reid, R.J., Jaber, J.R., Martin, V., Cunningham, A.A., Fernández, A., 2003. Gas bubble lesions in stranded cetaceans: was sonar responsible for a spate of whale deaths after an Atlantic military exercise? Nature 425, 575–576.

Johnson, M., Tyack, P.L., 2003. A digital acoustic recording tag for measuring the response of wild marine mammals to sound. IEEE J. Ocean. Eng. 28, 3–12.

Johnson, M., Madsen, P.T., Zimmer, W.M.X., Aguilar de Soto, N., Tyack, P.L., 2004. Beaked whales echolocate on prey. Proc. R. Soc. Lond. Ser. B 271, S383–S386.

Kerem, D., Hadar, N., Goffman, O., Scheinin, A., Kent, R., Boisseau, O., Schattner, U., 2012. Update on the cetacean fauna of the Mediterranean Levantine Basin. Open Mar. Biol. J. 6, 6–27.

Kovačić, I., Gomerčić, M.D., Gomerčić, H., Lucić, H., Gomerčić, T., 2010. Stomach contents of two Cuvier's beaked whales (*Ziphius cavirostris*) stranded in the Adriatic Sea. Mar. Biodiv. Rec. 3, e19. http://dx.doi.org/10.1017/S1755267210000059.

Lanfredi, C., 2014. Risk assessment of underwater sounds emissions: impact on marine environment. PhD thesis, Politecnico di Milano, Environmental Engineering Department, Milano, Italy.

Lefkaditou, E., Poulopoulos, Y., 1998. Cephalopod remains in the stomach-content of beaked whales, *Ziphius cavirostris* (Cuvier, 1823), from the Ionian Sea. Rapp. Comm. Int. Mer Medit. 35, 460–461.

Littardi, V., Rosso, M., Wurtz, M., 2004. Enquêtes historiques (1900–1966) sur les échouages de *Ziphius cavirostris* G. Cuvier, en Mer Ligure. Rapp. Comm. Int. Mer Médit. 37, 388.

LMIU (Lloyd's Marine Intelligence Unit), 2008. Study of Maritime Traffic Flows in the Mediterranean Sea. Regional Marine Pollution Emergency Response Centre for the Mediterranean Sea (REMPEC), Valletta.

MacLeod, C.D., 2000. Distribution of beaked whales of the genus *Mesoplodon* in the North Atlantic (Order: Cetacea, Family: Ziphiidae). Mammal Rev. 30, 1–8.

MacLeod, C.D., D'Amico, A., 2006. A review of knowledge about behavior and ecology of beaked whales in relation to assessing and mitigating potential impacts from anthropogenic noise. J. Cetacean Res. Manag. 7 (3), 211–221.

MacLeod, C.D., Mitchell, G., 2006. Known key areas for beaked whales around the world. J. Cetacean Res. Manag. 7 (3), 309–322.

MacLeod, C.D., Zuur, A.F., 2005. Habitat utilisation by Blainville's beaked whales off Great Abaco, Northern Bahamas, in relation to seabed topography. Mar. Biol. 174, 1–11.

MacLeod, C.D., Santos, M.B., Pierce, G.J., 2003. Review of data on diets of beaked whales: evidence of niche separation and geographic segregation. J. Mar. Biol. Assoc. U. K. 83, 651–665.

MacLeod, C.D., Perrin, W.F., Pitman, R., Barlow, J., Balance, L., D'Amico, A., Gerrodette, I., Joyce, G., Mullin, K.D., Palka, D.L., Waring, G.T., 2006. Known and inferred distributions of beaked whale species (Cetacea: Ziphiidae). J. Cetacean Res. Manag. 7, 271–286.

Malakoff, D., 2002. Suit ties whale deaths to research cruise. Science 298, 722–723.

Maldini, D., Mazzucca, L., Atkinson, S., 2005. Odontocete stranding patterns in the main Hawaiian islands (1937-2002): how do they compare with live animal surveys? Pac. Sci. 59 (1), 55–67.

Mannocci, L., Dorémus, G., Kiszka, J., Laran, S., Martinez, L., Van Canneyt, O., Ridoux, V., 2011. Habitat preferences of Ziphiids: an analysis of physiographic variables associated to beaked whales sightings in French temperate and tropical waters. International Whaling Commission, SC/63/SM12, 15 pp.

Marini, L., Consiglio, C., Angradi, A.M., Sanna, A., 1992. Four sightings of Ziphiidae (Cetacea, Odontoceti) in the central Tyrrhenian Sea. Ital. J. Zool. 4, 85–89.

Martín Martel, V., 2002. Especial varamiento de cetáceos–Viceconsejería de Medio Ambiente. Government of the Canary Islands, Gran Canaria. www.gobcan.es/medioambiente/varamientos.

Martín, V., Servidio, A., García, S., 2004. Mass strandings of beaked whales in the Canary Islands. In: Evans, P., Miller, L. (Eds.), Proceedings of the Workshop on Active Sonar and Cetaceans Held at the European Cetacean Society Held at the 17th Annual Conference Las Palmas, Gran Canaria, pp. 33–36. 8 March 2003. ECS Newsletter 42.

MEDACES, 2015. Mar. Zoology Unit, University of Valencia, Available at: http://medaces.uv.es/ (accessed 31 December 2015).

Moulins, A., Rosso, M., Nani, B., Wurtz, M., 2007. Aspects of the distribution of Cuvier's beaked whale (*Ziphius cavirostris*) in relation to topographic features in the Pelagos Sanctuary (north-western Mediterranean Sea). J. Mar. Biol. Assoc. U. K. 87, 177–186.

Moulins, A., Rosso, M., Ballardini, M., Würtz, M., 2008. Partitioning of the Pelagos Sanctuary (north-western Mediterranean Sea) into hotspots and coldspots of cetacean distributions. J. Mar. Biol. Assoc. U. K. 88, 1273–1281.

New, L.F., Moretti, D.J., Hooker, S.K., Costa, D.P., Simmons, S.E., 2013. Using energetic models to investigate the survival and reproduction of beaked whales (family Ziphiidae). PLoS One 8 (7), e68725. http://dx.doi.org/10.1371/journal.pone.0068725.

Northridge, S.P., 1994. World review of interactions between marine mammals and fisheries. Fisheries Technical paper 251, Food and Agriculture Organization of the United Nations, Rome.

Notarbartolo di Sciara, G., Birkun Jr., A., 2010. Conserving whales, dolphins and porpoises in the Mediterranean and Black Seas: an ACCOBAMS status report, 2010. ACCOBAMS, Monaco. 212 pp.

Orsi Relini, L., Garibaldi, F., 2005. Diversità dei cefalopodi mesopelagici del santuario dei cetacei in base a campionamenti diretti e osservazioni sull'alimentazione dello zifio, *Ziphius cavirostris*. Biol. Mar. Medit. 12 (1), 106–115.

Öztürk, A.A., Tonay, A.M., Dede, A., 2011. Strandings of the beaked whales, Risso's dolphins, and a minke whale on the Turkish coast of the Eastern Mediterranean Sea. J. Black Sea/Mediterranean Env. 17 (3), 269–274.

Pedà, C., Battaglia, P., Scuderi, A., Voliani, A., Mancusi, C., Andaloro, F., Romeo, T., 2015. Cephalopod prey in the stomach contents of odontocete cetaceans stranded in the western Mediterranean Sea. Mar. Biol. Res. 11, 593–602. http://dx.doi.org/10.1080/17451000.2014.966724.

Peltier, H., Ridoux, V., 2015. Marine megavertebrates adrift: a framework for the interpretation of stranding data in perspective of the European Marine Strategy Framework Directive and other regional agreements. Environ. Sci. Pol. 54, 240–247.

Peltier, H., Dabin, W., Daniel, P., Van Canneyt, O., Dorémus, G., Huona, M., Ridoux, V., 2012. The significance of stranding data as indicators of cetacean populations at sea: modelling the drift of cetacean carcasses. Ecol. Indic. 18, 278–290.

Pirotta, E., Milor, R., Quick, N., Moretti, D., Di Marzio, N., Tyack, P., Boyd, I., Hastie, G., 2012. Vessel noise affects beaked whale behavior: results of a dedicated acoustic response study. PLoS One 7 (8), e42535. http://dx.doi.org/10.1371/journal.pone.0042535.

Podestà, M., Magnaghi, L., 1989. Unusual number of cetacean by-catches in the Ligurian Sea. Eur. Res. Cetaceans 3, 67–70.

Podestà, M., Meotti, C., 1991. The stomach contents of a Cuvier's beaked whale *Ziphius cavirostris* and Risso's dolphin *Grampus griseus* stranded in Italy. Eur. Res. Cetaceans 5, 58–61.

Podestà, M., Cagnolaro, L., Cozzi, B., 2005. First record of a stranded Gervais' beaked whale, *Mesoplodon europaeus* (Gervais, 1855), in the Mediterranean waters. Atti Soc. Ital. Sci. Nat. Museo Civ. Stor. Nat. Milano 146 (I), 109–116.

Podestà, M., D'Amico, A., Pavan, G., Drougas, A., Komnenou, A., Portunato, N., 2006. A review of Cuvier's beaked whale strandings in the Mediterranean Sea. J. Cetacean Res. Manag. 7 (3), 251–261.

Pyenson, N.D., 2010. Carcasses on the coastline: measuring the ecological fidelity of the cetacean stranding record in the eastern North Pacific Ocean. Paleobiology 36, 453–480.

Pyenson, N.D., 2011. The high fidelity of the cetacean stranding record: insights into measuring diversity by integrating taphonomy and macroecology. Proc. R. Soc. B 278, 3608–3616.

Rosso, M., 2010. Population size, residency patterns and energy demand of Cuvier's beaked whales (*Ziphius cavirostris*) in the north western Mediterranean sea. PhD thesis, University of Basilicata, Potenza, Italy.

Rosso, M., Ballardini, M., Moulins, A., Würtz, M., 2011. Natural markings of Cuvier's beaked whale *Ziphius cavirostris* in the Mediterranean Sea. Afr. J. Mar. Sci. 33 (1), 45–57.

Schorr, G.S., Falcone, E.A., Moretti, D.J., Andrews, R.D., 2014. First long-term behavioral records from Cuvier's beaked whales (*Ziphius cavirostris*) reveal record-breaking dives. PLoS One 9 (3), e92633. http://dx.doi.org/10.1371/journal.pone.0092633.

Taylor, B.L., Baird, R., Barlow, J., Dawson, S.M., Ford, J., Mead, J.G., NotarbartolodiSciara, G., Wade, P., Pitman, R.L., 2008. *Ziphius cavirostris*. In: The IUCN Red List of Threatened Species. e.T23211A9429826.

Tepsich, P., Rosso, M., Halpin, P.N., Moulins, A., 2014. Habitat preferences of two deep-diving cetacean species in the northern Ligurian Sea. Mar. Ecol. Prog. Ser. 508, 247–260.

Tortonese, E., 1963. Insolita comparsa di cetacei (*Ziphius cavirostris* G. Cuv.) nel golfo di Genova. Natura 54, 120–122 (in Italian).
Tyack, P.L., Johnson, M., Aguilar Soto, N., Sturlese, A., Madsen, P.T., 2006. Extreme diving of beaked whales. J. Exp. Biol. 209, 4238–4253.
Tyack, P.L., Zimmer, W.M.X., Moretti, D., Southall, B.L., Claridge, D.E., Durban, J.W., Clark, C.W., D'Amico, A., DiMarzio, N., Jarvis, S., McCarthy, E., Morrissey, R., Ward, J., Boyd, I.L., 2011. Beaked whales respond to simulated and actual navy sonar. PLoS One 6 (3), e17009. http://dx.doi.org/10.1371/journal.pone.0017009.
Waring, G.T., Hamazaki, T., Sheenan, D., Wood, G., Baker, S., 2001. Characterization of beaked whale (Ziphiidae) and sperm whale (*Physeter macrocephalus*) summer habitat in shelf-edge and deeper waters off the northeast US. Mar. Mamm. Sci. 17 (4), 703–717.
Whitehead, H., 2009. SOCPROG programs: analyzing animal social structures. Behav. Ecol. Sociobiol. 63, 765–778.
Wimmer, T., Whitehead, H., 2004. Movements and distribution of northern bottlenose whales, *Hyperoodon ampullatus*, on the Scotian Slope and in adjacent waters. Can. J. Zool. 82, 1782–1794.
Woodside, J.M., David, L., Frantzis, A., Hooker, S.K., 2006. Gouge marks on deep-sea mud volcanoes in the eastern Mediterranean: caused by Cuvier's beaked whales? Deep Sea Res. I 53, 1762–1771.
Zimmer, W.M.X., Tyack, P.L., 2007. Repetitive shallow dives pose decompression risk in deep-diving beaked whales. Mar. Mamm. Sci. 23 (4), 888–925.
Zimmer, W.M.X., Johnson, M.P., Madsen, P.T., Tyack, P.L., 2005. Echolocation clicks of free-ranging Cuvier's beaked whales (*Ziphius cavirostris*). J. Acoust. Soc. Am. 117 (6), 3919–3927.

CHAPTER FIVE

Conservation Status of Killer Whales, *Orcinus orca*, in the Strait of Gibraltar

R. Esteban*[,1], P. Verborgh*, P. Gauffier*, D. Alarcón*, J.M. Salazar-Sierra*, J. Giménez[†], A.D. Foote[‡], R. de Stephanis*

*CIRCE (Conservation, Information and Research on Cetaceans), Pelayo-Algeciras, Cádiz, Spain
[†]Estación Biológica de Doñana (EBD-CSIC), Sevilla, Spain
[‡]CMPG, Institute of Ecology and Evolution, University of Bern, Bern, Switzerland
[1]Corresponding author: e-mail address: ruthesteban@gmail.com

Contents

Abstract

Killer whales (*Orcinus orca*) in the Mediterranean Sea are currently restricted to the Strait of Gibraltar and surrounding waters. Thirty-nine individuals were present in 2011, with a well-differentiated social structure, organized into five pods. Killer whale occurrence in the Strait is apparently related to the migration of their main prey, Atlantic bluefin tuna

Advances in Marine Biology, Volume 75
ISSN 0065-2881
http://dx.doi.org/10.1016/bs.amb.2016.07.001

(*Thunnus thynnus*). In spring, whale distribution was restricted to shallow waters off the western coast of the Strait where all pods were observed actively hunting tuna. In summer, the whales were observed in the shallow central waters of the Strait. A relatively new feeding strategy has been observed among two of the five pods. These two pods interact with an artisanal drop-line fishery. Pods depredating the fishery had access to larger tuna in comparison with pods that were actively hunting. The Strait of Gibraltar killer whales are socially and ecologically different from individuals in the Canary Islands. Molecular genetic research has indicated that there is little or no female-mediated gene migration between these areas. Conservation threats include small population size, prey depletion, vessel traffic, and contaminants. We propose the declaration of the Strait of Gibraltar killer whales as an endangered subpopulation. A conservation plan to protect the Strait of Gibraltar killer whales is urgently needed, and we recommend implementation of a seasonal management area where activities producing underwater noise are restricted, and the promotion of bluefin tuna conservation.

1. INTRODUCTION

Ancient naturalists described marine creatures in the Mediterranean Sea, but the identification of these creatures as marine mammals was uncertain because they were believed to be sea monsters. Moreover, the descriptions were usually symbolic and inclusive of mythology. Around the first century, Pliny the Elder produced what is likely the earliest published account of killer whales (*Orcinus orca*) (Notarbartolo di Sciara, 1987). In *Naturalis Historia* (Book IX.5), Pliny described whales that were the enemies of large whales because they fed on pregnant females and calves in the Gulf of Cadiz. These were likely to have been killer whales. In mid-winter, large whales were apparently abundant in the waters off Cadiz, but in summer they sought refuge in sheltered coastal areas to give birth, and it was there that killer whales attacked them. In the second century, Strabo, in *Geografia* (Book II 2,7), described Turdetania, a region of the southern Iberian Peninsula, and included what were likely killer whales among the marine life. Strabo referred to these whales as '*Oryges*' (Latin for gazelle or goat). In the same century, Aelianus, in *De Natura Animalium* (Book XV. 2), noted the presence of what were probably killer whales in the Strait of Bonifacio, between Corsica and Sardinia in the Mediterranean Sea. Aelianus referred to these whales as '*Marine Arietes*' (Latin for marine ram), and described their hunting techniques, noting that they were capable of 'intentional stranding' to penetrate the caves where seals were resting. These seals were most likely monk seals (*Monachus monachus*), which were abundant and widespread throughout the Mediterranean Sea, the North-African Atlantic coast, and

Macaronesian islands in that era (Brasseur et al., 1997; Israëls, 1992; King, 1956; Ronald and Healey, 1978).

In the 16th century, Horozco (1598) reported that tuna (Atlantic bluefin tuna, *Thunnus thynnus*) entered the Santi Pretri River (Cadiz, southern Spain), escaping from killer whales. Killer whales were reported to be common along the Atlantic coast of the Iberian Peninsula (Graells, 1889), and more particularly in the Gulf of Cadiz (Cornide, 1788; De Huerta, 1624), where it was said to be the most common marine mammal (Machado y Nuñez, 1869). Nevertheless, historically, the species was considered to be rare or absent in the Mediterranean Sea (Borrell et al., 2000; Cabrera, 1914; Casinos and Vericad, 1976; Cornide, 1788; Cuvier, 1836; Di Natale and Mangano, 1983; Duguy and Cyrus, 1973; Duguy et al., 1983; Graells, 1889; Graus et al., 1986; Machado y Nuñez, 1869; Notarbartolo di Sciara, 1987; Raga et al., 1985; Van Beneden, 1888).

Potential food resources for killer whales in the Mediterranean Sea were depleted throughout history by human activities such as the massive hunting of monk seals (Brasseur et al., 1997; Sal'nikov, 1959) and fin whales (*Balaenoptera physalus*). Hunting led to the near complete disappearance of the fin whale Mediterranean subpopulation south-west of the Strait of Gibraltar in the Atlantic Ocean (Aguilar, 1985, 2006; Clapham et al., 2008; Sanpera and Aguilar, 1992). It has been suggested that the decline of killer whales in the Atlantic Ocean may have been due to a shortage of prey, such as humpback (*Megaptera novaeangliae*) and other baleen whales (Duguy and Robineau, 1973). It is possible that the current killer whale diet, which is based on bluefin tuna (García-Tiscar, 2009), may be the result of adaptation in response the decline of the marine mammal populations that were apparently a part of their historical diet. However, another hypothesis is that two killer whale ecotypes (one specializing in predation of marine mammals and the other predating fish) inhabited the Mediterranean and nearby areas of the Atlantic historically (Esteban, 2015). The coexistence of these two killer whale ecotypes has been described in other areas such as the North Atlantic (Foote et al., 2009), and the Pacific Ocean (Ford et al., 1998).

From 1999 to 2011, we conducted research to investigate the distribution, ecology, social structure, population structure, and demography of killer whales in the Strait of Gibraltar. Here we have reviewed research on the Strait of Gibraltar killer whales and include a discussion of conservation threats, management actions, and current conservation. We conclude with recommendations regarding measures for the future protection of the Strait of Gibraltar killer whales.

2. KILLER WHALE DISTRIBUTION

Killer whales have a wide distribution throughout all oceans and seas, from polar waters to the equator (Forney and Wade, 2007; Heyning and Dahlheim, 1988; Leatherwood and Dalheim, 1978). They are known to be common in many coastal areas, particularly at high latitudes, but they are also present in high seas and tropical waters (Forney and Wade, 2007; Leatherwood and Dalheim, 1978 Reeves and Mitchell, 1988). Seasonal movement is associated with prey availability (e.g. Kuningas et al., 2013; Sigurjónsson and Leatherwood, 1988). Killer whales are distributed across the North Atlantic, with increasing density at higher latitudes where productivity is also higher (Forney and Wade, 2007; Hammond and Lockyer, 1988). Although killer whales were sometimes seen in the Mediterranean Sea prior to the 2000s (Table 1 and Fig. 1), there is no evidence of their continued regular occurrence within the Basin. Even with some dedicated survey effort searching for cetaceans in the Mediterranean Sea, there has been a lack of killer whale sightings since the 2000s (Table 1 and Fig. 1).

Killer whales are regularly seen in the Strait of Gibraltar, which is the natural passage between the Atlantic Ocean and the Mediterranean Sea (de Stephanis et al., 2008; Esteban et al., 2013; Guinet et al., 2007). Killer whale distribution in the southern Iberian Peninsula is related to the migration of its main prey, eastern Atlantic bluefin tuna (Esteban et al., 2013) (Fig. 2). This migratory pelagic tuna undertakes annual reproductive migrations, entering the Mediterranean Sea from the Atlantic Ocean during late spring, in search of suitable spawning areas (Sella, 1929). Esteban et al. (2013) used cetacean survey data (2002–2012, 322 killer whale sightings and 10,952 sightings of other cetaceans used as pseudo-absences of killer whales) combined with predictive generalized additive models to investigate probable factors affecting killer whale distribution in the Strait of Gibraltar, Gulf of Cadiz, and Alborán Sea. Models indicated that the killer whales were associated with the probable distribution of bluefin tuna migrating through the study area, thus restricting whale distribution to the shallow waters of the outer coastal Strait of Gibraltar and Gulf of Cadiz region in spring and to the shallow waters of the Strait of Gibraltar in spring and summer (de Stephanis et al., 2008; Esteban et al. 2013), when shoals of tuna migrate through the area (de la Serna et al., 2004).

Table 1 Sighting, Stranding, and Capture Information of Killer Whales (*Orcinus orca*) in the Mediterranean Sea and Surrounding Waters

Date (Day/Month/ Year)	Latitude	Longitude	Type of Observation	References
22/06/1787	43.31	6.75	Capture	Van Beneden (1881)
1886	39.86	0.66	Sighting from land	Brú (1913)
13/12/1890	43.46	−3.78	Stranding	Casinos and Vericad (1976)
22/06/1902	36.33	−5.15	Capture	Casinos and Vericad (1976)
1914	43.46	−3.78	Stranding	Casinos and Vericad (1976)
26/12/1941	39.7	3.45	Stranding	Casinos and Vericad (1976)
1966	40.07	4.09	Stranding	Hammond and Lockyer (1988)
26/05/1970	36.56	−6.20	Stranding	Valverde (2006)
Before 1973	43.33	3.58	Stranding	Duguy and Cyrus (1973)
Before 1973	42.41	3.18	Stranding	Duguy and Cyrus (1973)
01/05/1973	40.08	0.14	Stranding	Casinos and Filella (1975)
1978–1982	39.63	10.48	Sighting at sea	Di Natale and Mangano (1983)
1978–1982	38.46	12.80	Sighting at sea	Di Natale and Mangano (1983)
1978–1982	38.98	15.57	Sighting at sea	Di Natale and Mangano (1983)
1978–1982	36.73	18.18	Sighting at sea	Di Natale and Mangano (1983)
12/09/1982	38.00	25.00	Sighting at sea	Hammond and Lockyer (1988)
09/03/1983	37.30	−0.55	Sighting at sea	Hammond and Lockyer (1988)
17/06/1983	36.11	−4.03	Sighting at sea	Hammond and Lockyer (1988)
27/06/1984	39.00	9.42	Sighting at sea	Raga et al. (1985)
27/06/1984	39.00	9.70	Sighting at sea	Hammond and Lockyer (1988)
14/08/1985	43.55	8.5	Sighting at sea	Notarbartolo di Sciara (1987)
16/08/1985	43.53	7.77	Sighting at sea	Notarbartolo di Sciara (1987)
1/10/1985	43.86	8.4	Sighting at sea	Notarbartolo di Sciara (1987)
02/02/1986	37.95	24.75	Sighting at sea	Hammond and Lockyer (1988)
07/08/1990	36.75	−3.83	Stranding	Borrell et al. (2000)

Continued

Table 1 Sighting, Stranding, and Capture Information of Killer Whales (*Orcinus orca*) in the Mediterranean Sea and Surrounding Waters—cont'd

Date (Day/Month/Year)	Latitude	Longitude	Type of Observation	References
09/12/1990	36.73	−3.58	Stranding	Borrell et al. (2000)
08/05/1991	36.50	−4.86	Stranding	Borrell et al. (2000)
05/08/1991	36.50	−4.86	Stranding	Borrell et al. (2000)
01/03/1992	39.86	39.86	Sighting from land	Belenguer and Kersting (2011)
1997	36.48	−6.25	Stranding	Gutiérrez-Expósito et al. (2012)
06/06/2004	36.95	−7.20	Sighting at sea	Baez et al. (2007)
07/06/2004	36.99	−7.12	Sighting at sea	Baez et al. (2007)
08/06/2004	36.97	−8.01	Sighting at sea	Baez et al. (2007)
10/06/2004	36.92	−8.05	Sighting at sea	Baez et al. (2007)
12/06/2004	36.93	−7.50	Sighting at sea	Baez et al. (2007)
12/04/2012	36.28	−4.34	Sighting from land	T. Todorov (Iberian Peninsula, Spain, personal communication)
14/04/2012	36.27	−4.38	Sighting from land	T. Todorov (Iberian Peninsula, Spain, personal communication)

3. THE STRAIT OF GIBRALTAR KILLER WHALES

In the Strait of Gibraltar, 47 individual killer whales were identified between 1999 and 2011 (Esteban et al., 2016a). Nine individuals were considered to have died over the study period (one female stranded in 2006, known as 'Vega'; two juveniles; five calves that were considered dead, as they were not seen with their mother or within their pod for at least three consecutive years, and another adult individual that was not observed within its pod after 2007—recognizing that this individual may have left the area). After 2005, all newly identified individuals were calves (nine individuals) from known females of existing pods (Esteban et al., 2016b).

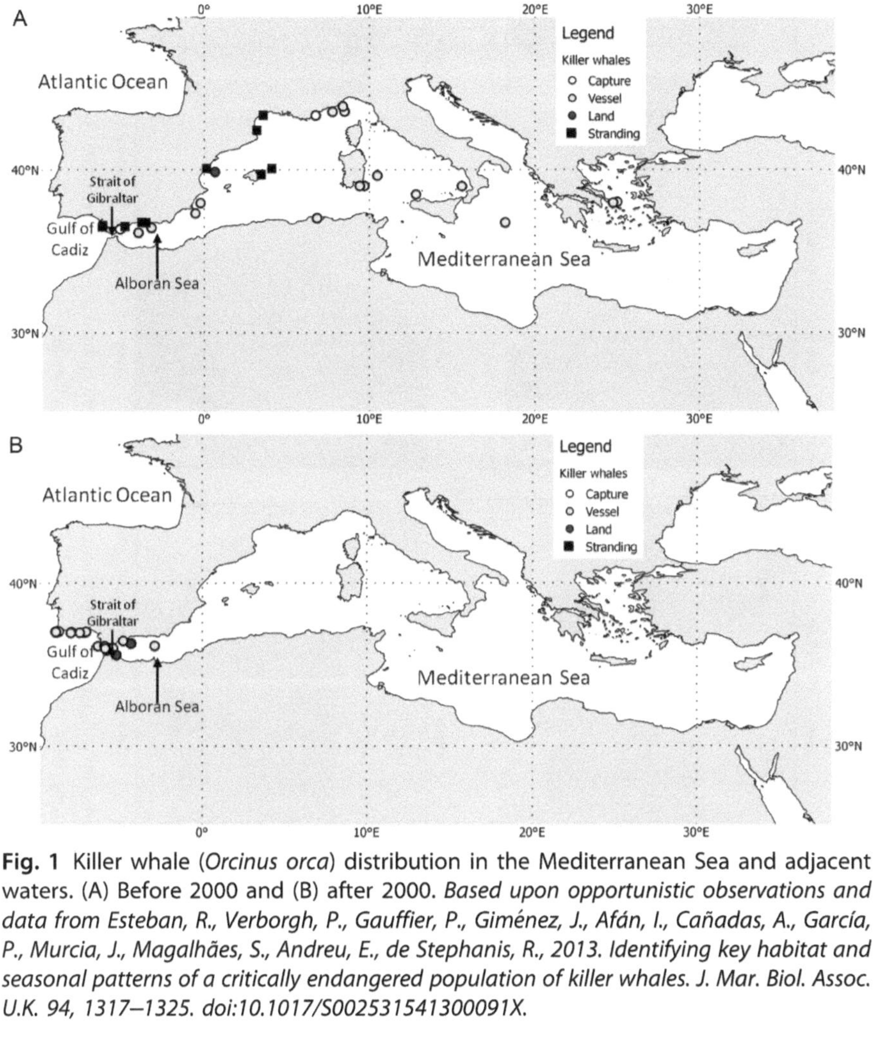

Fig. 1 Killer whale (*Orcinus orca*) distribution in the Mediterranean Sea and adjacent waters. (A) Before 2000 and (B) after 2000. *Based upon opportunistic observations and data from Esteban, R., Verborgh, P., Gauffier, P., Giménez, J., Afán, I., Cañadas, A., García, P., Murcia, J., Magalhães, S., Andreu, E., de Stephanis, R., 2013. Identifying key habitat and seasonal patterns of a critically endangered population of killer whales. J. Mar. Biol. Assoc. U.K. 94, 1317–1325. doi:10.1017/S002531541300091X.*

The first cumulative total abundance count (calculated in 2005) was 32 individuals. The count remained fairly constant after that time because most of the calves born subsequently died during this period, except for three new unidentified individuals which were added to the catalogue after 2005. In 2011, four births increased the count of that year to 39 individuals (Esteban et al., 2016a) (see also Section 6).

As has been well described for this species (e.g. Bigg et al., 1990), killer whales in the Strait of Gibraltar have a matrilineal social structure, and

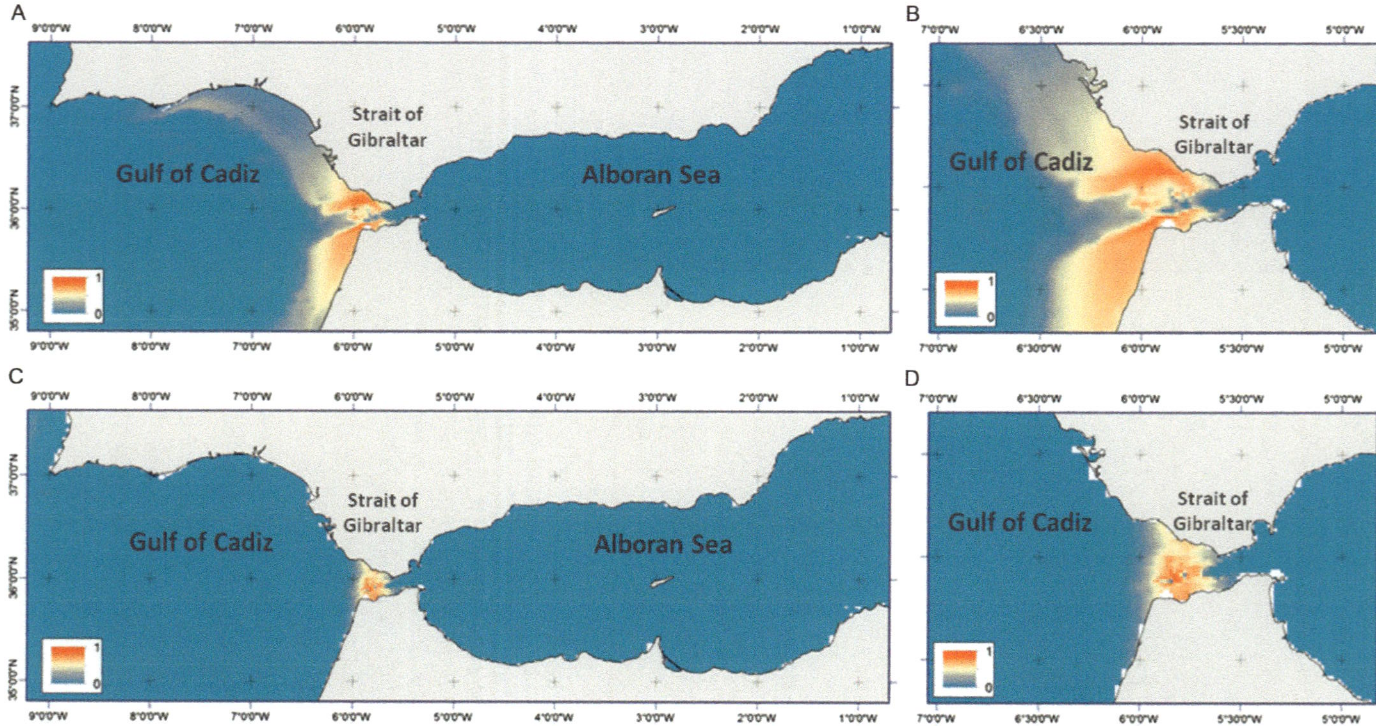

Fig. 2 Predictive density surface map of killer whales (*Orcinus orca*) based on spatial modelling: in spring, best predicted model (A) at southern Iberian Peninsula and (B) in the Strait of Gibraltar; in summer, best predicted model (C) at southern Iberian Peninsula and (D) in the Strait of Gibraltar. *Esteban, R., Verborgh, P., Gauffier, P., Giménez, J., Afán, I., Cañadas, A., García, P., Murcia, J., Magalhães, S., Andreu, E., de Stephanis, R., 2013. Identifying key habitat and seasonal patterns of a critically endangered population of killer whales. J. Mar. Biol. Assoc. U.K. 94, 1317–1325. doi:10.1017/S002531541300091X.*

certain individuals tend to have strong and enduring associations with specific individuals (Esteban et al., 2016b,c). These killer whales were originally observed, during the early 2000s, in four identifiable pods (known as A, B, C, and D pods). All individuals remained in the pod in which they were sighted for the first time. However in 2006 (due to the birth of one individual and the death of another), pod A underwent fission to form two socially differentiated pods (A1 and A2), resulting in a total of five identifiable pods (A1, A2, B, C, and D pods) (see Esteban et al. (2016b) (see also Section 5).

Of the 39 individual Strait of Gibraltar killer whales in 2011, 18 individuals were from pod A, nine from pod B, six from pod C, and six from pod D. At the start of our research, in 1999, 11 individuals were identified in pod A. Individuals from pods A1 and A2 were sighted every year from 1999 to 2011 (1999–2005 as members of pod A). Between 1999 and 2005, five viable calves were born (and one juvenile died), increasing the number of whales in pod A to 15 individuals by 2005. In 2011, three births occurred in pod A1, and one in pod B (see Esteban et al., 2016a).

4. ECOLOGY

4.1 Diet

Killer whales are apex predators, consuming a wide range of prey species, including squid, octopus, bony and cartilaginous fish, sea turtles, seabirds, sea and river otters, dugongs, pinnipeds, and cetaceans (Heyning and Dahlheim, 1988). Nevertheless, individual populations of killer whales regularly appear to specialize in particular types of prey (Ford et al., 1998).

Killer whales in the Strait of Gibraltar feed seasonally mainly on bluefin tuna (Esteban et al., 2016b). The diet of killer whales in the Strait of Gibraltar has been investigated using stable isotope analysis of skin samples (collected from May to July) (García-Tiscar, 2009). Isotope data collected from the female killer whale that stranded in 2006 ('Vega'), indicated that the animal had fed on other coastal fish species (García-Tiscar, 2009). Interestingly, this individual may have belonged to a migrant lineage, due to ancestry shared with individuals from both the Canary Islands and the Strait of Gibraltar (Esteban et al., 2016c). It is also possible that those killer whales opportunistically feed on coastal fish species in and around the Strait of Gibraltar.

4.2 Feeding Strategies

Killer whales in the Strait of Gibraltar use at least two different hunting techniques to feed on tuna. In spring, whales have been observed to chase tuna for up to 30 min at a relatively high speed (3.7 ± 0.2 m s^{-1}) until they captured them (Guinet et al., 2007). This has been described as an endurance-exhaustion technique, wherein killer whales drive tuna beyond their aerobic limits until the fish are exhausted and can be captured. Using this technique killer whales apparently capture small to medium size (0.8–1.5 m) tuna (Guinet et al., 2007). Killer whales have also been observed actively hunting in summer (Esteban et al., 2013). However, the whales were more frequently observed to be interacting with the artisanal drop-line fishery (Esteban et al., 2016a). The fishery began operating in 1995 (July–August) in the central waters of the Strait of Gibraltar (Srour, 1994). Killer whales have been interacting with the fishery since at least 1999 at the start of our observations in the Strait, and their distribution appears to be linked to the presence of drop-line boats in the area (de Stephanis et al., 2008). The whales patrol the vicinity of the boats until they find a tuna hooked on a line, and then depredate the fish before fishermen can bring it to the surface. Not all the killer whales sighted in the Strait have been observed to interact with the fishery (Esteban et al., 2013, 2016a).

4.3 Energy Requirements

Recently, Esteban et al. (2016a) investigated killer whale foraging behaviour and estimated energy requirements of 39 killer whales actively hunting and/or interacting with the drop-line fishery in the Strait of Gibraltar between 1999 and 2011. Esteban et al. (2016a) classified the whales into two categories: 'interacting individuals' observed interacting with the fishery at least once (INT; pods A1 and A2), and 'not interacting individuals' that were never observed interacting (NOT; pods B, C, and D). Energy requirements were estimated as tons of tuna needed for the entire subpopulation of 39 whales. Esteban et al. (2016a) estimated that (in 2011) the killer whales required approximately 1600 tons of tuna to meet their annual energy requirements. This estimate was converted to the number of fish needed, and to tons needed, to support killer whales in the Strait for each of the two feeding strategies (Esteban et al., 2016a). The interaction with the long-line fishery provided killer whales with larger tuna (2 m), compared to individuals that only hunted actively (tuna $\leq$ 0.8–1.5 m). Consequently INT whales needed fewer tuna to meet their energy requirements while interacting (8 tuna per day) compared to those actively hunting (21–141 tuna per day) (Esteban et al., 2016a).

4.4 Interspecies Interactions

In the Strait of Gibraltar, long-finned pilot whales (*Globicephala melas*) have repeatedly been observed apparently 'mobbing' killer whales and causing them to move out of the area (de Stephanis et al., 2015). However, isotopic analysis indicated different feeding niches for the two species. Interestingly, satellite tracking showed no overlap in the distribution of one pilot whale and one killer whale tagged for 21 days simultaneously (de Stephanis et al., 2015). Consequently, neither resource competition nor interspecific predation could explain the mobbing behaviour. The authors suggested that a possible historical presence of mammal-eating killer whales in the area could have provoked this behaviour as an antipredator defence that was maintained over time (de Stephanis et al., 2015).

5. SOCIAL STRUCTURE

Social structure in killer whales has been widely described and investigated and is typically characterized by stable hierarchically structured social units and strong natal philopatry (e.g. Bigg et al., 1990). However, several studies have described disparity in sociality among populations of killer whales (e.g. in mammal-eating 'transient' and fish-eating 'resident' killer whales off southern Vancouver Island) (Baird and Whitehead, 2000). A recent study of killer whales in the Northeast Atlantic suggested that sociality is plastic and can be modified depending on local ecological conditions (Beck et al., 2012). Furthermore, in the Northeast Pacific, group size of killer whales was related to prey abundance and the Pacific Decadal Oscillation (Lusseau et al., 2004). Anthropogenic activities have also been proposed to influence the social structure and behaviour of killer whales. Williams and Lusseau (2006) have suggested that targeted removals of individuals for dolphinaria may have altered the social structure into isolated groups in north-eastern Pacific killer whale populations. These populations are also thought to be experiencing social fragmentation related to low abundance of their primary prey, Chinook salmon (*Oncorhynchus tshawytscha*) (Foster et al., 2012; Parsons et al., 2009). Thus, disparity in social structure within populations can be affected by anthropogenic activities and changes in local ecological conditions.

Esteban et al. (2016b) found that foraging strategies are different between pods, and all members within each pod participate in the same strategy. All individuals from A1 and A2 pods were observed actively hunting tuna as well as interacting with the drop-line fishery annually during the summer. They were also seen actively hunting in spring. Pods B, C, and D were only observed actively hunting. Pods C and D were only seen actively hunting in spring, and pod B was seen hunting in both spring and summer (Fig. 3).

Such foraging strategies could arise through a combination of vertical transmission within matrilineal pods or by vertical and horizontal transmission with individuals aligning their behaviour with that of other pod members (Boyd and Richerson, 1985). Notably, all members of A pod were interacting with the drop-line fishery as early as 1999, indicating that this behaviour spread throughout A pod in less than a generation after the fishery began in 1995. Consequently, the transmission of this new behaviour could not have occurred only vertically from mother to calves. This within-pod uniformity and between-pod differentiation in a recently derived behaviour is consistent with, but not proof of, selective transmission via social learning (e.g. Galef, 1992).

Esteban et al. (2016b) proposed that social structure in these whales is shaped by maternal kinship and foraging behaviour. In the early 2000s, only

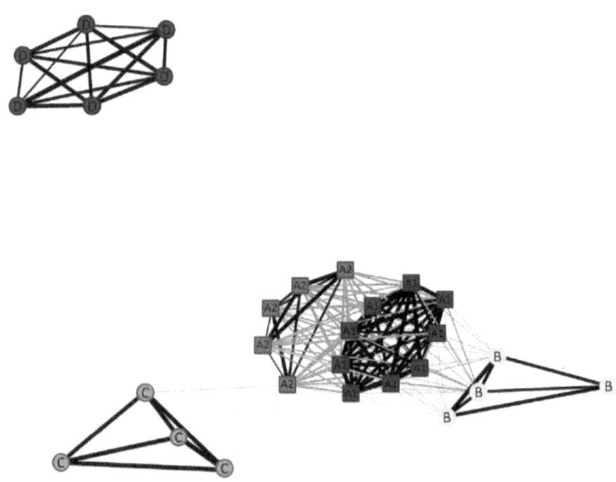

Fig. 3 Social network of the five killer whale (*Orcinus orca*) pods in the Strait of Gibraltar. *Squares* indicate interacting individuals and *circles* noninteracting individuals. *Black edges* are within-pod associations, and *gray edges* are between-pod associations. *Esteban, R., Verborgh, P., Gauffier, P., Giménez, J., Foote, A., de Stephanis, R., 2016b. Maternal kinship and fisheries interaction influences killer whale social structure. Behav. Ecol. Sociobiol. 70, 111–122. doi:10.1007/s00265-015-2029-3.*

one cohesive pod (A) interacted with the drop-line fishery, but after pod A underwent fission to form two socially differentiated pods (A1 and A2), each of these pods interacted with the fishery. Hence, social structure can influence the spread of novel foraging behaviours and may drive population fragmentation in this already small population. Observations of social changes at the earliest stages of diversification in foraging behaviour and social segregation may provide insights into the processes that ultimately result in the formation of socially isolated discrete ecotypes in killer whales (Esteban et al., 2016b).

6. POPULATION STRUCTURE AND MANAGEMENT UNITS

Genetic analysis of tissue samples provided information on population structuring in the north-eastern Atlantic Ocean, where three significantly differentiated populations were identified (Foote et al., 2011). Genetic analyses of skin samples collected from 85 killer whales included genotyping with 17 microsatellite loci and the sequencing of 989 bp of the mitochondrial control region for all tissue samples and the sequencing of complete mitochondrial genomes (mitogenomes; 16,386–16,392 bp) (Foote et al., 2011; and see Morin et al., 2010). The first population included individuals sampled in Norway, Iceland, and the North Sea; the second included individuals sampled in the North Sea and west coast of Iceland; and the third included individuals from the Strait of Gibraltar (10 free-ranging individuals and one stranded individual collected between 2006 and 2010) and the Canary Islands (nine individuals). The third population, which inhabits lower latitudes and was represented by a smaller sample size, consisted of maternal lineages from three highly divergent clades (Foote et al., 2011). For all three populations, all individuals sampled within a pod, shared the same mitochondrial DNA haplotype suggesting that the pods have a matrifocal social structure.

Recently Esteban et al. (2016c) adopted a multidisciplinary approach to assess ecological differences among killer whales. The authors combined photo-identification, molecular genetic analysis of individual genotypes, and isotopic biomarkers to assess contemporary gene flow and migration between killer whales in the Strait of Gibraltar and the Canary Islands. Strait of Gibraltar killer whales ($n=39$) appeared to be reproductively, socially, and ecologically differentiated from the individuals sampled in the Canary Islands ($n=16$) (Esteban et al., 2016c). Indeed, none of the individuals identified in the Canary Islands had ever been sighted in the Strait of Gibraltar (Fig. 4A). Additionally, genetic analyses indicated that only about 1–9% of

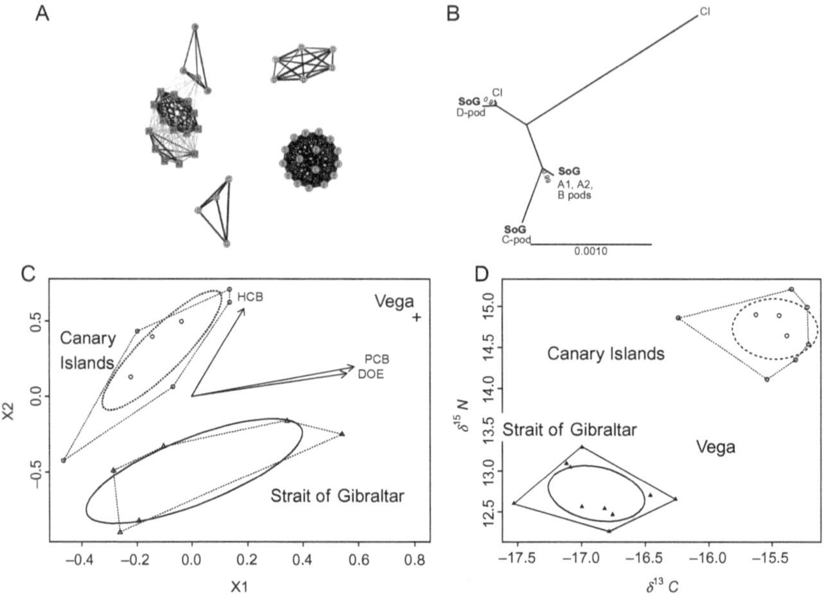

Fig. 4 Multidisciplinary study of killer whales (*Orcinus orca*) from the Strait of Gibraltar (SoG) and the Canary Islands (CI) (Esteban et al., 2016c): (A) Network diagram of killer whales including pods A1, A2, B, C, and D (including the female known as 'Vega') observed in the Strait of Gibraltar, and pod E observed in the Canary Islands. *Black edges* are within-pod associations, and *gray edges* are between-pod associations. In the SoG, *squares* indicate individuals known to interact with the drop-line fishery (INT) and *circles* indicate noninteracting (NOT) individuals. (B) Maximum-likelihood reconstruction of the genetic relationships between pods based upon complete mitochondrial genome sequences. Tip labels indicate population. Node labels indicate approximate likelihood ratio test (aLRT) support values. (C) Two-dimensional nonmetric multidimensional scaling configuration of similarities among study areas regarding 25 individual polychlorobiphenil (PCB) congeners, dichlorodiphenyldichloroethylene (*pp*′-DDE), and hexachlorobenzene (HCB) (Stress = 0.003), *dotted lines* represent convex hulls and *solid lines* the standard ellipses. (D) Stable isotopes biplot of carbon and nitrogen, *dotted lines* represent convex hulls and *solid lines* the standard ellipses. *Esteban, R., Verborgh, P., Gauffier, P., Giménez, J., Martín, V., Pérez-Gil, M., Tejedor, M., Almunia, J., Jepson, P., García-Tíscar, S., Barrettt-Lennard, L., Guinet, C., Foote, A., de Stephanis, R., 2016c. Using a multidisciplinary approach to identify a critically endangered killer whale management unit. Ecol. Indic. 66, 291–300. http://dx.doi.org/10.1016/j.ecolind.2016.01.043.*

the Strait of Gibraltar individuals were derived from the same subpopulation as Canary Islands individuals, while no complete mitogenome haplotypes were shared and no close kinship were detected between the two areas (Fig. 4B). The low genetic differentiation in combination with low levels of migration and the lack of any recent bottleneck signal are consistent with a very recent vicariant population split (Esteban et al., 2016c). Isotopic values and pollutant loads also suggested ecological differences between study areas (Fig. 4C and D). Interestingly, the Strait of Gibraltar's pod D had never been seen in association with any of the other four pods (A1, A2, B, and C) from the Strait (see Fig. 4A). The only individual sampled from pod D was the female ('Vega') that stranded in 2006. This female was identified via genetic analysis as having a potential migrant lineage and had intermediate isotopic values and contaminant between the Strait and the Canary Islands (see Fig. 4B–D) (Esteban et al., 2016c).

Esteban et al. (2016c) concluded that based on their genetic, social, and ecological differences, killer whales from the Canary Islands and the Strait of Gibraltar should be considered as two different subpopulations and management units. Furthermore, it was noted that pod D in the Strait of Gibraltar should be considered as a third management unit, due to ecological, social, and genetic differentiation from the other Strait of Gibraltar pods (Esteban et al., 2016c).

7. DEMOGRAPHIC PARAMETERS

7.1 Abundance

The abundance of the Strait of Gibraltar killer whales (39 as of 2011) is extraordinarily low compared to most other well-studied killer whale populations (Table 2). Moreover, the low numbers of individuals found in the Strait of Gibraltar are similar to sub-Antarctic southern Indian Ocean populations that have been considered to be in decline in the Crozet Archipelago (Poncelet et al., 2010), and around Marion Island, South Africa, where inshore abundance data indicated seasonal movement offshore and/or increases in range (Reisinger et al., 2011).

7.2 Survival Rate

Esteban et al. (2016a) estimated survival rate for the two categories (INT, NOT) of killer whales in the Strait of Gibraltar, using multistate mark-recapture models (Brownie et al., 1993; Hestbeck et al., 1991). Adult survival rate for INT whales (pods A1 and A2) was estimated at 0.991 (SE: 0.014; 95% CI: 0.837–1.000) (Fig. 5). It was similar to the survival rate estimated in the northern resident population, British Columbia (Olesiuk et al., 1990, 2005), which was considered to be a stable population (Table 3).

Adult survival for NOT whales (pods B, C, and D) was estimated at 0.901 (SE: 0.067; 95% CI: 0.672–0.980) (Fig. 5). This low rate was similar

Table 2 Abundance Estimates of Killer Whale (*Orcinus orca*) Populations in Regions Worldwide from Which Estimates Are Available (See Also Forney and Wade, 2007)

Population	Abundance	References
Norway	*731	Kuningas et al. (2013)
Iceland (east coast)	*300	Beck et al. (2012) and Foote et al. (2010)
Northern Isles and northeast coast of Scotland	*50	Beck et al. (2012) and Foote et al. (2010)
West coast of Scotland, Ireland, and Wales	*10	Beck et al. (2013)
Strait of Gibraltar	*38	Esteban et al. (2016c)
North-west Atlantic	*>67	Lawson and Stevens (2013)
Eastern Arctic	*53	Young et al. (2011)
Chilean Patagonia	*49	Häussermann and Acevedo (2013)
Southern resident (North Pacific)	*88	Ellis et al. (2011)
Northern resident (North Pacific)	*260	Ellis et al. (2011)
South Alaskan resident	>700	Matkin and Durban (2011)
Aleutian Islands, Bering Sea, and Kamchatka	*, **>1500	Hoelzel et al. (2007), Ivkovich et al. (2010), Matkin and Barrett-Lennard (2007), and Matkin and Durban (2011)
West coast transient	*250	Matkin et al. (2008)
Gulf of Alaska transient	*>100	Matkin and Durban (2011)
Alaska (AT1 transient)	*7	Barrett-Lennard and Heise (2006) and Matkin et al. (2008)
New Zealand	*115	Visser (2000)
Marion Island	*37	Reisinger et al. (2011)
Crozet Archipelago	*37	Poncelet et al. (2010)

*Estimate based on photo-identification catalogues; **Estimate based on distance sampling.

to Crozet Archipelago whales (Poncelet et al., 2010), where the most reasonable explanation for the high losses was direct killing by Patagonian toothfish (*Dissostichus eleginoides*) fishers during the late 1990s early 2000s (Guinet et al., 2014) (Table 3). Juvenile survival rate could only be estimated

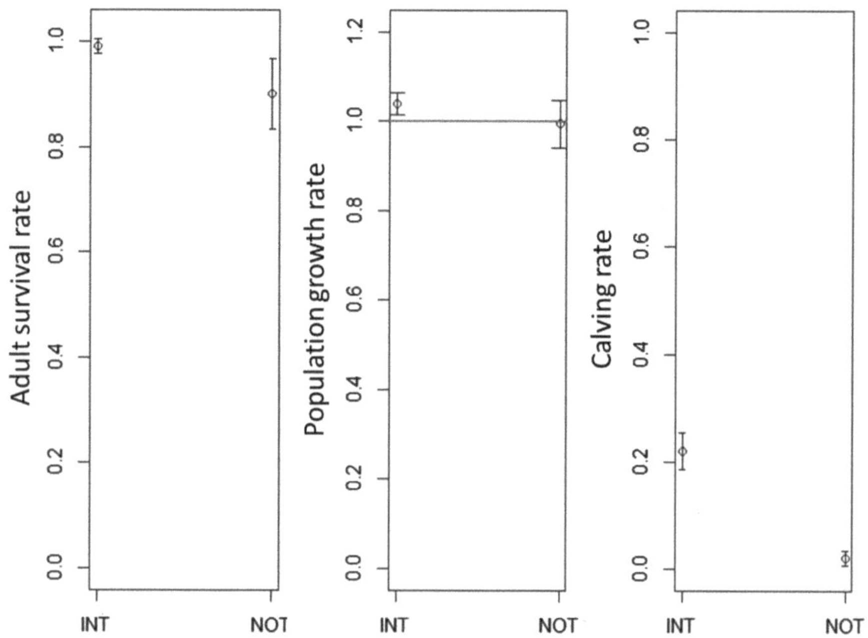

Fig. 5 Box plot demographic parameters compilation between Strait of Gibraltar killer whale (*Orcinus orca*) individuals known to interact with the drop-line fishery (INT) and noninteracting (NOT) individuals. *Red line* in population growth rate *box plot* indicates the threshold of growth of the population. *Circles* represent mean values and *lines* represent the standard error bars.

for INT whales and was 0.966 (SE: 0.031; 95% CI: 0.819–0.994). For INT whales, calf survival was one in 1999–2005 as all individuals survived their first year of life during this first period, and zero in 2006–2010, as no calf survived during this second period (Esteban et al., 2016a). Simultaneously, from 2005 to 2011 drop-line catches declined dramatically, reflecting the decline of tuna abundance in the Strait. Differences in survival rates between the two groups could be explained by the fact that depredation is a good opportunity for killer whales to feed in summer, when tuna are leaving the Strait for the Atlantic (Wilson and Block, 2009). These tuna use deep waters to cross the Strait and would likely be unavailable to killer whales (Aranda et al., 2013). However, drop-line fisheries bring tuna to the surface and could be acting as an accumulator of fish, locally increasing prey availability. This is likely especially important for killer whales when abundance of tuna is low. Food provisioning through depredation could have positively affected INT (pods A1 and A2) killer whale life history parameters in the years in which the tuna stock was at its lowest. This relationship between calf survival and tuna catches (as a proxy of tuna abundance) highlights

Table 3 Comparison of Demographic Parameters Among Killer Whale (*Orcinus orca*) Populations from the Strait of Gibraltar, North Pacific (Northern Resident), and Crozet Archipelago, Norway

	Strait of Gibraltar		**Northern Resident**		**Crozet Archipelago**		**Norway**
Parameter	1999–2011 (Esteban et al., 2016a)		1973–1995 (Olesiuk et al., 1990)	1996–2004 (Olesiuk et al., 2005)	1988–1989 (Poncelet et al., 2010)	1998–2000 (Poncelet et al., 2010)	1986–2003 (Kuningas et al., 2013)
	INT	NOT					
Adult survival rate	0.991 (SE: 0.014)	0.901 (SE: 0.067)	0.984 females (SE: 0.004)	0.971 females (SE: 0.007)	0.942 females (95% CI: 0.844–0.980)	0.901 females (95% CI: 0.742–0.966)	0.977 females (SE: 0.009)
			0.959 males (SE: 0.08)	0.909 males (SE: 0.017)	0.935 males (95% CI: 0.817–0.979)	0.895 males (95% CI: 0.746–0.961)	0.971 males (SE: 0.008)
Juvenile survival rate	0.966 (SE: 0.031)	–	0.984 (SE: 0.003)	0.946 (SE: 0.006)			0.816 (SE: 0.167)
Newborn survival rate	1	0	0.975 (SE: 0.013)	0.914 (SE: 0.033)			
Population growth rate	1.039 (SE: 0.025)	0.995 (SE: 0.053)	1.029	–	0.94 (95% CI: 0.90–0.99)		
Calving rate	0.219 (SE: 0.034)	0.020 (SE: 0.013)					
Calving interval	6–8 (mean = 7)	–	4.9 years (SE: 0.18)	5.5 years			5.06 years (SE: 0.722)
Fecundity rate	0.14	–		0.111 (SE: 0.086)	0.159 (SE: 0.14)		

the importance of artificial food provisioning in times when prey is scarce (Esteban et al., 2016a).

7.3 Other Parameters

Esteban et al. (2016a) estimated a growth rate of 1.039 (SE: 0.025; 95% CI: 0.986–1.091) for INT individuals (pods A1 and A2), which is equivalent to a 3.9% growth rate. In comparison, southern Alaska resident killer whale population growth rate was estimated to be 3.5%, and the population was thought to have reached a maximum growth rate thanks to the increasing return of their main prey, Chinook and coho (*Oncorhynchus kisutch*) salmon (Matkin et al., 2014). On the other hand, growth rate was almost stable for NOT pods (pods B, C, and D) at 0.995 (SE: 0.053; 95% CI: 0.832–1.159) (Fig. 5 and Table 3), with almost no recruitment over 12 years and lower adult survival, putting them a greater risk due to the already very low number of animals (18 individuals). Their situation could be similar to other populations currently on the verge of extinction such as the AT1 transient killer whales in Alaska, a group of seven individuals in which no recruitment had been reported since the Exxon Valdez oil spill in 1989 (Matkin et al., 2008, 2012), and to the group of nine killer whales around north-west Scotland and western Ireland, in which no calves had been reported for 19 years (Beck et al., 2013). Mean calving rate estimated for INT whales in the Strait of Gibraltar was 0.219 (SE: 0.034) and was higher than the rate of 0.020 estimated for the NOT whales (SE: 0.013) (Esteban et al., 2016a) (Fig. 5).

Only two births, from different females, were documented within the Strait of Gibraltar NOT whales (pods B, C, and D), and calving interval and fecundity rate were not estimated. Within the INT group (pods A1 and A2), 13 births were documented. However, only two intervals between viable calves of two females were observed and ranged from six to eight (mean = seven) years, producing a fecundity rate of 0.14 calves per year (Esteban et al., 2016a). These values were within the same ranges as for other populations available in the literature (Table 3).

8. CONSERVATION THREATS AND MANAGEMENT ACTIONS

8.1 Depletion of Bluefin Tuna Stock

Conservation of the Strait of Gibraltar killer whales depends on the adequate availability of tuna. Depletion of prey may force animals to spend more of their time foraging, which could lead to energetic and nutritional stress and subsequent reduction of reproductive rates and increased mortality rates. There is evidence to suggest that northern and southern resident killer whales are already prey limited, due to natural and anthropogenic factors affecting their preferred prey (Ford et al., 2010; Ward, 2009; Williams et al., 2011).

The eastern stock of North Atlantic bluefin tuna has been exploited for centuries, but in 1960 the industrial purse-seine and longline fleets replaced traditional fisheries in many areas of the Mediterranean Sea and the Atlantic Ocean (Fromentin and Powers, 2005). A decade later, the market development of sushi and sashimi had made it a highly lucrative fishery (Fromentin and Ravier, 2005). These new factors have caused a decline of tuna since the 1960s (Taylor et al., 2011). The International Convention for the Conservation of Atlantic Tunas (ICCAT) routinely establishes quotas for tuna. It was estimated that the established quotas were exceeded by 44% between 2005 and 2011 since most of the catches were not being declared (Gagern et al., 2013; ICCAT, 2014). This caused Atlantic bluefin tuna to be designated as Endangered on the International Union for Conservation of Nature (IUCN) Red List of Threatened Species in 2011. Between 2009 and 2011, the ICCAT fully endorsed scientific committee advice and enforced a low total allowable catch (TAC) around 18,500–22,000 tons (ICCAT, 2014). As a result, the most recent assessments show signs of biomass increase (ICCAT, 2014). Assessment of tuna indicated that spawning stock biomass peaked over 300,000 tons in the late 1950s and early 1970s, then declined to about 150,000 tons until the mid-2000s and stabilized around TAC levels established by the ICCAT at the end of the 2010s.

Conservation of the eastern Atlantic stock of bluefin tuna is essential for the future of killer whales in the Strait of Gibraltar, as they have a highly specialized diet (García-Tiscar, 2009). Similarly, in the north-eastern Pacific, the fecundity of the southern resident population has been strongly related to the abundance of their main prey, and a decrease in Chinook salmon populations caused a reduced calving rate (Ward, 2009). Moreover, Ford et al. (2010) showed that increased mortality relating to decreased prey availability was a greater factor than reduced calving rate, in driving the population decline of resident killer whales there. In the Strait of Gibraltar, the amount of tuna captured by drop-line fishery (as a proxy for tuna abundance) has been related to the survival of killer whale calves (Esteban et al., 2016a). Any decrease in the abundance of tuna could put the population of killer whales in the Strait of Gibraltar at greater risk. The 1600 tons estimated as the Strait of Gibraltar killer whale energy requirement in 2011 (see Section 4.3) represents 9% of the quota (18,500 tons) established by the

ICCAT for the eastern stock Atlantic bluefin tuna. Esteban et al. (2016a) proposed the use of ecosystem based management (see Pikitch et al., 2004) to allocate a specific annual quota for killer whales based upon their energy requirements, as was previously suggested for North Eastern Pacific southern resident killer whales under the Pacific Salmon Treaty (see Williams et al., 2011). No direct injury or retaliation by drop-line fishermen has ever been observed or reported. As a short-term urgent action, it has also been proposed to transfer a higher percentage of the quota to the artisanal drop-line fishery in the Strait from the industrial purse-seine fisheries in the Mediterranean Sea, at least until the Strait of Gibraltar killer whale community shows clear signs of stability and recovery (Esteban et al., 2016a).

8.2 Maritime Traffic

The Strait of Gibraltar is an area of heavy maritime traffic, due to the increase in commercial vessels, whale watching, ferries, and sports fishing boats in recent decades. Whale watching has become a major tourist industry in many places around the world since the 1980s (Hoyt, 2001, 2002). In addition to boosting the economy of coastal communities and providing an economic reason for the conservation of marine mammals stocks, whale watching has also proved beneficial to increase public awareness of marine mammals and the environmental problems that they face (Duffus and Dearden, 1993; Lien, 2001; Tilt, 1986). Andalusia, located at the southern end of the Iberian Peninsula, along with the Canary Islands, represent the main Spanish regions where this activity is being undertaken. In the Strait of Gibraltar, and especially Tarifa, whale watching began in the mid-1990s and in a few years has become a profitable activity in the tourism sector, with a clear upward trend: from 400 visitors in 1998 to 26,228 in 2007 (Carbó-Penche et al., 2007; Martín and Urquiola, 2001; Urquiola and de Stephanis, 2000).

In 2007, the Spanish Royal Decree for the protection of cetaceans (R.D. 1727/2007) was promulgated to prevent or minimize the impact of whale watching activities for tourist, scientific, recreational, or educational purposes (or any other circumstances in which people might interact with cetaceans). This Decree defines minimum approach distance (60 m) and maximum speed (4 kn) that must be met so as not to injure or disturb the cetaceans. However, in that same year 47% of manoeuvres displayed by commercial whale watching boats around cetaceans did not meet these standards (Salazar-Sierra et al., 2008).

Several studies have linked changes in the short-term behaviour of killer whales with boats manoeuvres (Foote et al., 2004; Kruse, 1991; Williams et al., 2002, 2006), due to the presence and activity of vessels, the sounds produced by vessels, or a combination of these factors. These reactions have been associated with a decrease in killer whale energy efficiency, whereby individuals need to spend more time foraging to capture the same amount of food than in the absence of boats (Kriete, 1994; Williams et al., 2002, 2006). However, individuals may react in different ways to boats. In some cases no response to the vessels was observed. Evasive tactics often differed between encounters with vessels (and sex of the whale), with the number of vessels and vessel proximity, activity, size, and the noise all affecting the killer whale responses (Williams et al., 2002, 2006).

In the Strait of Gibraltar, whale watching companies organize special trips to observe killer whales during July and August, when whales spend most of the time interacting with the drop-line fishery. The increased presence of vessels around the whales may affect foraging efficiency, making it harder for whales to obtain food in a noisier environment. This difficulty may have been exacerbated from 2005 to 2011, when tuna stocks were at their lowest and tuna were likely less available. In the past few years, whale watching companies in the Strait of Gibraltar have started to consider extending their activity to watching killer whales in spring, when whales are mainly actively hunting. In this area, at the north-western most part of the Strait of Gibraltar, vessel presence is low because the shallow coastal waters are outside of the main maritime traffic lanes. However, because whales are actively hunting, they may be more affected by vessel noise, which could mask detection of the tuna by the whales using passive listening and thus could cause a reduction in feeding efficiency.

Every year in spring, military exercises occur in the main distribution area of killer whales in the Strait. For example, in 2015 the North Atlantic Treaty Organization (NATO) coordinated the *Trident Juncture* exercise in that area, with 36,000 personnel of 30 different countries.

Two seasonal management areas for the Strait of Gibraltar were proposed by the Spanish Ministry of Environment to regulate possible disturbance activities such as commercial and recreational whale watching and military exercises in the main habitat of killer whales in spring and summer (Esteban et al., 2016a; Fig. 6).

8.3 Contamination

Organochlorine contaminants such as polychlorinated biphenyls (PCBs) pose a constant health threat to animals and humans. These components

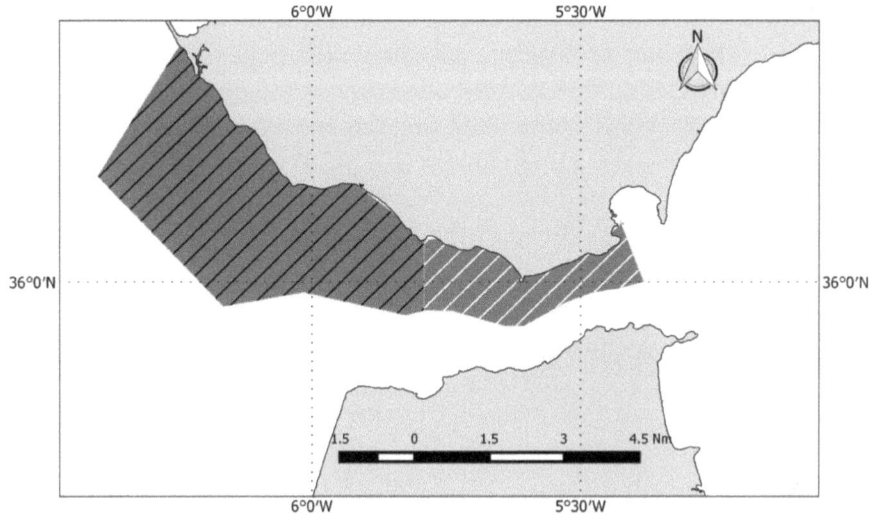

Fig. 6 Seasonal management areas proposed in the Spanish Ministry of Environment conservation plan for the killer whales (*Orcinus orca*) occurring in the Strait of Gibraltar and the Gulf of Cadiz. *Black cross lines* delimit the proposed spring management area, and *white cross lines* delimit the summer management area.

are lipophilic and persistent thus accumulate in the blubber of marine mammals. They were banned in developed countries in the 1970s and 1980s but can be still found in the blubber of animals, including marine mammals such as killer whales (e.g. Jepson et al., 2016). In cetaceans, concentrations of PCBs increase with age in males, but are reduced in females during lactation (Ross et al., 2000). Haraguchi et al. (2006) found higher concentrations of PCBs and polybrominated diphenyl ethers (PBDEs) in calves of lactating females, indicating that large amounts of halogenated organic compounds are transferred from female to offspring via lactation. In their recent meta-analysis of European data on summed 18–25 chlorobiphenyl congeners ($\sum$PCB) concentrations (mg/kg lipid weight (lw)) in odontocete cetaceans that were stranded or biopsied, Jepson et al. (2016) included tissue samples from 24 killer whales. Mean $\sum$PCB lipid concentrations in killer whales from the Northeast Atlantic were among the highest recorded in cetaceans globally. Killer whales of the Strait of Gibraltar ($n=7$ biopsied; $n=1$ stranded—the female, 'Vega') had some of the highest $\sum$PCB concentrations among cetaceans sampled (mean concentrations of 243.43 mg/kg lipid for males, and 186.74 mg/kg lipid for females of all ages sampled), markedly exceeding all known PCB toxicity thresholds for marine mammals (Jepson et al., 2016). Adult female killer whales in the Strait of Gibraltar had mean

$\sum$PCB concentrations of 215.43 mg/kg lipid (Jepson et al., 2016). Tuna sampled in the Strait of Gibraltar also presented high levels of pollutants (Sprague et al., 2012). The transfer of organic compounds from female to calf can negatively affect calf survival, and this may have been a factor when bluefin tuna stocks were at their lowest (2005–2011), as food deprivation in the lactating females could have promoted the metabolism of lipid stores, releasing sequestered pollutants into circulation (Aguilar et al., 1999; O'Shea, 1999).

9. CONSERVATION STATUS

Killer whales in the Strait of Gibraltar were proposed as 'Critically Endangered' by the Agreement on the Conservation of Cetaceans in the Black Sea Mediterranean Sea and Contiguous Atlantic Area (ACCOBAMS) to the IUCN (Cañadas and de Stephanis, 2006), and the International Whaling Commission (IWC) recommended the implementation of a conservation plan. In 2011, the Spanish Ministry of Environment catalogued them as vulnerable (R.D. 139/2011), and a conservation plan was drafted (and is currently under revision).

Ambiguity regarding the degree of genetic isolation between Strait of Gibraltar and Canary Islands killer whales was identified as a key question in confirming their status. However, Esteban et al. (2016c) demonstrated that whales from the Strait of Gibraltar belong to a separate population that presented social, genetic, and ecological differences in comparison with the Canary Islands individuals. The IUCN defines a subpopulation as 'geographically or otherwise distinct groups in the population between which there is little demographic or genetic exchange (typically one successful migrant individual or gamete per year or less)' (IUCN, 2012). Based on this new information, it is strongly recommended that killer whales of the Strait of Gibraltar should be considered as a subpopulation for conservation purposes.

Moreover, the population of killer whales of the Strait of Gibraltar would likely be categorized as Endangered under IUCN Red List of Endangered Species criteria D (small population size), if formally assessed, because its population size is fewer than 250 mature individuals. Regional level assessments should still consider the possibility of exchange with surrounding populations. However, there are differences in behaviour, ecology, and genetics between the Strait of Gibraltar population and the Canary Islands individuals (Esteban et al. 2016c). In the Spanish Catalogue of Endangered

Species, this population of killer whales should also be considered as endangered under criteria C1 (small population size), as actual population size of the killer whale population in the Strait was 39 individuals in 2011 (Esteban et al., 2016a). This is less than 16% of the abundance of other killer whale populations considered stable or increasing in size, such as in Norway, where 731 individuals were identified (Kuningas et al., 2013), or the northern residents of the Pacific Ocean with 260 individuals (Ellis et al., 2011). In addition, based upon factors reviewed here, the Strait of Gibraltar whales should also be considered as endangered in the Spanish Catalogue of Endangered Species under criteria D (expert judgment). These factors including the major decline of killer whale primary prey, which is still overexploited, low rate of recruitment, and dependence upon catches of the artisanal dropline fishery, all endanger the Strait of Gibraltar killer whale population with the risk of extinction (Esteban et al., 2016a).

ACKNOWLEDGEMENTS

We would like to especially thank CIRCE volunteers and research assistants that helped in the field work of CIRCE and EBD-CSIC projects. This work was funded by Loro Parque Foundation, CEPSA, Ministerio de Medio Ambiente, Fundación Biodiversidad, LIFE+ Indemares (LIFE07NAT/E/000732) and LIFE 'Conservación de Cetáceos y tortugas de Murcia y Andalucía' (LIFE02NAT/E/8610), and 'Plan Nacional I+D+I ECOCET' (CGL2011-25543) of the Spanish 'Ministerio de Economía y Competitividad'. Thanks to IFAW for providing the free software Logger 2000. Last, but not least, we would like to thank John K.B. Ford and an anonymous reviewer, and the Editors, Giuseppe Notarbartolo di Sciara, Michela Podestà, and Barbara E. Curry for greatly improving the quality of this manuscript.

REFERENCES

Aguilar, A., 1985. Biología y dinámica poblacional del rorcual común ("*Balaenoptera physalus*") en las aguas atlánticas ibéricas. Ph.D. dissertation, Universitat de Barcelona.

Aguilar, A., 2006. Catches of fin Whales Around the Iberian Peninsula: Statistics and Sources. International Whaling Commission, Cambridge.

Aguilar, A., Borrell, A., Pastor, T., 1999. Biological factors affecting variability of persistent pollutant levels in cetaceans. J. Cetacean Res. Manag. 1, 83–116.

Aranda, G., Abascal, F.J., Varela, J.L., Medina, A., 2013. Spawning behaviour and post-spawning migration patterns of Atlantic bluefin tuna (*Thunnus thynnus*) ascertained from satellite archival tags. PLoS One 8, e76445. http://dx.doi.org/10.1371/journal.pone.0076445.

Baez, J., Camiñas, J., Torreblanca, D., 2007. Analysis of spatial distribution for the marine bird and mammal in the Gulf of Cadiz (South-west of the Iberian Peninsula) during spring period. Bol. R. Soc. Esp. Hist. Nat. Sec. Biol. 102, 93–97.

Baird, R.W., Whitehead, H., 2000. Social organization of mammal-eating killer whales: group stability and dispersal patterns. Can. J. Zool. 78, 2096–2105. http://dx.doi.org/10.1139/cjz-78-12-2096.

Barrett-Lennard, L., Heise, K., 2006. The natural history and ecology of killer whales. In: Estes, J.A., DeMaster, D.P., Doak, D.F., Williams, T.M., Brownell, R.L. (Eds.), Whales, Whaling, and Ocean Ecosystems. University of California Press, Oakland, CA, pp. 163–173.
Beck, S., Kuningas, S., Esteban, R., Foote, A.D., 2012. The influence of ecology on sociality in the killer whale (*Orcinus orca*). Behav. Ecol. 23, 246–253. http://dx.doi.org/10.1093/beheco/arr151.
Beck, S., Foote, A.D., Kötter, S., Harries, O., Mandleberg, L., Stevick, P.T., Whooley, P., Durban, J.W., 2013. Using opportunistic photo-identifications to detect a population decline of killer whales (*Orcinus orca*) in British and Irish waters. J. Mar. Biol. Assoc. U.K. 94, 1327–1333. http://dx.doi.org/10.1017/S0025315413001124.
Belenguer, R., Kersting, D., 2011. Cetáceos en la Reserva Marina de las Islas Columbretes (Mediterráneo noroccidental): 20 años de avistamientos oportunistas. Marten Kwinkelenberg.
Bigg, M.A., Olesiuk, P.F., Ellis, G.M., Ford, J.K.B., Balcomb, K.C., 1990. Social organization and genealogy of resident killer whales (*Orcinus orca*) in the coastal waters of British Columbia and Washington State. Rep. Int. Whal. Comm. 12, 383–405.
Borrell, A., Vila, À., Forcada, J., 2000. Varamiento de cetáceos en las costas españolas del Mediterráneo durante el período 1989–1992. Miscel lània Zool. 23, 53–67.
Boyd, R., Richerson, P., 1985. Culture and the Evolutionary Process. University of Chicago Press, Chicago.
Brasseur, S., Reijnders, P., Verriopoulos, G., 1997. Mediterranean monk seal, *Monachus monachus*. In: Reijnders, P., Verriopoulos, G., Brasseur, S. (Eds.), Status of Pinnipeds Relevant to the European Union. Institute for Forestry and Nature Research, Wageningen, The Netherlands, pp. 1–195.
Brownie, C., Hines, J., Nichols, J., 1993. Capture-recapture studies for multiple strata including non-Markovian transitions. Biometrics 49, 1173–1187.
Brú, F., 1913. Notas de Caza. De José Guix, Valencia.
Cabrera, A., 1914. Fauna Ibérica; mamíferos. Museo Nacional de Ciencias Naturales, Madrid.
Cañadas, A., de Stephanis, R., 2006. Killer whale, or orca *Orcinus orca* (strait of Gibraltar subpopulation). In: Reeves, R.R., Notarbartolo di Sciara, G. (Eds.), The Status and Distribution of Cetaceans in the Black Sea and Mediterranean Sea. IUCN, Centre for Mediterranean Cooperation, Malaga, Spain, pp. 34–38.
Carbó-Penche, M., Salazar-Sierra, J.M., De Stephanis, R., Esteban, R., 2007. Socio-economic analysis of the whale-watching industry in Andalucia, Spain. In: Proceedings of the 21st Annual Meeting of the European Cetacean Society, San Sebastian, April 22–25, p. 140.
Casinos, A., Filella, S., 1975. Primer recull anual (1973) de la comissió de cetologia de la institució catalana d'Historia Natural. Bull. Inst. Catalana Hist. Nat. 39, 5–26.
Casinos, A., Vericad, J.R., 1976. The cetaceans of the Spanish coasts: a survey. Mammalia 40, 267–290.
Clapham, P., Aguilar, A., Hatch, L., 2008. Determining spatial and temporal scales for management: lessons from whaling. Mar. Mamm. Sci. 24, 183–201.
Cornide, J., 1788. Ensayo de una historia de los peces otras producciones marinas de la costa de galicia, arreglado al sistema del caballero Carlos Linneo.
Cuvier, F., 1836. De l'histoire naturelle des cétaces. Imprimerie de Félix Malteste et Cie, Paris.
De Huerta, G., 1624. Historia natural de Cayo Plinio Segundo. Traducida por el licenciado Geronimo de Huerta. Santo Oficio de la Inquisición, Médico y Familiar del.
de la Serna, J., Alot, E., Majuelos, E., Rioja, P., 2004. La migración trófica post reproductiva del atún rojo (*Thunnus thynnus*) a través del estrecho de Gibraltar. Collect. Vol. Sci. Pap. ICCAT 56, 1196–1209.

de Stephanis, R., Cornulier, T., Verborgh, P., Salazar-Sierra, J., Pérez-Gimeno, N., Guinet, C., Salazar Sierra, J., Pérez Gimeno, N., Guinet, C., 2008. Summer spatial distribution of cetaceans in the Strait of Gibraltar in relation to the oceanographic context. Mar. Ecol. Prog. Ser. 353, 275–288. http://dx.doi.org/10.3354/meps07164.

de Stephanis, R., Giménez, J., Esteban, R., Gauffier, P., García-Tiscar, S., Sinding, M.-H., Verborgh, P., 2015. Mobbing-like behavior by pilot whales towards killer whales: a response to resource competition or perceived predation risk? Acta Ethol. 18, 69–78. http://dx.doi.org/10.1007/s10211-014-0189-1.

Di Natale, A., Mangano, A., 1983. Killer whale, *Orcinus orca* (Linnaeus) and false killer whale *Pseudorca crassidens* owen, in the Italian seas. Rapp. Comm. Int. Explor. Sci. Mer. Médit. 28, 219–222.

Duffus, D., Dearden, P., 1993. Recreational use, valuation, and management, of killer whales (*Orcinus orca*) on Canada's Pacific coast. Environ. Conserv. 20, 149–156.

Duguy, R., Cyrus, J., 1973. Note préliminaire a l'étude des cétacés des côtes françaises de Méditerranée. Rev. Trav. Inst. Pêches Marit. 37, 151–158.

Duguy, R., Robineau, D., 1973. Cétacés et phoques des côtes de France. Société des sciences naturelles de la Charente-Maritime, La Rochelle, France.

Duguy, R., Casinos, A., Di Natale, A., 1983. Répartition et frequence des mammifères marins en Méditerranée. Rapp. Comm. Int. Mer. Médit. 28, 223–230.

Ellis, G., Towers, J., Ford, J., 2011. Northern Resident Killer Whales of British Columbia: Photo-Identification Catalogue and Population Status to 2010. Fisheries and Oceans Canada, Science Branch, Pacific Region.

Esteban, R., 2015. Ecology of killer whales (*Orcinus orca*) at southern Iberian Peninsula. Ph.D. dissertation, University of Cadiz, Spain.

Esteban, R., Verborgh, P., Gauffier, P., Giménez, J., Afán, I., Cañadas, A., García, P., Murcia, J., Magalhães, S., Andreu, E., de Stephanis, R., 2013. Identifying key habitat and seasonal patterns of a critically endangered population of killer whales. J. Mar. Biol. Assoc. U.K. 94, 1317–1325. http://dx.doi.org/10.1017/S002531541300091X.

Esteban, R., Verborgh, P., Gauf, P., Giménez, J., Guinet, C., de Stephanis, R., 2016a. Dynamics of killer whale, bluefin tuna and human fisheries in the Strait of Gibraltar. Biol. Conserv. 194, 31–38. http://dx.doi.org/10.1016/j.biocon.2015.11.031.

Esteban, R., Verborgh, P., Gauffier, P., Giménez, J., Foote, A., de Stephanis, R., 2016b. Maternal kinship and fisheries interaction influences killer whale social structure. Behav. Ecol. Sociobiol. 70, 111–122. http://dx.doi.org/10.1007/s00265-015-2029-3.

Esteban, R., Verborgh, P., Gauffier, P., Giménez, J., Martín, V., Pérez-Gil, M., Tejedor, M., Almunia, J., Jepson, P., García-Tíscar, S., Barrettt-Lennard, L., Guinet, C., Foote, A., de Stephanis, R., 2016c. Using a multi-disciplinary approach to identify a critically endangered killer whale management unit. Ecol. Indic. 66, 291–300. http://dx.doi.org/10.1016/j.ecolind.2016.01.043.

Foote, A.D., Osborne, R.W., Hoelzel, A.R., 2004. Environment: whale-call response to masking boat noise. Nature 428, 910. http://dx.doi.org/10.1038/428910a.

Foote, A.D., Newton, J., Piertney, S.B., Willerslev, E., Gilbert, M.T.P., 2009. Ecological, morphological and genetic divergence of sympatric North Atlantic killer whale populations. Mol. Ecol. 18, 5207–5217.

Foote, A.D., Similä, T., Víkingsson, G.a., Stevick, P.T., 2010. Movement, site fidelity and connectivity in a top marine predator, the killer whale. Evol. Ecol. 24, 803–814. http://dx.doi.org/10.1007/s10682-009-9337-x.

Foote, A.D., Vilstrup, J.T., De Stephanis, R., Verborgh, P., Abel Nielsen, S.C., Deaville, R., Kleivane, L., Martín, V., Miller, P.J.O., Oien, N., Pérez-Gil, M., Rasmussen, M., Reid, R.J., Robertson, K.M., Rogan, E., Similä, T., Tejedor, M.L., Vester, H., Víkingsson, G.A., Willerslev, E., Gilbert, M.T.P., Piertney, S.B., Nielsen, S.C.A., Øien, N., Pasmussen, M., 2011. Genetic differentiation among North Atlantic killer

whale populations. Mol. Ecol. 20, 629–641. http://dx.doi.org/10.1111/j.1365-294X.2010.04957.x.
Ford, J.K.J., Ellis, G.G.M., Barrett-Lennard, L.G., Morton, A.B., Palm, R.S., Balcomb III, K.C., 1998. Dietary specialization in two sympatric populations of killer whales (*Orcinus orca*) in coastal British Columbia and adjacent waters. Can. J. Zool. 76, 1456–1471. http://dx.doi.org/10.1139/z98-089.
Ford, J.K.B., Ellis, G.M., Olesiuk, P.F., Balcomb, K.C., 2010. Linking killer whale survival and prey abundance: food limitation in the oceans' apex predator? Biol. Lett. 6, 139–142. http://dx.doi.org/10.1098/rsbl.2009.0468.
Forney, K.A., Wade, P., 2007. Worldwide distribution and abundance of killer whales. In: -Estes, J.A., DeMaster, D.P., Doak, D.F., Williams, T.M., Brownell Jr., R.L. (Eds.), Whales, Whaling and Ocean Ecosystems. University of California Press, Berkeley, California, pp. 145–162.
Foster, E.A., Franks, D.W., Morrell, L.J., Balcomb, K.C., Parsons, K.M., van Ginneken, A., Croft, D.P., van Ginneken, A., 2012. Social network correlates of food availability in an endangered population of killer whales, *Orcinus orca*. Anim. Behav. 83, 731–736. http://dx.doi.org/10.1016/j.anbehav.2011.12.021.
Fromentin, J.J.-M., Powers, J.J.E., 2005. Atlantic bluefin tuna: population dynamics, ecology, fisheries and management. Fish Fish. 6, 281–306. http://dx.doi.org/10.1111/j.1467-2979.2005.00197.x.
Fromentin, J., Ravier, C., 2005. The East Atlantic and Mediterranean bluefin tuna stock: looking for sustainability in a context of large uncertainties and strong political pressures. Bull. Mar. Sci. 76, 353–361.
Gagern, A., van den Bergh, J., Sumaila, U.R., 2013. Trade-based estimation of bluefin tuna catches in the Eastern Atlantic and Mediterranean, 2005–2011. PLoS One 8, e69959. http://dx.doi.org/10.1371/journal.pone.0069959.
Galef, B., 1992. The question of animal culture. Hum. Nat. 3, 151–178.
García-Tiscar, S., 2009. Interacciones entre delfines mulares y orcas con pesqeuerías en el Mar de Alborán y Estrecho de Gibraltar. Ph.D. dissertation, Universidad Autónoma de Madrid.
Graells, M., 1889. Las ballenas en las costas oceánicas de españa. Noticias recogidas e investifaciones hechas por el Dr. M.P. Graelss. Imprenta de Don Luis Aguado, Madrid, Spain.
Graus, E., Filella, S., Raga, J.A., Raduán, A., 1986. Cetáceos varados en las costas del Mediterráneo Ibérico, durante los años 1980–1981. Miscel lània Zool. 10, 353–358.
Guinet, C., Domenici, P., de Stephanis, R., Barrett-Lennard, L., Ford, J.K.B., Verborgh, P., 2007. Killer whale predation on bluefin tuna: exploring the hypothesis of the endurance-exhaustion technique. Mar. Ecol. Prog. Ser. 347, 111–119. http://dx.doi.org/10.3354/meps07035.
Guinet, C., Tixier, P., Gasco, N., 2014. Long-term studies of Crozet Island killer whales are fundamental to understanding the economic and demographic consequences of their depredation behaviour on the Patagonian toothfish fishery. ICES J. Mar. Sci. 72, 1587–1597.
Gutiérrez-Expósito, C., Rivilla, J.C., Alís, S., Máñez, M., Garrido, H., Jiménez, F.J., Cobo, M.D., 2012. Veinticinco años (1986–2011) de monitorización de varamientos de mamíferos marinos en el litoral de Doñana (Huelva, SO España). Galemys: Span. J. Mamm. 24, 1–5. http://dx.doi.org/10.7325/Galemys.2012.N09.
Hammond, P.S., Lockyer, C., 1988. Distribution of killer whales in the eastern North Atlantic. Rit. Fiskid. 11, 24–41.
Haraguchi, K., Hisamichi, Y., Endo, T., 2006. Bioaccumulation of naturally occurring mixed halogenated dimethylbipyrroles in whale and dolphin products on the Japanese market. Arch. Environ. Contam. Toxicol. 51, 135–141. http://dx.doi.org/10.1007/s00244-005-1140-2.

Häussermann, V., Acevedo, J., 2013. Killer whales in Chilean Patagonia: additional sightings, behavioural observations, and individual identifications. Rev. Biol. Mar. Oceanogr. 48, 73–85. http://dx.doi.org/10.4067/S0718-19572013000100007.

Hestbeck, J., Nichols, J., Malecki, R., 1991. Estimates of movement and site fidelity using mark-resight data of wintering Canada geese. Ecology 72, 523–533.

Heyning, J.E., Dahlheim, M.E., 1988. Orcinus orca. Mamm. Species 304, 1–9.

Hoelzel, A.R., Hey, J., Dahlheim, M.E.M., Nicholson, C., Burkanov, V., Black, N., 2007. Evolution of population structure in a highly social top predator, the killer whale. Mol. Biol. Evol. 24, 1407–1415. http://dx.doi.org/10.1093/molbev/msm063.

Horozco, A., 1598. Historia de la ciudad de Cádiz. El Excmo. Ayuntamiento de esta M.N. M. L. y M. H. Ciudad.

Hoyt, E., 2001. Whale Watching 2001: Worldwide Tourism Numbers, Expenditures, and Expanding Socioeconomic Benefits. International Fund for Animal Welfare, Yarmouth Port, MA.

Hoyt, E., 2002. Whale watching. In: Perrin, W.F., Wursig, B., 'Hans' Thewissen, J.G.M. (Eds.), Encyclopedia of Marine Mammals. Academic Press, San Diego, California, pp. 1305–1310.

ICCAT, 2014. Report of the 2014 Atlantic Bluefin Tuna Stock Assessment Session. ICCAT, Madrid, Spain.

Israëls, L., 1992. Thirty Years of Mediterranean Monk Seal Protection, a Review. Instituut voor Taxonomische Zoölogie/Zoölogische Museum, Universiteit van Amsterdam, Amsterdam, The Netherlands.

IUCN, 2012. Red List Categories and Criteria: Version 3.1, second ed. IUCN, Gland, Switzerland and Cambridge, UK.

Ivkovich, T., Filatova, O.a., Burdin, A.M., Sato, H., Hoyt, E., 2010. The social organization of resident-type killer whales (*Orcinus orca*) in Avacha Gulf, Northwest Pacific, as revealed through association patterns and acoustic similarity. Mamm. Biol.—Z. Säugetierkd. 75, 198–210. http://dx.doi.org/10.1016/j.mambio.2009.03.006.

Jepson, P., Deaville, R., Barber, J., Aguilar, A., Borrell, A., Murphy, S., Barry, J., Brownlow, A., Barnett, J., Berrow, S., Cunningham, A., Davison, N., ten Doeschate, M., Esteban, R., Ferreira, M., Foote, A., Genov, T., Giménez, J., Loveridge, J., Llavona, A., Martin, V., Maxwell, D., Papachlimitzou, A., Penrose, R., Perkins, M., Smith, B., de Stephanis, R., Tregenza, N., Verborgh, P., Fernandez, A., Law, R., 2016. PCB pollution continues to impact populations of orcas and other dolphins in European waters. Sci. Rep. 6, 18573. http://dx.doi.org/10.1038/srep18573.

King, J., 1956. The monk seals (genus Monachus). Bull. Br. Mus. Zool. Lond. 3, 203–256.

Kriete, B., 1994. Bioenergetics in the Killer Whale, *Orcinus orca*. Ph.D. dissertation, University of British Columbia, British Columbia, Vancouver, Canada.

Kruse, S., 1991. The interactions between killer whales and boats in Johnstone Strait, BC. In: Pryor, K., Norris, K.S. (Eds.), Dolphin Societies: Discoveries and Puzzles. University of California Press, Berkeley, California, pp. 148–159.

Kuningas, S., Similä, T., Hammond, P.S., 2013. Population size, survival and reproductive rates of northern Norwegian killer whales (*Orcinus orca*) in 1986–2003. J. Mar. Biol. Assoc. U.K. 94, 1277–1291. http://dx.doi.org/10.1017/S0025315413000933.

Lawson, J.W., Stevens, T.S., 2013. Historic and current distribution patterns, and minimum abundance of killer whales (*Orcinus orca*) in the north-west Atlantic. J. Mar. Biol. Assoc. U.K. 94, 1253–1265. http://dx.doi.org/10.1017/S0025315413001409.

Leatherwood, J.S., Dalheim, M.E., 1978. Worldwide distribution of pilot whales and killer whales. Nav. Ocean Syst. Cent. Tech. Rep. 443, 1–39.

Lien, J., 2001. The conservation basis for the regulation of whale watching in Canada by the Department of Fisheries and Oceans: a precautionary approach.

Lusseau, D., Williams, R., Wilson, B., Grellier, K., Barton, T., Hammond, P., Thompson, P., 2004. Parallel influence of climate on the behaviour of Pacific killer whales and Atlantic bottlenose dolphins. Ecol. Lett. 7, 1068–1076. http://dx.doi.org/10.1111/j.1461-0248.2004.00669.x.

Machado y Nuñez, A., 1869. Catálogo metódico y razonado de los mamíferos de andalucia, clasificados segun el sistema del Dr. Enrique Schinz. Imprenta de Gironés y Orduña, Sevilla, Spain.

Martín, V., Urquiola, E., 2001. La observación de cetáceos. In: Naturaleza de Las Islas Canarias. Ecología Y Conservación. Turquesa, Spain, Santa Cruz de Tenerife, pp. 289–295.

Matkin, C., Barrett-Lennard, L., 2007. Ecotypic variation and predatory behavior among killer whales (*Orcinus orca*) off the eastern Aleutian Islands, Alaska. Fish. Bull. 105, 74–87.

Matkin, C., Durban, J., 2011. Killer whales in Alaskan waters. J. Am. Cetacean Soc. 40, 24–29.

Matkin, C., Saulitis, E., Ellis, G., Olesiuk, P., Rice, S.D.S., 2008. Ongoing population-level impacts on killer whales *Orcinus orca* following the 'Exxon Valdez' oil spill in Prince William Sound, Alaska. Mar. Ecol. Prog. Ser. 356, 269–281. http://dx.doi.org/10.3354/meps07273.

Matkin, C.O., Durban, J.W., Saulitis, E.L., Andrews, R.D., Straley, J.M., Matkin, D.R., Ellis, G.M., 2012. Contrasting abundance and residency patterns of two sympatric populations of transient killer whales (*Orcinus orca*) in the northern Gulf of Alaska. Fish. Bull. 110, 143–155.

Matkin, C.O., Testa, J.W., Ellis, G.M., Saulitis, E.L., 2014. Life history and population dynamics of southern Alaska resident killer whales (*Orcinus orca*). Mar. Mamm. Sci. 30, 460–479.

Morin, P.A., Archer, F.I., Foote, A.D., Vilstrup, J., Allen, E.E., Wade, P., Durban, J., Parsons, K., Pitman, R., Li, L., Bouffard, P., Abel Nielsen, S.C., Rasmussen, M., Willerslev, E., Gilbert, M.T.P., Harkins, T., 2010. Complete mitochondrial genome phylogeographic analysis of killer whales (*Orcinus orca*) indicates multiple species. Genome Res. 20, 908–916. http://dx.doi.org/10.1101/gr.102954.109.

Notarbartolo di Sciara, G., 1987. Killer whale, *Orcinus orca*, in the Mediterranean Sea. Mar. Mamm. Sci. 3, 356–360.

O'Shea, T., 1999. Environmental contaminants and marine mammals. Sci. Total Environ. 154, 141–151.

Olesiuk, P.F., Bigg, M.A., Ellis, G., 1990. Life history and population dynamics of northern resident killer whales (*Orcinus orca*) in British Columbia. Rep. Int. Whal. Comm. (Special Issue 12), 209–244.

Olesiuk, P.F., Ellis, G.M., Ford, J.K.B., 2005. Life History and Population Dynamics of Northern Resident Killer Whales (*Orcinus orca*) in British Columbia. Fisheries and Oceans Canada, Nanaimo, BC.

Parsons, K.M.M., Balcomb, K.C.C.I., Ford, J.K.B., Durban, J.W., 2009. The social dynamics of southern resident killer whales and conservation implications for this endangered population. Anim. Behav. 77, 963–971. http://dx.doi.org/10.1016/j.anbehav.2009.01.018.

Pikitch, E., Santora, C., Babcock, E., 2004. Ecosystem-based fishery management. Science 305, 346–347.

Poncelet, E., Barbraud, C., Guinet, C., 2010. Population dynamics of killer whales (*Orcinus orca*) in Crozet Archipelago, southern Indian Ocean: a mark-recapture study from 1977 to 2002. J. Cetacean Res. Manag. 11, 41–48.

Raga, J.A., Raduán, A., Blanco, C., 1985. Contribución al estudio de la distribución de cetàceos en el Mediterráneo y Atlantico Ibérico. Miscel lània Zool. 9, 361–366.

Reeves, R.R., Mitchell, E., 1988. Killer whale sightings and takes by American pelagic whalers in the North Atlantic. Rit. Fiskid. 11, 7–23.

Reisinger, R., de Bruyn, P., Bester, M., 2011. Abundance estimates of killer whales at subantarctic Marion Island. Aquat. Biol. 12, 177–185. http://dx.doi.org/10.3354/ab00340.
Ronald, K., Healey, P., 1978. The monk seal (*Monachus monachus*). In: FAO Advisory and Committee on Marine Resources Research (Ed.), Mammals in the Seas. Food and Agriculture Organization of the United Nations, Rome. ISBN: 92-5-100514-1, p. 534.
Ross, P.S., Ellis, G.M., Ikonomou, M.G., Barrett-Lennard, L.G., Addison, R.F., 2000. High PCB concentrations in free-ranging Pacific killer whales, *Orcinus orca*: effects of age, sex and dietary preference. Mar. Pollut. Bull. 40, 504–515. http://dx.doi.org/10.1016/S0025-326X(99)00233-7.
Sal'nikov, M., 1959. Novi dani pro tyulenya-monakha v Chernomu Mori. Nauch. Zap. Odes. Biol., St. AN URSR.
Salazar-Sierra, J.M., Torres Barranco, F.J., Cabaleiro Mora, E., 2008. Estudio de los aspectos socioeconómicos y de conservación de la actividad de avistamiento de cetáceos. Fundación Migres, Algeciras, Spain.
Sanpera, C., Aguilar, A., 1992. Modern whaling off the Iberian Peninsula during the 20th century. Rep. Int. Whal. Comm. 42, 723–730.
Sella, M., 1929. Biologia e pesca del tonno (*Thunnus thynnus* L.). Atti Conv. Biol. Mar. Appl. Pesca, Messina, Italy, June 1928, pp. 1–32.
Sigurjónsson, J., Leatherwood, S., 1988. The Icelandic live-capture fishery for killer whales, 1976–1988. Rit. Fiskid. 11.
Sprague, M., Dick, J., Medina, A., Tocher, D., 2012. Lipid and fatty acid composition, and persistent organic pollutant levels in tissues of migrating Atlantic bluefin tuna (*Thunnus thynnus*, L.) broodstock. Environ. Pollut. 171, 61–71.
Srour, A., 1994. Développement de la nouvelle pêcherie artisanale au thon rouge dans la région de Ksar sghir. Note d'information ISPM 26, 10–11.
Taylor, N.G.N., McAllister, M.M.K., Lawson, G.G.L., Carruthers, T., Block, B.A., 2011. Atlantic bluefin tuna: a novel multistock spatial model for assessing population biomass. PLoS One 6, e27693. http://dx.doi.org/10.1371/journal.pone.0027693.
Tilt, W., 1986. Whale Watching in California: Survey of Knowledge and Attitudes. Yale School of Forestry and Environmental Studies, Yale University, New Haven, Connecticut.
Urquiola, R., de Stephanis, R., 2000. Growth of whale watching in Spain. The success of the platforms in south mainland, new rules. In: Evans, P., Pitt-Aiken, R., Rogan, R. (Eds.), 14th Annual Conference of the European Cetacean Society, Cork, Ireland, pp. 198–204.
Valverde, J., 2006. Memorias de un Biólogo heterodoxo. Pesca, ballenas, barcas. Reflexiones y cuaderno de dibujos. Quercus, Madrid, Spain.
Van Beneden, M., 1881. Mémoire sur les orques observés dans les Mers D'Europe. L'Académie, Bruselles.
Van Beneden, M., 1888. Histoire naturelle des delphinides des mers d'Europe. In: Mémoires Couronnés et Autres Mémoires. l'Académie royale de Belgique, Bruselles, p. 256.
Visser, I., 2000. Orca (*Orcinus orca*) in New Zealand waters. Ph.D. dissertation, University of Auckland, Auckland, New Zealand.
Ward, E., 2009. Quantifying the effects of prey abundance on killer whale reproduction. J. Appl. Ecol. 46, 632–640.
Williams, R., Lusseau, D., 2006. A killer whale social network is vulnerable to targeted removals. Biol. Lett. 2, 497–500. http://dx.doi.org/10.1098/rsbl.2006.0510.
Williams, R., Bain, D., Ford, J.K.B., Trites, A.W., 2002. Behavioural responses of male killer whales to a leapfrogging vessel. J. Cetacean Res. Manag. 4, 305–310.
Williams, R., Lusseau, D., Hammond, P.S., 2006. Estimating relative energetic costs of human disturbance to killer whales (*Orcinus orca*). Biol. Conserv. 133, 301–311. http://dx.doi.org/10.1016/j.biocon.2006.06.010.

Williams, R., Krkošek, M., Ashe, E., Branch, T.A., Clark, S., Hammond, P.S., Hoyt, E., Noren, D.P., Rosen, D., Winship, A., 2011. Competing conservation objectives for predators and prey: estimating killer whale prey requirements for Chinook salmon. PLoS One 6, e26738. http://dx.doi.org/10.1371/journal.pone.0026738.

Wilson, S., Block, B., 2009. Habitat use in Atlantic bluefin tuna *Thunnus thynnus* inferred from diving behavior. Endanger. Species Res. 10, 355–367. http://dx.doi.org/10.3354/esr00240.

Young, B.G., Higdon, J.W., Ferguson, S.H., 2011. Killer whale (*Orcinus orca*) photo-identification in the eastern Canadian Arctic. Polar Res. 30, 1–7. http://dx.doi.org/10.3402/polar.v30i0.7203.

CHAPTER SIX

Conservation Status of Long-Finned Pilot Whales, *Globicephala melas*, in the Mediterranean Sea

P. Verborgh*,[1], P. Gauffier*, R. Esteban*, J. Giménez*, A. Cañadas†, J.M. Salazar-Sierra*, R. de Stephanis*

*CIRCE (Conservation, Information and Research on Cetaceans), Pelayo-Algeciras, Cádiz, Spain
†ALNILAM Research and Conservation, Navacerrada, Madrid, Spain
[1]Corresponding author: e-mail address: philippe@circe.info

Contents

Advances in Marine Biology, Volume 75
ISSN 0065-2881
http://dx.doi.org/10.1016/bs.amb.2016.07.004

Abstract

Mediterranean Sea long-finned pilot whales (*Globicephala melas*) are currently classified as Data Deficient on the International Union for the Conservation of Nature (IUCN) Red List. Multiple lines of evidence, including molecular genetic and photo-identification mark-recapture analyses, indicate that the Strait of Gibraltar population (distributed from 5.8°W longitude to west of Djibouti Bank and Alborán Dorsal in the Alborán Sea) is differentiated from the Mediterranean Sea population (east of Djibouti Bank and the Alborán Dorsal up to the Ligurian Sea). There is low genetic diversity within the Mediterranean population, and recent gene flow with the Strait of Gibraltar population is restricted. Current total abundance estimates are lacking for the species in the Mediterranean. Pilot whales in the Alborán Sea region were negatively affected by a morbillivirus epizootic from 2006 to 2007, and recovery may be difficult. The Strait of Gibraltar population, currently estimated to be fewer than 250 individuals, decreased by 26.2% over 5 years after the morbillivirus epizootic. Population viability analyses predicted an 85% probability of extinction for this population over the next 100 years. Increasing maritime traffic, increased contaminant burdens, and occasional fisheries interactions may severely impair the capacity of the Strait of Gibraltar population to recover after the decline due to the pathogen.

1. INTRODUCTION

Long-finned pilot whales, *Globicephala melas* (Traill, 1809), are medium-sized delphinid cetaceans that have an antitropical distribution (see Davies, 1963), and are found in cold and temperate waters (Olson and Reilly, 2002). Two subspecies are recognized as *G. m. edwardii* (Smith, 1834) living in the southern oceans, and *G. m. melas* (Traill, 1809) in the North Atlantic. Current gene flow between the two subspecies appears to be highly restricted (Oremus et al., 2009). The global population of long-finned pilot whales is currently classified as Data Deficient on the International Union for the Conservation of Nature (IUCN) Red List (Taylor et al., 2008). The Mediterranean Sea subpopulation is also designated as Data Deficient by the IUCN (Cañadas, 2012). Although data needed to sufficiently estimate worldwide population trends are lacking, about 778,000 (CV: 0.295) individuals have been estimated to occur in the north-eastern Atlantic (Buckland et al., 1993), approximately 31,000 (CV: 0.27) were estimated in the western North Atlantic (Waring et al., 2006), and 200,000 individuals (CV: 0.350) were estimated to occur in the southern oceans, south of the Antarctic Convergence, in the summer months (Kasamatsu and Joyce, 1995). Molecular genetic and stable isotope analyses indicate population structure in the North Atlantic (Fullard et al., 2000; Monteiro et al., 2015).

Risk of extinction for a species or specific population should be motivation for nations to prioritize efforts required for conservation and preservation of the ecosystem on which they depend (Hooker et al., 2002; Santos and Pierce, 2015). Cetaceans are top predators in the marine food web and have been used as ocean and biodiversity health indicators (e.g. Azzellino et al., 2014). For example, they are considered to be a good pollution indicator in the marine environment (e.g. Fossi et al., 2013), since they are long-lived animals and tend to accumulate contaminants throughout their life (e.g. Barón et al., 2015). Additionally, cetacean demographic parameters can be indicators of environmental change, as they are often both directly and indirectly affected by such change. Thus, the conservation status of cetacean populations can be used as an indicator of environmental status as required by the European Union Marine Strategy Framework Directive (European Union, 2008).

Here, we review and synthesize current scientific knowledge regarding long-finned pilot whales (hereafter, pilot whales) in the Mediterranean Sea, in an effort to evaluate their conservation status. We provide information on the distribution, abundance, population ecology, social structure and general ecology of pilot whales. Importantly, we present information that indicates there are two distinct pilot whale populations in the Mediterranean region, a Mediterranean population and a Strait of Gibraltar population. Next, we present an overview of the conservation threats impacting pilot whales in the region. We conclude with a summary of the conservation status of these delphinids in the Mediterranean Sea.

2. DISTRIBUTION

In the Mediterranean Sea, pilot whales are found almost exclusively in the western Basin (Abend and Smith, 1999; Boisseau et al., 2010; Notarbartolo di Sciara and Birkun, 2010; Notarbartolo di Sciara et al., 1993). Over the last century, they have not been observed during surveys in deep waters east of Italy (Boisseau et al., 2010; Frantzis et al., 2003; Kerem et al., 2012), although the region provides suitable habitat (deep water areas, known prey species) (Bello, 1991). There are confirmed records of the species around Malta (Notarbartolo di Sciara and Mifsud, 2002). In Croatia, one male pilot whale was reportedly caught in a tuna trap net on the island of Rab in August 1920 (Hirtz, 1922). The occurrence of this species around Greece is still uncertain (Frantzis et al., 2003).

Interestingly, a group of three short-finned pilot whales (*Globicephala macrorhynchus*) was sighted in the Adriatic Sea in 2010. Species identification was confirmed by photographs based on criteria developed by Rone and Pace (2012) (M. Affronte, Fondazione Cetacea, personal communication, 25 April 2016).

In recent decades, long-finned pilot whales have apparently been very rare or absent from the Tyrrhenian Sea (Arcangeli et al., 2013; Azzellino et al., 2014; Campana et al., 2015; Mussi et al., 2000; Notarbartolo di Sciara et al., 1993; Santoro et al., 2015) (Fig. 1). Density of pilot whales is highest in the Strait of Gibraltar, Alborán Sea and Gulf of Vera, where they are present throughout the year, generally in waters deeper than 500 m (Cañadas et al., 2005; de Stephanis et al., 2008a) (Fig. 1). Density is apparently low in other regions of the western Mediterranean Sea, with few sightings recorded over multiyear surveys around the central Mediterranean Sea (Gómez de Segura et al., 2006), the Balearic Sea (Raga and Pantoja, 2004), and in the Provençal Basin, and Ligurian Sea (Azzellino et al., 2012; Panigada et al., 2011; Praca and Gannier, 2008).

In the north-western Mediterranean Basin, pilot whales were generally found over waters more than 2000 m deep (deeper than waters occupied by pilot whales in the Alborán Sea) and were rarely associated with the continental shelf (Azzellino et al., 2008; Laran et al., 2012). Pilot whales were

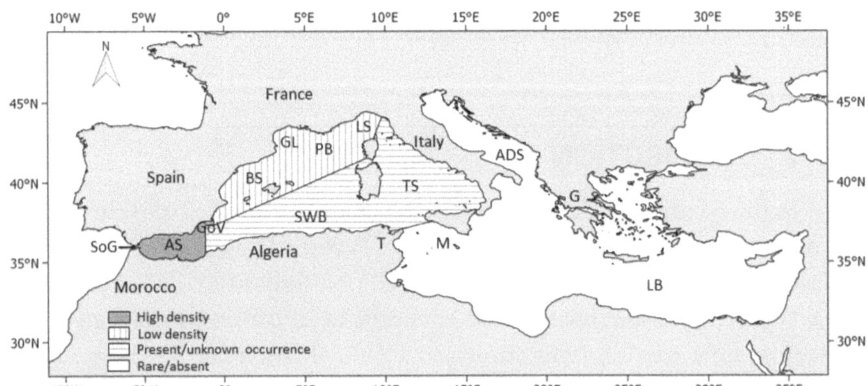

Fig. 1 Approximate distribution and density of long-finned pilot whales (*Globicephala melas*) in the Mediterranean Sea. *AS*, Alborán Sea; *ADS*, Adriatic Sea; *BS*, Balearic Sea; *G*, Greece; *GL*, Gulf of Lion; *GoV*, Gulf of Vera; *LB*, Levantine basin; *LS*, Ligurian Sea; *M*, Malta; *PB*, Provençal Basin; *SoG*, Strait of Gibraltar; *SWB*, south-western Mediterranean Basin; *T*, Tunisia; *TS*, Tyrrhenian Sea. *Note*: GL, PB and LS together form the north-western Mediterranean Basin).

initially thought to be present only in summer in the Ligurian Sea (Laran, 2005), which suggested temporary migration outside of this area during other seasons (Laran et al., 2010). However, visual surveys and acoustic investigation conducted during fall and winter of 2011 confirmed a year-round presence of this species in the north-western Mediterranean Basin (Giorli et al., 2016; Pettex et al., 2014).

Although most survey efforts have been concentrated in the northern Mediterranean Sea, pilot whales have also been described from a few sightings at sea and strandings in Morocco (Bayed, 1996; Masski and de Stephanis, 2015), Algeria (Bouslah, 2012; Boutiba, 1994) and Northern Tunisia (Attia El Hili et al., 2010; Karaa et al., 2012). These records indicate that pilot whales are present in the entire Western Mediterranean Sea, although their density is unknown in southern areas where further research should be carried out (Fig. 1).

3. POPULATION ECOLOGY

3.1 Genetic Structure

Verborgh (2015; and see Verborgh et al., 2010) used molecular analysis (11 microsatellites and 800 base pairs of the mitochondrial DNA control region) to investigate potential genetic differences among pilot whales sampled from the north-east Atlantic ($n=78$), the Strait of Gibraltar ($n=90$) and the Mediterranean Sea ($n=80$). Significant F_{ST} values, and recent migration rates under 10% (Hastings, 1993), were observed. The putative Strait of Gibraltar and Mediterranean populations appeared to be genetically differentiated (and potentially isolated) from individuals sampled in the north-east Atlantic and had lower within population genetic diversity (Verborgh, 2015; Verborgh et al., 2010) (Fig. 2). The study also indicated that pilot whales from the Alborán Sea, Gulf of Vera, Provençal Basin and Ligurian Sea form a single population (Verborgh, 2015; Verborgh et al., 2010) (Fig. 2). However, evidence of isolation by distance was found between pilot whales sampled in the Alborán and Ligurian seas, suggesting further structuring within the Mediterranean Sea (Verborgh, 2015). Small sample sizes from the Alborán and Ligurian seas necessitate further research to sufficiently address the potential differentiation between these two regions.

Seven mitochondrial DNA haplotypes were described in the North Atlantic and the Mediterranean Sea, and one was unique to the Mediterranean Sea and Strait of Gibraltar (found in 62.5% and 31.3% of sampled individuals, respectively), further suggesting isolation from Atlantic

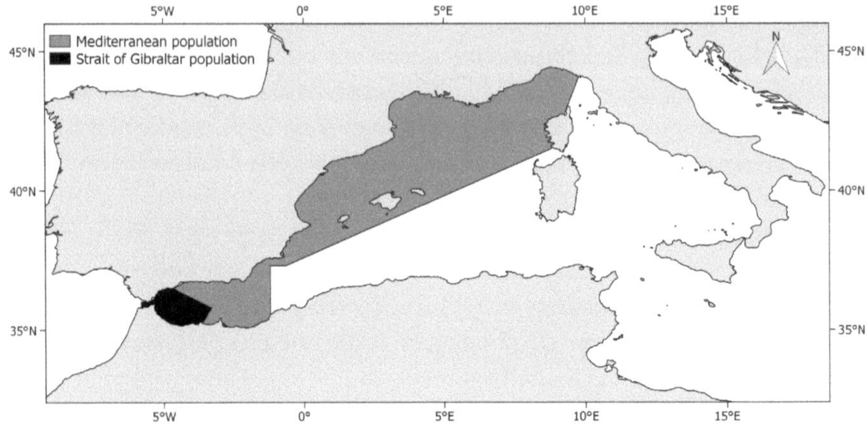

Fig. 2 Potential distribution of Strait of Gibraltar and Mediterranean populations of long-finned pilot whales (*Globicephala melas*) based on genetic structure and individual movements (see also Fig. 3). *From Verborgh, P., 2015. Demografía y estructura de las poblaciones de calderones comunes* (Globicephala melas) *en el Mediterráneo español. PhD Thesis, Universidad de Las Palmas de Gran Canaria, Las Palmas de Gran Canaria, Spain* (in Spanish).

individuals (Verborgh, 2015; and see Monteiro, 2013; Monteiro et al., 2015; Oremus et al., 2009; Verborgh et al., 2010).

Thus, there are apparently two populations of pilot whales in the Mediterranean region. One, the Mediterranean pilot whale population, is ranging from the east of Djibouti Bank and the Alborán Dorsal up to the Ligurian Sea (Verborgh, 2015). This population has a low genetic diversity compared to the Atlantic population (Verborgh, 2015; Verborgh et al., 2010) and similar to isolated populations of other cetacean species (Natoli et al., 2005) (see Figs 2 and 3).

The second pilot whale population identified in the Mediterranean region is the Strait of Gibraltar pilot whale population (defined geographically from 5.8°W longitude to the west of Djibouti Bank and the Alborán Dorsal in the Alborán Sea), which is genetically distinct from the Mediterranean and Atlantic populations (Verborgh, 2015). The Strait of Gibraltar population apparently had higher levels of inbreeding than did the Mediterranean and North Atlantic populations (likely due to small population size) (Verborgh, 2015; Verborgh et al., 2010), which may potentially make it more vulnerable to disease (Valsecchi et al., 2004).

3.2 Individual Movements

Movements of pilot whales have been studied via satellite tagging and photo-identification. Four satellite tags were deployed on pilot whales belonging to

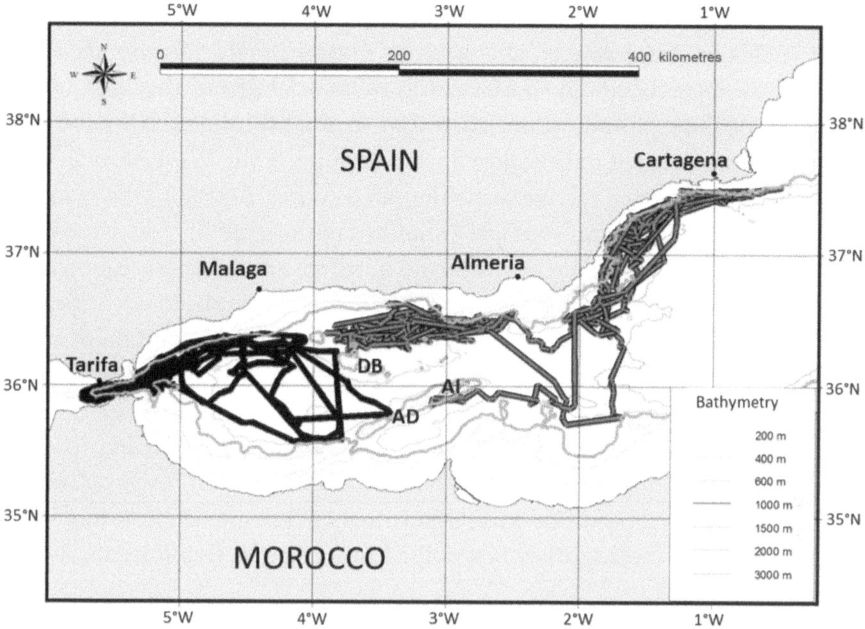

Fig. 3 Tracking from satellite tags deployed on four long-finned pilot whales (*Globicephala melas*) in the Strait of Gibraltar, south of Tarifa (in *black*), two individuals in the eastern Alborán Sea south of Almeria and two in the Gulf of Vera south of Cartagena (in *grey*). *AD*, Alborán Dorsal; *AI*, Alborán Island; *DB*, Djibouti Bank. *From Verborgh, P., 2015. Demografía y estructura de las poblaciones de calderones comunes* (Globicephala melas) *en el Mediterráneo español. PhD Thesis, Universidad de Las Palmas de Gran Canaria, Las Palmas de Gran Canaria, Spain* (in Spanish).

different social units in the Strait of Gibraltar, two in the eastern Alborán Sea and two in the Gulf of Vera (Fig. 3) (Verborgh, 2015). Tags transmitted positions for an average of 36 days (range: 22–49 days), which probably represented movements of pilot whale social units (Verborgh, 2015). Between 1992 and 2014, a total of 1569 individuals were photo-identified in the eastern Alborán Sea and Gulf of Vera (based on analysis of 14,912 high-quality dorsal fin images) (Morata et al., 2013; Verborgh, 2015; Wierucka et al., 2014). In the Strait of Gibraltar, 374 individuals were identified (from analyses of 62,964 high-quality dorsal fin images) between 1999 and 2011 (Verborgh, 2015; Verborgh et al., 2009). No individual that was photo-identified or tagged in the Strait of Gibraltar has ever been observed in the eastern Alborán Sea and Gulf of Vera, nor have any individuals from those areas been observed in the Strait (Morata et al., 2013). This may be further confirmation of the isolation suggested by the genetic analyses between pilot whales inhabiting in these regions (see Figs 2 and 3). Most pilot whales from the Strait of

Gibraltar are highly resident with few movements observed outside of the Strait itself (de Stephanis et al., 2015; Verborgh, 2015; Verborgh et al., 2009). One tagged individual belonging to a social group that was usually observed for short periods of time (i.e. days to weeks) inside the Strait before leaving the area seemed to be following the currents of the West Alborán Gyre to the deeper waters of the western Alborán Sea (Fig. 3). Pilot whales tagged and photo-identified from south of Almeria and the Gulf of Vera have made movements over larger areas between the northern and southern Alborán Sea (Verborgh, 2015; Verborgh et al., 2012) (Fig. 3). Photo-identification detected social group movements between the Gulf of Vera and the Alborán Sea for 24% of the identified individuals (Morata et al., 2013; Verborgh, 2015; Wierucka et al., 2014).

In the Ligurian Sea, some degree of site fidelity was revealed through photo-identification, with 40% of 92 animals having been resighted in different years over a 21-year period (Greco, 2011). The number of resightings per individual in the Ligurian Sea ranged from one to five (Greco, 2011).

3.3 Demographic Parameters

3.3.1 Abundance

Estimates of pilot whale total abundance are not available for the Mediterranean Sea. During the 1970s, pilot whales were apparently frequently encountered in the north-western Mediterranean Basin. In the Ligurian Sea, a group of 200 individuals was observed in 1974, and 50 opportunistic sightings were made mainly from ferries in 1975 (Duguy and Vallon, 1977; Vallon et al., 1977). Although no measure of effort is available for comparison, these observations contrast with the 32 sightings of a total 522 individuals between 1981 and 1988 (Podestà and Magnaghi, 1988), and the 47 sightings reported from dedicated vessel surveys (in addition to 87 opportunistic sightings) between 1994 and 2008, in the same area (Laran et al., 2012). This apparent decline is corroborated by a decrease in stranding rates along the Mediterranean French coast since the early 1990s (Van Canneyt et al., 2014). Since the 1960s, pilot whales have generally been less frequently observed in Italian and French waters than along the Spanish coast and especially the Strait of Gibraltar (Aloncle, 1964; Casinos and Vericad, 1976; Duguy and Cyrus, 1973; Notarbartolo di Sciara et al., 1993).

There is no current population estimate for the entire Mediterranean Sea; however, some areas have been studied and local abundance estimates are available. In the Strait of Gibraltar, where a resident population lives year-round, a mean abundance of 213 individuals (95% CI: 142–352) was estimated between 1999 and 2005 (Verborgh et al., 2009), and 374 individuals were identified through photo-identification between 1999 and 2011 (Verborgh, 2015). In the Alborán Sea and the Gulf of Vera, 1569 individuals were identified between 1992 and 2009 (Morata et al., 2013; Verborgh, 2015). All catalogues are available at https://www.cetidmed.com. Additionally, 2888 individuals (95% CI: 2565–3270) were estimated to be using the Strait of Gibraltar, northern Alborán Sea and Gulf of Vera between 1992 and 2009 (A. Cañadas, unpublished data) using spatial modelling as described in Cañadas and Hammond (2006).

In the north-western Mediterranean Basin, an abundance of 391 (95% CI: 270–565) pilot whales in summer and 369 pilot whales (95% CI: 128–1108) in winter was estimated based on aerial surveys conducted in 2012 (Pettex et al., 2014). However, these estimates should be taken with caution because they are based on very few sightings. A minimum population size of 184 individuals was estimated from photo-identification in the Ligurian Sea between 1991 and 2010 (Greco, 2011), confirming the low abundance of this species in that area.

3.3.2 Survival Rate

Survival rates have been estimated using mark-recapture models based on photo-identification data in the Strait of Gibraltar and the Alborán Sea. In the Strait, pilot whales had a high, constant survival rate around 0.982 (95% CI: 0.955–0.993) from 1999 to 2005 (Verborgh et al., 2009). In the Alborán Sea, annual survival rate estimates ranged from 0.919 (95% CI: 0.854–0.956) to 0.995 (95% CI: 0.952–0.999) among social groups during the time period of 1992–2009 (Wierucka et al., 2014).

Additionally, survival rates were estimated for different age classes in the Strait of Gibraltar from 1999 to 2006 and indicated a value of 0.629 (95% CI: 0.409–0.805) for calves within their first year (Gauffier et al., 2013), which is low compared to 0.862 estimated for calves in the Faroe Islands (Bloch et al., 1993). Juveniles (age range from 1 to 6.5 years) in the Strait had a higher survival rate, with a value of 0.869 (95% CI: 0.758–0.934), than did calves within their first year. As would be expected, adults (age >6.5 years) had a higher value, 0.972 (95% CI: 0.953–0.983), when compared to younger age classes in the Strait (Gauffier et al., 2013).

During October 2006 to April 2007, an unusually higher number of pilot whale strandings occurred along the coast of Spain in the Mediterranean Sea

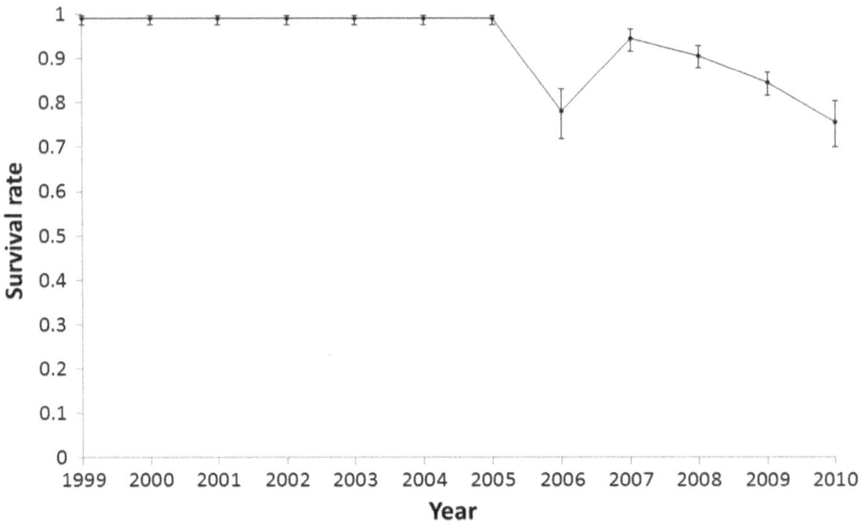

Fig. 4 Annual survival rate and 95% confidence intervals bars of long-finned pilot whales (*Globicephala melas*) in the Strait of Gibraltar, including effects of the 2006–07 morbillivirus epizootic and the negative trend in survival rate observed during the years following the event. *From Verborgh, P., 2015. Demografía y estructura de las poblaciones de calderones comunes (*Globicephala melas*) en el Mediterráneo español. PhD Thesis, Universidad de Las Palmas de Gran Canaria, Las Palmas de Gran Canaria, Spain* (in Spanish).

due to a morbillivirus epizootic event (Fernández et al., 2008). In the Alborán Sea, of 11 social groups studied, only 3 (27.3%), exhibited a decrease in survival rate, from 0.919 (95% CI: 0.854–0.956) in 1992–2005 to 0.547 (95% CI: 0.185–0.866) in 2006–08 (Wierucka et al., 2014).

In the Strait of Gibraltar, the adult survival rate decreased to 0.779 (95% CI: 0.717–0.830) for 2006–07 and then followed a 4-year negative trend, declining to 0.754 (95% CI: 0.698–0.803) for 2010–11 (Verborgh, 2015) (Fig. 4). The difference in survival rates between pilot whales studied in the Alborán Sea and those in the Strait of Gibraltar could be due to environmental stress (contaminant levels) and human impacts (whale watching, maritime traffic), which are both higher in the Strait than in the Alborán Sea (Verborgh, 2015).

3.3.3 Other Parameters

Birth interval for viable calves (i.e. calves that survive their first year of life) was estimated to be 4.5 years (range: 2–7 years) in the Strait of Gibraltar (Verborgh, 2015). Although newborn individuals are observed all year round in the Strait, there is a birth peak in spring (Aloncle, 1964; Verborgh, 2015). Breeding aggregations have been observed in summer in the Alborán Sea (Cañadas and Sagarminaga, 2000). In the Ligurian Sea, calves were encountered more frequently in August and September (Greco, 2011).

Mature individuals represented 77% of the pilot whales encountered in the Strait of Gibraltar from 1999 to 2011 (Verborgh, 2015), which is high compared to the 51% suggested by Taylor et al. (2007) for a pilot whale population experiencing a stable population growth rate.

4. SOCIAL STRUCTURE

Groups of pilot whales in the Mediterranean Sea are generally composed of around 10–50 individuals (Cañadas and Sagarminaga, 2000; Cañadas et al., 2005; de Stephanis et al., 2008a,c; Greco, 2011; Laran et al., 2012; Raga and Pantoja, 2004), but groups of several hundred individuals have been observed to aggregate (possibly for breeding) in the Alborán Sea (Cañadas and Sagarminaga, 2000; Cañadas et al., 2005; de Stephanis, 2008). A hierarchical social system was suggested by de Stephanis et al. (2008c) wherein the population is formed of "clans", each one composed of various pods. These pods are the equivalent of the "grinds" described in the Faroe Islands (Amos et al., 1993; Fullard, 2000), and of the multiple matrilines that were observed together in 12 mass strandings that occurred in New Zealand and Australia between 1992 and 2006 (bringing the existence of "extended matrilines" as was proposed by Amos et al., 1993, into question) (Oremus et al., 2013). Paternal males are not expected to be found within pods, because reproduction is thought to occur between individuals belonging to unrelated pods (Amos et al., 1993). Each pod is thus formed by stable, small matrilineal units (de Stephanis et al., 2008c; Mussi et al., 2000; Ottensmeyer and Whitehead, 2003), as suggested by the fact that individuals sampled within units in the Strait of Gibraltar shared the same haplotype (Verborgh, 2015).

5. ECOLOGY

5.1 Diving Behaviour

Pilot whales radio tagged in the Ligurian Sea were recorded diving to 824 m deep searching for prey in areas of greater than 2000 m depth (Airoldi et al., 2003; Baird et al., 2002). Deep foraging dives occurred mainly at night when the deep scattering layer moves closer to the surface and therefore makes prey more available (Airoldi et al., 2003; Baird et al., 2002; Giorli et al., 2016). Foraging dives lasted an average 14 min (maximum 15 min 50 s) (Airoldi et al., 2003). Pilot whales adjusted their foraging activity with night length, foraging longer during the longer winter nights (Giorli et al., 2016). In the Strait of Gibraltar pilot whales occur in waters 600–800 m deep (de Stephanis et al., 2008a), which means they can easily exploit the sea floor and feed at any time during the day or night. The same is probably true for pilot whales in the Alborán Sea although they have been encountered in waters of up to 2000 m deep in some areas (Cañadas and Sagarminaga, 2000; Cañadas et al., 2005).

5.2 Diet

In general, pilot whale diet is characterized mainly by mesopelagic prey living in oceanic waters such as cephalopods and some pelagic fish, or by neritic prey living in the ocean bottom (Olson and Reilly, 2002; Relini and Garibaldi, 1992; Spitz et al., 2011). In the Mediterranean Sea, they feed mainly on certain species of cephalopods such as angel squid (*Ancistroteuthis lichtensteini*), Chiroteuthiids, bobtail squid (*Heteroteuthis dispar*), umbrella squid (*Histioteuthis bonnellii*), elongated jewel squid (*H. elongata*), jewel squid (*H. reversa*), Octopoteuthiids, Ommastrephiids, European flying squid (*Todarodes sagittatus*) and fish such as members of the Gadidae and blue whiting (*Micromesistius poutassou*) (Astruc, 2005; Cañadas and Sagarminaga, 2000; Praca et al., 2011). In an examination of the stomach contents of pilot whales ($n = 10$) from French Mediterranean waters, flying squid had an Index of Relative Importance between 40% and 50%, and umbrella and jewel squid ranked between 10% and 20% (Astruc, 2005).

Mediterranean pilot whales feed at a lower trophic level than do other teuthophagous species, such as sperm whales (*Physeter macrocephalus*) and Risso's dolphins (*Grampus griseus*), in the north-western Mediterranean Basin (Praca and Gannier, 2008; Praca et al., 2011). These three species also exploit different ecological niches, with Risso's dolphins occurring on the upper slope, sperm whales on the lower part of the slope and pilot whales over the deeper regions (Azzellino et al., 2008; Praca and Gannier, 2008). Therefore, competition for resources is unlikely to occur amongst these species.

In the Strait of Gibraltar, pilot whales also feed at lower trophic levels than do common bottlenose dolphins (*Tursiops truncatus*), sperm whales and killer whales (*Orcinus orca*), but at higher levels than short-beaked common dolphins (*Delphinus delphis*) (Barón et al., 2015; de Stephanis, 2008; de Stephanis et al., 2015). Pilot whales in the Strait exhibit higher δ^{15}N and δ^{13}C values than those in the north-western Mediterranean Sea, suggesting that the two populations may have different food sources (Praca et al., 2011). As suggested by de Stephanis et al. (2008b), there appears to be some level of habitat segregation (perhaps dictated by specialization in different prey species) among pilot whales groups within the Strait of Gibraltar.

6. CONSERVATION THREATS

Little is known about the threats affecting pilot whales in the Mediterranean Sea. Few individuals strand each year in sufficient condition for proper necropsy; nevertheless, those that can be necropsied very rarely show signs of death caused directly by humans. Cause of death in the Mediterranean region has generally been attributed to either natural causes, mainly perinatal pathologies and morbillivirus infection, or unknown causes (Consejería de Medio Ambiente y Ordenación del Territorio, 2014; Fernández Maldonado, 2015; Fernández et al., 2008; Keck et al., 2010; Van Canneyt et al., 2014).

6.1 Bycatch

A number of fisheries have caused pilot whale mortality, mostly through incidental bycatch. The most important of these was the driftnet fishery (prohibited by the European Union since 2002), which incidentally caught an estimated 111 and 132 pilot whales, in 1990 and 1991, respectively, in the Tyrrhenian Sea (Di Natale, 1995). Driftnets were also responsible for the bycatch of nine pilot whales in the Ligurian Sea between 1983 and 1989 (Podestà and Magnaghi, 1989), and at least one in the Gulf of Lion from 2000 to 2003 (Bănaru et al., 2010). However, these numbers are probably underestimated because many of the catches were not declared, and the carcasses naturally sunk or were sunk on purpose by fishermen to avoid penalties (Notarbartolo di Sciara, 1990). The density of pilot whales was low in the Tyrrhenian Sea during this time (Notarbartolo di Sciara et al., 1993) and the species is no longer found in the Tyrrhenian Sea, so the few reported catches could have had a significant impact on the population. The Mediterranean Moroccan driftnet fishing fleet bycaught primarily common dolphins and striped dolphins (*Stenella coeruleoalba*) in the 1990s, although an unknown number of pilot whales were also incidentally caught (Tudela et al., 2003). Purse seiners also occasionally caught pilot whales as

Fig. 5 Evidence of long-finned pilot whale (*Globicephala melas*) interactions with fisheries in the Strait of Gibraltar. (A) Fishing gear entangled in the caudal peduncle of a pilot whale (3 September 2008). (B) Scars thought to have been caused by entanglement in fishing gear (28 July 2004). (C) Scars attributed to trolling lines (8 August 2010). *Photographs: CIRCE.*

bycatch in the 1980s (Di Natale, 1990). The Spanish long-line fleet, still operating in 2015 in the Western Mediterranean Sea, also bycaught pilot whales on a few occasions (Macías López et al., 2012).

In 2010 a pilot whale entangled in a net stranded near Almeria (Alborán Sea), but the fishery responsible could not be identified (Consejería de Medio Ambiente, 2010). In the Strait of Gibraltar, several free-ranging pilot whales were observed entangled in fishing gear (Fig. 5A). Injuries and scars most likely resulted from entanglement in fishing nets, lines or hooks (Fig. 5B and C). The most probable fisheries involved were likely the Spanish and Moroccan longline fleets, or sport fishing boats observed on many occasions passing, with trolling lines, near pilot whale groups (Vázquez et al., 2014).

6.2 Marine Traffic and Whale Watching

In general, there has been little evidence of collisions between boats and small cetaceans in the Mediterranean Sea. Pesante et al. (2002) examined 87 stranded individuals among various Mediterranean small cetacean species and found that only one pilot whale had died due to vessel collision. Worldwide, there have been only two known collisions with sailboats involving long-finned pilot whales (Ritter, 2012; Sodebo, 2011). This species inhabits pelagic waters and usually occurs far from the coast, in areas where maritime traffic intensity is lower. However, in the Strait of Gibraltar and the Alborán Sea, their distribution overlaps with merchant ships and ferries (de Stephanis et al., 2005). de Stephanis and Urquiola (2006) described a collision between a pilot whale and a ship in the Strait of Gibraltar in 2003, and since that time, six living individuals bearing marks associated with ship strikes have been observed in the area (Fig. 6).

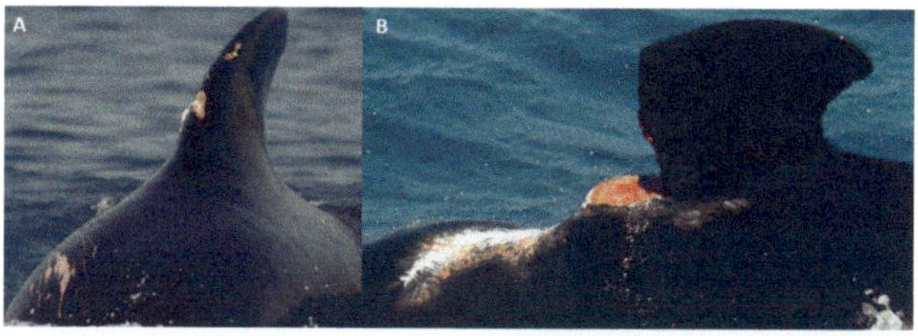

Fig. 6 Examples of long-finned pilot whales (*Globicephala melas*) with wounds likely incurred from collisions with vessels. (A) An individual with superficial wounds observed in 2010. (B) An individual with scars from a cut in front of the dorsal fin in 2008, and a fresh wound cutting the dorsal fin (observed in 2011). *Photographs: CIRCE.*

Collision risks can also come from whale watching operations. Mayol et al. (2014) found that the area where these conflicts were most likely to arise was in French Mediterranean waters near Toulon. However, these vessels must respect national legislation that prohibits the pursuit or harassment of marine mammals (Legifrance, 2011). France has also recently adopted the use of a voluntary "High Quality Whale Watching®" label, created by the Pelagos Sanctuary and the Agreement on the Conservation of Cetaceans in the Black Sea Mediterranean Sea and Contiguous Atlantic Area (ACCOBAMS).

Whale watching companies also operate in Italy and Spain, but pilot whales are only a target species in the Gulf of Vera and the Strait of Gibraltar due to their presence near the shore in those areas (Elejabeitia et al., 2012). Between 1999 and 2005, the number of whale watching boats in the Strait of Gibraltar increased and so did the numbers of individuals in the resident population of pilot whales in the area, suggesting that the activity had little effect on pilot whale survival rate (Verborgh et al., 2009). Andreu et al. (2009) observed that the pilot whale groups approached by whale watching vessels demonstrated evasive behaviour in only 2–5% of 774 encounters between 2003 and 2007. However, Salazar Sierra et al. (2008) reported that 47% of the manoeuvres undertaken by whale watching boats did not comply with the Spanish Royal Decree for the Protection of Cetaceans (Real Decreto 1727/2007), therefore potentially stressing the animals. Unfortunately, several years later, no measures have been undertaken by Spanish authorities to enforce this Royal Decree that applies to both commercial and recreational whale watching activities. Additionally, in 2010, pilot whales increased breathing synchrony after less than 10 min in the presence of a boat less than 100 m away, a possible sign of stress (Senigaglia et al., 2012). In 2011, pilot whales were the target species of from seven to eight

Fig. 7 Long-finned pilot whales (*Globicephala melas*) approaching whale watching boats in the Strait of Gibraltar. *Photographs: CIRCE.*

whale watching boats that operated from Tarifa and attracted 39,000 tourists (Fig. 7) (Elejabeitia et al., 2012). The decrease in population size from 324 individuals (95% CI: 302–359) in 2006 to 239 individuals (95% CI: 236–247) in 2011 (Verborgh, 2015) is likely to have increased the encounter rate of specific groups with whale watching boats. For example, in 2012, two whale watching boats from the same company approached the same two social groups on 70% ($n=237$) of their 339 total trips (Andréu Cazalla and Martín Bernal, 2012).

Disturbances to cetaceans by whale watching operations have been observed in other parts of the world, sometimes causing decreases in abundance or disrupting foraging behaviours (e.g. Bejder et al., 2006; Lusseau et al., 2009; Pirotta et al., 2015). Extra care should be taken, and general rules of good conduct including speed, angle of approach and distance restriction should be implemented and properly enforced in order to decrease any detrimental effects to individuals and populations.

6.3 Contaminants and Other Forms of Pollution

The Mediterranean Sea has high contaminant levels (UNEP/MAP, 2012), and these are likely to have negative effects on cetaceans (e.g. Jepson et al.,

2016). Average levels of organochlorine contaminants dichlorodiphenyl-trichloroethanes (DDTs) and polychlorinated biphenyls (PCBs) were five times higher for pilot whales in the Strait of Gibraltar than in the north-western Mediterranean Basin (Lauriano et al., 2014; Pinzone et al., 2015; Praca et al., 2011), which in turn were 5–10 times higher than for pilot whales from the North Atlantic Ocean (Dam and Bloch, 2000). Levels of contaminants found in pilot whales from the Strait of Gibraltar and the Mediterranean Sea have been well above the threshold considered harmful to the immune and reproductive systems in aquatic mammals (Jepson et al., 2005; Kannan et al., 2000). Furthermore, most of these contaminants can be transmitted to their offspring during lactation (Borrell et al., 1995) and may affect calf survival (Hoydal et al., 2015). High contaminant load may increase the risk of infections by viral and nonviral pathogens and probably increased mortality during the Mediterranean morbillivirus epizootic of 2006–07 (Fernández et al., 2008; Lauriano et al., 2014), as was also suggested during the 1990–92 morbillivirus epizootic in Mediterranean striped dolphins (Aguilar and Borrell, 1994). Levels of several flame retardants, such as polybrominated diphenyl ethers (PBDEs), found in pilot whales of the Strait of Gibraltar were noted to be well below the threshold associated with thyroid endocrine disruption in grey seals (*Halichoerus grypus*) (Barón et al., 2015).

The Mediterranean Sea was recently reported to contain a higher proportion of large plastic debris than that typically found in open oceanic gyres, a reflection of both its closer proximity to pollution sources and the low possibility for such debris to leave the semienclosed basin (Cózar et al., 2015). Few data on plastic debris ingestion by pilot whales are available (Baulch and Perry, 2014). However, other teuthophagous species in the Mediterranean, such as sperm whales (de Stephanis et al., 2013), Cuvier's beaked whales (*Ziphius cavirostris*) (Poncelet et al., 2000) or Risso's dolphin (Shoham-Frider et al., 2002), have been found with large quantities of plastics in their stomachs. Along the French Atlantic coast, plastic ingestion may have been the cause of death of three stranded pilot whales (Poncelet et al., 2000).

6.4 Underwater Noise

The Mediterranean Sea and Strait of Gibraltar host intense maritime traffic, causing high noise levels (Ross, 2005). Underwater noise can affect cetacean behaviour, energetics and physiology (Jepson et al., 2003; Nowacek et al., 2007). Seismic surveys, military sonar and other loud underwater sounds have affected pilot whales in different ways. Loud underwater noise has been documented as causing individuals to stop vocalizing in southern oceans

(Bowles et al., 1994) and to increase vocalizations in the Mediterranean Sea (Rendell and Gordon, 1999). In addition, evasive behaviours were observed during military sonar playback experiments in Norway (Miller et al., 2012). The Gulf of Vera and the Strait of Gibraltar are situated next to Cartagena, a busy port and naval base, and to Gibraltar's submarine harbours from which navy exercises take place regularly. This proximity may put pilot whales in the region at risk as a result of noise and military activities. For example, a recent mass stranding event of pilot whales in Scotland was reported to have been related to acoustic impairment, or a behavioural response, to a series of underwater military explosions (Brownlow et al., 2015).

Maritime traffic has been increasing in the Strait of Gibraltar over recent decades (e.g. the Moroccan port of Tangier-Med opened in 2007, and traffic in the area increased by 21% between 2006 and 2011 (Verborgh, 2015)).[1] This increase could have a direct impact on pilot whales by decreasing their ability to communicate or to find prey due to potential masking effects of ship noise. It could also affect prey species that have shown responses to low-frequency sounds, such as cephalopods (e.g. common cuttlefish, *Sepia officinalis*; European squid, *Loligo vulgaris*; longfin squid, *L. pealeii*; common octopus, *Octopus vulgaris*; broadtail shortfin squid, *Illex coindetii*) (Mooney et al., 2010; Packard et al., 1990). Low-frequency sounds between 1 Hz and 10 kHz, such as those emitted by commercial vessels crossing the Strait of Gibraltar (Ross, 2005), have been shown to alter cephalopod breathing rate and movements, and in some cases induced permanent acoustic trauma (André et al., 2011; Kaifu et al., 2008; Mooney et al., 2010; Packard et al., 1990; Solé et al., 2013).

6.5 Natural Mortality

Pilot whales have no significant predators in the Mediterranean Sea; however, killer whales are known to attack and eat marine mammals, including pilot whales, in other areas (Weller, 2008). In the Strait of Gibraltar, interactions between these species have been observed, but consisted in killer whales being chased away by pilot whales (de Stephanis et al., 2015). As these species do not compete for resources, the possible historical presence of marine mammal-eating killer whales in the area could explain the anti-predator defence behaviour (mobbing behaviour) of pilot whales in response to killer whales (de Stephanis et al., 2015; and see Esteban et al., 2016).

[1] For examples of realtime maritime traffic conditions in and around the Strait of Gibraltar, see http://www.marinevesseltraffic.com/2013/07/marine-traffic-gibraltar-strait-dual.html.

Mass strandings of pilot whales are not rare (Oremus et al., 2013). The reasons for these strandings are currently unknown; however, most of the animals are usually in healthy condition which has led to the suggestion that strong social structure may play a major role (Norris and Schilt, 1988). Mass strandings of pilot whales were described in the Mediterranean Sea in the 19th century: 72 individuals stranded in September 1827 in Calvi, Corsica; 150 in Majorca, Spain, in December 1860, and 17 in La Nouvelle, France, in February 1864 (Bompar, 2000). Since then, no other mass strandings of this species have been recorded in the Mediterranean Sea (Dhermain et al., 2002). Due to the low density of this species in the north-western Basin, a mass stranding could have a major impact on the local population and such events might have reduced substantially the Mediterranean population in the past.

The outbreak of the 2006–07 morbillivirus epizootic that affected the pilot whale populations of the Strait of Gibraltar (Verborgh, 2015) and the Alborán Sea (Wierucka et al., 2014) started in the Strait and then extended to the Alborán Sea and the Balearic Islands (Fernández et al., 2008), ending in France during 2007 (Keck et al., 2010). Thirty stranded individuals were determined to have died due to morbillivirus infection (Fernández et al., 2008; Keck et al., 2010). Other pathogens can affect the health of pilot whales. Steele et al. (2009) used blow samples to investigate the presence of respiratory pathogens in pilot whales in the Strait of Gibraltar and found bacteria in 20 samples, including: *Mycobacteria* (40%); *Streptococcus equi* (35%); *Staphylococcus aureus* (30%); *Streptococcus phocae* (25%); b-haemolytic streptococci (15%); *Streptococci* spp. (15%) and *Brucella* spp. (5%). Evidence of *Haemophilus influenzae*, *Cryptococcus neoformans* and *Mycoplasma* spp. were also found, the two latter causing a potential risk to human health, because these pathogens could be transmitted from cetaceans to humans.

6.6 Climate Change

The Intergovernmental Panel on Climate Change (IPCC) and European Environmental Agency (EEA) have warned of the impact of rising temperatures on many ecological factors including changes in the composition of phytoplankton blooms and changes in the northern boundary of distribution of warm water species (EEA, 2008; IPCC, 2007). Global climate change is affecting the marine ecosystem, including cetaceans (Kaschner et al., 2011; MacLeod, 2009). For the last 300 years, pilot whale distribution in the

North Atlantic has been highly influenced by changes in the subpolar gyre, which have an impact on primary production and the food chain (Hátún et al., 2009). Some pilot whale prey, such as squid species, show sensitivity to temperature and ocean acidification, which could influence their distribution and early development (Brodziak and Hendrickson, 1999; Lacoue-Labarthe et al., 2011, 2016). This could in turn affect pilot whale distribution and could probably even affect population structure (Fullard et al., 2000). The Mediterranean Sea has already been affected by climate change (Lejeusne et al., 2010), and this will likely cause an indirect impact on pilot whales in the future.

7. CONSERVATION STATUS

7.1 Mediterranean Population

There are fewer data available to make an assessment of the conservation status of this population. Nevertheless, pilot whales were apparently common in the 1970s in the north-western Mediterranean Basin (Vallon et al., 1977). Though, during the 1980s and 1990s, hundreds were probably bycaught in driftnets (Di Natale, 1995), which may have caused the species disappearance from the Tyrrhenian Sea. Although there is no total population estimate, it is likely that the number of mature individuals is less than 2500 individuals. Additionally, it has also suffered from the morbillivirus epizootic of 2006–07 (Fernández et al., 2008). For some social groups (27.3%), annual survival rate dropped from 0.919 (95% CI: 0.854–0.956) in 1992–2006 to 0.547 (95% CI: 0.185–0.866) in 2007–09 (Wierucka et al., 2014). This 37.2% decrease in survival rate potentially represents a 10.1% decrease for the entire Mediterranean population. Therefore, we recommend maintaining its current state of "vulnerable" in the Spanish Catalogue of Endangered Species.

7.2 Strait of Gibraltar Population

Population viability analyses estimated that the Strait of Gibraltar population have an 85% probability of extinction in the next 100 years when using the lower bound of 95% confidence intervals survival estimates for three age classes (0.41 for calves less than 1 year old, 0.76 for juveniles less than 6.5 years old and 0.95 for adults), and a calving interval of 4.5 years (Gauffier et al., 2013). In addition, although mean survival estimates predicted a null probability of extinction over the next 100 years, it is important to note that the population was predicted to follow a negative trend (decline) over the entire period. These analyses did not take into account the effects of the 2006–07 morbillivirus epizootic event, or the low demographic parameters estimated for the population during the years following the epizootic (Verborgh, 2015). Therefore, additional population viability analyses are needed to properly assess the risk of extinction using this newly available information.

The Strait of Gibraltar pilot whale population has a small size of less than 250 mature individuals, which is apparently continuing to decline. A decline of 26.2% of the population was observed over a 5-year period, wherein the population decreased from 324 individuals (95% CI: 302–359) in 2006 to 239 individuals (95% CI: 236–247) in 2011 (Verborgh, 2015) (see Section 3).

The decline in habitat quality in the Strait of Gibraltar due to the increased maritime traffic over the main pilot whale distribution area is negatively affecting this population (de Stephanis et al., 2008a; Verborgh, 2015). Moreover, the combined effects of pathogens and pollutants are impacting this population (Fernández et al., 2008; Lauriano et al., 2014). Contaminant levels reported for this population were well above the thresholds set in marine mammals from which the reproduction and immune system can be affected (Lauriano et al., 2014), suggesting that their resilience in response to the epizootic could have been compromised.

In Spain, the Strait of Gibraltar population is currently classified together with the rest of the Spanish Mediterranean pilot whale population as "vulnerable". Here, we suggest that the Strait of Gibraltar pilot whales are a distinct population that is "critically endangered". Urgent measures are required to protect the long-finned pilot whale in the Strait of Gibraltar.

ACKNOWLEDGEMENTS

We would like to thank CIRCE, Alnitak, Alnilam and ANSE volunteers, research assistants and staff that helped in the field work and abled the long-term studies mentioned here. This work was funded by Ministerio de Medio Ambiente, Fundación Biodiversidad, LIFE 'Conservación de Cetáceos y tortugas de Murcia y Andalucía' (LIFE02NAT/E/8610) and Loro Parque Foundation and CEPSA. We would also like to thank an anonymous reviewer and the Editors, Giuseppe Notarbartolo di Sciara, Michela Podestà and Barbara E. Curry for greatly improving the quality of this manuscript.

REFERENCES

Abend, A.G., Smith, T.D., 1999. Review of distribution of the long-finned pilot whale (*Globicephala melas*) in the North Atlantic and Mediterranean. NOAA Tech. Memo. NMFS-NE 117. 22 pp.

Aguilar, A., Borrell, A., 1994. Abnormally high polychlorinated biphenyl levels in striped dolphins (*Stenella coeruleoalba*) affected by the 1990-1992 Mediterranean epizootic. Sci. Total Environ. 154, 237–247. http://dx.doi.org/10.1016/0048-9697(94)90091-4.

Airoldi, S., Fasano, D., Gavazzi, N., Miragliuolo, A., Mussi, B., Mariani, M., 2003. Diving behaviour and reactions to tagging of long-finned pilot whales in the central Mediterranean Sea. In: Evans, P.G.H., Anderwald, P., André, M. (Eds.), Proceedings of the Seventeenth Annual Conference of the European Cetacean Society. European Cetacean Society, Kiel, pp. 347–348.

Aloncle, H., 1964. Premières observations sur les petits cétacés des côtes marocaines. Bull. l'inst. Pech. Marit. Maroc 12, 21–42.

Amos, B., Schlötterer, C., Tautz, D., 1993. Social structure of pilot whales revealed by analytical DNA profiling. Science 260, 670–672. http://dx.doi.org/10.1126/science.8480176.

André, M., Solé, M., Lenoir, M., Durfort, M., Quero, C., Mas, A., Lombarte, A., van der Schaar, M., López-Bejar, M., Morell, M., Zaugg, S., Houégnigan, L., 2011. Low-frequency sounds induce acoustic trauma in cephalopods. Front. Ecol. Environ. 9, 489–493. http://dx.doi.org/10.1890/100124.

Andréu Cazalla, E., Martín Bernal, C., 2012. Influencia del avistamiento de cetáceos sobre los grupos sociales de calderón común en el Estrecho. Biol. Mar. Aljaranda 87, 23–27.

Andreu, E., Gallego, P., Cervera, J.L., 2009. Long-finned pilot whale (*Globicephala melas*) response to whale-watching vessels in the Strait of Gibraltar: Report of the Scientific Committee, International Whaling Commission SC/61/WW5. .

Arcangeli, A., Marini, L., Crosti, R., 2013. Changes in cetacean presence, relative abundance and distribution over 20 years along a trans-regional fixed line transect in the Central Tyrrhenian Sea. Mar. Ecol. 34, 112–121. http://dx.doi.org/10.1111/maec.12006.

Astruc, G., 2005. Exploitation des chaînes trophiques marines de Méditerranée par les populations de cétacés. PhD Thesis, Ecole Pratique des Hautes Etudes, Montpellier, France.

Attia El Hili, H., Cozzi, B., Ben Salah, C., Podestà, M., Ayari, W., Ben Amor, N., Mraouna, R., 2010. A survey of cetaceans stranded along the northern coast of Tunisia: recent findings (2005–2008) and a short review of the literature. J. Coast. Res. 26, 982–985. http://dx.doi.org/10.2112/JCOASTRES-D-09-00010.1.

Azzellino, A., Gaspari, S., Airoldi, S., Nani, B., 2008. Habitat use and preferences of cetaceans along the continental slope and the adjacent pelagic waters in the western Ligurian Sea. Deep Sea Res. Part I Oceanogr. Res. Pap. 55, 296–323. http://dx.doi.org/10.1016/j.dsr.2007.11.006.

Azzellino, A., Panigada, S., Lanfredi, C., Zanardelli, M., Airoldi, S., Notarbartolo di Sciara, G., 2012. Predictive habitat models for managing marine areas: spatial and temporal distribution of marine mammals within the Pelagos Sanctuary (Northwestern Mediterranean Sea). Ocean Coast. Manag. 67, 63–74. http://dx.doi.org/10.1016/j.ocecoaman.2012.05.024.

Azzellino, A., Fossi, M.C., Gaspari, S., Lanfredi, C., Lauriano, G., Marsili, L., Panigada, S., Podestà, M., 2014. An index based on the biodiversity of cetacean species to assess the environmental status of marine ecosystems. Mar. Environ. Res. 100, 94–111. http://dx.doi.org/10.1016/j.marenvres.2014.06.003.

Baird, R.W., Borsani, J.F., Hanson, M.B., Tyack, P.L., 2002. Diving and night-time behaviour of long-finned pilot whales in the Ligurian Sea. Mar. Ecol. Prog. Ser. 237, 301–305.

Bănaru, D., Dekeyser, I., Imbert, G., Laubier, L., 2010. Non-target and released alive by-catch distributions observed during French driftnet fishery in the Northwestern Mediterranean Sea (2000-2003 database). J. Oceanogr. Res. Data 3, 33–45.

Barón, E., Giménez, J., Verborgh, P., Gauffier, P., de Stephanis, R., Eljarrat, E., Barceló, D., 2015. Bioaccumulation and biomagnification of classical flame retardants, related halogenated natural compounds and alternative flame retardants in three delphinids from Southern European waters. Environ. Pollut. 203, 107–115. http://dx.doi.org/10.1016/j.envpol.2015.03.041.

Baulch, S., Perry, C., 2014. Evaluating the impacts of marine debris on cetaceans. Mar. Pollut. Bull. 80, 210–221. http://dx.doi.org/10.1016/j.marpolbul.2013.12.050.

Bayed, A., 1996. First data on the distribution of cetaceans along the Moroccan coasts. In: Evans, P.G.H. (Ed.), Proceedings of the Tenth Annual Conference of the European Cetacean Society. European Cetacean Society, Lisbon, p. 106.

Bejder, L., Samuels, A., Whitehead, H., Gales, N., Mann, J., Connor, R., Heithaus, M., Watson-Capps, J., Flaherty, C., Krützen, M., 2006. Decline in relative abundance of bottlenose dolphins exposed to long-term disturbance. Conserv. Biol. 20, 1791–1798. http://dx.doi.org/10.1111/j.1523-1739.2006.00540.x.

Bello, G., 1991. Role of cephalopods in the diet of the swordfish, *Xiphias gladius*, from the eastern Mediterranean Sea. Bull. Mar. Sci. 49, 312–324.

Bloch, D., Lockyer, C., Zachariassen, M., 1993. Age and growth parameters of the long-finned pilot whale off the Faroe Islands. In: Donovan, G.P., Lockyer, C.H., Martin, A.R. (Eds.), Biology of Northern Hemisphere Pilot Whales. International Whaling Commission, Cambridge, pp. 163–208.

Boisseau, O., Lacey, C., Lewis, T., Moscrop, A., Danbolt, M., Mclanaghan, R., 2010. Encounter rates of cetaceans in the Mediterranean Sea and contiguous Atlantic area. J. Mar. Biol. Assoc. U. K. 90, 1589–1599. http://dx.doi.org/10.1017/S0025315410000342.

Bompar, J.-M., 2000. Les cétacés de Méditerranée, Edisud. ed. La Calade, Aix-en-Provence.

Borrell, A., Bloch, D., Desportes, G., 1995. Age trends and reproductive transfer of organochlorine compounds in long-finned pilot whales from the Faroe Islands. Environ. Pollut. 88, 283–292.

Bouslah, Y., 2012. Bilan actuel des échouages de cétacés sur le littoral occidental algérien. MSc Thesis, Université d'Oran, Oran, Algeria.

Boutiba, Z., 1994. Bilan de nos connaissances sur la présence des cétacés le long des côtes algériennes. Mammalia 58, 613–622.

Bowles, A.E., Smultea, M., Wursig, B., DeMaster, D.P., Palka, D., 1994. Relative abundance and behaviour of marine mammals exposed to transmission from the Heard Island feasibility test. J. Acoust. Soc. Am. 96, 2469–2484.

Brodziak, J., Hendrickson, L., 1999. An analysis of environmental effects on survey catches of squids *Loligo pealei* and *Illex illecebrosus* in the northwest Atlantic. Fish. Bull. 97, 9–24.

Brownlow, A., Baily, J., Dagleish, M., Deaville, R., Jensen, S., Krupp, E., Law, R., Penrose, R., Read, F., Jepson, P., 2015. Investigation into the long-finned pilot whale mass stranding event, Kyle of Durness, 22nd July 2011. Report to Defra and Marine Scotland, Inverness.

Buckland, S.T., Block, D., Cattanach, K.L., Gunnlaugsson, T., Hoydal, K., Lens, S., Sigurjónsson, J., 1993. Distribution and abundance of long-finned pilot whales in the North Atlantic, estimated from NASS-87 and NASS-88 data. In: Donovan, G., Lockyer, C.H., Martin, A.R. (Eds.), Biology of Northern Hemisphere Pilot Whales. International Whaling Commission, Cambridge, pp. 33–50.

Campana, I., Crosti, R., Angeletti, D., Carosso, L., David, L., Di-Méglio, N., Moulins, A., Rosso, M., Tepsich, P., Arcangeli, A., 2015. Cetacean response to summer maritime

traffic in the Western Mediterranean Sea. Mar. Environ. Res. 109, 1–8. http://dx.doi.org/10.1016/j.marenvres.2015.05.009.
Cañadas, A., 2012. *Globicephala melas* (Mediterranean subpopulation) [WWW Document]. IUCN Red List Threat. Species 2012. doi:10.2305/IUCN.UK.2012-1.RLTS.T16376479A16376495.
Cañadas, A., Hammond, P., 2006. Model-based abundance estimates for bottlenose dolphins off southern Spain: implications for conservation and management. J. Cetacean Res. Manag. 8, 13–27.
Cañadas, A., Sagarminaga, R., 2000. The northeastern Alboran Sea, an important breeding and feeding ground for the long-finned pilot whale (*Globicephala melas*) in the Mediterranean Sea. Mar. Mam. Sci. 16, 513–529.
Cañadas, A., Sagarminaga, R., de Stephanis, R., Urquiola, E., Hammond, P.S., 2005. Habitat preference modelling as a conservation tool: proposals for marine protected areas for cetaceans in southern Spanish waters. Aquat. Conservat. Mar. Freshwat. Ecosyst. 15, 495–521. http://dx.doi.org/10.1002/aqc.689.
Casinos, A., Vericad, J.R., 1976. The cetaceans of the Spanish coasts: a survey. Mammalia 40, 267–290.
Consejería de Medio Ambiente, 2010. Apoyo Técnico a la Gestión Sostenible del Medio Marino. Informe regional 2010, Junta de Andalucía, Sevilla.
Consejería de Medio Ambiente y Ordenación del Territorio, 2014. Programa de Gestión Sostenible del Medio Marino Andaluz. Informe final de resultados 2014. Junta de Andalucía, Sevilla.
Cózar, A., Sanz-Martín, M., Martí, E., González-Gordillo, J.I., Ubeda, B., Gálvez, J.Á., Irigoien, X., Duarte, C.M., 2015. Plastic accumulation in the Mediterranean Sea. PLoS One 10e0121762http://dx.doi.org/10.1371/journal.pone.0121762.
Dam, M., Bloch, D., 2000. Screening of mercury and persistent organochlorine pollutants in long-finned pilot whale (*Globicephala melas*) in the Faroe Islands. Mar. Pollut. Bull. 40, 1090–1099. http://dx.doi.org/10.1016/S0025-326X(00)00060-6.
Davies, J.L., 1963. The antitropical factor in cetacean speciation. Evolution 17, 107–116.
de Stephanis, R., 2008. Estrategias de alimentación en función del tiempo de los diferentes grupos de calderón común (*Globicephala melas*) en el Estrecho de Gibraltar en primavera verano. PhD Thesis, Universidad de Cádiz, Cadiz, Spain.
de Stephanis, R., Urquiola, E., 2006. Collisions between ships and cetaceans in Spain: Report of the Scientific Committee, International Whaling Commission SC/58/BC5.
de Stephanis, R., Verborgh, P., Pérez Gimeno, N., Sánchez Cabanes, A., Pérez Jorge, S., Esteban Pavo, R., Seller, N., Urquiola, E., Guinet, C., 2005. Impactos producidos por el tráfico marítimo en las poblaciones de cetáceos en el Estrecho de Gibraltar. Situación actual y previsiones de futuro: Report to Dirección General para la Biodiversidad del Ministerio de Medio Ambiente. .
de Stephanis, R., Cornulier, T., Verborgh, P., Salazar Sierra, J., Gimeno, N., Guinet, C., 2008a. Summer spatial distribution of cetaceans in the Strait of Gibraltar in relation to the oceanographic context. Mar. Ecol. Prog. Ser. 353, 275–288. http://dx.doi.org/10.3354/meps07164.
de Stephanis, R., García-Tíscar, S., Verborgh, P., Esteban-Pavo, R., Pérez, S., Minvielle-Sebastia, L., Guinet, C., 2008b. Diet of the social groups of long-finned pilot whales (*Globicephala melas*) in the Strait of Gibraltar. Mar. Biol. 154, 603–612. http://dx.doi.org/10.1007/s00227-008-0953-8.
de Stephanis, R., Verborgh, P., Pérez, S., Esteban, R., Minvielle-Sebastia, L., Guinet, C., 2008c. Long-term social structure of long-finned pilot whales (*Globicephala melas*) in the Strait of Gibraltar. Acta Ethol. 11, 81–94. http://dx.doi.org/10.1007/s10211-008-0045-2.
de Stephanis, R., Giménez, J., Carpinelli, E., Gutierrez-Exposito, C., Cañadas, A., 2013. As main meal for sperm whales: plastics debris. Mar. Pollut. Bull. 69, 206–214. http://dx.doi.org/10.1016/j.marpolbul.2013.01.033.

de Stephanis, R., Giménez, J., Esteban, R., Gauffier, P., García-Tiscar, S., Sinding, M.-H.S., Verborgh, P., 2015. Mobbing-like behavior by pilot whales towards killer whales: a response to resource competition or perceived predation risk? Acta Ethol. 18, 69–78. http://dx.doi.org/10.1007/s10211-014-0189-1.

Dhermain, F., Soulier, L., Bompar, J.-M., 2002. Natural mortality factors affecting cetaceans in the Mediterranean Sea. In: Notarbartolo di Sciara, G. (Ed.), Cetaceans of the Mediterranean and Black Seas: State of Knowledge and Conservation Strategies, pp. 1–14. A Report to the ACCOBAMS Secretariat. Monaco.

Di Natale, A., 1990. Marine mammals interactions in Scombridae fishery activities: the Mediterranean case. Collect. Vol. Sci. Pap. ICCAT 33, 140–142.

Di Natale, A., 1995. Drifnets impact on protected species: observer data from the Italian fleet and proposal for a model to assess the number of cetacean in the bycatch. Collect. Vol. Sci. Pap. ICCAT 44, 255–263.

Duguy, R., Cyrus, J.L., 1973. Note préliminaire a l'étude des cétacés des côtes françaises de Méditerranée. Rev. Trav. l'Inst. Pêches Marit. 37, 151–158.

Duguy, R., Vallon, D., 1977. Le Rorqual commun (*Balaenoptera physalus*) en Méditerranée occidentale: Etat actuel des observations. Rapp. la Commision Int. Mer Mediterr. 24, 21–24.

EEA, 2008. Impacts of Europe's Changing Climate—2008 Indicator-Based Assessment. European Commission, European Environment Agency, World Health Organization, Europe.

Elejabeitia, C., Urquiola, E., Verborgh, P., de Stephanis, R., 2012. Towards a sustainable whale-watching industry in the Mediterranean Sea. In: Rosalino, L.M., Silva, A., Abreu, A. (Eds.), New Trends Towards Mediterranean Tourism Sustainability. Nova Science Publishers Inc., New York, pp. 1–32.

Esteban, R., Verborgh, P., Gauffier, P., Alarcón, A., Salazar-Sierra, J.M., Giménez, J., Foote, A.D., de Stephanis, R., 2016. Conservation status of killer whales, *Orcinus orca*, in the Strait of Gibraltar. In: Notarbartolo di Sciara, G., Podestà, M., Curry, B.E. (Eds.), Mediterranean Marine Mammals Ecology and Conservation. Advances in Marine Biology, vol. 75. Elsevier, Amsterdam, pp. 141–172.

European Union, 2008. Directive 2008/56/EC of the European Parliament and of the Council of 17 June 2008 establishing a framework for community action in the field of marine environmental policy (Marine Strategy Framework Directive). Off. J. Eur. Un. L. 164, 19–40.

Fernández Maldonado, C., 2015. Patología y causas de la muerte de los cetáceos varados en Andalucía (años: 2011–2014). PhD Thesis, Universidad de Las Palmas de Gran Canaria, Las Palmas de Gran Canaria, Spain.

Fernández, A., Esperon, F., Herraèz, P., Espinosa de los Monteros, A., Clavel, C., Bernabé, A., Sánchez-Vizcaino, J., Verborgh, P., de Stephanis, R., Toledano, F., Bayón, A., 2008. Morbillivirus and pilot whale deaths, Mediterranean Sea. Emerg. Infect. Dis. 14, 792–794.

Fossi, M.C., Casini, S., Maltese, S., Panti, C., Spinsanti, G., Marsili, L., 2013. An "ex vivo" model to evaluate toxicological responses to mixtures of contaminants in cetaceans: integumentum biopsy slices. Environ. Toxicol. 29, 1107–1121. http://dx.doi.org/10.1002/tox.

Frantzis, A., Alexiadou, P., Paximadis, G., Politi, E., Gannier, A., Corsini-Foka, M., 2003. Current knowledge of the cetacean fauna of the Greek Seas. J. Cetacean Res. Manag. 5, 219–232.

Fullard, K.J., 2000. Microsatellite analysis of long-finned pilot whales. PhD Thesis, Cambridge University, Cambridge, UK.

Fullard, K.J., Early, G., Heide-Jorgensen, M.P., Bloch, D., Rosing-Asvid, A., Amos, W., 2000. Population structure of long-finned pilot whales in the North Atlantic: a correlation with sea surface temperature? Mol. Ecol. 9, 949–958.

Gauffier, P., Verborgh, P., Esteban, R., Giménez, J., de Stephanis, R., 2013. Estimating life history parameters of long-finned pilot whale (*Globicephala melas*) through mark-recapture models and population viability analysis. In: Twentieth Biennal Conference on Marine Mammals. Society for Marine Mammalogy, Dunedin, p. 77.

Giorli, G., Au, W.W.L., Neuheimer, A., 2016. Differences in foraging activity of deep sea diving odontocetes in the Ligurian Sea as determined by passive acoustic recorders. Deep. Res. Part I Oceanogr. Res. Pap. 107, 1–8. http://dx.doi.org/10.1016/j.dsr.2015.10.002.

Gómez de Segura, A., Crespo, E.A., Pedraza, S.N., Hammond, P.S., Raga, J.A., 2006. Abundance of small cetaceans in waters of the central Spanish Mediterranean. Mar. Biol. 150, 149–160.

Greco, M., 2011. The occurrence, distribution and social structure of long-finned pilot whales (*Globicephala melas*) in the Ligurian Sea. MSc Thesis, Università di Pisa, Pisa, Italia.

Hastings, A., 1993. Complex interactions between dispersal and dynamics: lessons from coupled logistic equations. Ecology 74, 1362–1372.

Hátún, H., Payne, M.R., Beaugrand, G., Reid, P.C., Sandø, A.B., Drange, H., Hansen, B., Jacobsen, J.A., Bloch, D., 2009. Large bio-geographical shifts in the north-eastern Atlantic Ocean: from the subpolar gyre, via plankton, to blue whiting and pilot whales. Prog. Oceanogr. 80, 149–162. http://dx.doi.org/10.1016/j.pocean.2009.03.001.

Hirtz, M., 1922. Kit debeloglavac, *Globicephalus melas* (Traill) u vodama Hrvatske. Glas. Hrvat. Naravosl. Društva 34, 84–89.

Hooker, S.K., Whitehead, H., Gowans, S., 2002. Ecosystem consideration in conservation planning: energy demand of foraging bottlenose whales (*Hyperoodon ampullatus*) in a marine protected area. Biol. Conserv. 104, 51–58.

Hoydal, K.S., Letcher, R.J., Blair, D.A.D., Dam, M., Lockyer, C., Jenssen, B.M., 2015. Legacy and emerging organic pollutants in liver and plasma of long-finned pilot whales (*Globicephala melas*) from waters surrounding the Faroe Islands. Sci. Total Environ. 520, 270–285. http://dx.doi.org/10.1016/j.scitotenv.2015.03.056.

IPCC, 2007. Climate Change 2007: Impacts, Adaptation and Vulnerability. Contribution of Working Group II to the Fourth Assessment Report of the Intergovernmental Panel on Climate Change. Cambridge University Press, Cambridge.

Jepson, P.D., Arbelo, M., Deaville, R., Patterson, I.A.P., Castro, P., Baker, J.R., Degollada, E., Ross, H.M., Herráez, P., Pocknell, A.M., Rodríguez, F., Howie, F.E., Espinosa, A., Reid, R.J., Jaber, J.R., Martin, V., Cunningham, A.A., Fernández, A., 2003. Gas-bubble lesions in stranded cetaceans. Nature 425, 575–576. http://dx.doi.org/10.1038/425575a.

Jepson, P.D., Bennett, P.M., Deaville, R., Allchin, C.R., Baker, J.R., Law, R.J., 2005. Relationships between polychlorinated biphenyls and health status in harbor porpoises (*Phocoena phocoena*) stranded in the United Kingdom. Environ. Toxicol. Chem. 24, 238–248. http://dx.doi.org/10.1897/03-663.1.

Jepson, P.D., Deaville, R., Barber, J.L., Aguilar, À., Borrell, A., Murphy, S., Barry, J., Brownlow, A., Barnett, J., Berrow, S., Cunningham, A.A., Davison, N.J., ten Doeschate, M., Esteban, R., Ferreira, M., Foote, A.D., Genov, T., Giménez, J., Loveridge, J., Llavona, Á., Martin, V., Maxwell, D.L., Papachlimitzou, A., Penrose, R., Perkins, M.W., Smith, B., de Stephanis, R., Tregenza, N., Verborgh, P., Fernandez, A., Law, R.J., 2016. PCB pollution continues to impact populations of orcas and other dolphins in European waters. Sci. Rep. 6, 18573. http://dx.doi.org/10.1038/srep18573.

Kaifu, K., Akamatsu, T., Segawa, S., 2008. Underwater sound detection by cephalopod statocyst. Fish. Sci. 74, 781–786. http://dx.doi.org/10.1111/j.1444-2906.2008.01589.x.

Kannan, K., Blankenship, A.L., Jones, P.D., Giesy, J.P., 2000. Toxicity reference values for the toxic effects of polychlorinated biphenyls to aquatic mammals. Hum. Ecol. Risk. Assess. 6, 181–201.

Karaa, S., Nejmeddine Bradai, M., Jribi, I., Attia El Hili, H., Bouain, A., 2012. Status of cetaceans in Tunisia through analysis of stranding data from 1937 to 2009. Mammalia 76, 21–29. http://dx.doi.org/10.1515/mamm.2011.100.

Kasamatsu, F., Joyce, G.G., 1995. Current status of Odontocetes in the Antarctic. Antarct. Sci. 7, 365–379.

Kaschner, K., Tittensor, D.P., Ready, J., Gerrodette, T., Worm, B., 2011. Current and future patterns of global marine mammal biodiversity. PLoS One 6, 19653.

Keck, N., Kwiatek, O., Dhermain, F., Dupraz, F., Boulet, H., Danes, C., Laprie, C., Perrin, A., Godenir, J., Micout, L., Libeau, G., 2010. Resurgence of Morbillivirus infection in Mediterranean dolphins off the French coast. Vet. Rec. 166, 654–655. http://dx.doi.org/10.1136/vr.b4837.

Kerem, D., Hadar, N., Goffman, O., Scheinin, A., Kent, R., Boisseau, O., Schattner, U., 2012. Update on the cetacean fauna of the Mediterranean Levantine Basin. Open Mar. Biol. J. 6, 6–27. http://dx.doi.org/10.2174/1874450801206010009.

Lacoue-Labarthe, T., Réveillac, E., Oberhänsli, F., Teyssié, J.L., Jeffree, R., Gattuso, J.P., 2011. Effects of ocean acidification on trace element accumulation in the early-life stages of squid *Loligo vulgaris*. Aquat. Toxicol. 105, 166–176. http://dx.doi.org/10.1016/j.aquatox.2011.05.021.

Lacoue-Labarthe, T., Nunes, P.A.L.D., Ziveri, P., Cinar, M., Gazeau, F., Hall-Spencer, J.M., Hilmi, N., Moschella, P., Safa, A., Sauzade, D., Turley, C., 2016. Impacts of ocean acidification in a warming Mediterranean Sea: an overview. Reg. Stud. Mar. Sci. 5, 1–11. http://dx.doi.org/10.1016/j.rsma.2015.12.005.

Laran, S., 2005. Variations spatio-temporelles du peuplement de cétacés en Mer Ligure (Méditerranée Nord-Occidentale) et relations avec les conditions environnementales. PhD Thesis, Free University of Brussels, Brussels, Belgium.

Laran, S., Joiris, C., Gannier, A., Kenney, R.D., 2010. Seasonal estimates of densities and predation rates of cetaceans in the Ligurian Sea, northwestern Mediterranean Sea: an initial examination. J. Cetacean Res. Manag. 11, 31–40.

Laran, S., Delacourtie, F., Di Fulvio, T., David, L., Di-Meglio, N., Monestiez, P., 2012. Synthèse sur la distribution des cétacés dans le Sanctuaire PELAGOS et les eaux adjacentes, mise en relation avec leur environnement. Sci. Rep. Port-Cros Natl. Park 26, 119–147.

Lauriano, G., Di Guardo, G., Marsili, L., Maltese, S., Fossi, M.C., 2014. Biological threats and environmental pollutants, a lethal mixture for mediterranean cetaceans? J. Mar. Biol. Assoc. U. K. 94, 1221–1225. http://dx.doi.org/10.1017/S0025315413000714.

Legifrance, 2011. Arrêté du 1er juillet 2011 fixant la liste des mammifères marins protégés sur le territoire national et les modalités de leur protection. JORF 0171□: 12708. NOR□: DEVL1110724A [WWW Document]. https://www.legifrance.gouv.fr/affichTexte.do?cidTexte=JORFTEXT000024396902 (accessed 3.23.16).

Lejeusne, C., Chevaldonné, P., Pergent-Martini, C., Boudouresque, C.F., Pérez, T., 2010. Climate change effects on a miniature ocean: the highly diverse, highly impacted Mediterranean Sea. Trends Ecol. Evol. 25, 250–260. http://dx.doi.org/10.1016/j.tree.2009.10.009.

Lusseau, D., Bain, D.E., Williams, R., Smith, J.C., 2009. Vessel traffic disrupts the foraging behavior of southern resident killer whales *Orcinus orca*. Endanger. Species Res. 6, 211–221.

Macías López, D., Barcelona, S.G., Báez, J.C., De la Serna, J.M., Ortiz de Urbina, J.M., 2012. Marine mammal bycatch in Spanish Mediterranean large pelagic longline fisheries, with a focus on Risso's dolphin (*Grampus griseus*). Aquat. Living Resour. 25, 321–331. http://dx.doi.org/10.1051/alr/2012038.

MacLeod, C., 2009. Global climate change, range changes and potential implications for the conservation of marine cetaceans: a review and synthesis. Endanger. Species Res. 7, 125–136. http://dx.doi.org/10.3354/esr00197.

Masski, H., de Stephanis, R., 2015. Cetaceans of the Moroccan coast: information from a reconstructed strandings database. J. Mar. Biol. Assoc. U. K., 1–9. http://dx.doi.org/10.1017/S0025315415001563.
Mayol, P., Di-Meglio, N., David, L., Couvat, J., Labach, H., 2014. Le whale-watching en Méditerranée française : état des lieux et recommandations. Sci. Rep. Port-Cros Natl. Park 28, 133–143.
Miller, P.J.O., Kvadsheim, P.H., Lam, F.-P.A., Wensveen, P.J., Antunes, R., Alves, A.C., Visser, F., Kleivane, L., Tyack, P.L., Sivle, L.D., 2012. The severity of behavioral changes observed during experimental exposures of killer (*Orcinus orca*), long-finned pilot (*Globicephala melas*), and sperm (*Physeter macrocephalus*) whales to naval sonar. Aquat. Mamm. 38, 362–401. http://dx.doi.org/10.1578/AM.38.4.2012.362.
Monteiro, S.R. da S., 2013. Population ecology of long-finned pilot whale (*Globicephala melas*) off the western coast of the Iberian Peninsula. PhD Thesis, Universidade do Minho, University of Aberdeen, Aberdeen, UK.
Monteiro, S.S., Méndez-Fernandez, P., Piertney, S., Moffat, C.F., Ferreira, M., Vingada, J.V., López, A., Brownlow, A., Jepson, P., Mikkelsen, B., Niemeyer, M., Carvalho, J.C., Pierce, G.J., 2015. Long-finned pilot whale population diversity and structure in Atlantic waters assessed through biogeochemical and genetic markers. Mar. Ecol. Prog. Ser. 536, 243–257. http://dx.doi.org/10.3354/meps11455.
Mooney, T.A., Hanlon, R.T., Christensen-Dalsgaard, J., Madsen, P.T., Ketten, D.R., Nachtigall, P.E., 2010. Sound detection by the longfin squid (*Loligo pealeii*) studied with auditory evoked potentials: sensitivity to low-frequency particle motion and not pressure. J. Exp. Biol. 213, 3748–3759. http://dx.doi.org/10.1242/jeb.048348.
Morata, A., Gauffier, P., Esteban, R., Murcia, J.L., García, P., Cañadas, A., Debons, E., Brévart, C., Carbou, M., Meade, R., Wierucka, K., Colmant, L., de Stephanis, R., Verborgh, P., 2013. Definición de unidades de gestión de calderón común (*Globicephala melas*) en el Mediterráneo occidental. In: SEC (Ed.), VI Congreso de La Sociedad Española de Cetáceos. Sociedad Española de Cetáceos, Tarifa, p. 35.
Mussi, B., Miragliuolo, A., Díaz López, B., 2000. Social structure and male parental care in a long finned pilot whale (*Globicephala melas*) pod off Ventotene Island (southern Tyrrhenian Sea, Italy). In: Evans, P.G.H., Pitt-Aiken, R., Rogan, E. (Eds.), European Research on Cetaceans 14: Fourteenth Annual Conference of the European Cetacean Society. European Cetacean Society, Cork, pp. 141–145.
Natoli, A., Birkun, A., Aguilar, A., Lopez, A., Hoelzel, A.R., 2005. Habitat structure and the dispersal of male and female bottlenose dolphins (*Tursiops truncatus*). Proc. R. Soc. Lond. B 272, 1217–12226.
Norris, K.S., Schilt, C.R., 1988. Cooperative societies in three-dimensional space: on the origins of aggregations, flocks, and schools, with special reference to dolphins and fish. Ethol. Sociobiol. 9, 149–179. http://dx.doi.org/10.1016/0162-3095(88)90019-2.
Notarbartolo di Sciara, G., 1990. A note on the incidental catch in the Italian driftnet swordfish fishery, 1986–1988. Rep. Int. Whaling Commn. 40, 459–460.
Notarbartolo di Sciara, G., Birkun Jr., A., 2010. Conserving whales, dolphins and porpoises in the Mediterranean and Black Seas: an ACCOBAMS status report. ACCOBAMS, Monaco.
Notarbartolo di Sciara, G., Mifsud, C., 2002. Action plan for the conservation of cetaceans in Maltese waters. RAC/SPA (Contract no 41/02) under the SAP/BIO framework, in cooperation with Environment and Planning Authority, Malta.
Notarbartolo di Sciara, G., Venturino, M.C., Zanardelli, M., Bearzi, G., Borsani, F.J., Cavalloni, B., 1993. Cetaceans in the central Mediterranean Sea: distribution and sighting frequencies. Ital. J. Zool. 60, 131–138.
Nowacek, D., Thorne, L.H., Johnston, D.W., Tyack, P.L., 2007. Responses of cetaceans to anthropogenic noise. Mammal Rev. 37, 81–115.

Olson, P.A., Reilly, S.B., 2002. Pilot whales *Globicephala melas* and *G. macrorhynchus*. In: Perrin, W.F., Würsig, B., Thewissen, J.G.M. (Eds.), Encyclopedia of Marine Mammals. Academic Press, San Diego, pp. 898–903.

Oremus, M., Gales, R., Dalebout, M.L., Funahashi, N., Endo, T., Kage, T., Steel, D., Baker, S., 2009. Worlwide mitochondrial DNA diversity and phylogeography of pilot whales (*Globicephala* spp.). J. Linn. Soc. 98, 729–744.

Oremus, M., Gales, R., Kettles, H., Baker, C.S., 2013. Genetic evidence of multiple matrilines and spatial disruption of kinship bonds in mass strandings of long-finned pilot whales, *Globicephala melas*. J. Hered. 104, 301–311. http://dx.doi.org/10.1093/jhered/est007.

Ottensmeyer, C.A., Whitehead, H., 2003. Behavioural evidence for social units in long-finned pilot whales. Can. J. Zool. 81, 1327–1338. http://dx.doi.org/10.1139/z03-127.

Packard, A., Karlsen, H.E., Sand, O., 1990. Low frequency hearing in cephalopods. J. Comp. Physiol. A 166, 501–505. http://dx.doi.org/10.1007/BF00192020.

Panigada, S., Lauriano, G., Burt, L., Pierantonio, N., Donovan, G., 2011. Monitoring winter and summer abundance of cetaceans in the Pelagos sanctuary (Northwestern Mediterranean Sea) through aerial surveys. PLoS One 6, e22878. http://dx.doi.org/10.1371/journal.pone.0022878.

Pesante, G., Collet, A., Dhermain, F., Frantzis, A., Panigada, S., Zanardelli, M., 2002. Review of collisions in the Mediterranean Sea. In: Pesante, G., Panigada, S., Zanardelli, M. (Eds.), Proceedings of the Workshop Collisions between Cetaceans and Vessels: Can We Find Solutions? ECS Newsletter No. 40—Special Issue. European cetacean Society, Rome, pp. 5–12.

Pettex, E., Lambert, C., Laran, S., Ricart, A., Virgili, A., Falchetto, H., Authier, M., Monestiez, P., Van Canneyt, O., Dorémus, G., Blanck, A., Toison, V., Ridoux, V., 2014. Suivi aérien de la mégafaune marine en France métropolitaine. Université de La Rochelle UMS 3462, La Rochelle. doi:10.13140/2.1.2698.5287.

Pinzone, M., Budzinski, H., Tasciotti, A., Ody, D., Lepoint, G., Schnitzler, J., Scholl, G., Thomé, J.-P., Tapie, N., Eppe, G., Das, K., 2015. POPs in free-ranging pilot whales, sperm whales and fin whales from the Mediterranean Sea: influence of biological and ecological factors. Environ. Res. 142, 185–196. http://dx.doi.org/10.1016/j.envres.2015.06.021.

Pirotta, E., Merchant, N.D., Thompson, P.M., Barton, T.R., Lusseau, D., 2015. Quantifying the effect of boat disturbance on bottlenose dolphin foraging activity. Biol. Conserv. 181, 82–89. http://dx.doi.org/10.1016/j.biocon.2014.11.003.

Podestà, M., Magnaghi, L., 1988. Sightings of pilot whales, *Globicephala melaena* (Traill, 1809) in the Ligurian Sea, 1981-1988. Atti Soc. Ital. Sci. Nat. Mus. Civ. Stor. Nat. Milano, 129, pp. 478–482.

Podestà, M., Magnaghi, L., 1989. Unusual number of cetacean bycatches in the Ligurian Sea. Eur. Res. Cetac. 3, 67–70.

Poncelet, E., Van Canneyt, O., Boubert, J.-J., 2000. Considerable amount of plastic debris in the stomach of a Cuvier's beaked whale (*Ziphius cavirostris*) washed ashore on the French Atlantic coast. In: Evans, P.G.H., Pitt-Aiken, R., Rogan, E. (Eds.), European Research on Cetaceans 14: Fourteenth Annual Conference of the European Cetacean Society. European Cetacean Society, Cork, pp. 44–47.

Praca, E., Gannier, A., 2008. Ecological niches of three teuthophageous odontocetes in the northwestern Mediterranean Sea. Ocean Sci. 4, 49–59.

Praca, E., Laran, S., Lepoint, G., Thomé, J.-P., Quetglas, A., Belcari, P., Sartor, P., Dhermain, F., Ody, D., Tapie, N., Budzinski, H., Das, K., 2011. Toothed whales in the northwestern Mediterranean: insight into their feeding ecology using chemical tracers. Mar. Pollut. Bull. 62, 1058–1065. http://dx.doi.org/10.1016/j.marpolbul.2011.02.024.

Raga, J.A., Pantoja, J., 2004. Proyecto mediterráneo. Zonas de especial interés para la conservación de los cetáceos en el Mediterráneo español. Ministerio de Medio Ambiente organismo autónomo Parques Nacionales, Madrid.
Relini, L.O., Garibaldi, F., 1992. Feeding of the pilot whale, *Globicephala melas*, in the Ligurian sea. A preliminary note. In: Evans, P.G.H. (Ed.), European Research on Cetaceans 6. Proceedings of the VI Annual Conference of the European Cetacean Society. European Cetacean Society, San Remo, pp. 142–145.
Rendell, L.E., Gordon, J.C.D., 1999. Vocal response of long-finned pilot whales (*Globicephala melas*) to military sonar in the Ligurian Sea. Mar. Mam. Sci. 15, 198–204.
Ritter, F., 2012. Collisions of sailing vessels with cetaceans worldwide: first insights into a seemingly growing problem. J. Cetacean Res. Manag. 12, 119–127.
Rone, B.K., Pace, R.M.I., 2012. A simple photograph-based approach for discriminating between free-ranging long-finned (*Globicephala melas*) and short-finned (*G. macrorhynchus*) pilot whales off the east coast of the United States. Mar. Mam. Sci. 28, 254–275. http://dx.doi.org/10.1111/j.1748-7692.2011.00488.x.
Ross, D., 2005. Ship sources of ambient noise. IEEE J. Ocean. Eng. 30, 257–261. http://dx.doi.org/10.1109/Joe.2005.850879.
Salazar Sierra, J.M., Torres Barranco, F.J., Cabaleiro Mora, E., 2008. Estudio de los aspectos socioeconómicos y de conservación de la actividad de avistamiento de cetáceos. Report to Fundación Migres, Sevilla.
Santoro, R., Sperone, E., Tringali, M.L., Pellegrino, G., Giglio, G., Tripepi, S., Arcangeli, A., 2015. Summer distribution, relative abundance and encounter rates of cetaceans in the Mediterranean waters off southern Italy (western Ionian Sea and southern Tyrrhenian Sea). Mediterr. Mar. Sci. 16, 613–620. http://dx.doi.org/10.12681/mms.1007.
Santos, M.B., Pierce, G.J., 2015. Marine mammals and good environmental status: science, policy and society; challenges and opportunities. Hydrobiologia 750, 13–41. http://dx.doi.org/10.1007/s10750-014-2164-2.
Senigaglia, V., de Stephanis, R., Verborgh, P., Lusseau, D., 2012. The role of synchronized swimming as affiliative and anti-predatory behavior in long-finned pilot whales. Behav. Process. 91, 8–14. http://dx.doi.org/10.1016/j.beproc.2012.04.011.
Shoham-Frider, E., Amiel, S., Roditi-Elasar, M., Kress, N., 2002. Risso's dolphin (*Grampus griseus*) stranding on the coast of Israel (eastern Mediterranean). Autopsy results and trace metal concentrations. Sci. Total Environ. 295, 157–166. http://dx.doi.org/10.1016/S0048-9697(02)00089-X.
Sodebo, 2011. 2011-TOUR DU MONDE. https://www.sodebo.com/fr/voile/actualites/article/3eme-tentative-record (accessed 2.27.16).
Solé, M., Lenoir, M., Durfort, M., López-Bejar, M., Lombarte, A., André, M., 2013. Ultrastructural damage of *Loligo vulgaris* and *Illex coindetii* statocysts after low frequency sound exposure. PLoS One 8, e78825. http://dx.doi.org/10.1371/journal.pone.0078825.
Spitz, J., Cherel, Y., Bertin, S., Kiszka, J., Dewez, A., Ridoux, V., 2011. Prey preferences among the community of deep-diving odontocetes from the Bay of Biscay, Northeast Atlantic. Deep Sea Res. Part I Oceanogr. Res. Pap. 58, 273–282. http://dx.doi.org/10.1016/j.dsr.2010.12.009.
Steele, J., de Stephanis, R., Macgregor, S., Acevedo-Whitehouse, K., 2009. Preliminary characterisation of respiratory bacteria in long-finned pilot whales, *Globicephala melas*, from the Strait of Gibraltar. In: 18th Biennial Conference on the Biology of Marine Mammals. Society for Marine Mammalogy, Quebec.
Taylor, B.L., Chivers, S.J., Larese, J., Perrin, W.F., 2007. Generation length and percent mature estimates for IUCN assessments of cetaceans: Report to National Marine Fisheries Service. Southwest Fisheries Science Center, La Jolla.
Taylor, B.L., Baird, R., Barlow, J., Dawson, S.M., Ford, J., Mead, J.G., Notarbartolo di Sciara, G., Wade, P., Pitman, R.L., 2008. *Globicephala melas*. IUCN Red List Threat.

Species. Version 2015.2. www.iucnredlist.org. http://dx.doi.org/10.2305/IUCN.UK.2008.RLTS.T9250A12975001.en.
Tudela, S., Guglielmi, P., El Andalossi, M., Kai Kai, A., Maynou, F., 2003. Biodiversity impact of the Moroccan driftnet fleet operating in the Alboran Sea (SW Mediterranean). A case study of the harmful effects inflicted by current IUU large-scale driftnet fleets in the Mediterranean on protected and vulnerable species. WWF Mediterranean Programme, Rome.
UNEP/MAP, 2012. State of the Mediterranean marine and coastal environment. United Nations Environment Programme/Mediterranean Action Plan (UNEP/MAP), Barcelona Convention, Athens.
Vallon, D., Guigo, C., Duguy, R., 1977. Le globicéphale noir (*Globicephala melaena* Traill 1809) en Méditerranée occidentale. Rapp. Commision Int. Mer Mediterr. 24, 25–26.
Valsecchi, E., Amos, W., Raga, J.A., Podestà, M., Sherwin, W., 2004. The effects of inbreeding on mortality during a morbillivirus outbreak in the Mediterranean striped dolphin (*Stenella coeruleoalba*). Anim. Conserv. 7, 139–146. http://dx.doi.org/10.1017/S1367943004001325.
Van Canneyt, O., Kerric, A., Authier, M., Dabin, W., Demaret, F., Dorémus, G., Spitz, J., 2014. Les échouages de mammifères marins sur le littoral français en 2013, Rapport UMS—Observatoire PELAGIS pour le Ministère de l'Ecologie, du Développement Durable et de l'Energie, Direction de l'eau et de la biodiversité. Programme Observatoire du Patrimoine Naturel, La Rochelle.
Vázquez, J.A., Cañadas, A., Martínez-Cedeira, J., López, A., Tejedor, M., Gauffier, P., Gazo, M., Brotons, J.M., 2014. Documento técnico sobre la incidencia de la captura accidental de especies de cetáceos amenazadas en artes de pesca. Ministerio de Agricultura, Alimentación y Medio Ambiente, Madrid.
Verborgh, P., 2015. Demografía y estructura de las poblaciones de calderones comunes (*Globicephala melas*) en el Mediterráneo español. PhD Thesis, Universidad de Las Palmas de Gran Canaria, Las Palmas de Gran Canaria, Spain.
Verborgh, P., de Stephanis, R., Pérez, S., Jaget, Y., Barbraud, C., Guinet, C., 2009. Survival rate, abundance, and residency of long-finned pilot whales in the Strait of Gibraltar. Mar. Mam. Sci. 25, 523–536. http://dx.doi.org/10.1111/j.1748-7692.2008.00280.x.
Verborgh, P., de Stephanis, R., Gauffier, P., García Tiscar, S., Esteban, R., Minvielle-Sebastia, L., Ridoux, V., Dabin, W., Llavona, Á., Marcos Ipiña, E., Monteiro, S., Ferreira, M., Monaghan, N.T., Berrow, S., Fossi, M.C., Marsili, L., Laran, S., Praca, E., Cañadas, A., Sagarminaga, R., Murcia, J.L., García, P., 2010. Population structure of long-finned pilot whales in Europe. In: Verfuß, U., Osterrieder, S., Vincent, C., Jabbusch, M., Gallus, A., Subklew, J., Bräger, S., Brandecker, A., Dähne, M., Großer, S., Hanke, F., Hanke, W., Hermann, A., Krügel, K., Meyer, T., Müller, S., Sundermeyer, J., Zehl, L. (Eds.), 24th Annual Meeting European Cetacean Society. European Cetacean Society, Stralsund, pp. 59–60.
Verborgh, P., de Stephanis, R., García, P., Murcia, J.L., Andrews, R., Cañadas, A., Gauffier, P., Giménez, J., Esteban, R., 2012. The territorial long-finned pilot whales of the Alboran Sea. In: McGovern, B., Berrow, S., McKeogh, E., O'Connor, I. (Eds.), 26th Annual Conference of the European Cetacean Society. European Cetacean Society, Galway, p. 91.
Waring, G.T., Josephson, E., Fairfield, C.P., Maze-Foley, K., Editors, 2006. U. S. Atlantic and Gulf of Mexico Marine Mammal Stock Assessments—2005. NOAA Technical Memorandum NMFS-NE-194.
Weller, D.W., 2008. Predation on marine mammals. In: Perrin, W.F., Würsig, B., Thewissen, J.G.M. (Eds.), Encyclopedia of Marine Mammals, second ed. Academic Press, New York, p. 1355.
Wierucka, K., Verborgh, P., Meade, R., Colmant, L., Gauffier, P., Esteban, R., de Stephanis, R., Cañadas, A., 2014. Effects of a morbillivirus epizootic on long-finned pilot whales *Globicephala melas* in Spanish Mediterranean waters. Mar. Ecol. Prog. Ser. 502, 1–10. http://dx.doi.org/10.3354/meps10769.

CHAPTER SEVEN

Risso's Dolphin, *Grampus griseus*, in the Western Ligurian Sea: Trends in Population Size and Habitat Use

A. Azzellino*,†,1, S. Airoldi†, S. Gaspari‡, C. Lanfredi*,†, A. Moulins§, M. Podestà¶, M. Rosso§, P. Tepsich§,‖

*Politecnico di Milano, University of Technology, Milano, Italy
†Tethys Research Institute, Milano, Italy
‡National Research Council (CNR), Institute of Marine Sciences (ISMAR), Ancona, Italy
§CIMA Research Foundation, Savona, Italy
¶Museum of Natural History of Milan, Milano, Italy
‖University of Genoa, Genoa, Italy
[1]Corresponding author: e-mail address: arianna.azzellino@polimi.it

Contents

Abstract

This paper provides a summary of 25 years of research on Risso's dolphins (*Grampus griseus*) in the western Ligurian Sea. Seasonal variations in abundance, distribution and habitat use were observed. Photographic mark–recapture methods provided a

Advances in Marine Biology, Volume 75
ISSN 0065-2881
http://dx.doi.org/10.1016/bs.amb.2016.08.003

population size estimate for the period from 1998 to 2012, of about 100 individuals (95% CI of 60–220 individuals). The same methods detected a decline in population size from an average of about 120–150 from 2000 to 2005, to an average of 70–100 during 2010 to 2014. Species occurrence appeared to be significantly decreasing in coastal and continental slope areas, while it seemed to be stable in the most pelagic area. In addition, a dramatic change was observed in the local primary production, which was analysed based on time series of sea surface temperature and chlorophyll data from 1990 to 2014. Based on fisheries landings, there may have been a general decrease in fishery productivity, both in the western Ligurian Sea and in adjacent regions. Environmental variability, depletion of resources by fisheries and possibly interspecies competition may all have contributed to cause changes in Risso's dolphin habitat use and occurrence in the western Ligurian Sea.

1. INTRODUCTION

Risso's dolphins, *Grampus griseus* (G. Cuvier, 1812) (Fig. 1), occur worldwide in tropical and temperate waters, with a strong preference for mid-temperate waters of the continental shelf and slope (Jefferson et al., 2014). The species, well known from the major ocean basins of the world (Baird, 2009; Kruse et al., 1999), is also relatively widespread, although apparently not abundant, in the Mediterranean Sea (Bearzi et al., 2011; Boisseau et al., 2010; Gaspari and Natoli, 2006; Reeves and Notarbartolo di Sciara, 2006). There is no baseline of abundance for Risso's dolphin in the Mediterranean Sea. However, a recent aerial survey conducted over

Fig. 1 Adult and juvenile Risso's dolphins (*Grampus griseus*) in the western Ligurian Sea. *Photograph: S. Airoldi, Tethys Research Institute.*

the north-western portion of the Mediterranean Basin (approximately 200,000 km^2), used distance sampling methods to estimate a total 2550 individuals (95% CI: 849–7658) in winter and 1783 individuals (95% CI: 849–7658) during the summer season (Pettex et al., 2014). Currently, the Mediterranean subpopulation is classified on the International Union for the Conservation of Nature (IUCN) Red List, as Data Deficient (Gaspari and Natoli, 2012). The worldwide status is classified as Least Concern (Taylor et al., 2012). The occurrence and distribution of this species in the Mediterranean, as is true of other cetacean species in the region, are heterogeneous. Areas of occurrence include the Alborán Sea (Cañadas et al., 2002, 2005; Gannier, 2005), Ligurian Sea (Azzellino et al., 2008, 2012; Di Méglio et al., 1999; Gannier, 2005; Moulins et al., 2008; Notarbartolo di Sciara et al., 1993), Tyrrhenian Sea (Campana et al., 2015; Marini et al., 1996), Adriatic Sea (UNEP MAP-RAC/SPA, 2014), and the Ionian (Dimatteo et al., 2011; Frantzis et al., 2003) and Aegean basins (Frantzis et al., 2003), and very little is known about Levantine and North African waters (Kerem et al., 2012). Long-term monitoring studies concerning this species have also been extremely rare. De Boer et al. (2013) provided an exception and reported a regular seasonal occupancy (with some site fidelity) for this species in the waters off Bardsey Island, Wales, over a 10-year study period.

Here, we present information resulting from 25 years of survey work in the western Ligurian Sea, within a portion of the Pelagos Sanctuary for Mediterranean Marine Mammals (Fig. 2; and see Notarbartolo di Sciara et al., 2008). First, we provide background information on the research and study area, as well as details of the data and methods used for the analyses discussed here. Next, we present a discussion including information on Risso's dolphin occurrence, population size and structure in the western Ligurian Sea, and we examine the factors that may have caused changes in these parameters for Risso's dolphins in the area over the last 25 years.

2. BACKGROUND AND METHOD

During the 25-year study period from 1990 to 2014, the Tethys Research Institute (TRI) conducted opportunistic cetacean sighting surveys in the western Ligurian Sea (Azzellino and Lanfredi, 2015). These data were integrated with data collected from dedicated surveys conducted year-round from 2004 to 2013 in the area by the CIMA Research Foundation (CIMA). (Tepsich et al., 2014), dedicating a greater effort during the good weather

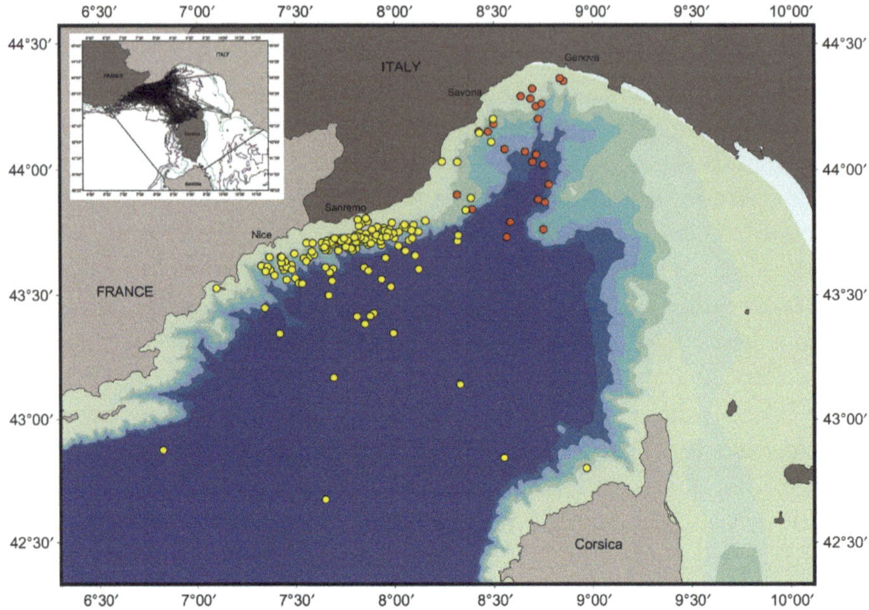

Fig. 2 Sighting data for Risso's dolphins (*Grampus griseus*) collected by Tethys Research Institute (TRI) (*yellow circle*) and the CIMA Research Foundation (CIMA) (*red circle*) in the north-western Ligurian Sea, Pelagos Sanctuary. Panel: Tethys Research Institute research effort (1990–2014). Boundaries of the Pelagos Sanctuary are shown.

months of spring and summer. From 1990 to 2014, 169 Risso's dolphin sightings were recorded by TRI, and photo-identification data were collected for 142 of these sightings. An additional 28 sightings with photo-identification data were collected by CIMA (2004–13).

The study area of about 25,000 km^2 within the Pelagos Sanctuary is characterized by a heterogeneous geomorphology, wherein a narrow continental shelf, deeply cut by several submarine canyons, leads to off-shore waters with a uniform abyssal plain that is greater than 2500 m in depth. The study area was divided into 3312 cells of 6.8×9.3 km grid size (about 4×5 nautical miles) (Azzellino et al., 2012), and grid cells were classified based on three different depth (D) zones: Dzone$_1$: $D < 300$ m; Dzone$_2$: $D = 300$–1500 m; Dzone$_3$: $D > 1500$ m.

2.1 Sighting Data

The main TRI sighting dataset was generated from shipboard surveys conducted between May and October from 1990 to 2014. Sailing vessels ranging in length from 15 to 21 m were used as research platforms. Data were collected roughly along North-to-South transects surveyed at a mean cruising speed of 9–11 km h^{-1}. Observations were made under good sea and weather conditions, defined as 'favourable conditions', and search effort ceased when wind exceeded Force 3 on the Beaufort scale (wind speed higher than 5.4 m s^{-1}). Two trained observers were positioned, one on each side of the vessel, at a height of about 3 m above the sea surface. The CIMA surveys were conducted at a speed of approximately 10–15 km h^{-1}, and at least three trained observers were positioned on the upper deck at about 4 m above sea level to scan 360 degrees around the vessel with and without binoculars. The same data collection protocol was used by both TRI and CIMA survey teams. Effort, environmental conditions and sighting data were recorded regularly using Logger2000 software or a portable GPS (see Azzellino et al., 2012; Tepsich et al., 2014). Search effort was evaluated as kilometres track line per grid cell unit: only the effort made in favourable conditions was considered. To minimize the effect of effort outliers in the analysis of encounter rates, only cell units with effort higher than 35 km and lower than 600 km (i.e. higher than the 10th and lower than the 90th percentile) were used in the analysis.

Variability of effort during the overall study period was tested using a factorial ANOVA to assess whether differences in the effort coverage of depth zones among years could significantly bias results in terms of occurrence and consequently of photographic effort. Although the effort varied significantly among years (F_{year}: 3.27; *P*: 0.007) and among depth zones ($F_{depthzone}$: 49.9; $P<0.001$), coverage of the study area was proportionally maintained since no difference was found in the depth zone coverage across the time series ($F_{year \times depthclass}$: 1.49; *P*: 0.140) supporting the hypothesis of comparability of the different field seasons over the entire time series.

2.2 Photo-Identification Data

A standardized photo-identification protocol was followed, including standardized photographic procedures, image preparation and assignation of scores for photographic quality and distinctiveness of individual marks (see Airoldi et al., 2015). Two photo-identification datasets were compiled: one for the dorsal fin right side and another for dorsal fin left side. The photo-identification catalogue resulted in a total of 326 photo-identified individuals for the right side and 283 identified individuals for the left side (Table 1).

Table 1 Summary of Photo-Identification Catalogue Information for Risso's Dolphins (*Grampus griseus*) in the Western Ligurian Sea: Number of Individual Dorsal Fin Right Side and Dorsal Fin Left Side Images

Source	Number of Photo-Identified Individuals	Time Period
Tethys Research Institute	258 right side 213 left side	1989–2014
CIMA Research Foundation/ Genoa University	103 right side 114 left side	2004–10
CIMA Research Foundation	40 right side 37 left side	2011–13

2.3 Stranding Data

Eighteen Risso's dolphin stranding events, collected between 1986 and 2014 along the Ligurian Coast also contributed to our knowledge of the species in the area (Banca Dati Spiaggiamenti, 2015; Podestà et al., 2006, 2009). Stranding data were retrieved from the Italian historical database of strandings, Banca Dati Italiana Spiaggiamenti (BDS) (http://mammiferimarini.unipv.it/index_en.php). The BDS provides information about each stranding, including the date, location (latitude and longitude), region, species, sex and length of individual, as well as the information about toxicological and parasitological investigations (when available), and samples collected/institutions where samples are housed.

2.4 Sea Surface Temperature, Chlorophyll and Fishery Landings Statistics

With the aim of identifying the effects of different potential drivers of change, the time series of sea surface temperature (SST) and chlorophyll data were analysed together with fishery landings. The monthly time series of remotely sensed SST (°C, AVHRR sensor) and chlorophyll-a (Chl-a) data (mg m^3; SeaWiFS sensor Ocean Colour and MODIS Aqua) were examined for the period from 1990 to 2014. Data were generated by the National Oceanic and Atmospheric Administration (NOAA) Earth and System Research Laboratory (ESRL) and were downloaded from www.esrl.noaa.gov/psd/data/gridded/.

In addition, the regional statistics on fishery landings, available from the Italian National Institute of Statistics (ISTAT),[1] for the Ligurian and the

[1] ISTAT Censimenti Nazionali Agricoltura, Zootecnia e Pesca.

Tyrrhenian Sea regions, and the corresponding statistics for the Gulf of Lion[2] were examined to provide a rough idea of the fishery pressure in these regions during the time period from 2004 to 2011. These statistics do not account for catch per unit effort (CPUE). Fishery registration data from the European Union Community Fishing Fleet Register[3] reflect the number of fishery vessel registrations corresponding to the main ports. These statistics can be evaluated by type of fishing gear (i.e. bottom trawl, purse seine, small-scale fishery, polyvalent, passive polyvalent and longlines), and thus may provide general information regarding the potential for direct (e.g. bycatch) and indirect impact on cetacean species.

2.5 Encounter Rate

The temporal variability of Risso's dolphin occurrence was investigated by analysing the encounter rate among years and months (Azzellino and Lanfredi, 2015). The encounter rate was calculated as the number of sightings per kilometre surveyed under favourable conditions within each grid cell unit. To account for the heterogeneity of the effort coverage during the 25 years time series, generalized linear models (GLMs, Rutherford, 2001) were used:

$$y_{ijk} = \mu + x_{effort} + x_{year} + \alpha_i + \beta_j + \alpha\beta_{ij} + \varepsilon_{ijk}$$

where y_{ijk} is the k_{th} encounter rate of the i_{th} year; μ true overall mean; x_{effort} covariate to account for the effort coverage; x_{year} covariate to account for the year; α_i incremental effect of month i, such that $\alpha_i = \mu_i \_\mu$ (factor A); β_j incremental effect of depth zone j, such that $\beta_j = \mu_j\mu$ (factor B); μ_i true population mean for the i_{th} level of factor A; μ_j true population mean for the j_{th} level of factor B; $\alpha\beta_{ij}$ month $\times$ depth zone interaction term; ε_{ijk} error for the k_{th} observation.

In addition, partial correlation analysis (Afifi and Clark, 1998) was used to investigate encounter rate trends in the different depth zones controlling for the effort.

2.6 Mark–Recapture Estimates

To apply mark–recapture (M/R) methods based on the individual photo-identification records, a dataset of capture histories per encounter was created (Azzellino and Lanfredi, 2015). A capture was defined as an individual

[2] FAO Fisheries and Aquaculture Department Yearbooks. Fishery and Aquaculture Statistics.

[3] Available at http://ec.europa.eu/fisheries/fleet/.

identification within an encounter. The capture frequency histogram (Fig. 3) shows that the number of photographic captures varied greatly among years as the percentage of recaptures. Only a few seasons had enough recaptures to allow robust abundance estimates with mark–recapture methods based on the field season as primary sampling interval. As a result, data were pooled based on the homogeneity of photographic effort, and the time interval between capture occasions was set in order to have sample sizes that were sufficient for analysis. Capture–recapture models included closed models, wherein the population is assumed to remain unchanged for the duration of the study, and open models, wherein the population may change through additions (births or immigration) and deletions (deaths or emigration) (Pollock et al., 1990). Estimates of population size were obtained assuming both closed and open population models. Closed population models were assumed when the consecutive primary periods of sampling was sufficiently short in time to assume that the population was stable. For closed population models, 'year' was the primary sampling interval (i.e. the field season), and the monthly interval was the secondary sampling unit within each field season. Capture histories were analysed using CAPTURE application run within MARK (mark and recapture parameter estimation) v.8.0 (White and Burnham, 1999). This application has 11 available models that test for three sources of variation in sighting probabilities (Otis et al., 1978): (i) a time response, which considers the sighting probability varying from sampling period to sampling period but assuming that all

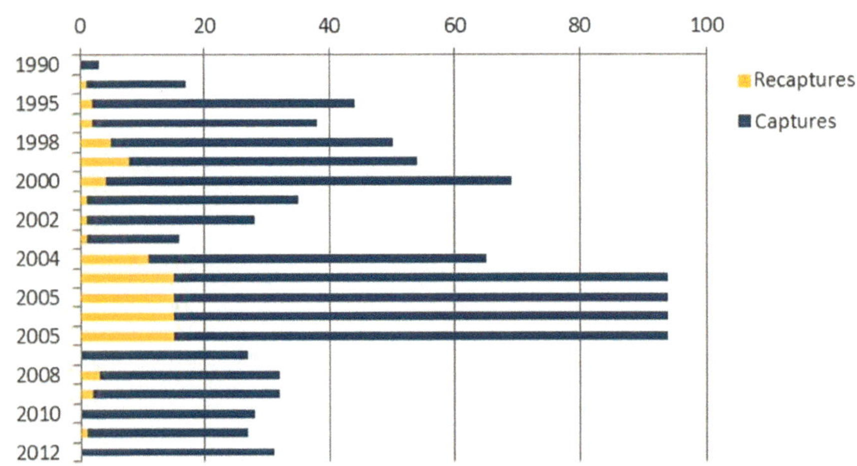

Fig. 3 Number of captures and recaptures of photo-identified Risso's dolphin (*Grampus griseus*) individuals by year during the period from 1990 to 2012.

animals within each sampling period have the same probability of being sighted (M_t), (ii) a behavioural response, where animals become either 'trap happy' or 'trap shy' after their first capture (M_b) and (iii) the individual heterogeneity response, where individuals vary in their capture probability (M_h). All models are based on these principles and/or combinations of the three (for example, M_{bh}, M_{th}, M_{tb}), plus one additional model wherein probability of capture remains constant (M_0). Open population models were also applied to support the estimates based on the closed population assumption and to provide population size estimates for field seasons (e.g. 2010–12) for which sample size could not support estimates based on the field season as primary time unit (see Table 2). Three different types of open population models were tested: (1) the standard Jolly–Seber (Model A), (2) the Jolly–Seber Model B, assuming constant survival rate per unit time and time-specific capture probabilities, and (3) Jolly–Seber Model D, assuming constant survival rate per unit time and constant capture probability. In these models, the primary sampling period was a three-field season interval, and the secondary sampling unit was the single field season.

Using MARK, a testing procedure allows for comparison of alternative models to determine which effects are operating and assesses the most appropriate model for estimating population size. For closed population models, the time model (M_t) (assuming that sighting probability varies from sampling period to sampling period but the same probability of being sighted for all animals captured within each sampling period), was selected as most appropriate in most of the situations. The Jolly–Seber Model B (assuming constant survival rate per unit time and time-specific capture probabilities) was chosen as best model for the estimates based on the open population assumption. Photographic sample size for the early 1990s field seasons (i.e. 1990–1997) was too low to allow robust estimates.

2.7 Association Indices

The proportion of co-occurrence of any two individuals can be measured by means of association indices. We conducted a preliminary analysis of the associations among Risso's dolphin individuals during the time period from 1990 to 1998, using SOCPROG (Whitehead, 2009). There were 128 individuals photo-identified for that time period, and only individuals sighted more than two times were included in the analyses ($n = 58$). The simple ratio (SR) index (Ginsberg and Young, 1992) was calculated for each individual dyad. The SR index ranges from zero (individuals never seen together) to

Table 2 Population Size Estimates ($\check{N}$) Based on Mark–Recapture Estimators From Photo-Identification of Risso's Dolphins (*Grampus griseus*) in the Western Ligurian Sea, 1998 to 2012

Time Period	Capture Occasions	Model	$\check{N}$	SE	95% CI Lower	95% CI Upper	Recaptures	Captures
1998	3	*M(t)*	89	26.3	60	175	5	45
1999	3	*M(t)*	77	16.8	58	129	8	46
2000	4	*M(0)*	426	197	198	1046	4	65
1998–99	3	*Jolly–Seber Model B*	92.7	20.6	52.3	133.1	16	91
1999–2000	3		127.9	41.9	45.8	201	23	112
1998–2000	3		110.3	41.7	28.6	192	39	118
1998–2000	3	*M(t)*	187	18.9	159	234	39	118
1998–2000	3	*M(t) Chao*	261	47	195	385	39	118
2001	3	*M(t)*	110	78.1	49	435	1	34
2005	6	*M(t)*	210	44	148	327	15	79
2005	3	*M(0)*	216	50.5	147	369	15	79
2005L	3	*M(0)*	213	93.8	107	513	4	48
2004–06	3	*M(t)*	189	15	167	227	42	132
2004–05	3	*Jolly–Seber Model B*	182.2	41.9	100	264	42	132

2005–06	3		172	51.9	70	273	42	132
2004–06	3		177.1	63.6	52	271	42	132
2008	3	*M(t)*	67	27.1	40	162	3	29
2010–12	3	*M(0)*	138	25.2	104	207	15	69
2010–11	3	*Jolly–Seber Model B*	103.3	52.5	1	206.1	15	69
2011–12	3		81.7	51.8	1	183.26	15	69
2010–12	3		92.5	68.8	1	227.3	15	69

The standard error (SE), and upper and lower bounds of the 95% confidence interval (95% CI) are also shown. The used model are *M*(0)—constant probability of capture; and *M*(*t*)—sighting probability varying from sampling period to sampling period, assuming the same probability of being sighted for all the individuals—under the closed population hypothesis and Jolly–Seber Model B, assuming constant survival rate per unit time and time-specific capture probabilities under the open population hypothesis.

one (individuals always seen together). The higher the value of the index, the greater the level of association between that pair of individuals (Ginsberg and Young, 1992).

3. OCCURRENCE AND DISTRIBUTION

Table 3 shows the GLM main effects and the corresponding *F* and significance levels resulting from modelling of Risso's dolphins encounter rates in the western Ligurian Sea. Although the effort is the most significant covariate ($P<0.01$), the year is also significant ($P<0.05$) as depth zone ($P<0.05$).

These results highlight a dramatic change in the species use of the habitat in the study area (see Fig. 4).

Species occurrence appeared to be significantly decreasing in coastal and continental slope areas, while it seemed to be stable in the most pelagic area. A partial correlation analysis controlling for effort confirmed these results, indicating a significant inverse correlation of encounter rates with year for both $Dzone_1$ (*r*: −0.696; df: 9; $P<0.05$) and $Dzone_2$ (*r*: −0.303; df: 60; $P<0.05$), and the lack of such correlation for $Dzone_3$ (*r*: 0.313; df: 25; $P>0.10$).

For month of the year during the summer season there was no significant effect on Risso's dolphin occurrence, although significant differences were

Table 3 Analysis of Risso's Dolphins (*Grampus griseus*) Encounter Rates (Dependent) From 1990 to 2014 in the Western Ligurian Sea Based on a Generalized Linear Model (GLM)

Source	Sum of Squares	df	Mean Square	*F*	Significance
Intercept	2.810E−5	1	2.810E−5	4.846	0.031
Effort (km)	0.00014	1	0.00014	24.251	*0.000*
Year	2.608E−5	1	2.608E−5	4.496	*0.037*
Month	9.824E−6	3	3.275E−6	0.565	0.640
Depth_zone	3.712E−5	2	1.856E−5	3.200	*0.046*
Month × Depth_zone	0.00012	6	1.924E−5	3.318	*0.006*
Error	0.00048	82	5.800E−6		
Total	0.0028	96			

$R^2=0.474$ (adjusted $R^2=0.390$).
Effort and year are covariates; month and depth zones are fixed factors. Significant terms are italicized.

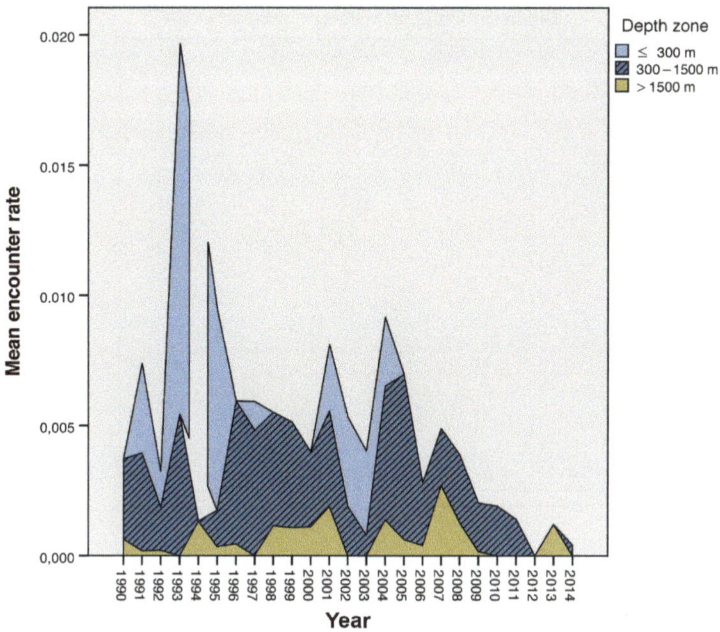

Fig. 4 Risso's dolphin (*Grampus griseus*) encounter rate decreasing trend.

found in the species' monthly pattern of use for the three depth zones ($P < 0.05$, Fig. 5).

The model also shows the overall higher occurrence of this species in the 300–1500 m depth zone ($Dzone_2$), and this is consistent with the previously reported preference of this species for outer-slope habitats, which is generally attributed to the distribution of their cephalopod prey (Azzellino et al., 2008; Cañadas et al., 2002).

Risso's dolphins feed primarily on deep-water cephalopods (Clarke and Pascoe, 1985; Cockroft et al., 1993; Pauly et al., 1998). Stomach contents were analysed of 14 Risso's dolphins stranded along the Spanish coasts (Blanco et al., 2006), five stranded on Italian coasts (Bello and Bentivegna, 1996; Pedà et al., 2015; Podestà and Meotti, 1991; Würtz et al., 1992) and specimens taken as bycatch by the eastern Mediterranean Sea swordfish driftnet fishery off the Turkish coast (Ozturk et al., 2007). Stomach contents of dolphins stranded on the Spanish coastline were mainly composed of greater argonaut, *Argonauta argo*. Contents from dolphins stranded along the Italian coasts and those from the waters off Turkey, were primarily composed of bathypelagic cephalopods belonging to the Histioteuthidae, Onychoteuthidae, Ommastrephidae and Crachiidae families.

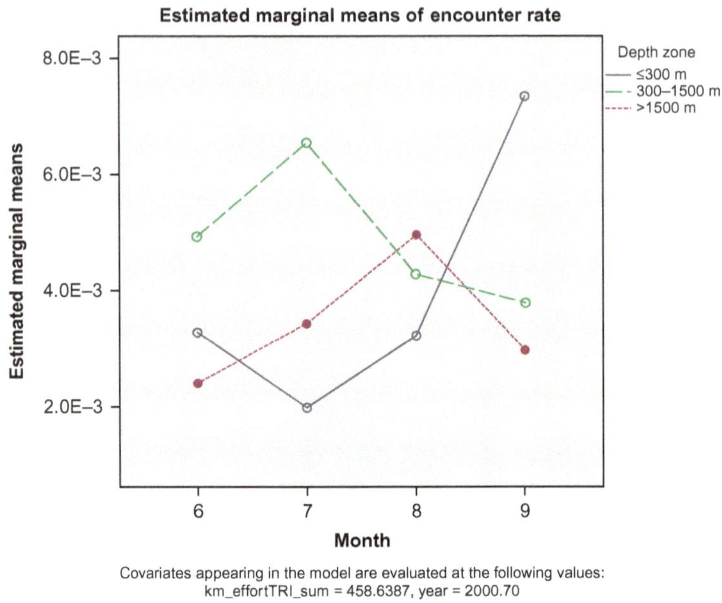

Fig. 5 Monthly pattern of Risso's dolphin (*Grampus griseus*) encounter rates in the western Ligurian Sea in the three depth zones (Dzone) ($Dzone_1$: $D < 300$ m; $Dzone_2$: $D = 300$–1500 m; $Dzone_3$: $D > 1500$ m).

The pattern of Risso's dolphin presence in the region has also been studied in terms of interval between sightings. Excluding same day encounters, the median interval between Risso's dolphin sightings in this area is eight days, with 50% of the sightings (i.e. interquartile range (IQR)) occurring between three and 19 days of a previous sighting (range: 1–69 days). This pattern suggests intermittent use of the study area, with Risso's dolphins moving into and out of the area. Azzellino et al. (2008) have suggested that this particular occurrence pattern might be a strategy for conveniently exploiting temporary formation of food resources induced by the zooplankton accumulation that is typical of the slope area (Macquart-Moulin and Patriti, 1996), and that the pattern might facilitate interspecific coexistence through reduction of overlap in resource exploitation.

The species stranding frequencies reflect the time pattern of the sighting frequencies (Fig. 6). It is interesting to note that there has not been any Risso's dolphin strandings reported in the area since 2008. The pattern of the stranding series is apparently consistent with the documented decrease in species occurrence in coastal and continental slope areas.

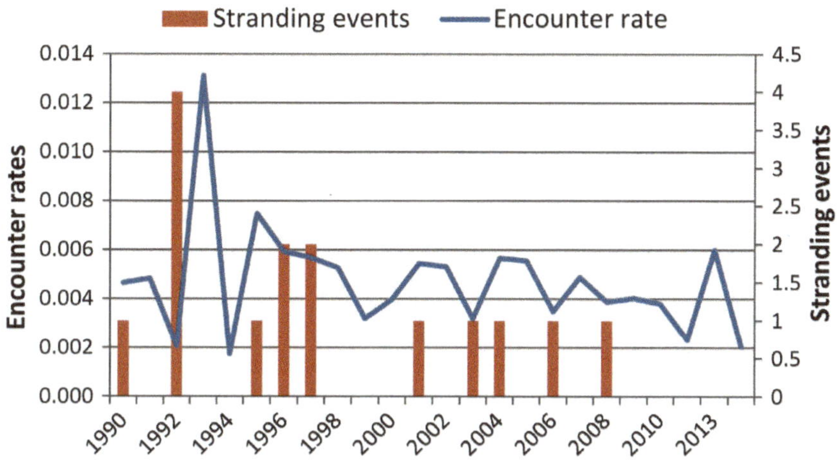

Fig. 6 Time series of Risso's dolphin (*Grampus griseus*) encounter rates and stranding events.

4. POPULATION SIZE

Mark–recapture models based on the closed population assumption indicated that Risso's dolphin population size in the period from 1998 to 2012, averaged approximately 100 individuals (95% CI: 60–220), with a peak of an estimated 212 individuals (95% CI: 146–345) in 2005. Population size estimates based on the open population assumption confirmed the estimates based on closed population, with an average abundance of 110 individuals (95% CI: 29–192), and a peak of 177 individuals (95% CI: 52–271) in 2005. Notably, the time pattern suggested by the population size estimates is coherent with the pattern of relative abundance indicated by the species encounter rates (Fig. 7).

Mark–recapture methods can be prone to serious negative biases (e.g. in relation to unaccounted for heterogeneity of photo-identification data) (Hammond, 2010). However, because the estimates of population size obtained assuming the closed population were based on independent field seasons, and those estimates are reciprocally consistent and comparable with the corresponding estimates obtained from the open population models (which were based on a different time interval), the possibility of a serious bias due to the heterogeneity of photographic effort might be ruled out. In addition, the existence of a significant correlation (r: 0.59, $P < 0.05$) between M/R population size estimates and encounter rates, indicates a

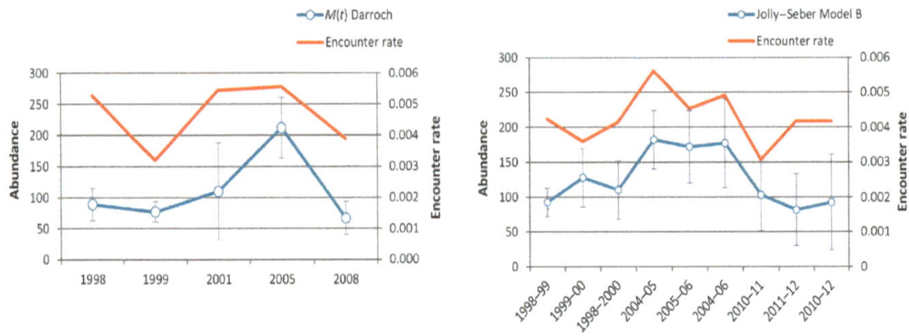

Fig. 7 Comparison between mark–recapture population size estimates based on closed population (*M*(*t*) after Darroch) and on open population assumption (Jolly–Seber Model B). *Error bars* are standard error (SE) of the estimates. Concurrent encounter rates are also shown.

proportionality between survey and photographic effort (supporting the consistency of the photographic effort).

The correlation between absolute (i.e. population size estimates) and relative abundance (i.e. encounter rate) allows for estimation of the population size in the early 1990s in which the bulk of the photographic records were not yet available. Encounter rates suggest that Risso's dolphin population size in the study area peaked in 1993, with a maximum of about 400 individuals, while a minimum size occurred in 1992, 1994, 2011 and 2014 (Fig. 8). Greatest survey effort in the early 1990s concentrated in the pelagic zone ($Dzone_3$), and thus the minimum size predictions for 2011 and 2014 could be more robust than the estimates for 1992 and 1994. In the last period of the time series, Risso's dolphin population size significantly decreased, from an average of about 120–150 individuals in the period from 2000 to 2005 to an average of 70–100 dolphins in the period from 2010 to 2014.

These local estimates of Risso's dolphin abundance roughly corresponded to 12,500 km^2 of optimal habitat for the species, approximately half of the overall survey coverage, and indicate a density estimate of 0.0096–0.012 dolphins per km^2. This density estimate is generally similar to estimates available from the aerial surveys conducted in the western central Mediterranean from 2001 to 2003, in which there were an estimated 493 Risso's dolphins (95% CI: 162–1498) over an area of 32,270 km^2 corresponding to a density of 0.0153 dolphins per km^2 (Gomez de Segura et al., 2006).

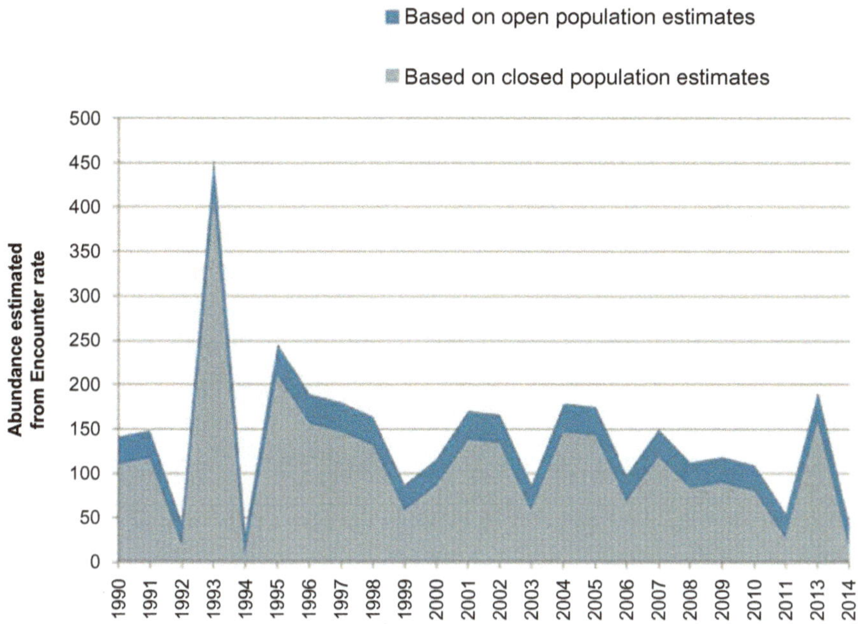

Fig. 8 Population size time series for Risso's dolphins (*Grampus griseus*) in the western Ligurian Sea. Population sizes for the period from 1990 to 1997 are estimated based on encounter rates relying on their correlation with mark–recapture estimates.

5. POPULATION STRUCTURE

5.1 Genetics

Although Risso's dolphins are distributed worldwide, little is known about their population structure. In particular, in the Mediterranean Sea there are very few published studies (Gaspari, 2004; Gaspari et al., 2007). Risso's dolphins generally occur in small, localized coastal populations. Therefore, the question of regional differentiation by genetic drift and overall diversity levels has been recognized as a conservation concern (Gaspari, 2004; Gaspari et al., 2007).

Molecular genetic analysis of nuclear and mitochondrial DNA comparing samples collected from Risso's dolphins in the Mediterranean Sea ($n=24$ live dolphins in the Ligurian Sea; $n=9$ from dead stranded dolphins in other regions of the Mediterranean Sea) to those collected from dolphins sampled in areas off the United Kingdom (UK) ($n=18$), revealed genetic differentiation between the Mediterranean and the Northeastern Atlantic

(Gaspari, 2004; Gaspari et al., 2007). Gaspari et al. (2007) suggested a possible subdivision within the Mediterranean samples, with apparent differentiation between samples collected form the western (Tyrrhenian Sea) and eastern (Adriatic Sea) coasts of Italy. Also, Risso's dolphins from the Mediterranean Sea had a higher level of genetic variability than those from the Northeastern Atlantic. Gaspari (2004) suggested that the Mediterranean population was presumably more recent and may have therefore retained a higher variability. The Northeastern Atlantic population was mainly represented by samples from Scotland, and that region represents a sort of extreme range limit for the species; it is therefore conceivable that adaptation to that particular environment could have led to the separation of the UK population driven by 'a local founder effect' at this range margin.

5.2 Social Structure

Risso's dolphins, like most delphinids, are very social and spend much time with conspecifics.

Association indices ($SR \geq 0.5$) for the time period between 1990 and 1998, showed that only 4.3% (48) of all possible pairwise interactions between dolphins (1624) were observed. Association indices of 0 (i.e. dyads never seen together) were dominant, while nonzero values mostly ranged between 0.1 and 0.2. Thirty-four per cent of the individual associations of dolphin pairs were found to be significantly different ($p < 0.01$) from random distributions derived from a randomization test (30,000 unrestricted permutations). In this type of study, the photo-identified individuals need to be assumed as representative of the overall population. The photo-identification data used, dated back to 1990, although specific effort on Risso's dolphin photo-identification began in 1997. This could have led to an underestimation of the number and strength of Risso's dolphin associations, and groups outlined in this preliminary analysis. Even with such a premise, there is the need to remark that most of the individual associations found were weak ($SR < 0.5$), although stronger relationships between individuals were found over periods of months and years. Only a few small groups showed high individual fidelity. Outside of these potentially stable small groups, the structure of Risso's dolphin society appeared to be labile. Pairs or small cliques may have such stable bonds that they form subunits of larger changeable groups. Kruse (1989) investigated Risso's dolphin ecology in Monterey Bay, California, and found that school composition was highly variable, while Hartman et al. (2008) found that Risso's dolphins in localized populations in the Azores, formed stable groups with long-term bonds. It is possible that the species may adapt its social organization depending upon environmental variables. Delphinid species have been documented to adopt different social strategies depending upon habitat and resource availability (Gowans et al., 2007). The fluid composition of the Risso's dolphin organization in the Ligurian Sea may also be the result of a gathering of Risso's dolphins coming from other areas in the Mediterranean Sea. This might also explain the high genetic variability found among Risso's dolphins sampled from the Ligurian Sea (Gaspari, 2004; Gaspari et al., 2007).

Among the few well-studied marine delphinid species, group stability appears to decrease with decreasing body size (Bräger, 1999). This stability becomes more fluid in smaller dolphins. Even though body size itself is unlikely to be a factor that determines group stability, it seems to be related to longevity and the duration of nursing in odontocetes (Perrin and Reilly, 1994). If this is true, we would expect Risso's dolphin social structure to be generally comparable to that of bottlenose dolphins. Associations of Risso's dolphins studied in the Ligurian Sea appear to be in agreement with this hypothesis, with a 'fission and fusion' model quite similar to the one reported for common bottlenose dolphins (*Tursiops truncatus*) (Wells, 1991). Limited data on Risso's dolphin subgroups from strandings in the Mediterranean region suggest that some of these cohesive subgroups may be same age, same sex animals. Risso's dolphins in our study area in the Ligurian Sea were frequently observed to gather in larger postfeeding aggregations and be engaged in social activities (Gaspari, 2004). Often small groups, typically four to eight individuals, appeared to be engaged in presumably feeding activity, and then aggregate with different members to rest and socialize in larger groups. The small groups may be stable over periods of years. A very similar behaviour was observed in Argentinian dusky dolphins (Würsig and Würsig, 1980). On the other hand, Hartman et al. (2008) found that Risso's dolphins off Pico Island in the Azores probably have a more complex social structure where most of the individuals belong to stable, long-term units or to a strongly associated pair, while others have no long-term associations. Although stability was not found for all age classes (e.g. young adults, females without calves), association patterns varied between age classes and were stable in adult males and most fluid in subadults (Hartman et al., 2008). Thus, according to Hartman et al. (2008), the society of Risso's dolphin in the Azores could be structurally based on preferred associations between individuals of the same age and sex class with highly stable pods of three to 12 individuals, formed by both adult males and adult females.

Data on group structure and cohesiveness of Risso's dolphins in the western Ligurian sea is not conclusive at this time. However, there is reason to suspect that social organization in these dolphins might be different from that observed in the Azores. Further studies are needed to shed light on the ecology of this species in the Mediterranean Sea area.

6. ENVIRONMENTAL DRIVERS AND ANTHROPOGENIC THREATS

Elucidating the ways in which human impacts may have interacted with natural variability to affect populations and ecosystems is particularly challenging. Humans have been altering the structure and function of marine ecosystems and food webs for centuries (Jackson et al., 2001; Lotze et al., 2006; Lotze and Worm, 2009). Moreover, there is no doubt that the increasing cumulative pressure deriving from human activities, interacting with ongoing climate change, will further destabilize marine organisms and ecosystems (Cury et al., 2008; Perry et al., 2009; Planque et al., 2009). Indeed, although there is general agreement about the global causes of such declines in marine biodiversity and degradation of the ecosystem, there is little understanding of how recovery might be achieved at local levels within individual ecosystems (Eero et al., 2011). Risso's dolphin occurrence and distribution in the western Ligurian Sea suggest that this population may be responding to both natural and anthropogenic drivers of change.

A number of anthropogenic threats are likely to have some impact on Risso's dolphins in the Mediterranean Sea. Underwater noise pollution is suspected to be a significant source of impact on the species, and exposure to high-intensity sonar has been found to be deleterious to Risso's dolphins (Jepson et al., 2005; Weilgart, 2007). Chemical contamination may also pose a threat to Risso's dolphins in the Mediterranean Sea (Fossi and Marsili, 2003; Marsili and Focardi, 1997; Shoham-Frider et al., 2002).

According to IUCN's Red List, the major recognized threat for Risso's dolphins in the Mediterranean Sea is fisheries bycatch (Gaspari and Natoli, 2012). Data collected from the Spanish longline fishery in the Western Mediterranean Sea from 2000 to 2009, showed that Risso's dolphins were the most frequent marine mammal bycatch (33–59%—of 56 total instances for four species) (Macías López et al., 2012). Moreover, the authors stated that the species was mainly caught over the continental shelf with Japanese longlines or an experimental home-based longline. This could have had a serious impact on Risso's dolphin populations in the area around the Balearic Islands and may have affected this species in the western Ligurian Sea, which, as we have noted, likely moves into and out of the area. Currently, longline fishing in the western Ligurian Sea has dwindled (no longline landings have been reported since 2006), while bottom trawl (19.5% of total landings) and purse seine (54.4% of total landings) are the dominant types of gear, together with the small-scale fishery (20.9% of total landings). Regional statistics indicated that overall fishery landings in the Ligurian Sea region are of the order of 4500 tonnes per year, which is less than the annual amount of adjacent regions (e.g. 9000 tonnes of landings in the upper Tyrrhenian Sea and 11,000 tonnes of landings in the Gulf of Lions). However, 44% (eight out of 18) of the Risso's dolphins stranded between 1986 and 2014, were reported in the BDS as bycatches or as having signs of net entanglement. Based on the high level of Risso's dolphin bycatch that was documented by Macías López et al. (2012), it seems that this species is highly susceptible to catch by some longline gear and that Risso's dolphins in the Ligurian Sea could have been impacted by the fishery.

In addition to bycatch, other fisheries effects (e.g. prey depletion) could be a cause of the population decrease observed for Risso's dolphins in the western Ligurian Sea. However, the absence of stranding reports of the species since 2008, concurrent to the reduction of the local longline fishery, and the fact that the local fishery pressure is about half of the fishery pressure in adjacent regions suggest that climate-driven environmental changes might be implied in the decline of these dolphins in the area. As shown in Fig. 9, local fishery landings in the Ligurian Sea have been significantly decreasing, as have the catches in the Gulf of Lion. No such trend is detectable for the statistics of fishery landings of the adjacent Tyrrhenian Sea (see Section 2.4).

Fig. 10 shows the summer average patterns of SST and chlorophyll-a local data series. Both SST and chlorophyll-a show a moderate decreasing trend. All of these elements suggest that the ecosystem in the western Ligurian Sea may be undergoing an environmental change. The period from 2003 to 2005 seems to constitute a sort of breakpoint for the time series. The summer of 2003 was characterized by the highest SST, and the summer of 2004 had the highest primary productivity. After these relative peaks the series showed a monotonic decreasing trend. The same pattern was apparently underscored by the *M/R* estimates of the Risso's dolphin population size, which indicated 2005 as a relative maximum of Risso's dolphin abundance, followed by seasons of relative minima. In this same period, some

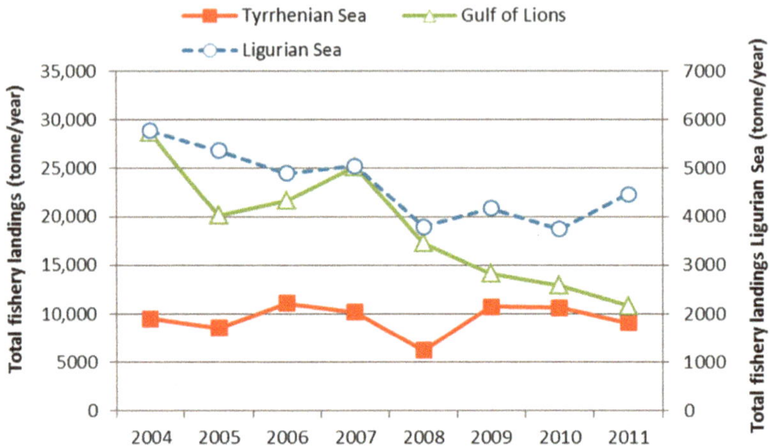

Fig. 9 Statistics for fishery landings in the Ligurian Sea region compared with Tyrrhenian Sea and Gulf of Lion. *Data source: Ligurian Sea, Tyrrhenian Sea—Italian National Institute of Statistics (ISTAT) Censimenti Nazionali Agricoltura, Zootecnia e Pesca. Gulf of Lion—FAO Fisheries and Aquaculture Department Yearbooks, Fishery and Aquaculture Statistics.*

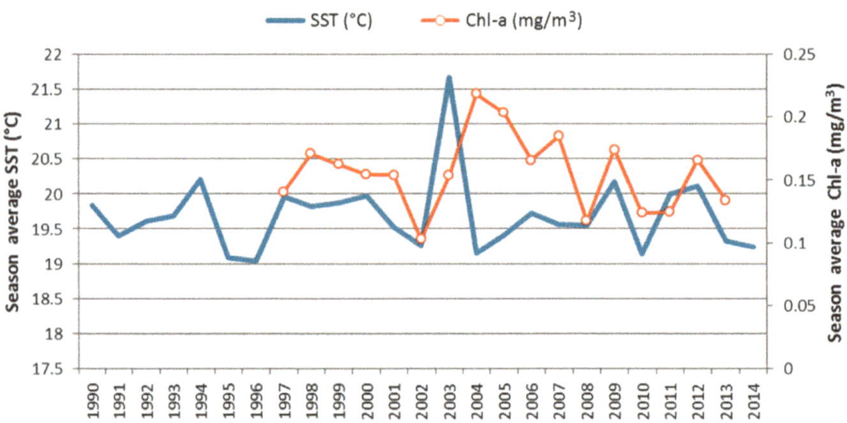

Fig. 10 Sea surface temperature and chlorophyll-a time series for the western Ligurian Sea for the period from 1990 to 2014. *Data generated by the National Oceanic and Atmospheric Administration (NOAA) Earth and System Research Laboratory (ESRL) and were downloaded from www.esrl.noaa.gov/psd/data/gridded/.*

changes in the occurrence and use of the habitat have been reported for other species (Azzellino et al., 2012; Azzellino and Lanfredi, 2015). In particular, striped dolphins (*Stenella coeruleoalba*), sperm whales (*Physeter macrocephalus*) and Cuvier's beaked whales (*Ziphius cavirostris*) have apparently intensified their presence in the study area, probably in order to maximize foraging effort. It is possible that Risso's dolphins may be responding to this change in a different way, leaving the area to find more productive, and possibly less competitive, foraging sites elsewhere. Interspecific competition may be an issue in conditions of decreasing resources (Begon et al., 1986). Azzellino et al. (2008) speculated about the possibility of a temporal and spatial partitioning of the habitat between Risso's dolphin, sperm whale and Cuvier's beaked whale, all of which are deep diving species and have a high degree of diet overlap. Based on the occurrence of cephalopod prey in the stomach contents of different marine mammals stranded along the coast of Tuscany, Pedà et al. (2015) suggested a possible partitioning of cephalopod resources among predator species. They also found overlap in the diet of striped dolphins and Risso's dolphins. According to Pedà et al. (2015), a large proportion of striped dolphin prey consisted of muscular squids belonging to the families Onychoteuthidae (*Onychoteuthis banksii* and *Ancistroteuthis lichtensteinii*) and Ommastrephidae (*Illex coindetii* and *Todarodes sagittatus*), with deep-water species forming a smaller proportion of the prey (Pedà et al., 2015). The findings of Pedà et al. (2015) also confirmed the dominance of mesopelagic (*O. banksii*, *A. lichtensteinii* and *Heteroteuthis dispar*) and deep-water (*Histioteuthis bonnellii* and *Galiteuthis armata*) cephalopods as prey items in Risso's dolphin diet as was described in previous studies (Bearzi et al., 2011; Blanco et al., 2006; Clarke, 1996; Würtz et al., 1992). Thus, in the context of declining ecological resources, Risso's dolphins in the western Ligurian Sea may be competing with both sperm whales and striped dolphins, which might have forced the species to change its use of the habitat.

Further studies that systematically address the ecological connections of this specific population with populations of other Mediterranean areas are urgently needed to fill the gap between our understanding of causes and effects concerning this particular case of study.

7. CONCLUSIONS

Long-term monitoring studies, such as those we have summarized here for Risso's dolphins in the western Ligurian Sea, are a valuable tool allowing scientists to understand the factors that influence ecosystem structure and dynamics at variable spatial and temporal scales. Our results seem to indicate that the occurrence of this cetacean species is responding to changes in environmental and anthropogenic factors. The observed changes in Risso's dolphin local distribution and relative abundance suggest that the population is declining in the region. The local Risso's dolphin population in the western Ligurian Sea was estimated to consist of about 100 individuals with a 95% CI of 60–220 individuals. Mark–recapture estimates confirmed the local decreasing trend within this population, which was also indicated by a decrease in encounter rates. Based on *M/R* estimates, the population size was significantly declining from an average of about 120–150 Risso's dolphin individuals, during 2000 to 2005, to an average of 70–100 dolphins during the period from 2010 to 2014.

The causes of decline in Risso's dolphin numbers in the area remain unclear. However, the observed shift in the local primary production and a possible general decrease of fishery productivity affecting the local area and the adjacent regions, suggest the ecosystem may be facing conditions of declining resources. The concurrent change of use (i.e. intensified occurrence in the study area) that was observed for other species that might compete with Risso's dolphins for prey resources in the western Ligurian Sea, adds support to this hypothesis. Although overfishing alone cannot be excluded, the most likely reasons for such ecosystem functional change might be attributed to environmental variability effects combined with fishery impacts. Further research to systematically address the ecological connections of this specific population with Risso's dolphin populations of other areas of the Mediterranean Sea will be useful to fill the gap between our understanding of causes and effects observed here.

ACKNOWLEDGEMENTS

Part of the findings included in this review derive from the study 'Analysis of the distribution and absolute/relative abundance of sperm whale (*P. macrocephalus*), Risso's dolphin (*G. griseus*) and Cuvier's beaked whale (*Z. cavirostris*) in the Pelagos Sanctuary in function of environmental changes and anthropogenic pressures' financed by the Italian Ministry for the Environment (N.0003302/PNM 19/02/2014). In addition, this work would not have been possible without the enthusiasm of all the people (researchers, research assistants, collaborators, volunteers, under- and postgraduate students) who participated in the surveys on board the research vessels by Tethys Research Institute. Many thanks to the skippers who have provided an invaluable support during fieldwork. Thanks to the International Fund of Animal Welfare (IFAW) for the use of the software LOGGER and to Marina degli Aregai and Portosole for logistic support. Many thanks to the anonymous reviewers whose comments allowed to significantly improving the manuscript.

REFERENCES

Afifi, A., Clark, V., 1998. Computer-aided multivariate analysis. In: Texts in Statistical Science, fourth ed. Chapman & Hall/CRC Press, London, UK.

Airoldi, S., Azzellino, A., David, L., Dhermain, F., Di-Meglio, N., Jimenez, V., Jourdan, J., Labach, H., Lanfredi, C., Remonato, E., Rosso, M., Roul, M., Scuderi, A., 2015.

Absolute abundance of Risso's dolphin (*Grampus griseus*) in the north western part of the Pelagos Sanctuary. Report Convention Pelagos Sanctuary No. 03/2014.

Azzellino, A., Gaspari, S., Airoldi, S., Nani, B., 2008. Habitat use and preferences of cetaceans along the continental slope and the adjacent pelagic waters in the western Ligurian Sea. Deep-Sea Res. I 55, 296–323.

Azzellino, A., Lanfredi, C., 2015. Analisi di distribuzione e abbondanza assoluta e relativa delle specie capodoglio (*Physeter macrocephalus*), grampo (*Grampus griseus*) e zifio (*Ziphius cavirostris*) nel Santuario Pelagos alla luce dei cambiamenti ambientali e dei fattori di pressione antropica. Report of Project financed by Ministero dell'Ambiente e della Tutela del Territorio e del Mare (N.0003302/PNM 19/02/2014) (In Italian).

Azzellino, A., Panigada, S., Lanfredi, C., Zanardelli, M., Airoldi, S., Notarbartolo di Sciara, G., 2012. Predictive models for managing marine areas: habitat use and temporal variability of marine mammal distribution within the Pelagos Sanctuary (northwestern Mediterranean Sea). Ocean Coast. Manag. 67, 63–74.

Baird, R.W., 2009. Risso's dolphin *Grampus griseus*. In: Perrin, W.F. , et al., (Eds.), Encyclopedia of Marine Mammals, second ed. Academic Press, San Diego, CA, pp. 975–976.

Bearzi, G., Reeves, R.R., Remonato, E., Pierantonio, N., Airoldi, S., 2011. Risso's dolphin *Grampus griseus* in the Mediterranean Sea. Mamm. Biol. 76, 385–400.

Begon, M., Harper, J.L., Townsend, C.R., 1986. Ecology: Individuals, Populations and Communities. Blackwell Scientific Publications, London.

Bello, G., Bentivegna, F., 1996. Cephalopod remains from the stomach of a Risso's dolphin (*Grampus griseus*), stranded along the eastern Tyrrhenian coast. Atti Soc. Ital. Sci. Nat. Museo Civ. Stor. Nat. Milano 135/1994 (2), 467.

Blanco, C., Raduán, M.Á., Raga, J.A., 2006. Diet of Risso's dolphin (*Grampus griseus*) in the western Mediterranean Sea. Sci. Mar. 70, 407–411.

Boisseau, O., Lacey, C., Lewis, T., Moscrop, A., Danbolt, M., Mclanaghan, R., 2010. Encounter rates of cetaceans in the Mediterranean Sea and contiguous Atlantic area. J. Mar. Biol. Assoc. UK 90 (8), 1589–1599.

Bräger, S., 1999. Association patterns in three populations of Hector's dolphin, *Cephalorhyncus hectori*. Can. J. Zool. 77, 13–18.

Campana, I., Crosti, R., Angeletti, D., Carosso, L., David, L., Di-Meglio, N., Moulins, A., Rosso, M., Tepsich, P., Arcangeli, A., 2015. Cetacean response to summer maritime traffic in the Western Mediterranean Sea. Mar. Environ. Res. 109, 1–8.

Cañadas, A., Sagarminada, R., Garcìa-Tiscar, S., 2002. Cetacean distribution related with depth and slope in the Mediterranean waters off southern Spain. Deep-Sea Res. I 49, 2053–2073.

Cañadas, A., Sagarminaga, R., De Stephanis, R., Urquiola, E., Hammond, P.S., 2005. Habitat preference modelling as a conservation took proposals for marine protected areas for cetaceans in southern Spanish waters. Aquat. Conserv. Mar. Freshwat Ecosyst 15 (5), 495–521.

Clarke, M.R., 1996. Cephalopods as prey. III. Cetaceans. Philos. Trans. R. Soc. B 351, 1053–1065.

Clarke, M., Pascoe, P., 1985. Stomach contents of a Risso's dolphin (*Grampus griseus*) stranded at Thurlestone, Devon. J. Mar. Biol. Assoc. UK 65, 983–986.

Cockroft, V.G., Haschick, S.L., Klages, N.T.W., 1993. The diet of Risso's dolphin, *Grampus griseus* (Cuvier, 1812) from the east coast of South Africa. Mamm. Biol. 58, 286–293.

De Boer, M.N., Clark, J., Leopold, M.F., Simmonds, M.P., Reijnders, P.J.H., 2013. Photo-identification methods reveal seasonal and long-term site-fidelity of Risso's dolphins (*Grampus griseus*) in shallow waters (Cardigan Bay, Wales). Open J. Mar. Sci. 3, 65–74.

Dimatteo, S., Siniscalchi, M., Esposito, L., Prunella, V., Bondanese, P., Bearzi, G., Quaranta, A., 2011. Encounters with pelagic and continental slope cetacean species near the northern shore of the Gulf of Taranto, Italy. Ital. J. Zool. 78 (1), 130–132.
Cury, P.M., Shin, Y.-J., Planque, B., Durant, J.M., Fromentin, J.-M., Kramer-Schadt, S., Stenseth, N.C., Travers, M., Grimm, V., 2008. Ecosystem oceanography for global change in fisheries. Trends Ecol. Evol. 23, 338–346.
Eero, M., Mackenzie, B.R., Köster, F.W., Gislason, H., 2011. Multi-decadal responses of a cod (*Gadus morhua*) population to human-induced trophic changes, fishing, and climate. Ecol. Appl. 21 (1), 214–226.
Fossi, M.C., Marsili, L., 2003. Effects of endocrine disruptors in aquatic mammals. Pure Appl. Chem. 75 (11–12), 2235–2247.
Di Méglio, N., David, L., Beaubrun, P., 1999. Spatio-temporal distribution of Risso's dolphin, *Grampus griseus* (Cuvier, 1812) in summer in the North-Western Mediterranean Sea. Eur. Res. Cetaceans 13, 195–200.
Frantzis, A., Alexiadou, P., Paximadis, G., Politi, E., Gannier, A., Corsini-Foka, M., 2003. Current knowledge of the cetacean fauna of the Greek Seas. J. Cetacean Res. Manag. 5 (3), 219–232.
Gannier, A., 2005. Summer distribution and relative abundance of delphinids in the Mediterranean Sea. Rev. Écol. (Terre Vie) 60, 223–238.
Gaspari, S., 2004. Social and population structure of striped and Risso's dolphins in the Mediterranean Sea. Ph.D. thesis, University of Durham, Durham, UK.
Gaspari, S., Natoli, A., 2006. Risso's dolphin *Grampus griseus* (Mediterranean sub-population). In: Reeves, R.R., Notarbartolo di Sciara, G. (Eds.), The Status and Distribution of Cetaceans in the Black Sea and Mediterranean Sea. IUCN Centre for Mediterranean Cooperation, Malaga, Spain, pp. 29–33.
Gaspari, S., Natoli, A., 2012. *Grampus griseus* (Mediterranean sub-population). In: The IUCN Red List of Threatened Species 2012. e.T16378423A16378453, http://dx.doi.org/10.2305/IUCN.UK.2012-1.RLTS.T16378423A16378453.en.
Gaspari, S., Airoldi, S., Hoelzel, R., 2007. Risso's dolphins (*Grampus griseus*) in UK waters are differentiated from a population in the Mediterranean Sea and genetically less diverse. Conserv. Genet. 8, 727–732.
Ginsberg, J.R., Young, T.P., 1992. Measuring association between individuals or groups in behavioural studies. Anim. Behav. 44, 377–379.
Gomez de Segura, A., Crespo, E.A., Pedraza, S.N., Hammond, P.S., Raga, J.A., 2006. Abundance of small cetaceans in the waters of the Central Spanish Mediterranean. Mar. Biol. 150, 149–160.
Gowans, S., Würsig, B., Karczmarski, L., 2007. The social structure and strategies of delphinids: predictions based on an ecological framework. Adv. Mar. Biol. 53, 195–294.
Hammond, P.S., 2010. Estimating the abundance of marine mammals. In: Boyd, I.L., Bowen, W.D., Iverson, S.J. (Eds.), Marine Mammal Ecology and Conservation. A Handbook of Techniques. Oxford University Press, Inc., New York, pp. 42–67
Hartman, K.L., Visser, F., Hendriks, A.J.E., 2008. Social structure of Risso's dolphins (*Grampus griseus*) at the Azores: a stratified community based on highly associated social units. Can. J. Zool. 86 (4), 294–306.
Jackson, J.B.C., Kirby, M.X., Berger, W.H., Bjorndal, K.A., Botsford, L.W., Bourque, B.J., Bradbury, R.H., Cooke, R., Erlandson, J., Estes, J.A., Hughes, T.P., Kidwell, S., Lange, C.B., Lenihan, H.S., Pandolfi, J.M., Peterson, C.H., Steneck, R.S., Tegner, M.J., Warner, R.R., 2001. Historical overfishing and the recent collapse of coastal ecosystems. Science 293, 629–638.
Jefferson, T.A., Weir, C.R., Anderson, C.R., Balance, L.T., Kenney, R.D., Kiszka, J.J., 2014. Global distribution of Risso's dolphin *Grampus griseus*: a review and critical evaluation. Mammal Rev. 44, 56–68.

Jepson, P.D., Deaville, R., Patterson, I.A.P., Pocknell, A.M., Ross, H.M., Baker, J.R., Howie, F.E., Reid, R.J., Colloff, A., Cunningham, A.A., 2005. Acute and chronic gas bubble lesions in cetaceans stranded in the United Kingdom. Vet. Pathol. 42, 291–305.

Kerem, D., Hadar, N., Goffman, O., Scheinin, A., Kent, R., Boisseau, O., Schattner, U., 2012. Update on the cetacean fauna of the Mediterranean Levantine Basin. Open J. Mar. Sci. 6, 6–27.

Kruse, S.L., 1989. Aspects of the biology, ecology, and behavior of Risso's dolphins (*Grampus griseus*) off the California coast. PhD thesis, University of California, Santa Cruz.

Kruse, S., Caldwell, D.K., Caldwell, M.C., 1999. Risso's dolphin *Grampus griseus* (G. Cuvier, 1812). In: Ridgway, S.H., Harrison, R. (Eds.), The Second Book of Dolphins and the Porpoises. In: Handbook of Marine Mammals vol. 6. Academic Press, London, UK, pp. 183–212.

Lotze, H.K., Lenihan, H.S., Bourque, B.J., Bradbury, R.H., Cooke, R.G., Kay, M.C., Kidwell, S.M., Kirby, M.X., Peterson, C.H., Jackson, J.B.C., 2006. Depletion, degradation, and recovery potential of estuaries and coastal seas. Science 312, 1806–1809.

Lotze, H.K., Worm, B., 2009. Historical baselines for large marine animals. Trends Ecol. Evol. 24, 254–262.

Macías López, D., García Barcelona, S., Baez, J.C., de la Serna, J.M., Ortiz de Urbina, J.M., 2012. Marine mammal bycatch in Spanish Mediterranean large pelagic longline fisheries, with a focus on Risso's dolphin (*Grampus griseus*). Aquat. Living Resour. 25, 321–331.

Marsili, L., Focardi, S., 1997. Chlorinated hydrocarbon (HCB, DDTs and PCBs) levels in cetaceans stranded along the Italian coasts: an overview. Environ. Monit. Assess. 45, 129–180.

Marini, L., Consiglio, C., Angradi, A.M., Catalano, B., Sanna, A., Valentini, T., Finoia, M.G., Villetti, G., 1996. Distribution, abundance and seasonality of cetaceans sighted during scheduled ferry crossings in the central Tyrrhenian Sea: 1989–1992. Ital. J. Zool. 63 (4), 381–388.

Macquart-Moulin, C., Patriti, G., 1996. Accumulation of migratory macronekton crustaceans over the upper slope and submarine canyons of the northwestern Mediterranean. Deep-Sea Res. I 43 (5), 579–601.

Moulins, A., Rosso, M., Ballardini, M., Würtz, M., 2008. Partitioning of the Pelagos Sanctuary (north-western Mediterranean Sea) into hotspots and coldspots of cetacean distributions. J. Mar. Biol. Assoc. UK 88, 1273–1281.

Notarbartolo di Sciara, G., Venturino, M.C., Zanardelli, M., Bearzi, G., Borsani, J.F., Cavalloni, B., 1993. Cetaceans in the central Mediterranean Sea: distribution and sighting frequencies. Boll. Zool. 60, 131–138.

Notarbartolo di Sciara, G., Agardy, T., Hyrenbach, D., Scorazzi, T., Van Klaveren, P., 2008. The Pelagos Sanctuary for Mediterranean marine mammals. Aquat. Conserv. 18, 367–391.

Otis, D.L., Burnham, K.P., White, G.C., Anderson, D.R., 1978. Statistical inference from capture data on closed animal populations. Wildl. Monogr. 62, 3–135.

Ozturk, B., Salman, A., Ozturk, A.A., Tonay, A., 2007. Cephalopod remains in the diet of striped dolphins (*Stenella coeruleoalba*) and Risso's dolphins (*Grampus griseus*) in the eastern Mediterranean Sea. Vie Milieu—Life Environ. 57 (1–2), 57–63.

Pauly, D., Trites, A.W., Capuli, E., Christensen, V., 1998. Diet composition and trophic levels of marine mammals. ICES J. Mar. Sci. 55, 467–481.

Pedà, C., Battaglia, P., Scuderi, A., Voliani, A., Mancusi, C., Andaloro, F., Romeo, T., 2015. Cephalopod prey in the stomach contents of odontocete cetaceans stranded in the western Mediterranean Sea. Mar. Biol. Res. 11 (6), 593–602.

Perrin, W.F., Reilly, S.B., 1994. Reproductive parameters of dolphins and small whales of family of Delphinidae. Report International Whaling Commission Spec. Issue No. 6, 97–133.

Perry, R.I., Cury, P., Brander, K., Jennings, S., Möllmann, C., Planque, B., 2009. Sensitivity of marine systems to climate and fishing: concepts, issues and management responses. J. Mar. Syst. 79 (3–4), 427–435.
Pettex, E., Lambert, C., Laran, S., Ricart, A., Virgili, A., Falchetto, H., Authier, M., Monestiez, P., Van Canneyt, O., Doremus, G., Blanck, A., Toison, V., Ridoux, V., 2014. Suivi Aerien de la Megafaune Marine en France metropolitaine. SAMM Rapport Finalpp. 1–169.
Planque, B., Fromentin, J.-M., Cury, P., Drinkwater, K., Jennings, S., Kifani, S., Perry, R.I., 2009. How does fishing alter marine populations and ecosystems sensitivity to climate? J. Mar. Syst. 79 (3–4), 403–417.
Podestà, M., Meotti, C., 1991. The stomach contents of a Cuvier's beaked whale *Ziphius cavirostris* and Risso's dolphin *Grampus griseus* stranded in Italy. Eur. Res. Cetaceans 5, 58–61.
Podestà, M., D'Amico, A., Pavan, G., Drougas, A., Komnenou, A., Portunato, N., 2006. A review of Cuvier's beaked whale strandings in the Mediterranean Sea. J. Cetacean Res. Manag. 7 (3), 251–261.
Podestà, M., Pavan, G., Bernuzzi, E., 2009. The online Italian stranding database. In: 23rd Annual Conference European Cetacean Society, Istanbul, Turkey.
Pollock, K.H., Nicholls, J.D., Brownie, C., Hines, J.E., 1990. Statistical inference for capture–recapture experiments. Wildl. Monogr. 107, 3–97.
Reeves, R.R., Notarbartolo di Sciara, G., 2006. The status and the distribution of cetaceans in the Black Sea and Mediterranean sea. IUCN Centre for Mediterranean Cooperation, Malaga, Spain. 137 pp.
Rutherford, A., 2001. Introducing Anova and Ancova: A GLM Approach Introducing Statistical Methods Series. SAGE Publications, London, UK.
Shoham-Frider, E., Amiel, S., Roditi-Elasar, M., Kress, N., 2002. Risso's dolphin (*Grampus griseus*) stranding on the coast of Israel (eastern Mediterranean). Autopsy results and trace metal concentrations. Sci. Tot. Environ. 295 (1–3), 157–166.
Taylor, B.L., Baird, R., Barlow, J., Dawson, S.M., Ford, J.K.B., Mead, J.G., Notarbartolo di Sciara, G., Wade, P., Pitman, R.L., 2012. *Grampus griseus*. The IUCN Red List of Threatened Species 2012: e.T9461A17386190.
Tepsich, P., Rosso, M., Halpin, P.N., Moulins, A., 2014. Habitat preferences of two deep-diving cetacean species in the northern Ligurian Sea. Mar. Ecol. Prog. Ser. 508, 247–260.
UNEP MAP-RAC/SPA, 2014. Status and conservation of cetaceans in the Adriatic Sea. By Holcer, D., Fortuna, C.M., Mackelworth, P.C. Draft internal report for the purposes of the Mediterranean Regional Workshop to Facilitate the Description of Ecologically or Biologically Significant Marine Areas, Malaga, Spain.
White, G.C., Burnham, K.P., 1999. Program MARK: survival estimation from populations of marked animals. Bird Study 46 (Suppl.), S120–S139.
Whitehead, H., 2009. SOCPROG programs: analysing animal social structures. Behav. Ecol. Sociobiol. 63, 765–778.
Weilgart, L.S., 2007. The impacts of anthropogenic ocean noise on cetaceans and implications for management. Can. J. Zool. 85 (11), 1091–1116.
Wells, R.S., 1991. The role of long-term study in understanding the social structure of a bottlenose dolphin community. In: Pryor, K., Norris, K.S. (Eds.), Dolphin Societies, Discoveries & Puzzles. University of California Press, Berkeley, CA, pp. 199–226.
Würsig, B., Würsig, M., 1980. Behaviour and ecology of dusky porpoises, *Lagenorhynchus obscurus*, in the South Atlantic. Fish. Bull. 77, 871–890.
Würtz, M., Poggi, R., Clarke, M.R., 1992. Cephalopods from the stomach of a Risso's dolphin (*Grampus griseus*) from the Mediterranean. J. Mar. Biol. Assoc. UK 72, 861–867.

CHAPTER EIGHT

The Rough-Toothed Dolphin, *Steno bredanensis*, in the Eastern Mediterranean Sea: A Relict Population?

D. Kerem[*,†,1], **O. Goffman**[*,†], **M. Elasar**[*,†], **N. Hadar**[†], **A. Scheinin**[*,†], **T. Lewis**[‡]

[*]School of Marine Sciences, University of Haifa, Haifa, Israel
[†]Israel Marine Mammal Research & Assistance Center (IMMRAC), Haifa, Israel
[‡]North Atlantic & Mediterranean Sperm Whale Catalogue (NAMSC), London, United Kingdom
[1]Corresponding author: e-mail address: dankerem@research.haifa.ac.il

Contents

Abstract

Only recently included among the cetacean species thought to regularly occur in the Mediterranean, the rough-toothed dolphin (*Steno bredanensis*) is an obscure and

Advances in Marine Biology, Volume 75
ISSN 0065-2881
http://dx.doi.org/10.1016/bs.amb.2016.07.005

enigmatic member of this ensemble. Preliminary genetic evidence strongly indicates an Atlantic origin, yet the Mediterranean distribution for this species is conspicuously detached from the Atlantic, with all authenticated records during the last three decades being east of the Sicilian Channel and most within the bounds of the Levantine Basin. These dolphins are apparently a small, relict population, probably the remnant of a larger one, contiguous with that in the Atlantic and nowadays entrapped in the easternmost and warmest province. Abundance data are lacking for the species in the Mediterranean. Configuring acoustic detection software to recognise the apparently idiosyncratic vocalisations of rough-toothed dolphins in past and future acoustic recordings may prove useful for potential acoustic monitoring. Evidence accumulated so far, though scant, points to seasonal occupation of shallow coastal waters. Vulnerability to entanglement in gill-nets, contaminants in the region, and the occurrence of mass strandings (possibly in response to anthropogenic noise), are major conservation concerns for the population in the Mediterranean Sea.

1. INTRODUCTION

The rough-toothed dolphin (*Steno bredanensis*, Lesson, 1828) is a monotypic species in the genus *Steno*. Its specific name was given in honour of J.G.S. van Breda, who, in 1825, first brought a fresh specimen and its skull to the attention of the scientific community. The species is distributed globally, mainly in tropical, subtropical and warm-temperate waters, where it is typically found in deep oceanic waters though in some areas it can be seen in shallow nearshore waters (Jefferson et al., 2015; Miyazaki and Perrin, 1994; West et al., 2011). It seems to favour the surrounds of steep volcanic islands such as French Polynesia, Hawaii and the Canary Islands, where deep waters reach close to the coast. At these locations the species is mostly sighted over depths between 500 and 2000 m and at distances of 2–5.5 km from shore (Baird et al., 2008; Gannier and West, 2005; Oremus et al., 2012; West et al., 2011). Animals can be solitary or in groups of between two and 50 individuals (West et al., 2011), but have been observed in larger groups. For example, Baird et al. (2008) observed groups of up to 90 individuals around the Hawaiian Islands.

At close quarters (e.g. when bow riding) the rough-toothed dolphin is easily distinguished from other small delphinids by the lack of demarcation separating the melon and beak, which results in a slanted forehead, gradually merging into a long and slender rostrum (Jefferson et al., 2015; West et al., 2011) (Fig. 1). Other distinguishing features include white lips, a relatively tall and erect dorsal fin, relatively large pectoral fins, light-coloured splotches

Fig. 1 A rough-toothed dolphin (*Steno bredanensis*) photographed 21 August 2013, southwest of Cyprus, showing the slanted forehead unique to the species. *Photograph: International Fund for Animal Welfare (IFAW)/Marine Conservation Research (MCR).*

Fig. 2 The tight formation of a group of rough-toothed dolphins (*Steno bredanensis*) photographed 22 March 2006, inside the Port of Haifa, Israel. *Photograph: M. Elasar.*

on the flanks and the tendency of small subgroups to swim in very tight formation (Addink and Smeenk, 2001; Götz et al., 2005; Ritter, 2002) (Fig. 2). All of these features may be less discernable and/or distinctive at longer ranges, such that this species could easily be confused with the common bottlenose dolphin (*Tursiops truncatus*) (Jefferson et al., 2015). It is for this reason that in the present communication, definite Mediterranean records of *S. bredanensis* are only considered to be valid if substantiated by photographs. For the sake of completeness, other records will be mentioned, but are regarded as 'unsubstantiated'.

Here, we provide an overview of information regarding the rough-toothed dolphin in the Mediterranean Sea. It is a story told in three parts: (1) The historic period, from the early 1820s up to 1985, throughout which the species was considered a rare/occasional visitor; (2) The period starting in 1985, when a spectacular sighting of an estimated 160 individuals aggregated in the Ionian Sea (Watkins et al., 1987), the largest ever reported for the species (West et al., 2011), marked a change of status to 'possibly regular'; (3) The period starting in 1997 and ongoing, during which sightings became frequent, and a 'regular' status of the species was accepted and scientific data about it began to accumulate. All sightings within this period occurred in the eastern Mediterranean and all except two were confined to the Levantine Basin. We also present information resulting from a preliminary genetic analysis of the species in the Mediterranean and discuss the potential for acoustic monitoring and its usefulness for collecting data regarding the occurrence and abundance of rough-toothed dolphins in the Mediterranean Sea. In addition, we discuss the ecology of rough-toothed dolphins in the region and provide a brief discussion of potential conservation threats to this species.

2. ROUGH-TOOTHED DOLPHIN DISTRIBUTION WITHIN THE MEDITERRANEAN SEA

2.1 The Historic Period

The few definite records from the Mediterranean Sea in the historic period (starting in the early 1820s), including museum holdings and sightings, were reviewed in Watkins et al. (1987). A remarkable item, on which, unfortunately, we could not get any more details, was a personal communication by René Guy Busnel to Anne Collet, stating that between 1950 and 1960, 'about 10' specimens were live-captured in the Mediterranean for acoustic studies in the 'Laboratoire de Physiologie Acoustique' in Jouy-en-Josas, France (Collet, 1984). The communication seems credible, as it also included a specimen captured near Madeira, at the same period and for the same laboratory, on which details have been published (Busnel and Dziedzic, 1966). The item is remarkable in the sense that being commissioned to live-capture 10 animals of a reportedly rare species in the Mediterranean, one would have to know where such animals could reliably be found. One possibility would be the southern Ionian Sea and Sicilian Channel, in which possible (Di Natale and Mangano, 1981) and definite (Boisseau et al., 2010) sightings were made, respectively, prior to and following the extraordinary sighting of an estimated 160 rough-toothed dolphins (in eight groups of approximately 20 individuals each) in the Ionian Sea, reported by Watkins et al. (1987).

2.2 The Period of 1985–1997

The sighting recounted by Watkins et al. (1987) prompted the authors to propose '...a more stable *Steno* population in the Mediterranean than was previously thought'. They made the proposal based on three factors: (1) the large number of dolphins observed (approximately 160 dolphins), (2) the size range of individuals (including adults of 2–2.5 m length and numerous cow-calf pairs), and (3) the considerable distance from Gibraltar (the sighting was made on 4 September 1985 in the Sicily Channel, about 170 km from Cape Passero, Sicily, and 130 km off southeast of Malta, in waters of about 500 m depth). As will be detailed later, it is based on these same indications that the regular presence of the species in the Mediterranean has recently been accepted (Kerem et al., 2012; Notarbartolo di Sciara and Birkun, 2010). During the 12-year period between 1985 and 1997, there were no additional records of the species.

2.3 The Period of 1997–Present

Since 1997, there have been 23 authenticated records of rough-toothed dolphins, including 12 sightings and 11 strandings/bycatches (Table 1). On one occasion, a large group of c.40 dolphins of various sizes, including calves, was documented for ~5 h inside Haifa Port. At the same time, port authorities and naval base observers reported the presence of a similarly sized group of approximately 40 dolphins on the seaward side of the port's breakwater (for a total of c.80 dolphins), which, although not photographed, was also likely to be rough-toothed dolphins (Kerem, 2005; Kerem et al., 2012). The even greater distance from Gibraltar of the majority of recent sightings (see Table 1 and Fig. 3) suggests a resident relict population apparently favouring the easternmost province of the Mediterranean. Despite the relatively high number of recent records, little research has been conducted to improve scientific knowledge about the species' biology and ecology in the Mediterranean. Being removed from the major cetacean research centres in the west, it remains the least studied and least known of the basin's regular species.

Table 1 Mediterranean Rough-Toothed Dolphin (*Steno bredanensis*) Records from 1985 to 2016

Date (d/m/y)[Source]	Location	Position	Depth	Area	Number of Individuals	Sex	Age Group	Record Type
4/9/1985[1]	130 km SE of Malta	35.47°N, 015.88°E	~500 m	Ion	~160	M + F	A, J + C	S (8 subgroups)
16/3/1997[2]	Dugit, Gaza	31.59°N, 034.49°E	—	LB	1	M	C	St
1/3/1998[2]	Jaffa, Israel	32.06°N, 034.76°E	—	LB	1	?	C	St
13/4/1998[2]	Atlit, Israel	32.70°N, 034.93°E	—	LB	1	F	J	BC
5/4/2001[2]	Shavey-Zion, Israel	32.98°N, 035.08°E	—	LB	1	F	C	St
16/2/2002[2]	Atlit, Israel	32.70°N, 034.94°E	—	LB	1	M	C	BC
5/4/2002[3]	Donnalucata, Sicily	36.76°N, 014.64°E	—	EM	6	4M + 2F	A + J	LSt (3 refloated)
9/3/2003[2]	Haifa, Israel	32.81°N, 034.95°E	—	LB	1	F	C	BC
20/3/2003[2]	Acko, Israel	32.94°N, 035.07°E	—	LB	1	M	C	St
24/9/2003[4]	~160 km west of Kephalonia, Greece	38.50°N, 018.80°E	3000 m	Ion	8	—	7A + C	S
3/5/2005[5]	Off Haifa	32.60°N, 034.02°E	1424 m	LB	2	—	A + J	S
3/5/2005[5]	Off Haifa	32.57°N, 034.23°E	1650 m	LB	4	—	A	S
15/3/2006[2]	Nahariya, Israel	33.01°N, 035.09°E	—	LB	1	F	C	St
22/3/2006[2]	Haifa Port, Israel	32.82°N, 034.99°E	5–10 m	LB	~40	—	A, J + C	S
21/5/2007[6]	Off Jounieh, Lebanon	34.01°N, 035.60° E*	?	LB	1(?2)	—	A	S
5/6/2007[2]	~100 km SW of Haifa, Israel	32.42°N, 033.87°E	1500 m	LB	4	—	A	S

17/6/2007[4]	Off N Cyprus	35.51°N, 032.66°E	2822 m	LB	9	—	A	S
1/7/2007[4]	Cyrenaica, Libya	32.90°N, 021.11°E	320 m	LB	10	—	A + C	S
4/3/2008[7]	Tyre, Lebanon		—	LB	2	F	A	BC (1 pregnant)
4/5/2008[2]	Netanya, Israel	32.33°N, 034.85°E	—	LB	1	?	C	St
15/11/2009[2]	En route Larnaca, Cyprus to Herzliya, Israel	33.32°N, 034.11°E	1800 m	LB	5	—	A	S
10/3/2010[8]	Laidy's Mile, Limassol Bay, Cyprus	34.62°N, 033.01°E	<5 m	LB	21	—	A + J	14 LS (and refloated)
21/8/2013[9]	Between Cyprus and Lebanon	34.11°N, 033.86°E	>2000 m	LB	5	—	A	S (+2 Grampus)
25/8/2013[9]	Between Cyprus and Lebanon	34.27°N, 033.44°E	>3000 m	LB	3	—	A	S

[1]Watkins, W.A., Tyack, P., Moore, K.E., Notarbartolo di Sciara, G., 1987. *Steno bredanensis* in the Mediterranean Sea. Mar. Mamm. Sci. 3, 78–82;
[2]Kerem, D., Hadar, N., Goffman O., Scheinin, A., Kent, R., Boisseau, O., Schattner, U., 2012. Update on the cetacean fauna of the Mediterranean Levantine Basin. Open Mar. Biol. J. 6, 6–27;
[3]Centro Studi Cetacei, 2004. Cetacei spiaggiati lungo le coste italiane. XVII Rendiconto 2002 (Mammalia). Atti Soc. It. Sci. Nat. Museo Civ. St. Nat. Milano 145, 155–169;
[4]Boisseau, O., Lacey, C., Lewis, T., Moscrop, A., Danbolt, M., McLanaghan, R., 2010. Encounter rates of cetaceans in the Mediterranean Sea and contiguous Atlantic area. J. Mar. Biol. Assoc. U. K. 90, 1589–1599;
[5]Eyre and Frizell (2012);
[6]Khalaf, G., National Center for Marine Sciences, Lebanon personal communication to G. Notarbartolo di Sciara, 22 May 2007 (with photographs);
[7]Gonzalvo, J. 2009. Action Plan for the Conservation of Cetaceans in Lebanon. Agreement on the Conservation of Cetaceans of the Black Sea, Mediterranean Sea and contiguous Atlantic Area (ACCOBAMS), Monaco, 42 pp;
[8]Demetropoulos, A., Cyprus Wildlife Society, Greece, personal communication to G. Notarbartolo di Sciara, 12 March 2010 (with photographs);
[9]Ryan, C., Cucknell, A.C., Romagosa, M., Boisseau, O., Moscrop, A., Frantzis, A., McLanaghan, R., 2014. Final report of a visual and acoustic survey for marine mammals in the eastern Mediterranean Sea during summer 2013. International Fund for Animal Welfare. Available at: http://www.marineconservationresearch.co.uk/.
Area: *EM*, eastern Mediterranean; *Ion*, Ionian Sea; *LB*, Levantine Basin (as defined in Kerem et al., 2012). Age-group: *A*, adult; *C*, calf; *J*, juvenile. Record type: *BC*, bycaught; *LSt*, live stranded; *S*, sighting; *St*, stranding. Position: *estimated position.

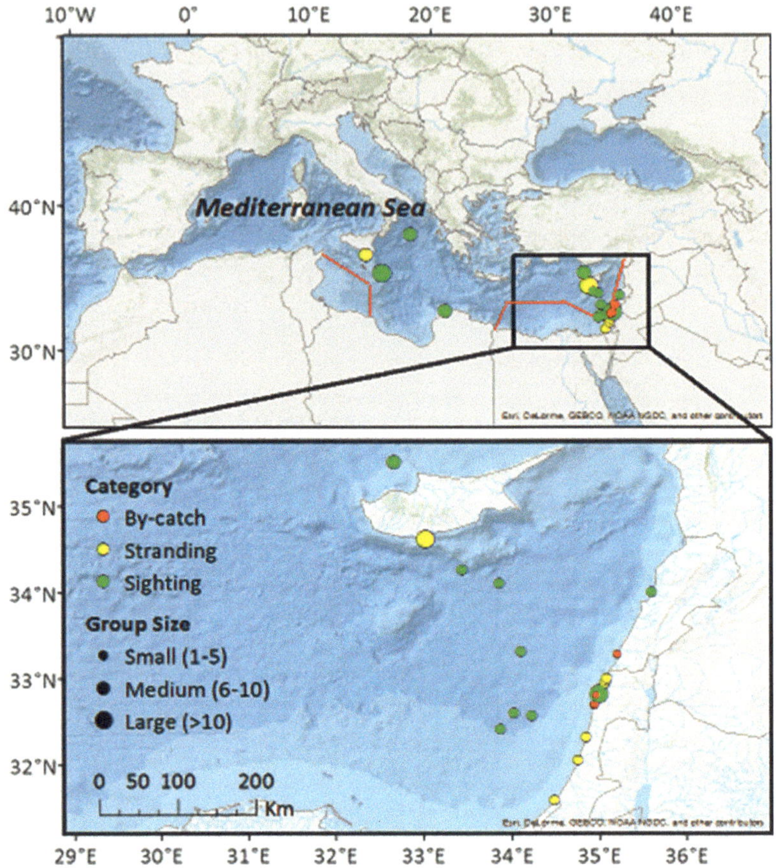

Fig. 3 Map of sightings, strandings and gill-net entanglements of rough-toothed dolphins (*Steno bredanensis*), 1985–2013, in the Mediterranean region. *Upper panel*: areas in the eastern Mediterranean delimited by *red* lines and coastline were not covered by International Fund for Animal Welfare/Marine Conservation Research surveys (see Boisseau et al., 2010; Ryan et al., 2014). *Map prepared by Marina Costa, Tethys Research Institute.*

3. RECENT OCCURRENCE OF ROUGH-TOOTHED DOLPHINS IN THE MEDITERRANEAN SEA

3.1 Spatial Patterns of Records

There are only a small number of recent records of the species in the Mediterranean (see Table 1). Geographic positions of these records are mapped in Fig. 3. Most records are anecdotal. The only published large-scale line-transect surveys using distance sampling methodology in the eastern

Mediterranean were performed by the International Fund for Animal Welfare (IFAW) in 2007 (Boisseau et al., 2010) and by IFAW and Marine Conservation Research (MCR) in 2013 (Ryan et al., 2014). The areas not covered by these surveys are shown in Fig. 3.

In addition, there are four unsubstantiated published records, all west of the Sicilian Channel: sightings (total of 19 individuals) during ferry crossings in Gibraltar Strait (between Algeciras, Spain and Ceuta, Spanish North Africa), during 1989 and 1990 (Hashmi and Adloff, 1991); a mummified body found on 6 June 1994 on the coast of Ibiza, Balearic Islands, reported by the Fundación Aspro Natura (MEDACES, 2011); an adult male found stranded in Santa Teresa di Gallura, Sardinia, in August 2004 (Centro Studi Cetacei, 2006), and a sighting of five individuals near the Pontine Islands, Tyrrhenian Sea, during ferry crossings between Catania and Civitavecchia, Italy, in 2011 (Santoro et al., 2015).

Based only upon the verified reports, the roughed-toothed dolphin appears to demonstrate a complementary distribution pattern to the long-finned pilot whale (*Globicephala melas*), staying east and west of the Sicilian Channel, respectively (see Fig. 3). Rough-toothed dolphins have not been sighted in the northern Alborán Sea during >60,000 km of survey effort, since 1992 (A. Cañadas, Alnilam, Madrid, Spain, personal communication, November 2012). Likewise, during 45,000 km of survey effort in the Gibraltar Strait between 1999 and 2012 and 20,000 km of effort in the Gulf of Cadiz between 2001 and 2012, the species has never been observed (R. de Stephanis, Conservation, Information and Research on Cetaceans (CIRCE), Algeciras, Spain, personal communication, November 2012). This apparent absence extends further north in the eastern North Atlantic, with no known records of the species off mainland Portugal (M. Sequeira, Instituto da Conservação da Natureza e das Florestas (ICNF), Lisbon, Portugal, personal communication, February 2016). Sighting reports from the Moroccan coast of the Alborán Sea are altogether lacking and sightings of the species from the Atlantic waters of Morocco and Western Sahara (Djiba et al., 2015) are also lacking. A compilation of 205 stranding records from the entire Moroccan coast between 1980 and 2009 (Masski and de Stephanis, 2015) revealed a single record in Tarfaya, an Atlantic coastal area closest to the Canary Islands. Taken together, this distribution pattern would indicate that recent entrance from or exchange with the Atlantic Ocean is extremely unlikely.

Within the Eastern Mediterranean, there are no records (strandings or sightings) from the Aegean and Adriatic seas, despite considerable survey effort and beach monitoring (e.g. Frantzis, 2009; Ryan et al., 2014;

UNEP-MAP-RAC/SPA, 2014). The only countries where authenticated strandings/incidental catches of rough-toothed dolphins have occurred are Israel, Lebanon, Cyprus and Italy (Table 1 and Fig. 3). The lack of sighting records from Egypt may be due to the lack of survey effort (Boisseau et al., 2010; Ryan et al., 2014; see Fig. 3).

3.2 Temporal Patterns of Records

An interesting seasonal pattern emerges when plotting records by months of the year (Fig. 4).

Temporal patterns of offshore, deepwater records may be biased by timing and locations of survey effort. However, all strandings (dead or live), bycaught animals and sightings in shallow nearshore waters (14 records in all: 10 from Israel, two from Lebanon and one each from Cyprus and Sicily) occurred between February and June (Table 1) (Kerem et al., 2012). Due to the thorough beach monitoring by the Israel Marine Mammal Research and Assistance Center (IMMRAC), it seems unlikely that strandings occurring during other seasons in Israel would have been missed. One explanation for the apparent seasonal pattern in occurrence of rough-toothed dolphins in the coastal areas of the Levantine Basin was proposed by Kerem et al. (2012), who suggested the existence of a resident deepwater population, that, in late

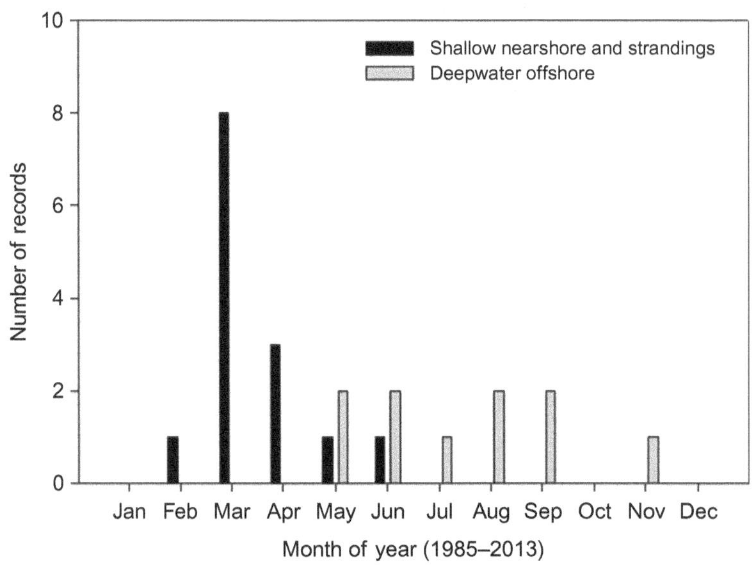

Fig. 4 Distribution of verified Mediterranean records of the rough-toothed dolphin (*Steno bredanensis*) by month of year.

winter-early spring, approaches coastal waters where individuals or groups are more likely to be encountered. The movement of rough-toothed dolphins from offshore, deepwater areas could be driven by the movement and schooling patterns of preferred prey species.

4. ACOUSTIC DETECTION

Acoustic detection and classification may be used to replace or augment visual surveys, while the latter component is absent or off-effort or when visibility is compromised. Acoustic monitoring may be particularly useful for species that are highly vocal and have distinct vocalisations (Adam and Samaran, 2013).

Regarding rough-toothed dolphins, Rankin et al. (2008) commented that '…during 7 yr of combined visual and acoustic surveys in the eastern tropical Pacific, this species was found to be one of the most vocal dolphin species'. The rough-toothed dolphin produces both broadband clicks and whistles, and the characteristics of rough-toothed dolphin whistles have been reported from various areas including the south-western Atlantic Ocean (e.g. Lima et al., 2012), the tropical eastern and central Pacific (e.g. Rankin et al., 2015) and the Mediterranean Sea (Busnel and Dziedzic, 1966; Watkins et al., 1987).

Rankin et al. (2015) summarised the distinguishing characteristics of rough-toothed dolphin whistles in the Pacific as: (1) relatively low frequency with a small frequency range, (2) most whistles containing at least one step (with a mean of 1.5 steps and maximum of eight steps per whistle) and (3) with individual components (steps) of the stepped whistles often being downswept.

In addition to the previously reported sound recordings of Mediterranean rough-toothed dolphins (Busnel and Dziedzic, 1966; Watkins et al., 1987), more recent recordings have been made from IFAW's research vessel, *Song of the Whale*, during surveys in the Ionian Sea in 2003 and in the Levantine Sea in 2007 and 2013 (Boisseau et al., 2010; Ryan et al., 2014). We hereby present results of preliminary analysis performed on segments from these recordings. The results show that a predominant whistle type (65% of 310 analysed whistles from one encounter) was a segmented-upswept whistle. Examples of segmented-upswept whistles recorded in 2003 are shown in Fig. 5.

These whistles contained segments of higher amplitude separated by dips or gaps in the whistle contour usually, but not always, appearing as steps in the frequency similar to those described by Watkins et al. (1987), Rankin

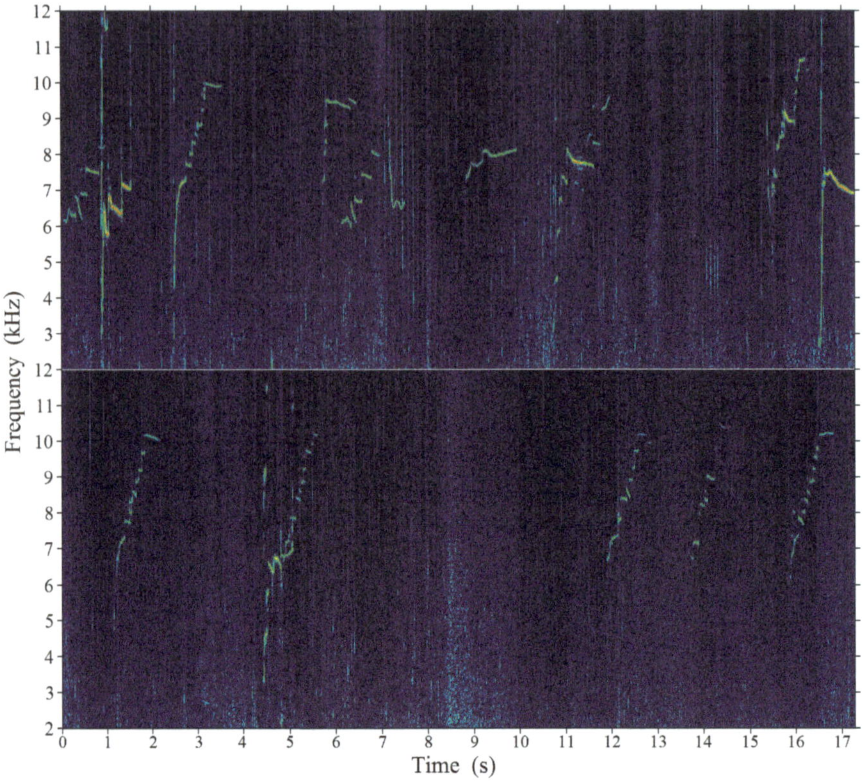

Fig. 5 Spectrograms of rough-toothed dolphin (*Steno bredanensis*) whistles, and some clicks, showing the predominance of characteristic segmented-upswept whistles. Recording made by International Fund for Animal Welfare (IFAW) on 14 September 2003 at 04:03:39 UTC, in the central Ionian Sea at 38°29.672′N, 018°47.418′E, water depth ≈3000 m (see Boisseau et al., 2010). 48 kHz sample rate, 2048 FFT, Hann window, 50% FFT overlap.

et al. (2015) and Lima et al. (2012) (e.g. lower-left spectrogram in Fig. 6 of the latter reference). The analysed segmented-upswept whistles comprised of between two and 12 segments (average = 5.7 and SD = 2.5 segments), with individual segments showing a variety of down-, up- and flat-frequency contours (Fig. 5). Frequencies of these whistles ranged from 2.4 to 10.7 kHz, with a maximum individual whistle sweep of 7.2 kHz (average = 3.0 and SD = 1.4 kHz), and had a relatively short duration with a maximum length of 1.8 s (average = 0.5 and SD = 0.3 s). These whistles often started with a long and steep upsweep and ended with a more gradual downsweep; additionally, the starts and ends of segments sometimes coincided with a click-like vertical component (examples in Fig. 5).

Although these results are from the preliminary analysis of a single encounter, they would suggest that this very distinctive whistle type may be more prevalent in Mediterranean rough-toothed dolphin than reported for the species in other geographic locations. For example, Lima et al. (2012) reported that only 28.5% of rough-toothed dolphin whistles in southeastern Brazil had breaks or steps in contour. Furthermore, whistles of Mediterranean rough-toothed dolphins may be more distinctive than those emitted by conspecifics from other locales. For instance, the mean number of steps for all analysed whistles reported here was over four steps per whistle, compared to 1.5 steps per whistle reported by Rankin et al. (2015) in the Pacific Ocean.

Classification parameters for the semiautomated real-time identification of whistles emitted by delphinids in the eastern tropical Pacific, including rough-toothed dolphin, have been identified and tested (Oswald et al., 2013), with a high success rate for classifying the latter species. Misclassifications of rough-toothed dolphin whistles mostly involved confusion with the false killer whale (*Pseudorca crassidens*), which also produces low-frequency whistles (Oswald et al., 2013). Within the Mediterranean, acoustic monitoring has mainly been used to alert surveyors to the presence of dolphin species rather than as a sole means of species identification (Boisseau et al., 2010; Ryan et al., 2014). Recently, attempts at acoustic classification of some small Mediterranean delphinids have been initiated, with correct classification scores for individual species ranging from 43.1% for short-beaked common dolphin (*Delphinus delphis*) to 69.7% for striped dolphin (*Stenella coeruleoalba*) (Azzolin et al., 2014). Rankin et al. (2015) noted that although many stepped whistles produced by rough-toothed dolphins are upswept, some or all of the stepped segments within a given whistle may be downswept. The authors stated that: 'In our experience, this type of whistle is rare in all dolphin species except *S. bredanensis* and could in itself be considered diagnostic' (Rankin et al., 2015). The combination of a low-frequency upswept whistle with downswept steps may thus be peculiar to this species. The limited past experience from the Mediterranean (Busnel and Dziedzic, 1966; Watkins et al., 1987), and the present analysis would support this. Indeed, tentative evidence suggests that this whistle type may be even more distinctive and prevalent within the Mediterranean, making automatic species identification even more reliable in this region. In the Mediterranean, this would not only enhance the capabilities of future acoustic surveys by allowing the detection and classification of rough-toothed dolphins that had not been visually confirmed, but would also permit retrospective analysis of recordings from previous surveys.

5. ORIGIN OF THE MEDITERRANEAN POPULATION

5.1 Geographic Considerations

The repopulation of the Mediterranean with cetaceans following its desiccation and hypersalination during the Messinian salinity crisis of c.5.6 million years ago (Hsü et al., 1973) could only have occurred from the Atlantic by way of the reopened Gibraltar Strait, as the connection of the Mediterranean to the Red Sea and the Indian Ocean via the Gulf of Suez had ceased to exist 14–10 million years ago (Bosworth et al., 2005). The speciation of *Steno* is estimated to have occurred shortly after the Messinian event (though not related to this event), at approximately five million years ago (Cunha et al., 2011). Climatic oscillations in the Mediterranean driven by the glacial/interglacial cycles of the quaternary period (up to and including the last glacial maximum (LGM) of 26.5–19 thousand years ago) (Clark et al., 2009), may have induced waves of migration of cetacean species into and out of the Mediterranean, with cold and warm water species moving in opposite directions. Post-LGM warming was invoked to explain the discontinuity of Atlantic and Black Sea populations of harbour porpoise (*Phocoena phocoena*) (Fontaine et al., 2010, 2014). Similarly, LGM cooling may have trapped and isolated rough-toothed dolphins in the Levantine Basin, a warm, saline and oligotrophic body of water, the shelf of which has been described as 'the closest analogue to a tropical shelf to be found anywhere in the Mediterranean' (Malanotte-Rizzoli et al., 1999).

The opportunity for Indian Ocean biota to invade and colonise the Mediterranean recurred after the excavation and opening of the Suez Canal in 1869 (Por, 1978). Rough-toothed dolphins in the Red Sea were very rarely documented (Notarbartolo di Sciara et al., 2007); however, this may be due to the scarcity of expert observers rather than to its actual rarity and an Indian Ocean origin for the Mediterranean rough-toothed dolphin cannot be ruled out.

5.2 Preliminary Genetic Evidence

Here we report the results of a preliminary molecular genetic analysis of tissue samples collected from three stranded *S. bredanensis* specimens from Israel and six specimens of *S. bredanensis* stranded on the Canary Islands. A 450 bp fragment of the mitochondrial DNA (mtDNA) control region was sequenced from each sample. Nine distinct haplotypes were revealed, one for each of the nine tissue samples (M. Behagen, University of Haifa, unpublished data). These samples were processed together with 21 other published *S. bredanensis* sequences from around the world (Oremus, 2008; GenBank). Sequences were aligned using Geneious v. 7.1.2 (http://www.geneious.com; Kearse et al., 2012). A maximum likelihood phylogeny was estimated using RaxML v. 8.0.24 (Stamatakis, 2014). The GTRGAMMA model of evolution was implemented, as selected in PartitionFinder v. 1.0.1 (Lanfear et al., 2012). A majority-rule consensus tree was generated from 500 bootstrap replicates. *Stenella attenuata* (Pantropical spotted dolphin) and *D. delphis* (short-beaked common dolphin) were specified as outgroups (Fig. 6).

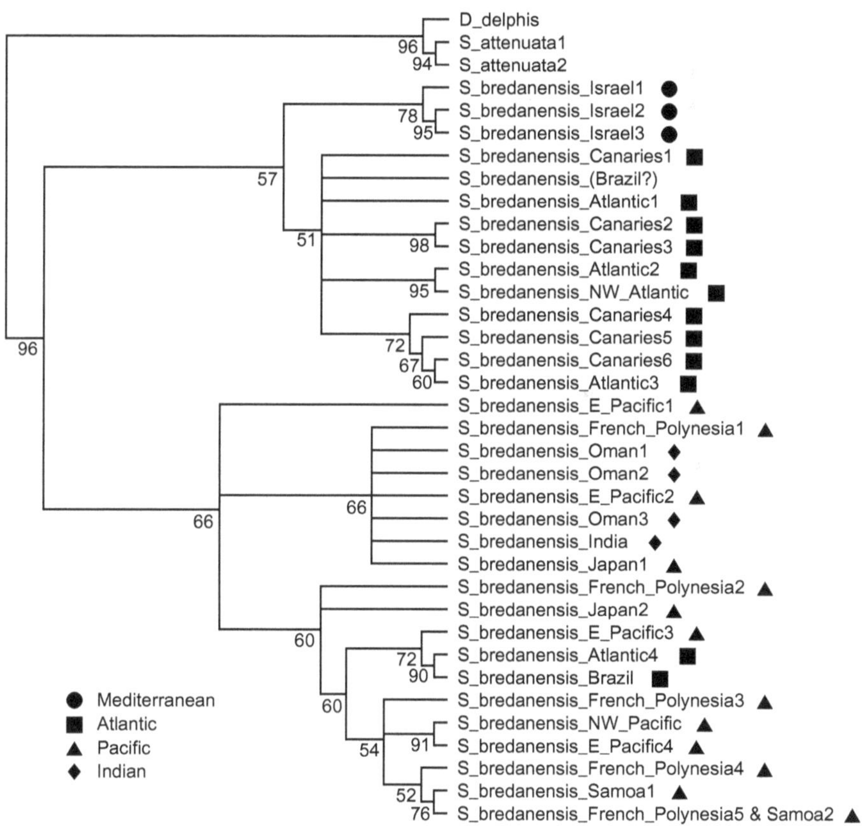

Fig. 6 Majority-rule consensus maximum likelihood tree of rough-toothed dolphin (*Steno bredanensis*) control region sequences, worldwide. Bootstrap support values indicated at nodes. *Stenella attenuata* (Pantropical spotted dolphin) and *Delphinus delphis* (short-beaked common dolphin) were specified as outgroups. Tree branch lengths are not to scale. *E*, east; *NW*, northwest.

The divergence of Atlantic and Indo-Pacific haplotypes was highly supported which concurred with the findings of Albertson (2014) and da Silva et al. (2015). All nine haplotypes sequenced for the present analysis clustered with the Atlantic haplotypes, but were separate from the Pacific and Indian Ocean (India and Oman) haplotypes, strongly suggesting an Atlantic origin of the Mediterranean rough-toothed dolphin. As evident in Fig. 6, within the Atlantic cluster, Mediterranean haplotypes are more basal, possibly indicative of long divergence and isolation.

6. ECOLOGY

6.1 Group Size

Observed group sizes for rough-toothed dolphins in various parts of their range are from one to 90 individuals, although larger aggregations have been sighted. The observed group sizes of sightings and live strandings in the Mediterranean Sea are comparable with published group sizes for this species in other regions (Table 2). Even within the unusually large aggregation of

Table 2 Group Sizes of Rough-Toothed Dolphins (*Steno bredanensis*) from Sightings in Various Regions Within Its Global Distributional Range

Region	Number of Individuals	Mean	Median	Standard Deviation	Range	Source
Mediterranean Sea	80	8.5	5	3.6	1–40	Kerem (2005), Kerem et al. (2012)
Canary Islands, Western North Atlantic Ocean	137	12.8	—	12.0	1–50	Ritter (2002)
French Polynesia, South Pacific Ocean	38	10.8	—	58.6	1–35	Gannier and West (2005)
Hawaiian Islands, North Pacific Ocean	202	11.1	7	12.1	1–90	Baird et al. (2013)
Honduras, Caribbean Sea	15	—	8–12*	—	5–30	Kuczaj and Yeater (2007)
Brazil, Eastern South Atlantic Ocean	32	4.6	—	4.4	1–18	Rossi-Santos et al. (2006)

Med, median; *SD*, standard deviation; *most common.

rough-toothed dolphins observed in the Sicilian Channel by Watkins et al. (1987), the dolphins appeared to be segregated into subgroups of approximately 20 individuals.

6.2 Occurrence in Mixed Groups

In other areas of its range, the species has been sighted in mixed groups with individuals from eight different species (Baird et al., 2008; West et al., 2011). In the Mediterranean, there has only been one such sighting reported so far, by observers aboard IFAW's R/V *Song of the Whale*, where a group of five adults was observed with Risso's dolphins (*Grampus griseus*) in August 2013 (Ryan et al., 2014). This particular interspecies association, to our knowledge, had not been reported previously (Fig. 7).

6.3 Abundance

Few estimates of abundance exist for this species globally. An abundance of 145,900 (CV=32%) individuals was estimated for the eastern tropical Pacific (Wade and Gerrodette 1993). In the northern Gulf of Mexico there are approximately 2746 (CV=36%) rough-toothed dolphins (Waring et al., 2008), which includes an estimated 1238 (CV=65%) individuals on the continental shelf (Fulling et al., 2003). The regional population around Hawaii was estimated to be 19,904 (CV=52%) (Carretta et al., 2006). The population size of rough-toothed dolphins within the Mediterranean Sea is not known.

Fig. 7 Rough-toothed dolphins (*Steno bredanensis*) in a mixed group with Risso's dolphins (*Grampus griseus*), photographed on 21 August 2013, southwest of Cyprus. *Photograph: International Fund for Animal Welfare (IFAW)/Marine Conservation Research (MCR).*

Existing worldwide survey sighting data may provide useful information on the relative abundance of rough-toothed dolphins in the Mediterranean. While there are differences in surveying platforms and methodology, uncorrected encounter rates (number of groups/individuals visually detected per unit survey effort) from regions where rough-toothed dolphins occur may allow rough assessment of relative abundance (Table 3). Existing data also provide information regarding relative encounter rates of rough-toothed dolphins among all on-effort visually encountered delphinid groups, by encounter frequency (the number of detected groups of species *x*, as a percentage of the total number of detected groups) (Table 3). Interestingly, in spite of the Levantine Basin's ultra-oligotrophicity (Herut et al., 2000), the encounter rate is only slightly lower than those in tropical and subtropical waters worldwide. Comparing individuals encountered per unit track length, the encounter rate of rough-toothed dolphins in the Mediterranean was similar to the Gulf of Mexico. The frequency of

Table 3 Worldwide Encounter Rates and Encounter Frequency Ranking of Rough-Toothed Dolphins (*Steno bredanensis*) During Dedicated Surveys

Region	Total Effort (km)	Groups/10^2 km (Individuals/10^2 km)	Rank (*n*; *f*)	Source
French Polynesia, South Pacific Ocean	6458	0.590 (12.4)	2 (7; 34%)*	Gannier (2000) and Gannier and West (2005)
Levantine Basin, Mediterranean Sea	5180	0.077 (0.5)	2 (6; 17%)	Boisseau et al. (2010) and Ryan et al. (2014)
Hawaiian Islands, North Pacific	84,758	0.239 (2.6)	4 (12; 14%)	Baird et al. (2013)
Western Tropical Indian Ocean	9784	0.123 (2.6)	5 (14; 6.5%)	Ballance and Pitman (1998)
Southern Sulu Sea, South Pacific Ocean	925	0.200 (1.6)	5 (9; 3%)	Dolar et al. (1997)
Gulf of Mexico	195,000	0.026 (0.39)	8 (15; 1.4%)	Roberts et al. (2016)
Eastern Tropical Pacific Ocean	135,324	0.100 (1.5)	9 (13; 4.5%)	Wade and Gerrodette (1993)

f, encounter frequency (% rough-toothed dolphin groups out of total detected groups); *n*, number of detected delphinid species; *data for the Society Islands.

encounter for rough-toothed dolphins was 17% and was surpassed only by that in French Polynesia (34%).

The ranking of rough-toothed dolphins among the encountered groups of delphinids in the Levantine Basin is relatively high (similar to that in French Polynesia) and is second only to striped dolphin. The rough-toothed dolphin outranks Risso's dolphin, false killer whale, common bottlenose dolphin and short-beaked common dolphin, in that order (Table 3; Boisseau et al., 2010; Ryan et al., 2014). One should, however, keep in mind that these results derive from very few encounters and that survey track lines in the Levantine Basin were almost entirely offshore (Boisseau et al., 2010; Ryan et al., 2014), with predominantly coastal species such as common bottlenose dolphin probably underrepresented.

An extensive and comprehensive cetacean density modelling and habitat-based distribution analysis was recently performed by Roberts et al. (2016) for United States Atlantic waters and the Gulf of Mexico. The rough-toothed dolphin in the Gulf of Mexico was unusual among all modelled species in that the only environmental predictor found to be marginally significant in explaining its distribution was bottom slope (Roberts et al., 2015). The year-round computed density in the Gulf of Mexico was 0.69 ind/100 km^2 (no seasonal effect). Extrapolation to the Levantine Basin is implausible given difference in the marine ecosystems, yet, since the number of animals encountered per unit track length was similar in the Levantine Basin to that in the Gulf of Mexico, it may provide a useful exercise for a very rough measure of possible abundance. This density, translated to the Levantine Basin, with an area of 320,000 km^2, would give an expected population abundance of 2208 rough-toothed dolphins. However, if unsubstantiated with actual data from the Mediterranean, such a number would not be useful as a population abundance estimate for management purposes.

6.4 Diet

The large group of rough-toothed dolphins sighted inside Haifa Port (approximately 40 individuals) in 2006, was observed actively feeding on flathead grey mullet (*Mugil cephalus*) that swarmed in the water (Kerem et al., 2012). Scant information from stomach contents of two calves, both bycatch from gill-net entanglement, revealed both fish and cephalopods as items in the diet. One stomach contained the head of a single, slightly digested, narrow-barred Spanish mackerel (*Scomberomorus commerson*)

Fig. 8 Rough-toothed dolphin (*Steno bredanensis*) calf, found floating off Carmel Beach, Haifa, Israel, 9 March 2003. Faint monofilament marks can be seen circling the head and crossing the blowhole. *Inset*: bitten-off Spanish mackerel (*Scomberomorus commerson*) head, found in the dolphin's oesophagus. *Photograph: O. Goffman, University of Haifa.*

(Fig. 8), while the other contained milk remains, unidentified fish remains and 21 histioteuthid and sepiid beaks (M. Sapir, Israel Marine Mammal Research & Assistance Center, IMMRAC, unpublished data).

7. CONSERVATION CONCERNS

With such limited information on rough-toothed dolphins in the Mediterranean Sea, identification of specific threats is speculative at best. The apparent seasonal occupation of nearshore shallow water areas puts individuals at greater risk from anthropogenic threats. Incidental catch in gill-nets appears to be a major threat. At least three of the eight stranded animals in Israel (two calves and a subadult) and two females in Lebanon, were determined to have been entangled in gill-nets. In comparison, out of 160 strandings of common bottlenose dolphin and 25 strandings of striped dolphin during the last three decades in Israel, only five bottlenose dolphins and one striped dolphin were determined to have been entangled by gill-nets (O. Goffman, IMMRAC, unpublished data).

Other potential threats include fisheries impacts on the limited resources of rough-toothed dolphin prey species. Limited dietary information indicate that rough-toothed dolphins feed on prey species of high commercial value, the standing crop of which in the Levantine Basin may well be dwindling (Edelist et al., 2011).

Heavy metal load has been determined for a single rough-toothed dolphin calf (0.5–1 years of age) from the coast of Israel (Shoham-Frider et al., 2014). Interestingly, liver mercury levels were >3 SDs higher than the mean level for common bottlenose dolphin ($n=3$) and striped dolphin ($n=1$) calves of similar age, from the same general region (Shoham-Frider et al., 2014).

Anthropogenic noise may impact rough-toothed dolphins. Mass strandings are highly unusual in the Mediterranean, but the rough-toothed dolphin presented two mass stranding events in a decade (see Table 1). The seafloor between Israel and Cyprus is subject to intense hydrocarbon prospecting and production activities. The mass stranding in Lady's Mile (34.62° N, 033.01°E), south of Limassol Harbour (Table 1), may have been linked to such activities. Twenty-one dolphins were first observed in shallow water (<5 m), of which 14 later stranded and were refloated. They then drifted slowly north and when nearing the harbour's breakwater, six to eight dolphins stranded again, and again were helped back to sea (A. Demetropoulos, personal communication to G. Notarbartolo di Sciara, 12 March, 2010). The location of the live stranding in Donnalucata, Sicily (36.76°N, 014.64°E) on 5 April 2002, in which three animals died and three were apparently successfully refloated, is in close proximity to waters where naval manoeuvres including antisubmarine warfare (and therefore the use of active military sonar) periodically occur. Such activities have included *Dynamic Manta* 2015, a North Atlantic Treaty Organization (NATO) exercise conducted in September 2015, and other activities described by Podestà et al. (2016).

The current global International Union for the Conservation of Nature (IUCN) Red List assignment for rough-toothed dolphin is 'Least Concern' (Hammond et al., 2012). Assessment of the conservation status of the Mediterranean 'subpopulation' is pending, and when assessed, it would most likely be Data Deficient, as population abundance estimation is lacking.

8. SUMMARY AND RECOMMENDATIONS

Knowledge about the rough-toothed dolphin in the Mediterranean Sea has accumulated slowly. The species has not been recorded in Israel since 2008, which hampers potential research. The species is, however, sighted whenever large-scale dedicated surveys are carried out in its apparently preferred habitat in the Levantine Basin (Boisseau et al., 2010; Ryan et al., 2014). A recently published request for tenders for a visual and acoustic survey to take place during 2016 in Cypriot exclusive economic zone (EEZ) has a potential for yielding more records and possibly preliminary regional abundance estimates. The adaptation and training of dedicated acoustic classification software for the identification of Mediterranean rough-toothed dolphin whistles and for its use in analysing existing recordings (e.g. IFAW's extensive acoustic archive) hold the potential to determine whether the species' known range should be extended. Strict adherence by known range states to the provisions of Agreement on the Conservation of Cetaceans in the Black Sea Mediterranean Sea and Contiguous Atlantic Area (ACCOBAMS) Resolution 4.17: 'Guidelines to address the impact of anthropogenic noise on cetaceans in the ACCOBAMS area' (ACCOBAMS, 2016), outlining best practices to be employed for each noise-producing human activity at sea, in particular high-power active sonar and seismic air-gun use, is urgently called for to protect this lesser known Mediterranean cetacean.

ACKNOWLEDGEMENTS

The authors wish to thank Antonio Fernandez for providing tissue samples of rough-toothed dolphin from the Canary Islands; the International Fund for Animal Welfare and Marine Conservation Research for providing photographic material and acoustic recordings; Marina Costa, Tethys Research Institute, for preparing the distribution map; and Ana Cañadas, Andreas Demetropoulos, Giuseppe Notarbartolo di Sciara, Michela Podestà, Marina Sequeira and Renaud de Stephanis for factual inputs. We would also like to thank the two anonymous reviewers whose remarks and suggestions did much to improve the manuscript and Barbara Curry for expert and insightful edits.

REFERENCES

ACCOBAMS, 2016. Anthropogenic noise. Available at: http://www.accobams.org/index.php?option=com_content&view=article&id=1091&Itemid=141. Accessed on 3 June 2016.

Adam, O., Samaran, F. (Eds.), 2013. Detection, Classification and Localization of Marine Mammals Using Passive Acoustics, 2003–2013: 10 Years of International Research. DIRAC NGO, Paris.

Addink, M.J., Smeenk, C., 2001. Opportunistic feeding behavior of rough-toothed dolphins *Steno bredanensis* off Mauritania. Zool. Verh. (Leiden) 334, 37–48.

Albertson, G.R., 2014. Worldwide phylogeography and local population structure of the rough-toothed dolphin (*Steno bredanensis*). PhD dissertation, Oregon State University, Corvallis, Oregon.

Azzolin, M., Gannier, A., Lammers, M.O., Oswald, J.N., Papale, E., Buscaino, G., Buffa, G., Mazzola, S., Giacoma, C., 2014. Combining whistle acoustic parameters to discriminate Mediterranean odontocetes during passive acoustic monitoring. J. Acoust. Soc. Am. 135, 502–512.

Baird, R.W., Webster, D.L., Aschettino, J.M., Schorr, G.S., McSweeney, D.J., 2013. Odontocete cetaceans around the main Hawaiian islands: habitat use and relative abundance from small-boat sighting surveys. Aquat. Mamm. 39, 253–269.

Baird, R.W., Webster, D.L., Mahaffy, S.D., McSweeney, D.J., Schorr, G.S., Ligon, A.D., 2008. Site fidelity and association patterns in a deep-water dolphin: rough-toothed dolphins (*Steno bredanensis*) in the Hawaiian Archipelago. Mar. Mamm. Sci. 24 (3), 535–553.

Ballance, L.T., Pitman, R.L., 1998. Cetaceans of the western tropical Indian Ocean: distribution, relative abundance, and comparisons with cetacean communities of two other tropical ecosystems. Mar. Mamm. Sci. 14, 429–459.

Boisseau, O., Lacey, C., Lewis, T., Moscrop, A., Danbolt, M., McLanaghan, R., 2010. Encounter rates of cetaceans in the Mediterranean Sea and contiguous Atlantic area. J. Mar. Biol. Assoc. U. K. 90, 1589–1599.

Bosworth, W., Huchon, P., McClay, K., 2005. The Red Sea and Gulf of Aden Basins. J. Afr. Earth Sci. 43, 334–378.

Busnel, R.G., Dziedzic, A., 1966. Caractéristiques physiques de certains signaux acoustiques du Delphinidé *Steno bredanensis*, Lesson. C. R. Acad. Sci. (Paris) 262, 143–146.

Carretta, J.V., Forney, K.A., Muto, M.M., Barlow, J., Baker, J., Hanson, B., Lowry, M.S., 2006. U.S. Pacific marine mammal stock assessments—2005. NOAA Technical Memorandum NOAA-TM-NMFS-SWFSC-388, 317 pp.

Centro Studi Cetacei, 2006. Cetacei spiaggiati lungo le coste italiane. XIX Rendiconto 2004 (Mammalia). Atti Soc. Ital. Sci. Nat. Museo Civ. Stor. Nat. Milano 147, 145–157.

Clark, P.U., Dyke, A.S., Shakun, J.D., Carlson, A.E., Clark, J., Wohlfarth, B., Mitrovica, J.X., Hostetler, S.W., McCabe, A.M., 2009. The last glacial maximum. Science 325, 710–714.

Collet, A., 1984. Live-capture of cetaceans for European institutions. Rep. Int. Whal. Commn. 34, 603–607.

Cunha, H.A., Moraes, L.C., Medeiros, B.V., Lailson-Brito Jr., J., da Silva, V.M.F., Solé-Cava, A.M., Schrago, C.G., 2011. Phylogenetic status and timescale for the diversification of *Steno* and *Sotalia* dolphins. PLoS One 6, e28297. http://dx.doi.org/10.1371/journal.pone.0028297.

da Silva, D.M.P., Azevedo, A.F., Secchi, E.R., Barbosa, L.A., Flores, P.A.C., Carvalho, R.R., Bisi, T.L., Lailson-Brito Jr., J., Cunha, H.A., 2015. Molecular taxonomy and population structure of the rough-toothed dolphin *Steno bredanensis* (Cetartiodactyla: Delphinidae). Zool. J. Linn. Soc. 175, 949–962. http://dx.doi.org/10.1111/zoj.12301.

Di Natale, A., Mangano, A., 1981. Report on the progress of "Project Cetacea" 6. July 1978–October 1981. Mem. Biol. Ocean. N. S. 9, 1–49. Supplemento speciale.

Djiba, A., Bamy, I.L., Samba Ould Bilal, A., Van Waerebeek, K., 2015. Biodiversity of cetaceans in coastal waters of Northwest Africa: new insights through platform of opportunity visual surveying in 2011–2013. In: Valdés, L., Déniz-González, I. (Eds.), In: Oceanographic and Biological Features in the Canary Current Large Marine Ecosystem. vol. 115. IOC-UNESCO, Paris, pp. 283–297. IOC Technical Series.

Dolar, M.L., Perrin, W.F., Yaptinchay, A.A.S.P., Jaaman, S.A.B.H., Santos, M.D., Alava, M.N., Suliansa, M.S.B., 1997. Preliminary investigation of marine mammal distribution, abundance and interactions with humans in southern Sulu Sea. Asian Mar. Biol. 14, 61–81.

Edelist, D., Sonin, O., Golani, D., Rilov, G., Spanier, E., 2011. Spatiotemporal patterns of catch and discards of the Israeli Mediterranean trawl fishery in the early 1990s: ecological and conservation perspectives. Sci. Mar. 75, 641–652.

Eyre, E.J., Frizell, J., 2012. A note on observations of cetaceans in the Indian Ocean Sanctuary, Australia to Israel, April 1995. J. Cetacean Res. Manag. 12, 277–285.

Fontaine, M.C., Tolley, K.A., Michaux, J.R., Birkun Jr., A., Ferreira, M., Jauniaux, T., Llavona, Á., Öztürk, B., Öztürk, A.A., Ridoux, V., Rogan, E., Sequeira, M., Bouquegneau, J.-M., Baird, S.J.E., 2010. Genetic and historic evidence for climate-driven population fragmentation in a top cetacean predator: the harbour porpoises in European water. Proc. R. Soc. B Biol. Sci. 277, 2829–2837.

Fontaine, M.C., Roland, K., Calves, I., Austerlitz, F., Palstra, F.P., Tolley, K.A., Ryan, S., Ferreira, M., Jauniaux, T., Llavona, A., Öztürk, B., Öztürk, A.A., Ridoux, V., Rogan, E., Sequeira, M., Siebert, U., Vikingsson, G.A., Borrell, A., Michaux, J.R., Aguilar, A., 2014. Postglacial climate changes and rise of three ecotypes of harbour porpoises, *Phocoena phocoena*, in western Palearctic waters. Mol. Ecol. 23, 3306–3321. http://dx.doi.org/10.1111/mec.12817.

Frantzis, A., 2009. Cetaceans in Greece: present status of knowledge: Technical report—initiative for the conservation of cetaceans in Greece. Athens, Greece, 94 pp.

Fulling, G.L., Mullin, K.D., Hubard, C.W., 2003. Abundance and distribution of cetaceans in outer continental shelf waters of the U.S. Gulf of Mexico. Fish. Bull. 101, 923–932.

Gannier, A., 2000. Distribution of cetaceans off the Society Islands (French Polynesia) as obtained from dedicated surveys. Aquat. Mamm. 26, 111–126.

Gannier, A., West, K.L., 2005. Distribution of the rough-toothed dolphin (*Steno bredanensis*) around the Windward Islands (French Polynesia). Pac. Sci. 59, 17–24.

Götz, T., Verfuss, U., Schnitzler, H., 2005. 'Eavesdropping' in wild rough-toothed dolphins (*Steno bredanensis*)? Biol. Lett. 2, 1–3.

Hammond, P.S., Bearzi, G., Bjørge, A., Forney, K.A., Karczmarski, L., Kasuya, T., Perrin, W.F., Scott, M.D., Wang, J.Y., Wells, R.S., Wilson, B., 2012. *Steno bredanensis*. The IUCN Red List of Threatened Species 2012. e.T20738A17845477, http://dx.doi.org/10.2305/IUCN.UK.2012.RLTS.T20738A17845477.en. Downloaded on 1 March 2016.

Hashmi, D., Adloff, B., 1991. Surface frequency of cetaceans in the Strait of Gibraltar. Eur. Res. Cetac. 5, 16–17.

Herut, B., Almogi-Labin, A., Jannink, N.T., Gertman, I., 2000. The seasonal dynamics of nutrient and chlorophyll a concentrations on the SE Mediterranean shelf-slope. Oceanol. Acta 23, 771–782.

Hsü, K.J., Ryan, W.B.F., Cita, M.B., 1973. Late Miocene desiccation of the Mediterranean. Nature 242, 240–244.

Jefferson, T.A., Webber, M.A., Pitman, R.L., 2015. Marine Mammals of the World. A Comprehensive Guide to Their Identification, second ed. Academic Press, London.

Kearse, M., Moir, R., Wilson, A., Stones-Havas, S., Cheung, M., Sturrock, S., Buxton, S., Cooper, A., Markowitz, S., Duran, C., Thierer, T., Ashton, B., Meintjes, P., Drummond, A., 2012. Geneious basic: an integrated and extendable desktop software platform for the organization and analysis of sequence data. Bioinformatics 28, 1647–1649.

Kerem, D., 2005. Rough-toothed dolphins "invading" the port of Haifa. FINS 2 (1), 19.

Kerem, D., Hadar, N., Goffman, O., Scheinin, A., Kent, R., Boisseau, O., Schattner, U., 2012. Update on the cetacean fauna of the Mediterranean Levantine Basin. Open Mar. Biol. J. 6, 6–27.

Kuczaj Jr., S.A., Yeater, D.B., 2007. Observations of rough-toothed dolphins (*Steno bredanensis*) off the coast of Utila, Honduras. J. Mar. Biol. Assoc. U. K. 87, 141–148.

Lanfear, R., Calcott, B., Ho, S.Y., Guindon, S., 2012. PartitionFinder: combined selection of partitioning schemes and substitution models for phylogenetic analyses. Mol. Biol. Evol. 29, 1695–1701.

Lima, I.M.S., Andrade, L.G., Carvalho, R.R., Lailson-Brito Jr., J., Azevedo, A.F., 2012. Characteristics of whistles from rough-toothed dolphins (*Steno bredanensis*) in Rio de Janeiro coast, southeastern Brazil. J. Acoust. Soc. Am. 131, 4173–4181.

Malanotte-Rizzoli, P., Manca, B.B., D'alcala, M.R., Theocharis, A., Brenner, S., Budillon, G., Ozsoy, E., 1999. The eastern Mediterranean in the 80s and in the 90s: the big transition in the intermediate and deep circulation. Dyn. Atmos. Oceans 29, 365–395.

Masski, H., de Stephanis, R., 2015. Cetaceans of the Moroccan coast: information from a reconstructed strandings database. J. Mar. Biol. Assoc. U. K. 1, 1–9. http://dx.doi.org/10.1017/S0025315415001563.

MEDACES, 2011. Mediterranean Database of Cetacean Strandings. Marine Zoology Unit, University of Valencia, Valencia, Spain. [updated on 3 March 2011]. Available from: http://medaces.uv.es. Accessed on 15 April 2011.

Miyazaki, N., Perrin, W.F., 1994. Rough-toothed dolphin *Steno bredanensis* (Lesson, 1928). In: Ridgway, S.H., Harrison, R. (Eds.), Handbook of Marine Mammals. Academic Press, San Diego, CA, pp. 1–21.

Notarbartolo di Sciara, G., Addink, M., Baldwin, R.M., Rudolph, P., Smeenk, C., 2007. A review of cetaceans from the Red Sea. In: Proceedings of the 21st Annual Conference of the European Cetacean Society, Donostia—San Sebastian (Spain), 23–25 April.

Notarbartolo di Sciara, G., Birkun Jr., A., 2010. Conserving Whales, Dolphins and Porpoises in the Mediterranean and Black Seas: An ACCOBAMS Status Report. ACCOBAMS and Department of External Relations, Principality of Monaco. 212 p.

Oremus, M., 2008. Genetic and demographic investigation of population structure and social system in four delphinid species. PhD dissertation, University of Auckland, Auckland, New Zealand.

Oremus, M., Poole, M.M., Albertson, G.R., Baker, C.S., 2012. Pelagic or insular? Genetic differentiation of rough-toothed dolphins in the Society Islands, French Polynesia. J. Exp. Mar. Biol. Ecol. 432, 37–46.

Oswald, J.N., Rankin, S., Barlow, J., Oswald, M., Lammers, M.O., 2013. Real-time Call Classification Algorithm (ROCCA): software for species identification of delphinid whistles. In: Adam, O., Samaran, F. (Eds.), Detection, Classification and Localization of Marine Mammals Using Passive Acoustics, 2003–2013: 10 Years of International Research. DIRAC NGO, Paris, pp. 245–266.

Podestà, M., Azzellino, A., Cañadas, A., Frantzis, A., Moulins, A., Rosso, M., Tepsich, P., Lanfredi, C., 2016. Cuvier's beaked whale, *Ziphius cavirostris*, distribution and occurrence in the Mediterranean Sea: High-Use areas and conservation threats. In: Notarbartolo di Sciara, G., Podestà, M., Curry, B.E. (Eds.), Mediterranean Marine Mammals Ecology and Conservation. Advances in Marine Biology, vol. 75. Elsevier, Amsterdam, pp. 103–140.

Por, F., 1978. Lessepsian Migration: The Influx of Red Sea Biota into the Mediterranean by Way of the Suez Canal. Springer Verlag, Berlin.

Rankin, S., Oswald, J.N., Barlow, J., 2008. Acoustic behavior of dolphins in the Pacific Ocean: implications for using passive acoustic methods for population studies. Can. Acoust. 36, 88–92.

Rankin, S., Oswald, J.N., Simonis, A.E., Barlow, J., 2015. Vocalizations of the rough-toothed dolphin, *Steno bredanensis*, in the Pacific Ocean. Mar. Mamm. Sci. 31, 1538–1548.

Ritter, F., 2002. Behavioral observations of rough-toothed dolphins (*Steno bredanensis*) off La Gomera, Canary Islands (1995–2000), with special reference to their interactions with humans. Aquat. Mamm. 28, 46–59.

Roberts, J.J., Best, B.D., Mannocci, L., Fujioka, E., Halpin, P.N., Palka, D.L., Garrison, L.P., Mullin, K.D., Cole, T.V.N., Khan, C.B., McLellan, W.A., Pabst, D.A., Lockhart, G.G., 2016. Habitat-based cetacean density models for the U.S. Atlantic and Gulf of Mexico. Sci. Rep. 6, 22615. http://dx.doi.org/10.1038/srep22615.

Roberts, J.J., Best, B.D., Mannocci, L., Fujioka, E., Halpin, P.N., Palka, D.L., Garrison, L.P., Mullin, K.D., Cole, T.V.N., Khan, C.B., McLellan, W.M., Pabst, D.A., Lockhart, G.G., 2015. Density Model for Rough-Toothed Dolphin (*Steno bredanensis*) for the U.S. Gulf of Mexico, Version 1.3, 2015-09-29, and Supplementary Report. Marine Geospatial Ecology Lab, Duke University, Durham, North Carolina.

Rossi-Santos, M., Wedekin, L.L., Sousa-Lima, R.S., 2006. Distribution and habitat use of small cetaceans off Abrolhos Bank, Eastern Brazil. Lat. Am. J. Aquat. Mamm. 5, 23–28.

Ryan, C., Cucknell, A.C., Romagosa, M., Boisseau, O., Moscrop, A., Frantzis, A., McLanaghan, R., 2014. Final report of a visual and acoustic survey for marine mammals in the eastern Mediterranean Sea during summer 2013. International Fund for Animal Welfare. Available at: http://www.marineconservationresearch.co.uk/.

Santoro, R., Sperone, E., Tringali, M.L., Pellegrino, G., Giglio, G., Tripepi, S., Arcangeli, A., 2015. Summer distribution, relative abundance and encounter rates of cetaceans in the Mediterranean waters off southern Italy (western Ionian Sea and southern Tyrrhenian Sea). Mediterr. Mar. Sci. 16, 613–620.

Shoham-Frider, E., Kerem, D., Roditi-Elasar, M., Goffman, O., Morick, D., Yoffe, O., Kress, N., 2014. Trace elements in tissues of cetacean species rarely stranded along the Israeli Mediterranean coast. Mar. Pollut. Bull. 83, 376–382.

Stamatakis, A., 2014. RAxML version 8: a tool for phylogenetic analysis and post-analysis of large phylogenies. Bioinformatics 30, 1312–1313.

UNEP-MAP-RAC/SPA, 2014. Status and conservation of cetaceans in the Adriatic Sea. Draft internal report prepared by Holcer, D., Fortuna, C.M., Mackelworth, P.C. for the purposes of the Mediterranean regional workshop to facilitate the description of ecologically or biologically significant marine areas, Malaga, Spain, 7–11 April 2014.

Wade, P.R., Gerrodette, T., 1993. Estimates of cetacean abundance and distribution in the eastern tropical Pacific. Rep. Int. Whal. Commn. 43, 477–493.

Waring, G.T., Josephson, E., Fairfield, C.P., Maze-Foley, K., 2008. U.S. Atlantic and Gulf of Mexico marine mammal stock assessments—2007. NOAA Technical Memorandum.

Watkins, W.A., Tyack, P., Moore, K.E., Notarbartolo di Sciara, G., 1987. *Steno bredanensis* in the Mediterranean Sea. Mar. Mamm. Sci. 3, 78–82.

West, K.L., Mead, J.G., White, W., 2011. *Steno bredanensis* (Cetacea: Delphinidae). Mamm. Species 43, 177–189.

CHAPTER NINE

The Gulf of Ambracia's Common Bottlenose Dolphins, *Tursiops truncatus*: A Highly Dense and yet Threatened Population

J. Gonzalvo*[,1], G. Lauriano[†], P.S. Hammond[‡], K.A. Viaud-Martinez[§], M.C. Fossi[¶], A. Natoli[||], L. Marsili[¶]

*Tethys Research Institute, Milan, Italy
[†]Institute for Environmental Protection and Research (ISPRA), Roma, Italy
[‡]Sea Mammal Research Unit, Gatty Marine Laboratory, University of St Andrews, Fife, Scotland, United Kingdom
[§]Illumina, Inc., San Diego, CA, United States
[¶]University of Siena, Siena, Italy
[||]UAE Dolphin Project, Dubai, United Arab Emirates
[1]Corresponding author: e-mail address: joan.gonzalvo@gmail.com

Contents

Advances in Marine Biology, Volume 75
ISSN 0065-2881
http://dx.doi.org/10.1016/bs.amb.2016.07.002

Abstract

The common bottlenose dolphin (*Tursiops truncatus*) is the only cetacean present in the semiclosed waters of the Gulf of Ambracia, Western Greece. This increasingly degraded coastal ecosystem hosts one of the highest observed densities in the Mediterranean Sea for this species. Photo-identification data and tissue samples collected through skin-swabbing and remote biopsy sampling techniques during boat-based surveys conducted between 2006 and 2015 in the Gulf, were used to examine bottlenose dolphin abundance, population trends, site fidelity, genetic differentiation and toxicological status. Bottlenose dolphins showed high levels of year-round site fidelity throughout the 10-year study period. Dolphin population estimates mostly fell between 130 and 170 with CVs averaging about 10%; a trend in population size over the 10 years was a decline of 1.6% per year (but this was not significant). Genetic differentiation between the bottlenose dolphins of the Gulf and their conspecifics from neighbouring populations was detected, and low genetic diversity was found among individuals sampled. In addition, pesticides where identified as factors posing a real toxicological problem for local bottlenose dolphins. Therefore, in the Gulf of Ambracia, high dolphin density does not seem to be indicative of favourable conservation status or pristine habitat.

1. INTRODUCTION

The common bottlenose dolphin (*Tursiops truncatus*), hereafter bottlenose dolphin, is the most common cetacean on the continental shelf of the Mediterranean Sea, where its distribution appears to be scattered and possibly fragmented into small units (Notarbartolo di Sciara, 2016). Despite a worldwide distribution and the capability of moving across long distances, bottlenose dolphins show marked population genetic structure in areas throughout their range (Natoli et al., 2004; Tezanos-Pinto et al., 2009). Coastal bottlenose dolphin populations, in particular, are often resident and prone to population fragmentation, sometimes exhibiting strong population differentiation even across a small geographic scale (Gaspari et al., 2015; Mirimin et al., 2011; Natoli et al., 2005; Richards et al., 2013). Identifying those population units and assessing their boundaries is crucial for the implementation of effective conservation measures to protect small resident populations and ensure the survival of this species across its range.

In 2006, the International Union for Conservation of Nature (IUCN) Red List Authority and the Agreement on the Conservation of Cetaceans in the Black Sea, Mediterranean Sea and Contiguous Atlantic Area (ACCOBAMS) agreed to list the Mediterranean subpopulation of common bottlenose dolphins as Vulnerable according to IUCN Red List criteria A2cde and based on a suspected population decline of at least 30% over the last 60 years (Bearzi et al., 2012). Intentional kills and extermination campaigns conducted until at least the early 1960s in portions of the basin, and recent and ongoing incidental mortality in fishing gear (i.e. bycatch) are considered to be among the main factors contributing to the decline, along with other ongoing threats derived from overfishing and from general habitat degradation caused by contamination and disturbance from marine traffic (Bearzi et al., 2008b). The common bottlenose dolphin is listed in the Appendix II (Mediterranean population) of the Convention on the Conservation of Migratory Species of Wild Animals (CMS), in the Appendix II (Strictly Protected Fauna Species) of the Convention on the Conservation of European Wildlife and Natural Habitats (Bern Convention), and in the Annexes II and IV of the EU Habitats Directive (Council Directive 92/43/EEC).

Greek waters still harbour a remarkable richness of cetacean fauna compared to the rest of the Mediterranean (Frantzis, 2009). And yet, such richness is progressively eroding due to rapidly growing degradation of the marine environment (Notarbartolo di Sciara and Bearzi, 2010). The bottlenose dolphin is the cetacean species that is most commonly found in Greek coastal waters and the second most abundant species after the striped dolphin (*Stenella coeruleoalba*) (Frantzis, 2009). Cetaceans living in coastal areas, and particularly in semiclosed inshore habitats, such as the Gulf of Ambracia (GA) in Western Greece, are exposed to risks from a variety of anthropogenic impacts and are especially vulnerable because they often have restricted geographic ranges, disjunct distributions and limited movements (Reeves et al., 2003). In this increasingly degraded Gulf, where the bottlenose dolphin is the only cetacean present at a relatively high density (an average density of 0.37 animals km^2), which is one of the highest observed densities in the Mediterranean for this species (Bearzi et al., 2008a), dolphins may be suffering significant physiological stress due to anthropogenic activities (Gonzalvo et al., 2015). Here, we present the results of our research to determine the conservation status of bottlenose dolphins in the GA (Fig. 1) by examining their abundance, population trends, site fidelity, toxicological status and genetic differentiation.

2. MATERIALS AND METHODS

2.1 Study Area

The GA, also known by its Greek name, Amvrakikòs Kolpos, is a shallow, semiclosed embayment of 405 km^2, whose only communication with the

Fig. 1 Common bottlenose dolphins (*Tursiops truncatus*) photographed in the Gulf of Ambracia showing the characteristic morphology of the species. *Bottom-right* image shows an adult bottlenose dolphin severely affected by a skin condition firstly reported by Gonzalvo et al. (2015). *Photographs courtesy of J. Gonzalvo/Tethys Research Institute.*

open Ionian Sea is through the Preveza Channel, a narrow (minimum width of 370 m) and shallow (2–12 m) 3-km-long corridor (Fig. 2). On average, the depth of the Gulf is approximately 30 m (maximum 60 m), and its bottom mostly consists of mud and sand (Ferentinos et al., 2010). It is characterized by abundant wildlife, and in addition to providing key habitat for bottlenose dolphins (Bearzi et al., 2008a; Gonzalvo et al., 2015), it is an important foraging ground for loggerhead sea turtles (*Caretta caretta*) (Rees et al., 2013), as well as a breeding site for Dalmatian pelicans (*Pelecanus crispus*) (Catsadorakis et al., 2015). Its northern side, a complex ecosystem, is composed of a double delta from the rivers Arachthos and Louros, and their associated marshes and lagoons are of particular importance for bird diversity. The GA is included in Natura 2000 and Ramsar sites and is protected by national, European, and international legislation. In 2008, it was designated as a 'National Park' in accordance with Greek national legislation (11989/08 KYA).

Due to the isolated character of the GA, water quality is strongly influenced by man-made processes; input of organic matter and pollutants comes from various sources (Friligos et al., 1997; Karras et al., 2007; Kountoura and

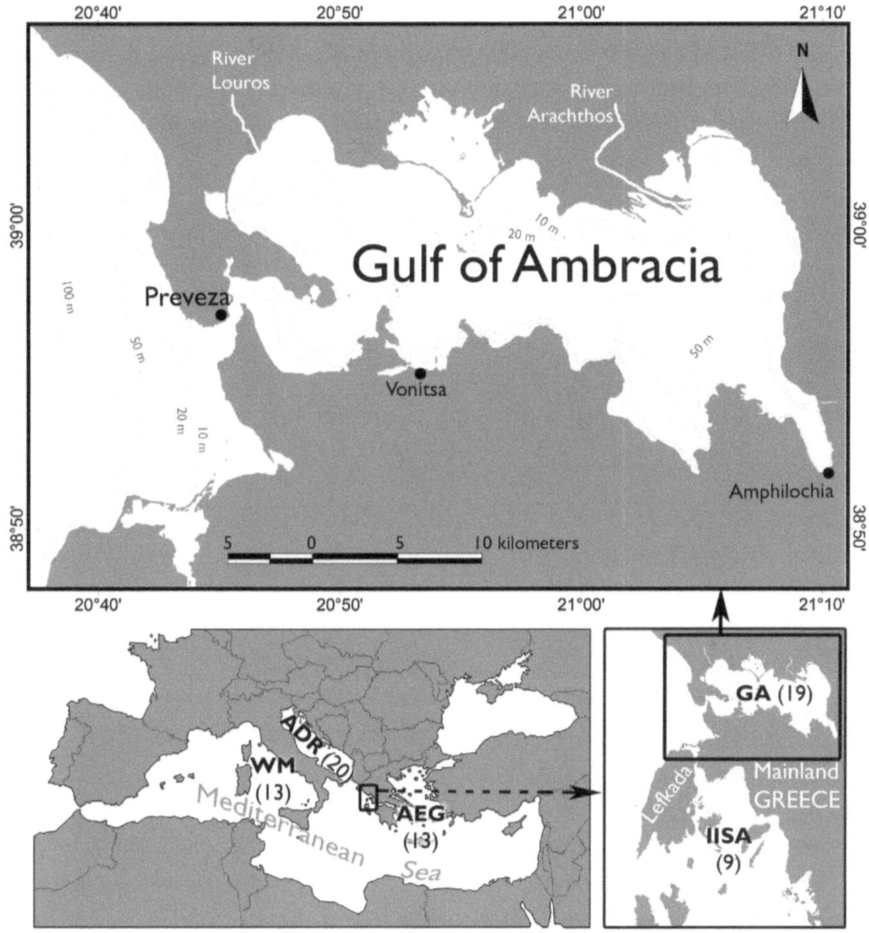

Fig. 2 (*Top*) Map of the Gulf of Ambracia. (*Bottom*) Geographic origin of common bottlenose dolphin (*Tursiops truncatus*) skin samples analysed using molecular genetic mtDNA analysis; *WM*, Western Mediterranean; *ADR*, Adriatic Sea; *IISA*, Inner Ionian Sea Archipelago; *GA*, Gulf of Ambracia and *AEG*, Aegean Sea (numbers indicate sample size).

Zacharias, 2013; Tsangaris et al., 2010). The Louros and Arachthos rivers are the main pathways bringing agricultural runoff (Tsabaris et al., 2015; Tsangaris et al., 2010). The use of dichlorodiphenyltrichloroethanes (DDTs), organochlorine pesticides, as well as of other xenobiotic compounds such as polychlorinated biphenyls (PCBs) and hexachlorobenzene (HCB), was banned by the Stockholm Convention in 2001 because of their environmental impacts. Nevertheless, in the GA, high levels of these compounds have been detected in eggs of Dalmatian pelicans as well as in eels

Anguilla anguilla (Albanis et al., 1995), mussels *Mytilus galloprovincialis* (Tsangaris et al., 2010) and various fishes (Hatzianestis et al., 2001). Although the use of these chemicals reached its peak between the decades of the 1940s and 1950s, and continued until the late 1970s, they remain priority contaminants in toxicological studies. Hence, they are often referred to as legacy contaminants, which persist in the environment, bioaccumulate through the food web and pose a risk to human health and the environment (Adams et al., 2014; Jepson et al., 2016).

Fish farms, agriculture, livestock and discharges of domestic sewage from coastal towns and villages contribute to the nutrient enrichment of the Ambracian waters (Ferentinos et al., 2010; Gonzalvo et al., 2014), which are rather murky and highly eutrophic, with Secchi disk readings often as small as 2 m (Bearzi et al., 2008a). The western part of the Gulf is seasonally hypoxic, while the eastern part is seasonally anoxic (Kountoura and Zacharias, 2013). The local active fishing fleet totals about 360 boats and is composed exclusively of small-scale fishing boats working primarily with set nets (i.e. trammel and gill nets), targeting mainly small pelagic/epipelagic fish and shrimp *Penaeus kerathurus* (Gonzalvo et al., 2014).

2.2 Boat-Based Surveys Data Collection

Boat surveys were conducted between 2006 and 2015, from a 5.80-m inflatable boat with fibreglass keel (Novamarine RH-580) powered by a 100 HP four-stroke outboard engine. Surveys were conducted using predefined routes (i.e. transects) designed for survey effort to provide uniform coverage of the entire GA on a monthly basis. In the years 2006 and 2007, monthly surveys were also conducted in the nearby open waters of the Ionian Sea. Survey conditions were considered as 'positive' under daylight and good visibility, sea state $\leq$3 Beaufort (large wavelets, crests beginning to break and scattered whitecaps), and with at least two observers scanning the sea surface looking for dolphins. When spotted, dolphin groups were approached at low speed, progressively converging with their routes and avoiding sudden changes of speed and directionality to minimize potential disturbance. As proposed by Mann (2000), when observing a dolphin group both dolphin activity and distance among individuals was considered; hence, a group was defined as 'that composed of dolphins within approximately 100 m radius of each other (Irvine et al., 1981) that were observed in apparent association, moving in the same direction and often, but not always, engaged in the same activity (Shane, 1990)'. Group size estimates and composition were recorded. The latter included four age categories (newborn, calf, juvenile and adult) based on visual assessment of size in relation to average adult size (Bearzi et al., 1997).

2.3 Photo-Identification

During each dolphin sighting, photo-identification effort was conducted to obtain as many good images as possible of every individual present in our focal group throughout the duration of the observation, using digital Canon SLR cameras equipped with Canon EF 70–200 mm f/2.8 L USM zoom lenses, and avoiding bias towards any particular individuals. Photo-identification was consistently based on long-term natural marks such as notches and nicks in the dolphins' dorsal fins (Wilson et al., 1997; Würsig and Jefferson, 1990; Würsig and Würsig, 1979), as well as on any additional marks in other body parts.

During processing, photos were cropped around the dorsal fin and visible part of the body. Selection of photographs followed recommendations provided by Read et al. (2003); therefore, it was exclusively based on photographic quality using consistent criteria (entire dorsal fin visible, fin perpendicular to camera, high sharpness and resolution, no water spray or other interference masking fin profile). The best images of every dolphin during each sighting were selected and compared with a catalogue of identified individuals, where dolphins were classified, depending on their degree of distinctiveness, into one of the following four categories: D1—highly marked (easily distinguishable, even with very poor quality photos), D2—marked (distinguishable even under poor light conditions), D3—moderately marked (small nicks/notches noticeable only with good quality photos) and D4—slightly marked (fins with unusual shapes or with tiny nicks and scars poorly distinguishable unless with high quality images). When a match was not found, the individual was given a unique identification code and added to the catalogue. The matching procedure was done twice, by two different experienced researchers who were working independently and using exactly the same protocol to minimize the number of matching errors. Identifications were used to construct individual sighting histories. In most cases, calves were recognized in the field primarily based on their regular association with an identifiable adult dolphin; therefore, only calves and juveniles with distinctive marks were included in the analysis.

2.4 Estimation of Population Size

Annual population size was estimated using the multisample closed population models available in CAPTURE, using MARK software (Otis et al., 1978; https://www.phidot.org/software/mark). The assumptions of closed models are well known and discussed in, for example, Hammond (2009). All closed population models assume that no births, death, permanent immigration or emigration occur within the study period. In addition, there are three fundamental assumptions related to the data: (1) marks are unique, (2) marks cannot be lost and (3) all marks are correctly recorded and reported. Other assumptions that relate to variation in capture probability depend on the particular mark–recapture model used. Failure to meet these assumptions can lead to bias in estimates of population size (Hammond, 2009, 2010). In this case, the assumptions were mostly addressed as described by Bearzi et al. (2008a), who produced the first population estimate for bottlenose dolphins living in the GA based on photo-identification effort conducted between 2002 and 2005.

Capture histories for each individual bottlenose dolphin identified in the GA were constructed for each year using a calendar month as a sampling occasion. A time frame of one month was considered a good compromise between three factors: having a reasonable number of sampling occasions per year; maximizing capture probability within each capture occasion and ensuring that the population mixes between sampling occasions (Genov et al., 2008; Hammond, 2010). For the years 2006–2014, five sampling occasions (May–September) were chosen for each year. In 2015, the season was shorter resulting in only four sampling occasions (June–September).

The models available in CAPTURE are: M_0, no variation in capture probability; M_t, variation in capture probability among sampling occasions (Darroch, 1958); M_h, variation in capture probability among individuals (Burnham and Overton, 1978) and M_b, variation in capture probability as a behavioural response to first capture (Otis et al., 1978). Models M_{th} (Chao et al., 1992), M_{tb} (Pollock, 1975) and M_{bh} (Norris and Pollock, 1996) combine these effects. In analysis, CAPTURE was applied first to determine the most appropriate of these models for each data subset D1+D2 and D1+D2+D3 in each year according to the model section criteria based on the goodness of fit tests conducted. Behavioural response models are not generally considered appropriate in a photo-identification study because there is no reason why dolphins should react differently leading to a change in capture probability following first capture. Nevertheless, behavioural responses were included to investigate whether the model selection would favour them in place of other models in any of the years. In addition, a consistent set of models was run on all years (M_t, M_h, M_{th}) with the aim of constructing consistent time series to investigate trends in population size over time.

Mark–recapture estimates refer to the population of animals that have appropriate marks, as defined earlier. To derive an estimate of the total population, the estimated number of marked individuals must be divided by the proportion of marked animals in the population (e.g. Wilson et al., 1999).

The proportion of marked individuals (θ) was estimated for each year as the mean of the proportion of 'marked' individuals in each encounter for the D1 + D2 and D1 + D2 + D3 data subsets. The variance of estimated total population size was estimated as:

$$\mathrm{var}\left(\hat{N}_{\mathrm{total}}\right) = \hat{N}^2_{\mathrm{total}}\left(\frac{\mathrm{var}\left(\hat{N}\right)}{\hat{N}^2} + \frac{\mathrm{var}(0)}{0^2}\right)$$

where N is the number of marked individuals in the population and N_{total} is total population size (Urian et al., 2015).

In analysis, two data subsets were considered. The first data set was restricted to the most distinctively marked individuals (D1 + D2). The second set also included moderately marked individuals (D1 + D2 + D3). Including moderately distinct animals maximizes sample size, but may cause positive bias because of missed matches between individuals. Using only the most distinct animals minimizes the possibility of bias resulting from missed matches, but the reduced sample size will result in poorer precision. Slightly marked individuals (D4) were not considered at all in analysis. The estimate of the number of marked animals in the D1 + D2 data subset is expected to be smaller than in the D1 + D2 + D3 data subset. However, the proportion of marked animals in the population will also be smaller for D1 + D2 subset than the D1 + D2 + D3 subsets and, theoretically, estimates of the total population from both analyses should be the same. Consideration of both sets of estimates allowed the trade-off between bias and precision to be investigated.

The power of the data to detect a trend was investigated using the approximate inequality (Gerrodette, 1987):

$$r^2 n^3 \geq 12\mathrm{CV}^2\left(z_{\alpha/2} + z_\beta\right)^2$$

where r is the annual rate of change in population size, n is the number of annual estimates of population size, CV is the coefficient of variation of population size, z is the value of a standard normal variable, α is the probability of making a Type I error and β is the probability of making a Type II error ($1-\beta=$power). In the calculation, r was set to the estimated annual rate of change in population size over the $n=10$ years study period and CV was set to the average CV of estimated population size. α was set to 0.05 and the value of β was determined from the inequality.

2.5 Tissue Sampling

Remote tissue sampling of free-ranging cetaceans is an inexpensive tool used to collect valuable data for conservation and management. Analyses of skin and blubber samples have provided information on population genetics, hormone levels, and contaminant loads (e.g. Gorgone et al., 2008). Between 2001 and 2007, 20 skin samples for use in molecular genetic analysis were collected from bow-riding dolphins with the use of a sterilized nylon scrub pad affixed to the round end of a wooden pole through 'skin swabbing'; a method, adapted from Harlin et al. (1999), that provides samples suitable for amplification and sequencing of mitochondrial DNA (mtDNA). Other skin samples were also opportunistically obtained from four dead individuals found stranded or floating adrift.

In July 2013, a one-week biopsy sampling campaign was conducted using standard procedures and equipment (Gorgone et al., 2008; Kiszka et al., 2010). Skin and blubber tissue samples for toxicological analysis were collected using a crossbow (Barnett Panzer V, 68 kg draw weight) and custom made bolts and tips (Ceta Dart, Copenhagen, Denmark). Tips were 25 mm in length and 7 mm in internal diameter. A high-pressure moulded stopper prevented the bolt from penetrating more than about 20 mm forcing it to bounce back. The dorsolateral area below the dorsal fin was the body region targeted. Sampling was only attempted on adult dolphins. No individuals accompanied by offspring (or in groups in which offspring were present) were sampled. In case of doubt, a precautionary approach was applied (i.e. no sampling attempted). Continuous photo-identification effort allowed identification of 15 out of the 16 individuals sampled.

2.6 Genetic Analysis

A total of 74 bottlenose dolphin samples from five different populations were included in this study (Fig. 2). From the 24 skin samples collected from bottlenose dolphins in the GA, only 19 were viable (Table 1). The samples collected from those 19 dolphins were sequenced for the current study and

Table 1 Details Regarding Skin Samples of Common Bottlenose Dolphin (*Tursiops truncatus*) Collected in the Gulf of Ambracia for Use in Genetic Analysis ($n = 20$ Obtained by Skin Swabbing from Bow-Riding Specimens and $n = 4$ Collected from Dead Animals)

Sample Id Code	Date (Day–Month–Year)	Time (hh:mm)	Sampling Method	Notes	Haplotype
1	4-8-2004	11:26	Skin swabbing	No photo-identified	IV
2	4-8-2004	11:40	Skin swabbing	Identified as 3022	IV
3	4-8-2004	11:50	Skin swabbing	No photo-identified	Failed sequencing
4	6-8-2004	10:42	Skin swabbing	No photo-identified	Failed sequencing
5	6-8-2004	11:21	Skin swabbing	No photo-identified	XXXXII
6	6-8-2004	13:09	Skin swabbing	No photo-identified	Failed sequencing
7	9-8-2004	11:32	Skin swabbing	Identified as 3036	Failed sequencing
8	15-8-2004	9:40	Skin swabbing	No photo-identified	Failed sequencing
9	14-8-2006	9:40	Skin swabbing	No photo-identified	IV
10	14-8-2006	11:50	Skin swabbing	No photo-identified	IV
11	8-9-2006	10:33	Skin swabbing	No photo-identified	IV
12	8-9-2006	11:11	Skin swabbing	No photo-identified	IV
13	8-9-2006	11:22	Skin swabbing	No photo-identified	IV
14	15-9-2006	10:40	Skin swabbing	No photo-identified	IV
15	28-9-2006	19:05	Dead stranded	No photo-identified (Juvenile, male, length = 210 cm)	IV

Continued

Table 1 Details Regarding Skin Samples of Common Bottlenose Dolphin (*Tursiops truncatus*) Collected in the Gulf of Ambracia for Use in Genetic Analysis ($n=20$ Obtained by Skin Swabbing from Bow-Riding Specimens and $n=4$ Collected from Dead Animals)—cont'd

Sample Id Code	Date (Day–Month–Year)	Time (hh:mm)	Sampling Method	Notes	Haplotype
16	20-6-2007	19:15	Dead stranded	No photo-identified (adult, male, length = 257 cm)	IV
17	26-6-2007	8:52	Skin swabbing	No photo-identified	IV
18	26-6-2007	9:04	Skin swabbing	No photo-identified	IV
19	27-6-2007	11:38	Skin swabbing	No photo-identified	IV
20	4-7-2007	11:47	Dead floating adrift	Male new born; length about 100 cm (mother identified as 03046)	IV
21	19-8-2007	11:01	Skin swabbing	No photo-identified	IV
22	19-8-2007	11:19	Skin swabbing	No photo-identified	IV
23	27-8-2007	10:45	Skin swabbing	No photo-identified	IV
24	8-9-2009	14:40	Dead stranded	No photo-identified (adult, male, length = 270 cm)	IV

were compared with previously analysed samples from the Western Mediterranean Sea (WM, $n=13$), Adriatic Sea (ADR, $n=20$), Inner Ionian Sea Archipelago (IISA, $n=9$) and the Aegean Sea (AEG, $n=13$) (Natoli et al., 2005; Viaud-Martinez et al., 2008). The methodology for tissue preservation, DNA isolation and the sequencing of the first 442 base pairs (bp) of the mtDNA control region was as described by Viaud-Martinez et al. (2008).

Sequence alignment was performed using 650 bp of mtDNA obtained from previous studies and samples collected from dolphins in the ADR ($n=6$), IISA ($n=9$) (Natoli et al., 2005), 442 bp of mtDNA from dolphins in the ADR ($n=4$) and AEG ($n=13$) (Viaud-Martinez et al., 2008). These sequences were aligned with new dolphin sequences obtained from the GA

($n = 19$). Sequence alignment was performed using Sequencher 3.1 (Gene Codes Corporation). Final haplotypes were reported for the 442-bp shared between these populations.

Genetic diversity within dolphin populations was measured using haplotype diversity (H_e), which is equivalent to expected heterozygosity after adjustment to population size, and nucleotide diversity (π; Nei, 1987) using Arlequin v.3.5.1.2 (Excoffier and Lischer, 2010). Analysis of molecular variance (AMOVA) was used to estimate the degree of subdivision among the five dolphin populations. The null hypothesis of no genetic divergence between the five populations was tested using Contingency or Exact Test in Arlequin v.3.5.1.2 (Excoffier and Lischer, 2010); this test has previously been demonstrated to be a robust measurement for identifying population boundaries (Waples and Gaggiotti, 2006). Genetic differentiation among pairwise population was inferred as F_{ST} (Weir and Cockerham, 1984). Tajima's D and Fu's Fs estimates for each populations were calculated with Arlequin v.3.5.1.2 (Excoffier and Lischer, 2010). Haplotype genealogy was reconstructed using Haplotype Viewer (Ewing, 2010) using maximum parsimony (Phylip v.3.69; Felsenstein, 1989), where the size of each node (circle) represents the frequency of a haplotype and the length of the links (or number of black nodes) represents the genetic divergence among haplotypes.

2.7 Toxicology Analysis

Of the 16 biopsy samples obtained, two were discarded from the analysis because the quantity of blubber obtained was too small. HCBs, DDTs and PCBs were measured in subcutaneous blubber of 14 bottlenose dolphins according to the EPA (Environmental Protection Agency) method 8081/8082 with modifications (Marsili and Focardi, 1997). The samples (about 0.1 g) were lyophilized in an Edwards freeze drier for two days and extracted with *n*-hexane for gas chromatography in a Soxhlet apparatus (Marsili et al., 2014). Capillary gas-chromatography revealed HCB, op′- and pp′-isomers of DDT and its derivatives DDD and DDE, and 30 PCB congeners. Total PCBs were quantified as the sum of individual congeners (IUPAC nos. 95, 101, 99,151, 144, 135, 149, 118, 146, 153, 141, 138, 178, 187, 183, 128, 174, 177, 156, 171, 202, 172, 180, 199, 170, 196, 201, 195, 194, 206; Ballschmiter and Zell, 1980), while total DDTs were calculated as the sum of pp′DDT, op′DDT, pp′DDE, op′DDE, pp′DDD and op′DDD. The detection limit was 0.1 ng/kg (ppt) for all the organochlorine compounds (OCs) analysed. The extracted organic matter (EOM%) from freeze-dried samples has been evaluated in all samples.

Data were processed using Statistica 5.0 (Statsoft). Normal distribution of values was tested using the Shapiro–Wilk test. Most of the variables were not normally distributed; so nonparametric tests should be used. However, due to the small number of samples no subsequent statistical elaboration was possible to evaluate differences between groups.

3. RESULTS

3.1 Boat-Based Surveys

Across 10 years of research in the GA, with 74 months spent in the field and a total of 770 daily surveys, more than 13,000 km of survey effort were covered under positive conditions, resulting in 631 bottlenose dolphin sightings and almost a thousand hours spent with dolphins (Table 2). In the Ionian Sea during 2006 and 2007, a total of 667 km of survey effort under favourable conditions produced only one sighting near the mouth of the GA, in which 15 bottlenose dolphins were photo-identified; none of them were ever observed inside the Gulf prior to, or after that time period.

3.2 Photo-Identification

A total of 185 dolphins were identified. The rate at which new individuals were photo-identified during the 10-year study period is shown in Fig. 3. This discovery curve rose sharply in 2006, coinciding with the start of the photo-identification work, and then increased more slowly, followed by an asymptotic pattern from 2007 onwards. The progressive flattening of the discovery curve and the high site fidelity shown by the dolphins (Fig. 4), indicate that the population was effectively geographically restricted (confined to the GA) during sampling seasons across the study period.

3.3 Population Size and Trend

The data subsets D1 + D2 and D1 + D2 + D3 included 108 and 148 individuals, respectively. The population estimates produced for both subsets are shown in Table 3. The model selected as most appropriate varied from year to year. Model M_{tb} was selected in 2009 for both data subsets but the estimates were considerably larger than any other and were very imprecise; they were therefore considered unreliable and excluded from further consideration. Estimates for the D1 + D2 + D3 data subset were larger and slightly more precise than for the D1 + D2 data subset in each year, as expected because they include additional less distinct animals and therefore

Table 2 Summary of Research Effort Conducted to Survey for Common Bottlenose Dolphins (*Tursiops truncatus*) in the Gulf of Ambracia (2006–2015)

							Dataset Used for Mark-Recapture Population Estimates (Sightings on Transect Between May–September)		
Year	Research Calendar (Months at Sea)		Days at Sea	Km Surveyed 'Positive'	Total Sightings	Time with Dolphins	Total Sightings	Photos Taken	Photos Selected
2006	Apr–Dec	9	107	1630	70	99 h 32 min	38	5872	2770
2007	Year-round	12	136	2776	101	139 h 45 min	25	4674	2656
2008	Year-round	12	105	1648	77	110 h 52 min	25	2466	1561
2009	May–Oct	6	82	1310	76	105 h 50 min	25	5478	2241
2010	Apr–Oct	7	88	1234	80	97 h 36 min	29	5478	2241
2011	Apr–Sep	6	43	785	35	46 h 56 min	23	4765	2045
2012	Apr–Sep	6	55	656	50	75 h 19 min	37	8873	3296
2013	Apr–Sep	6	60	667	55	85 h 17 min	37	8216	2992
2014	Apr–Sep	6	56	707	50	77 h 30 min	33	9695	3287
2015	Jun–Sep	4	38	412	37	54 h 10 min	30	5462	2674
Total	74 months		770	13,303	631	992 h 47 min	302	60,979	25,763

Framed dataset refers exclusively to the number of sightings and photo-id effort used for the population estimates.

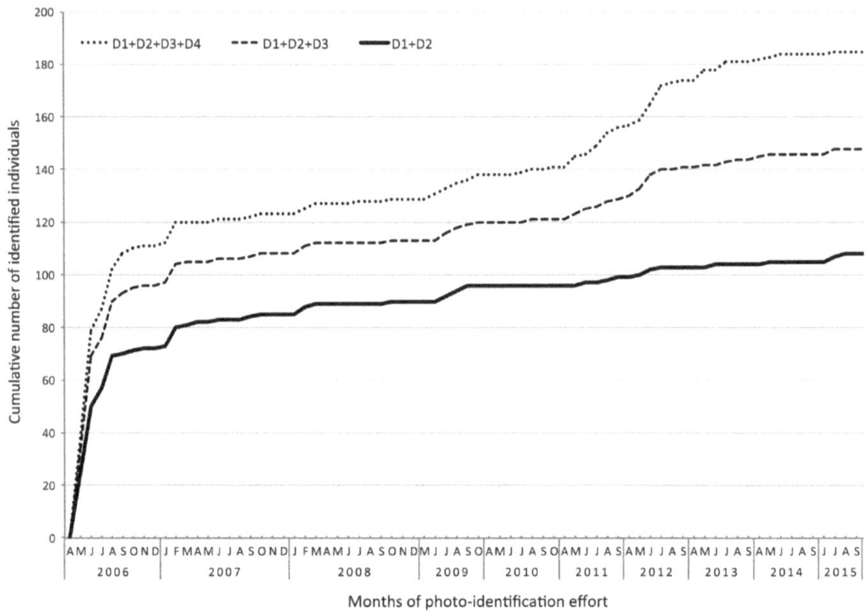

Fig. 3 Discovery curves, taking into account the degree of distinctiveness (from D1-highly marked to D4-poorly marked), for individually identified common bottlenose dolphins (*Tursiops truncatus*) from 2006 to 2015 in the Gulf of Ambracia.

larger sample sizes. The D1 + D2 + D3 data subset estimates were, however, more variable among years.

Estimates for each year did not vary greatly among models. Model M_0 was selected as the most appropriate in three of the 10 years for each data subset; in each case, the estimates were the same as for model M_t. Models including heterogeneity of capture probabilities, M_h and M_{th}, were selected as the most appropriate in four of the 10 years for each data subset. Estimates from these models were almost always larger than from model M_t, which is indicative of heterogeneity of capture probabilities in the data causing a negative bias if not accounted for. Estimates from model M_{th} were usually the least precise but CVs were generally less than 0.1.

The series of estimates from model M_{th} were identified as the most robust. These estimates from the same model were much more consistent over the 10-year study period than estimates chosen as most appropriate by CAPTURE. They are also likely to be the least biased estimates because they allow capture probability to vary by both time and individual.

Table 4 shows the estimates from model M_{th} corrected for the estimated proportion of marked animals in the population for each year and each data

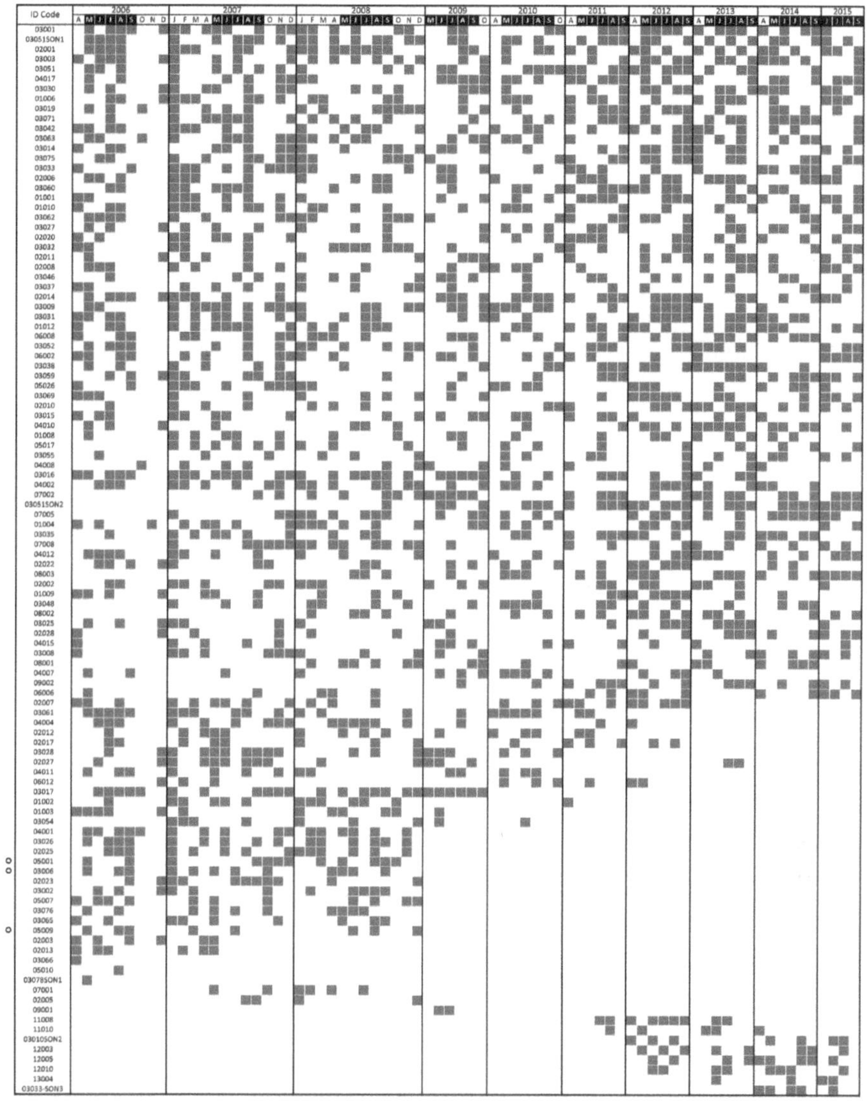

Fig. 4 Residency pattern of marked common bottlenose dolphin (*Tursiops truncatus*) individuals (D1–D2) in the Gulf of Ambracia. Grey cells indicate presence documented through digital photo-identification. Months included in the mark-recapture model are shown with white letters on black background. Three individuals with 'o' beside their ID-Code were last seen in our study area in 2008; they were identified in June 2010 in the Inner Ionian Sea Archipelago and two of them were spotted in the Gulf of Corinth in August that same year (T1, T2 and T3 in Bearzi et al., 2011).

Table 3 Mark-Recapture Population Estimates of the Number of Marked Common Bottlenose Dolphin (*Tursiops truncatus*) Individuals Generated by Different Models for Data Subsets D1 + D2 (Top) and D1 + D2 + D3 (Bottom)

Year	M(tb)	SE	CV	M(0)	SE	CV	M(t)	SE	CV	M(h)	SE	CV	M(th)	SE	CV
D1–D2															
2006							71	2.38	0.033	82	5.68	0.069	**78**	6.28	0.081
2007							**82**	3.58	0.044	93	6.33	0.068	85	7.07	0.083
2008				**74**	3.52	0.048	74	3.24	0.044	86	6.86	0.080	83	8.14	0.098
2009	**126**	244.63	1.942				78	5.25	0.067	101	11.48	0.114	98	15.12	0.154
2010				**66**	3.40	0.051	66	3.14	0.048	76	6.13	0.081	73	7.27	0.100
2011	**66**	4.05	0.061				66	2.54	0.038	75	5.40	0.072	73	6.24	0.085
2012							72	1.34	0.019	**78**	4.50	0.058	75	3.62	0.048
2013							73	2.20	0.030	80	9.80	0.123	**72**	2.91	0.040
2014							63	2.07	0.033	**69**	5.03	0.073	63	3.79	0.060
2015				**66**	3.23	0.050	66	3.03	0.050	77	6.66	0.090	80	10.33	0.130
D1–D2–D3															
2006							93	2.80	0.030	105	6.31	0.060	**99**	6.62	0.067
2007							104	4.21	0.040	119	7.67	0.064	**111**	8.93	0.080
2008							86	3.22	0.037	99	6.53	0.066	94	7.67	0.082
2009	**174**	391.10	2.248				94	6.06	0.064	114	11.57	0.101	107	14.31	0.134

2010				**84**	3.92	0.047	84	4.83	0.058	94	6.33	0.067	86	6.90	0.080
2011	**100**	15.27	0.153				94	3.64	0.039	112	9.33	0.083	111	9.99	0.090
2012							104	3.47	0.033	116	5.81	0.050	**111**	5.18	0.047
2013				**74**	3.03	0.041	73	2.81	0.038	82	5.53	0.067	76	5.45	0.072
2014							93	2.443	0.026	**103**	5.97	0.058	95	4.97	0.052
2015				**90**	3.96	0.044	90	3.75	0.040	100	6.23	0.060	95	8.81	0.090

Estimates from models identified as the most appropriate by CAPTURE (Otis et al., 1978; https://www.phidot.org/software/mark) are shown in bold. The model identified as most appropriate for 2008 data subset D1 + D2 + D3 was undefined. Shaded columns correspond to the results from M_t (variation in capture probability among sampling occasions), M_h (variation in capture probability among individuals) and M_{th} (combining these effects) models run for each year.

Table 4 Estimates of the Number of Marked Common Bottlenose Dolphin (*Tursiops truncatus*) Individuals from Model M_{th}, of the Proportion of Marked Animals in the Population, and of Total Population Size for Data Subsets *D1 + D2* and D1 + D2 + D3

Year	Subset	CAPTURE Estimate	HM θ	Corrected Estimate	CV	SE	95%CI
2006	*D1 + D2*	*78*	*0.552*	***141***	*0.094*	*13.30*	*117–170*
	D1 + D2 + D3	99	0.702	**141**	0.086	12.25	119–167
2007	*D1 + D2*	*85*	*0.582*	***145***	*0.121*	*17.72*	*115–185*
	D1 + D2 + D3	111	0.679	**163**	0.107	17.575	132–201
2008	*D1 + D2*	*83*	*0.645*	***128***	*0.118*	*15.20*	*102–161*
	D1 + D2 + D3	94	0.734	**127**	0.094	12.099	106–153
2009	*D1 + D2*	*98*	*0.560*	***174***	*0.167*	*29.34*	*126–242*
	D1 + D2 + D3	107	0.636	**168**	0.145	24.432	126–223
2010	*D1 + D2*	*73*	*0.535*	***136***	*0.117*	*15.95*	*108–171*
	D1 + D2 + D3	86	0.652	**131**	0.102	13.451	107–160
2011	*D1 + D2*	*73*	*0.469*	***155***	*0.133*	*20.73*	*119–201*
	D1 + D2 + D3	111	0.661	**167**	0.110	18.470	135–207
2012	*D1 + D2*	*75*	*0.516*	***145***	*0.077*	*11.29*	*124–169*
	D1 + D2 + D3	111	0.772	**143**	0.054	7.862	129–159
2013	*D1 + D2*	*72*	*0.541*	***133***	*0.077*	*10.26*	*114–154*
	D1 + D2 + D3	76	0.708	**107**	0.080	8.636	91–125
2014	*D1 + D2*	*63*	*0.432*	***145***	*0.119*	*17.42*	*115–184*
	D1 + D2 + D3	95	0.685	**138**	0.067	9.394	121–158
2015	*D1 + D2*	*80*	*0.582*	***137***	*0.153*	*21.12*	*101–185*
	D1 + D2 + D3	95	0.707	**134**	0.110	14.84	110–170

subset. The estimated proportion of marked animals in the population was larger in each year for data subset D1 + D2 + D3 than for data subset D1 + D2, as expected. The estimated proportions were consistent for data subset D1 + D2 (0.432–0.645) and highly consistent for data subset D1 + D2 + D3 (0.636–0.772). Estimates of total population size for each data subsets are very consistent in most years, differing by less than 10% except in 2007 (D1 + D2 higher) and 2013 (D1 + D2 + D3 higher). Thus, there was no evidence of any bias caused by including moderately distinctive animals in analysis. In addition, estimates of total population size for data subset D1 + D2 + D3 were consistently more precise than those from the D1 + D2 data subset (mean CV of 0.096 for D1 + D2, compared to 0.118 for D1 + D2 + D3).

The best series of population estimates was therefore considered to be the one using the M_{th} model applied to the D1 + D2 + D3 data subset (Fig. 5). A linear regression of log total population size on year generated an annual decline of 1.64% (SE = 1.54%; 95% CI = − 5.04–1.89%) but this was not significant ($p = 0.31$).

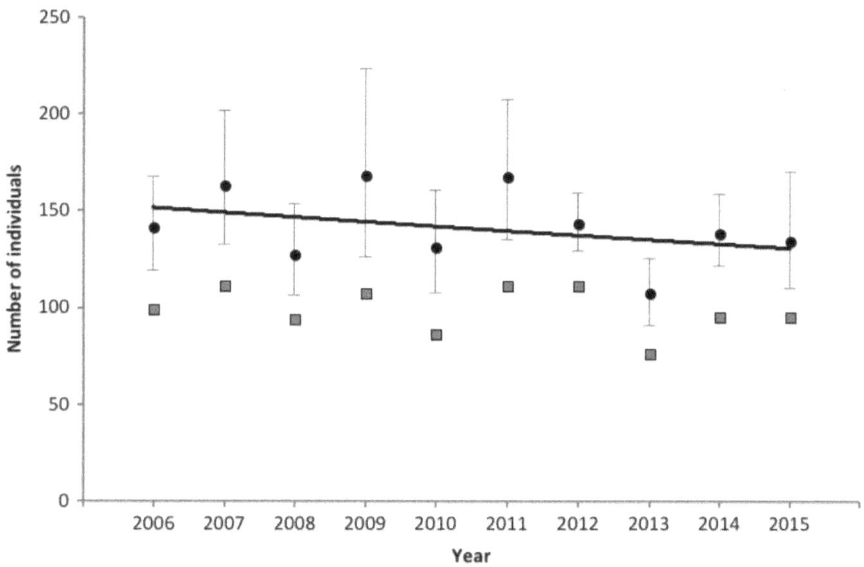

Fig. 5 Estimates of the number of identifiable common bottlenose dolphin (*Tursiops truncatus*) individuals D1 + D2 + D3 (*grey squares*) and total population size (*black circles*) within years using the closed model M_{th} Chao. The line shows the exponential decline at a rate of 1.6% (SE = 1.5%) per year derived from the linear regression of log total population size on year.

The power analysis indicated that the power of the data (10 annual estimates with $CV = 0.1$) to detect a rate of decline of this magnitude is only around 30%. For the power to increase to the typically accepted desirable level of 0.8, there would need to be an additional six years of survey with estimates at the same precision to detect a decline of 1.6% per year.

3.4 Genetic Analysis

The analysis of 74 sequences for 442 bp identified 16 unique haplotypes characterized by 22 polymorphic sites (Table 5; Genbank accession in progress). The GA population was characterized by only two haplotypes and one of those was observed in 18/19 individuals. Both haplotypes were unique to the GA population. The GA dolphins did not share any haplotype with the other four populations analysed. The haplotype network identifies a highly represented haplotype among all the populations except for the GA and Inner Ionian Sea Archipelago populations (Fig. 6). The two unique haplotypes that characterize the GA population were one and five mutation steps away from a haplotype observed in the ADR and AEG populations, respectively.

Both haplotype and nucleotide diversities were lower for the GA population ($H_e = 0.11$, $\pi = 0.003$) when compared to all the other populations ($0.74 \leq H_e \leq 0.92$, and $0.009 \leq \pi \leq 0.0016$). Tajima's D were not significant for all the populations except for GA (Tajima's $D = -2.273$, p value $= 0.001$), whereas Fu's values were all not significant including for the GA population.

The analysis of population differentiation showed highly significant differentiation (AMOVA $F_{ST} = 0.375$, $df = 73$, overall $p < 0.001$). Exact tests of haplotype distributions supported significant differences between the five putative populations sampled. The null hypothesis of no genetic divergence was rejected ($p < 0.001$). Genetic differentiation among pairwise populations is reported in Table 6. Significant genetic differentiation was observed among all populations, except between the ADR and AEG population. The GA population showed the highest F_{ST} values when compared with all the other populations analysed.

3.5 Toxicology

Organochlorine levels found in the subcutaneous blubber of the 14-sampled bottlenose dolphins, reported as dry weight and in lipid weight values (d.w. and l.w., respectively), together with the lipid percent (EOM%), water percent (H_2O%) and information on dolphin sex and sexual maturity, are

Table 5 Haplotype Affiliations Across Five Putative Common Bottlenose Dolphin (*Tursiops truncatus*) Populations in the Mediterranean Region
Populations

No.	Haplotype	WM	ADR	IISA	AEG	GA
1	I	2	11		6	
2	IV					18
3	XIV		3			
4	XV		2		3	
5	XVI	6		1		
6	XVII	1				
7	XVIII	2				
8	XIX		2	3	1	
9	XXII			3	2	
10	XXIII	1				
11	XXXVIII				1	
12	XXXXI	1	1			
13	XXXXII					1
14	XXXXIII		1			
15	XXXXIV			1		
16	XXXXV			1		
	$n=16$	$n=13$	$n=20$	$n=9$	$n=13$	$n=19$

Genetic data from previously analysed samples representing the Western Mediterranean Sea (WM, $n=13$), Adriatic Sea (ADR, $n=20$), Inner Ionian Sea Archipelago (IISA, $n=9$) and the Aegean Sea (AEG, $n=13$) (Natoli et al., 2005; Viaud-Martinez et al., 2008) were compared with skin samples collected from the Gulf of Ambracia (GA, $n=19$) and analysed in the current study. Shared haplotypes are shaded.

shown in Table 7. Sex of the individuals was established genetically. Males of at least 10 years of age and five-year-old females were considered sexually mature (Wells and Scott, 1999). Dolphin age was estimated based on individual photo-identification histories. For instance, a dolphin recorded as a newborn in September 2006, when first identified, was considered to be seven-years old when sampled in July 2013. Also, one individual, AMV12, was genetically identified as a male but, because it was not photo-identified during the sampling, this individual's age could not be determined.

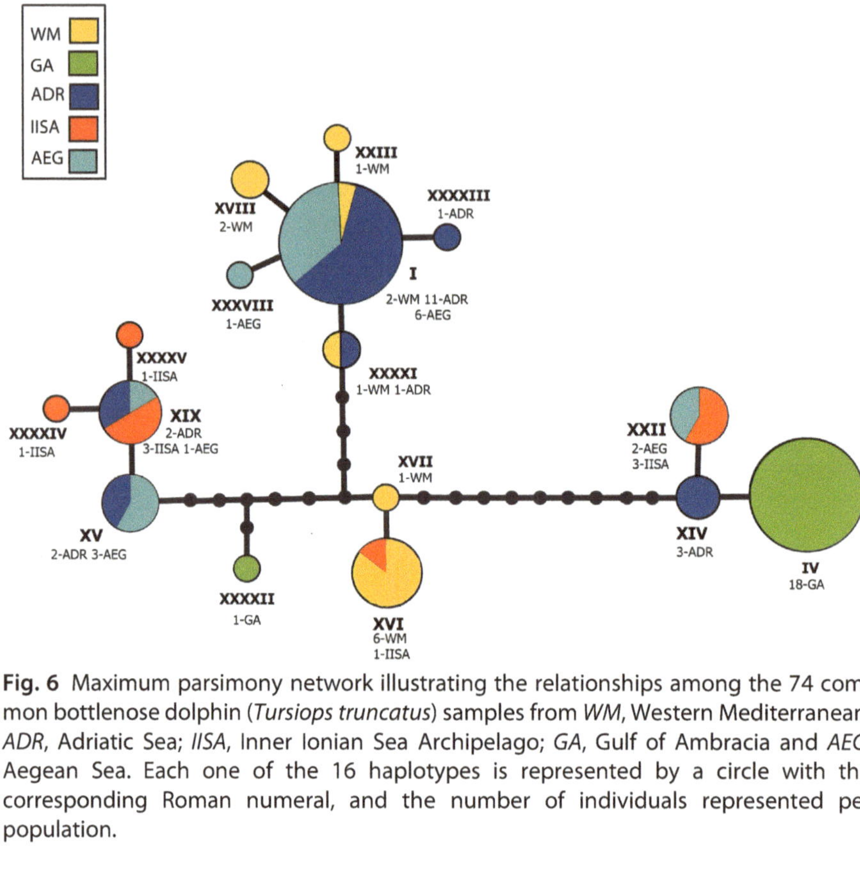

Fig. 6 Maximum parsimony network illustrating the relationships among the 74 common bottlenose dolphin (*Tursiops truncatus*) samples from *WM*, Western Mediterranean; *ADR*, Adriatic Sea; *IISA*, Inner Ionian Sea Archipelago; *GA*, Gulf of Ambracia and *AEG*, Aegean Sea. Each one of the 16 haplotypes is represented by a circle with the corresponding Roman numeral, and the number of individuals represented per population.

Table 6 Genetic Differentiation Among Common Bottlenose Dolphin (*Tursiops truncatus*) Populations Estimated with F_{ST} Based on mtDNA Data

n	Populations	WM	ADR	IISA	GA	AEG
13	WM					
20	ADR	0.199**				
9	IISA	0.150**	0.226**			
19	GA	0.600**	0.600**	0.627**		
13	AEG	0.172**	−0.005	0.141*	0.612**	

Statistical significance: $*F_{ST} > 0$ ($p < 0.05$), $**F_{ST} > 0$ ($p \leq 0.001$). *WM*, Western Mediterranean Sea; *ADR*, Adriatic Sea; *IISA*, Inner Ionian Sea Archipelago; *GA*, Gulf of Ambracia and *AEG*, Aegean Sea.

Table 7 Sex and Sexual Maturity, Lipid Percent (EOM%), Water Percent (H_2O%) and OCs Levels (i.e. HCB, DDT and PCB) Reported as d.w. (and l.w.) for Biopsy Samples Collected from 14 Common Bottlenose Dolphins (*Tursiops truncatus*) in the Gulf of Ambracia

Dolphin Identification	Sex	Sexual Maturity	Photo-Identification Information (to Estimate Minimum Age and Sexual Maturity)	EOM%	H_2O%	HCB ng/g d.w. (l.w.)	DDTs ng/g d.w. (l.w.)	PCBs ng/g d.w. (l.w.)
AMV01	M	No	First identified in September 2009 as a calf	30.3	55.0	6.45 (21.28)	25182.4 (83110.2)	7344.5 (24239.1)
AMV03	M	Yes	First identified in September 2004 as an adult	29.7	59.3	6.63 (22.32)	2830.0 (9528.7)	1312.6 (4419.6)
AMV04	M	Yes	First identified in June 2010 as an adult	14.0	55.2	5.80 (41.43)	1133.2 (8094.2)	1236.9 (8834.6)
AMV05	F	No	First identified in June 2012 as a juvenile	54.7	51.0	10.02 (18.33)	14249.5 (26050.3)	5628.2 (10289.2)
AMV06	M	Yes	First identified in September 2002 as an adult	28.7	58.9	6.11 (21.28)	14586.3 (50823.4)	5647.4 (19677.4)
AMV07	M	No	First identified in April 2011 as a calf	21.1	56.4	6.21 (29.41)	1593.3 (7551.2)	626.1 (2967.3)
AMV08	F	No	First identified in May 2011 as a juvenile	17.3	54.8	7.54 (43.56)	3072.1 (17757.7)	1348.3 (7793.7)
AMV09	M	Yes	First identified in January 2007 as an adult	37.8	52.5	11.79 (31.20)	11941.1 (31590.1)	7346.5 (19435.2)
AMV10	M	Yes	First identified in July 2003 as an adult	21.2	53.6	16.75 (79.00)	13025.2 (61439.7)	10302.3 (48595.5)

Continued

Table 7 Sex and Sexual Maturity, Lipid Percent (EOM%), Water Percent (H_2O%) and OCs Levels (i.e. HCB, DDT and PCB) Reported as d.w. (and l.w.) for Biopsy Samples Collected from 14 Common Bottlenose Dolphins (*Tursiops truncatus*) in the Gulf of Ambracia—cont'd

Dolphin Identification	Sex	Sexual Maturity	Photo-Identification Information (to Estimate Minimum Age and Sexual Maturity)	EOM%	H_2O%	HCB ng/g d.w. (l.w.)	DDTs ng/g d.w. (l.w.)	PCBs ng/g d.w. (l.w.)
AMV11	F	Yes	First identified in July 2003 as an adult	18.6	58.6	15.12 (81.29)	7123.5 (38298.1)	20218.9
(108704.0)								
AMV12	M	?	Unknown; no photo-identified	34.4	55.8	13.97 (40.62)	18066.6 (52519.1)	9859.1 (28660.1)
AMV13	F	Yes	First identified in June 2004 as an adult	22.5	56.6	12.81 (56.92)	16097.3 (71543.4)	6603.3 (29347.9)
AMV14	M	Yes	First identified in July 2001 as an adult	40.3	57.3	20.61 (51.14)	161091.2 (399730.0)	21584.2 (53558.9)
AMV15	F	Yes	First identified in July 2003 as an adult. Two offspring recorded (in 2005 and 2008)	30.0	55.2	12.72 (42.39)	6539.6 (21798.8)	2463.9 (8212.9)

To make comparisons between the different xenobiotic levels detected in each single individual and within the various sex/age classes, the results were considered on the basis of lipid content. Among the OCs analysed, HCB was the compound with the lowest levels and with relatively similar values among individuals (Table 7). Large differences were detected in the levels of DDTs and PCBs between individual blubber samples, which ranged from 7.6 to 399.7 μg/g l.w. and from 3.0 to 108.7 μg/g l.w., respectively. The DDT values were higher than HCB and PCB for all the analysed animals, except for dolphin AMV04, which had almost the same levels of the two xenobiotics, and for dolphin AMV11, which had three times more PCBs than the DDTs (Fig. 7).

In the Ambracian dolphins, the ΣDDT/ΣPCB ratio varied between 0.352 and 7.463, with a mean value of 2.43. In all bottlenose dolphins sampled, pp′DDE/pp′DDT ratio was higher than 10, reaching a maximum value of 147 in the blubber of dolphin AMV6. The values of ppDDE/DDTs ratios ranged from 0.81 to 0.96.

Sexually mature males had higher levels for all xenobiotics than immature dolphins of the same sex (Table 8). This difference was particularly high in the case of DDTs and was observed independently of the group in which

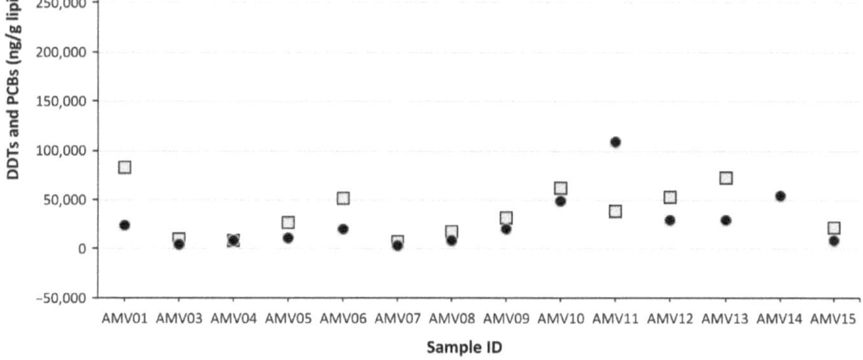

Fig. 7 Plot of DDT and PCB levels (ng/g lipid weight) in the subcutaneous blubber of the single common bottlenose dolphin (*Tursiops truncatus*) specimens collected from the Gulf of Ambracia.

Table 8 Arithmetic Mean Levels and Standard Deviation (in Brackets) of HCB, DDTs and PCBs in Common Bottlenose Dolphins (*Tursiops truncatus*) ($n = 14$) of the Gulf of Ambracia, Subdivided by Sex (Males and Females) and Sexual Maturity

	n	HCB ng/g l.w.	DDTs μg/g l.w.	PCBs μg/g l.w.
Males and females	14	41.44 (20.19)	62.85 (99.92)	26.77 (28.27)
Males	9	37.52 (18.70)	78.27 (123.4)	23.38 (18.00)
Females	5	48.50 (23.01)	35.09 (21.78)	32.87 (43.33)
Sexually mature males	6	41.06 (21.80)	93.53 (151.5)	25.75 (20.56)
No-sexually mature males	3	30.44 (9.71)	47.73 (38.01)	18.62 (13.74)
No-sexually mature males (*AMV12 not included*)	2	25.35 (5.75)	45.33 (53.43)	13.60 (15.04)
Sexually mature females	3	60.20 (19.66)	43.88 (25.34)	48.75 (52.98)
No-sexually mature females	2	30.94 (17.84)	21.90 (5.86)	9.04 (17.64)

male dolphin AMV12, whose sexual maturity was considered uncertain, was included. Higher levels of xenobiotics were also detected in sexually mature females. Mean value of DDTs was higher in males than in the females. Conversely, females had higher HCB and PCBs mean values than did males.

4. DISCUSSION

4.1 Site Fidelity

Bottlenose dolphins in the GA showed high levels of year-round site fidelity throughout the 10-year study period, which is consistent with previous findings (Bearzi et al., 2008a,b). Dolphin groups encountered in the southwest portion of the Gulf, in areas closer to the Preveza channel, were followed for periods of up to several hours, but they were never observed to enter the narrow and shallow corridor leading to open seawaters. Three individuals first photo-identified in the GA and regularly observed between 2003 and 2008 were subsequently found in the Inner Ionian Sea Archipelago and in the Gulf of Corinth (Bearzi et al., 2011), and were not observed in Ambracian waters after 2008 (J. Gonzalvo, unpublished data; see Fig. 4). Based on photographs of their genital area, all three animals were males. This is consistent with the hypothesis that males are more wide-ranging than females, and they may therefore be the primary vectors of genetic exchange (Bearzi et al., 1997; Wells et al., 1987). Such observations apparently indicate a small degree of emigration, but no immigration into the Gulf has been recorded to date. The observed extent of occurrence of the population, defined as 'the area contained within the shortest continuous imaginary boundary which can be drawn to encompass all the known, inferred or projected sites of present occurrence of a taxon, excluding cases of vagrancy', was about three quarters of the total area of the Gulf (i.e. approx. 300 km^2, excluding enclosed marshes and lagoons), would fall well within the range of values required (<5000 km^2) for classification as Endangered under criterion B1 of the IUCN Red List (IUCN, 2012).

4.2 Population Size and Trend

The approach taken to estimate population size was to use closed population models applied to monthly sampling occasions within years, which was considered more appropriate than using open population models applied to annual sampling occasions. Open models are best suited for estimation of survival rate and tend to be less robust than closed models for estimating population size; in particular, they are unable to take account of heterogeneity of capture probabilities (Hammond, 2009, 2010), which was shown to be influential in our case. An alternative could be to use the robust design analytical framework, which would combine the use of closed models based on monthly sampling occasions to estimate population size within years with open models based on annual data to estimate survival rate. Temporary emigration/immigration rates between years can also be estimated and this could be interesting to investigate here. However, robust design models are data-hungry and the small number of animals in this population may limit the utility of this approach (Kendall, 1999; Pollock, 1982).

Estimates of the number of marked individuals from the closed population models were variable from year to year if the model selected as the most appropriate was used in each year. There was much more consistency when a single model was used for each year. Models accounting for heterogeneity of capture probabilities among individuals consistently gave larger annual estimates than models without heterogeneity, indicating that even when these models were not chosen there was still likely some heterogeneity in the data. Thus, we are confident that choosing the time series of estimates from a single model with heterogeneity is justified as the best approach in this study. A similar approach was taken in a study of Northern Norwegian killer whales (Kuningas et al., 2014).

Analysing data subsets with and without moderately distinct individuals to explore the trade-off between bias and precision in estimates showed that including moderately distinct individuals increased precision, as expected because of larger sample size, but found no evidence that including them caused bias. Thus, our estimates based on the D1 + D2 + D3 data subset were the best estimates.

These population estimates over the 10-year study mostly fell between 130 and 170 with CVs averaging about 10%. Confidence that the true size of the population lies within this range is therefore high. The estimated trend in population size over the 10 years was a decline of 1.6% per year but this was not significant and there is insufficient power in the data to detect a decline of this magnitude over 10 years.

A population size estimated to number fewer than 250 mature individuals can be classified as Endangered under criterion D (IUCN, 2012). Our most robust estimates of the total number of bottlenose dolphins in the GA never exceeded 170 individuals and even the uppermost 95% confidence limit did not exceed 250. Further work may be useful to determine the extent of geographic and genetic differentiation between bottlenose dolphins in the GA and the Mediterranean bottlenose dolphin populations.

4.3 Genetics

Despite our relatively small sample sizes, results indicate that the dolphins of the GA are genetically differentiated from the surrounding Mediterranean populations. Furthermore, the genetic diversity observed for the GA population was the lowest when compared with the rest of populations analysed in this study and with previous studies found in the literature (Curry, 1997; Natoli et al., 2005; Viaud-Martinez et al., 2008) which suggests that the GA population is highly isolated.

Samples used in this study were obtained by different research teams through a variety of means and over the course of several years. In the case of GA samples (see Table 1), whenever a sample was taken, the research boat abandoned that particular group and headed to a different side of the Gulf in search of other dolphins, to minimize the possibility of samples taken the same day representing closely related individuals. Bottlenose dolphins from the neighbouring Inner Ionian Sea Archipelago area, which belong to an even smaller resident population, despite their smaller sample size, exhibited a higher genetic diversity. Nevertheless, despite such precautions, relatedness among any GA individuals, no matter how distantly sampled, cannot be ruled out.

The genetic uniqueness of the dolphins of the GA, as shown by low mtDNA diversity and the presence of unique haplotypes, demonstrates low connectivity with neighbouring Ionian and even Adriatic populations and could be explained by historical events, and/or low gene flow. Additionally, the fact that the Tajima's D estimate was negative and significant for the GA population indicates that the population may have undergone a recent reduction; however, Fu's Fs was not in support of this scenario. To estimate both historical and contemporary connectivity between the GA and the neighbouring populations, as well as understanding the ancestry of the GA population, the analysis of a larger sample set and the use of nuclear markers would be required.

4.4 Toxicology

Free-ranging bottlenose dolphins sampled in the GA were considered to be in good nutritional status (no signs of emaciation) and were frequently engaged in surface feeding. Nevertheless, lipid content in their subcutaneous blubber (represented by EOM%) resulted in 28.6% ($n=14$). This value is much lower than what was reported in bottlenose dolphins of the Sea of Cortez ($n=11$; 43.5%), the Strait of Gibraltar ($N=3$; 48.9%), the Strait of Sicily ($n=4$; 60.8%), and a single specimen from the Ionian Sea with 74.0%, but it was similar to EOM% from specimens of the Croatian Adriatic Sea ($n=14$; 24.5%) (L. Marsili, unpublished data; Marsili et al., 1996). Variables such as nutritional status, reproductive status, ontogeny, environmental temperature changes and pathologies have been observed to affect blubber composition (Dunkin et al., 2005; Struntz et al., 2004; Wells and Scott, 1999). Although some of those variables may have contributed to the variation in lipid content (range 14.0–54.7), a mean value as low as that detected in the present study is indication of a generalized poor health condition.

With respect to toxicological status in delphinids, it must be also considered that when lipids mobilize from the blubber, lipophilic contaminants redistribute leading to higher tissue concentrations. Consequently, specimens with a low EOM%, may have a higher risk of suffering exposure-related adverse health effects (Yordy et al., 2010). Generally, the ΣDDT/ΣPCB ratio tends to be <1 in waters near industrialized areas, and >1 in waters close to agricultural areas (Aguilar et al., 1999). Hence, our results indicate that contamination in the Gulf has a predominantly agricultural origin. This is in accordance with other studies, reporting a predominance of DDTs rather than PCBs, conducted in this part of the Mediterranean

(Fossi et al., 2004; Marsili, 2000; Marsili and Focardi, 1996; Marsili et al., 2014). A typical technical DDT is composed of pp′DDT (77.1%), op′DDT (14.9%), pp′DDD (0.3%), op′DDD (0.1%), pp′DDE (4.0%), op′DDE (0.1%) and unidentified compounds (3.5%) (WHO, 1979) and presents a pp′DDE/pp′DDT ratio of 0.05. Therefore, if the ratio pp′DDE/pp′DDT has high values, it can be deduced that the majority of the active substance (pp′DDT) has been degraded to pp′DDE, which means that no recent entries of insecticide occurred (Aguilar, 1984). pp′DDE/DDTs ratio, as well as having a similar meaning to the pp′DDE/pp′DDT ratio, can also indicate the efficiency of the metabolic processes of a population (Borrell and Aguilar, 1987). Values of pp′DDE/pp′DDT ratio >0.05 for all sampled dolphins showed that DDT contamination was not recent. Moreover, based on the values obtained of the pp′DDE/DDTs ratio, varying from 0.81 to 0.96, bottlenose dolphins of the GA manifested a very high metabolism of this pesticide.

Sex and sexual maturity are important variables when assessing delphinid toxicological status. Females lose up to 90% of their total body burden of OCs during pregnancy and lactation (Borrell and Aguilar, 2005; Tanabe et al., 1982). Consequently, whereas females are able to eliminate their OC load during their reproductive life, males accumulate persistent organic pollutants for their entire life, increasing in contamination levels with age. The mean value of DDTs found in bottlenose dolphins of the GA was higher in males than in females, but contrary to what we expected, the exact opposite was found for HCB and PCBs. Three sexually mature females had levels of average HCB, DDTs and PCBs higher than younger immature females. However, when looking at individual life history information (i.e. health, nutritional status, pregnancies/lactation), only female AMV15 (firstly identified in July 2003 as an adult) had been observed with offspring; one calf in 2005 and another in 2008 (both currently alive according to our photo-identification records). In fact this female had very low OC levels, contrary to the other two females, which were never observed in association with offspring and are unlikely to have reproduced.

Anthropogenic pollution may have important consequences for dolphin population dynamics (García-Alvarez et al., 2014). Most studies on the toxicological status from OCs in bottlenose dolphins in the Mediterranean Sea were exclusively based on stranded animals (Marsili and Focardi, 1997; Romanić et al., 2014; Shoham-Frider et al., 2009; Storelli and Marcotrigiano, 2003; Storelli et al., 2007; Wafo et al., 2005). A recent paper by Jepson et al. (2016), using samples from both stranded and free-living biopsied animals, reported high mean $\sum$PCB concentrations (>100.0 mg/kg l.w.) in bottlenose dolphins from the Western Mediterranean and the Northern Adriatic Sea. The present study is one of the few toxicology studies conducted in the Mediterranean Sea that is based entirely on free-ranging bottlenose dolphins and provides unprecedented information on the toxicological status of dolphins in the GA. The OCs levels found in bottlenose dolphins were similar to those reported by Fossi et al. (2003) for the same species in the neighbouring waters of the Ionian Sea for HCB and PCBs, while for DDTs, levels were four times higher in the Gulf, which indicates that the latter legacy pesticide poses a real toxicological problem for the dolphins of the GA.

5. CONCLUSIONS

Our findings indicate that bottlenose dolphins in the Gulf Ambracia constitute a geographically distinct and genetically differentiated unit with little demographic exchange. Further, these dolphins are exposed to high levels of pollution, mostly derived from local agriculture (i.e. pesticides). Based on photographic mark-recapture estimates, 134 animals (CV=0.11) resided in the Gulf in 2015. By applying standard criteria provided by the IUCN Red List of Threatened Species, we conclude that this 'subpopulation' would qualify as Endangered if formally assessed by the IUCN. While local density of dolphins is among the highest recorded anywhere in the Mediterranean Sea, this is not indicative of favourable conservation status or pristine habitat. On the contrary, the viability of bottlenose dolphins in the GA may be at risk due to their likely reproductive isolation, small population size and small extent of occurrence, as well as to the increasing acute anthropogenic impacts in their semiclosed shallow habitat. Management of human pressures is an obvious way of reducing such a risk, and is consistent with national and regional commitments to protect this coastal area and cetaceans generally.

ACKNOWLEDGEMENTS

This work was supported in part by OceanCare and UNEP's Regional Activity Centre for Specially Protected Areas (RAC-SPA). Special thanks to Stefano Agazzi, Giovanni Bearzi, Silvia Bonizzoni, Marina Costa, Tilen Genov and Ioannis Giovos. We are also grateful for the work carried out by participants in our citizen science program handling and cataloguing hundreds of digital images. The Milan Civic Aquarium and Hydrobiological Station, Lefkas Marina and Jimmy Pandazis provided logistical support. Comments by Tilen Genov and Aviad Scheinin helped improve the manuscript.

REFERENCES

Adams, J., Speakman, T., Zolman, E., Mitchum, G., Wirth, E., Bossart, G.D., Fair, P.A., 2014. The relationship between land use and emerging and legacy contaminants in an apex predator, the bottlenose dolphin (*Tursiops truncatus*), from two adjacent estuarine watersheds. Environ. Res. 135, 346–353.

Aguilar, A., 1984. Relationship of DDE/ΣDDT in marine mammals to the chronology of DDT input into the ecosystem. Can. J. Fish. Aquat. Sci. 41, 840–844.

Aguilar, A., Borrell, A., Pastor, T., 1999. Biological factors affecting variability of persistent pollutant levels in cetaceans. J. Cetacean Res. Manage. 1, 83–116.

Albanis, T., Hela, D.G., Hatzilacou, D., 1995. Organochlorine residues in eggs of *Pelecanus crispus* and its prey in wetlands of Amvrakikos Gulf, North-western Greece. Chemosphere 31, 4341–4349.

Ballschmiter, K., Zell, M., 1980. Analysis of polychlorinated biphenyls (PCB) by glass capillary gas chromatography. Fresen. Z. Anal. Chem. 302, 20–31.

Bearzi, G., Notarbartolo di Sciara, G., Politi, E., 1997. Social ecology of bottlenose dolphins in the Kvarneric (northern Adriatic Sea). Mar. Mamm. Sci. 13, 650–668.

Bearzi, G., Agazzi, S., Bonizzoni, S., Costa, M., Azzellino, A., 2008a. Dolphins in a bottle: abundance, residency patterns and conservation of common bottlenose dolphins *Tursiops truncatus* in the semi-closed eutrophic Amvrakikos Gulf. Greece. Aquat. Conserv. 18, 130–146.

Bearzi, G., Fortuna, C.M., Reeves, R.R., 2008b. Ecology and conservation of common bottlenose dolphins *Tursiops truncatus* in the Mediterranean Sea. Mamm. Rev. 39, 92–123.

Bearzi, G., Bonizzoni, S., Gonzalvo, J., 2011. Mid-distance movements of common bottlenose dolphins in the coastal waters of Greece. J. Ethol. 29, 369–374.

Bearzi, G., Fortuna, C., Reeves, R., 2012. *Tursiops truncatus* (Mediterranean subpopulation). The IUCN Red List of Threatened Species 2012: e.T16369383A16369386. http://dx.doi.org/10.2305/IUCN.UK.2012-1.RLTS.T16369383A16369386.en. http://www.iucnredlist.org/details/16369383/0.

Borrell, A., Aguilar, A., 1987. Variations in DDE percentage correlated with total DDT burden in the blubber of fin and sei whales. Mar. Pollut. Bull. 18, 70–74.

Borrell, A., Aguilar, A., 2005. Mother-calf transfer of organochlorine compounds in the common dolphin (*Delphinus delphis*). Bull. Environ. Contam. Toxicol. 75, 149–156.

Burnham, K.P., Overton, W.S., 1978. Estimating the size of a closed population when capture probabilities vary among animals. Biometrika 65, 625–633.

Catsadorakis, G., Onmus, O., Bugariu, S., Gül, O., Hatzilacou, D., Hatzofe, O., Malakou, M., Michev, T., Naziridis, T., Nikolaou, H., Rudenko, A., Saveljic, D., Shumka, S., Sıkı, M., Crivelli, A.J., 2015. Current status of the Dalmatian pelican and the great white pelican populations of the Black Sea/Mediterranean flyway. Endanger. Spec. Res. 27, 119–130.

Chao, A., Lee, S.M., Jeng, S.L., 1992. Estimating population size for capture–recapture data when capture probabilities vary by time and individual animal. Biometrics 48, 201–216.

Curry, B.E., 1997. Phylogenetic Relationships Among Bottlenose Dolphins (Genus Tursiops) in a World-Wide Context. Ph.D. dissertation, Texas A&M University, Texas, USA, p. 138.

Darroch, J.N., 1958. The multiple recapture census: I. Estimation of a closed population. Biometrika 45, 343–359.

Dunkin, R.C., McLellan, W.A., Blum, J., Pabst, D.A., 2005. Ontogenetic changes in the thermal properties of Atlantic bottlenose dolphin (*Tursiops truncatus*) blubber. J. Exp. Biol. 208, 1469–1480.

Ewing, G., 2010. Haplotype Viewer. http://www.cibiv.at/~greg/haploviewer.

Excoffier, L., Lischer, H.E.L., 2010. Arlequin suite ver 3.5: a new series of programs to perform population genetics analyses under Linux and Windows. Mol. Ecol. Resour. 10, 564–567.

Felsenstein, J., 1989. PHYLIP-phylogeny inference package (version 3.2). Cladistics 5, 164–166.
Ferentinos, G., Papatheodorou, G., Geraga, M., Iatrou, M., Fakiris, E., Christodoulou, D., Dimitriou, E., Koutsikopoulos, C., 2010. Fjord water circulation patterns and dysoxic/anoxic conditions in a Mediterranean semi-enclosed embayment in the Amvrakikos Gulf, Greece. Estuar. Coast. Shelf Sci. 88, 473–481.
Fossi, M.C., Marsili, L., Neri, G., Natoli, A., Politi, E., Panigada, S., 2003. The use of a non-lethal tool for evaluating toxicological hazard of organochlorine contaminants in Mediterranean cetaceans: new data 10 years after the first paper published in MPB. Mar. Pollut. Bull. 46, 972–982.
Fossi, M.C., Marsili, L., Lauriano, G., Fortuna, C., Canese, S., Ancora, S., Leonzio, C., Romeo, T., Merino, R., Abad, E., Jiménez, B., 2004. Assessment of toxicological status of a SW Mediterranean segment population of striped dolphin (*Stenella coeruleoalba*) using skin biopsy. Mar. Environ. Res. 58, 269–274.
Frantzis, A., 2009. Cetaceans in Greece: Present Status of Knowledge. Initiative for the Conservation of Cetaceans in Greece, Athens, Greece.
Friligos, N., Balopoulos, E.T., Psyllidou-Giouranovits, R., 1997. Eutrophication and hydrography in the Amvrakikos Gulf Ionian Sea. Fresenius Environ. Bull. 6, 21–26.
García-Alvarez, N., Martín, V., Fernández, A., Almunia, J., Xuriach, A., Arbelo, M., Tejedor, M., Boada, L.D., Zumbado, M., Luzardo, O.P., 2014. Levels and profiles of POPs (organochlorine pesticides, PCBs, and PAHs) in free-ranging common bottlenose dolphins of the Canary Islands, Spain. Sci. Total Environ. 493, 22–31.
Gaspari, S., Scheinin, A., Holcer, D., Fortuna, C., Natali, C., Genov, T., Frantzis, A., Chelazzi, G., Moura, A.E., 2015. Drivers of population structure of the bottlenose dolphin (Tursiops truncatus) in the Eastern Mediterranean Sea. Evol. Biol. 42, 177–190.
Genov, T., Kotnjek, P., Lesjak, J., Hace, A., Fortuna, C.M., 2008. Bottlenose dolphins (*Tursiops truncatus*) in Slovenian and adjacent waters (northern Adriatic Sea). Ann. Ser. Hist. Nat. 18, 227–244.
Gerrodette, T., 1987. A power analysis for detecting trends. Ecology 68, 364–372.
Gonzalvo, J., Giovos, I., Moutopoulos, D.K., 2014. Fishermen's perception on the sustainability of small-scale fisheries and dolphin-fisheries interactions in two increasingly fragile coastal ecosystems in western Greece. Aquat. Conserv. Mar. Freshw. Ecosyst. 25, 91–106.
Gonzalvo, J., Giovos, I., Mazzariol, S., 2015. Prevalence of epidermal conditions in common bottlenose dolphins (*Tursiops truncatus*) in the Gulf of Ambracia, western Greece. J. Exp. Mar. Biol. Ecol. 463, 32–38.
Gorgone, A.M., Haase, P.A., Griffith, E.S., Hohn, A.A., 2008. Modeling response of target and nontarget dolphins to biopsy darting. J. Wildl. Manage. 72, 926–932.
Hammond, P.S., 2009. Mark-recapture. In: Perrin, W.F., Würsig, B., Thewissen, J.G.M. (Eds.), Encyclopedia of Marine Mammals, second ed. Elsevier, San Diego, pp. 705–709.
Hammond, P.S., 2010. Estimating the abundance of marine mammals. In: Boyd, I.L., Bowen, W.D., Iverson, S. (Eds.), Marine Mammal Ecology and Conservation: A Handbook of Techniques. Oxford University Press, Oxford and New York, pp. 42–67.
Harlin, A.D., Würsig, B., Baker, C.S., Markowitz, T.M., 1999. Skin swabbing for genetic analysis: application to dusky dolphins (*Lagenorhynchus obscurus*). Mar. Mamm. Sci. 15, 409–425.
Hatzianestis, I., Sklivagou, E., Georgakopoulou, E., 2001. Levels of chlorinated compounds in marine organisms from Greek waters. In: 7th International Conference on Environmental Science and Technology Ermoupolis, Syros Island, Greece, pp. 310–317.

Irvine, A.B., Scott, M.D., Wells, R.S., Kaufmann, J.H., 1981. Movements and activities of the Atlantic bottlenose dolphin, *Tursiops truncatus*, near Sarasota, Florida. Fish. Bull. 79, 671–688.

IUCN, 2012. IUCN Red List Categories and Criteria: Version 3.1, second ed. IUCN, Gland, Switzerland/Cambridge, UK, p. iv+32.

Jepson, P.D., Deaville, R., Barber, J.L., Aguilar, A., Borrell, A., Murphy, S., Barry, J., Brownlow, A., Barnett, J., Berrow, S., Cunningham, A.A., Davison, N.J., Ten Doeschate, M., Esteban, R., Ferreira, M., Foote, A.D., Genov, T., Giménez, J., Loveridge, J., Llavona, Á., Martin, V., Maxwell, D.L., Papachlimitzou, A., Penrose, R., Perkins, M.W., Smith, B., de Stephanis, R., Tregenza, N., Verborgh, P., Fernández, A., Law, R.J., 2016. PCB pollution continues to impact populations of orcas and other dolphins in European waters. Sci. Rep. 6, 18573. http://dx.doi.org/10.1038/srep18573.

Karras, G., Barelos, D., Karra, A., Patakioutas, G., Gizas, G., Albanis, T., 2007. Occurrence of pesticide persistent in Amvrakikos Gulf—N.W. Greece. In: Proceedings of the 10th International Conference Environmental Science Technology, Kos Island, Greece, 5–7 September, pp. 640–647.

Kendall, W.L., 1999. Robustness of closed capture-recapture methods to violations of the closure assumption. Ecology 80, 2517–2525.

Kiszka, J., Simon-Bouhet, B., Charlier, F., Pusineri, C., Ridoux, V., 2010. Individual and group behavioural reactions of small delphinids to remote biopsy sampling. Anim. Welf. 19, 411–417.

Kountoura, K., Zacharias, I., 2013. Trophic state and oceanographic conditions of Amvrakikos Gulf: evaluation and monitoring. Desalin. Water Treat. 51, 2934–2944.

Kuningas, S., Similä, T., Hammond, P.S., 2014. Population size, survival and reproductive rates of northern Norwegian killer whales (*Orcinus orca*) in 1986–2003. J. Mar. Biol. Assoc. U.K. 94, 1277–1291.

Mann, J., 2000. Unraveling the dynamics of social life: long-term studies and methods. In: Mann, J., Connor, R., Tyack, P., Whitehead, H. (Eds.), Cetacean Societies: Field Studies of Dolphins and Whales. The University of Chicago Press, Chicago, pp. 45–64.

Marsili, L., 2000. Lipophilic contaminants in marine mammals: review of the results of ten years work at the Department of Environmental Biology, Siena University (Italy). Int. J. Environ. Pollut. 13, 416–452.

Marsili, L., Focardi, S., 1996. Organochlorine levels in subcutaneous blubber biopsies of fin whale (*Balaenoptera physalus*) and striped dolphin (*Stenella coeruleoalba*) from the Mediterranean sea. Environ. Pollut. 91, 1–9.

Marsili, L., Focardi, S., 1997. Chlorinated hydrocarbons (HCB, DDTs and PCBs) levels in cetaceans stranded along the Italian coasts, an overview. Environ. Monit. Assess. 45, 129–180.

Marsili, L., Fossi, M.C., Notarbartolo Di Sciara, G., Zanardelli, M., Focardi, S., 1996. Organochlorine levels and mixed function oxidase activity in skin biopsy specimens from Mediterranean cetaceans. Fresenius Environ. Bull. 5, 723–728.

Marsili, L., Maltese, S., Coppola, D., Carletti, L., Mazzariol, S., Fossi, M.C., 2014. Ecotoxicological status of seven sperm whales (*Physeter macrocephalus*) stranded along the Adriatic coast of Southern Italy. Aquat. Conserv. Mar. Freshw. Ecosyst. 24, 103–118.

Mirimin, L., Miller, R., Dillane, E., Berrow, S.D., Ingram, S., Cross, T.F., Rogan, E., 2011. Fine-scale population genetic structuring of bottlenose dolphins in Irish coastal waters. Anim. Conserv. 14, 342–353.

Natoli, A., Peddemors, V.M., Hoelzel, A.R., 2004. Population structure and speciation in the genus Tursiops based on microsatellite and mitochondrial DNA analyses. J. Evol. Biol. 17, 363–375.

Natoli, A., Birkun, A., Aguilar, A., Lopez, A., Hoelzel, A.R., 2005. Habitat structure and the dispersal of male and female bottlenose dolphins (*Tursiops truncatus*). Proc. R. Soc. Lond. B 272, 1217–1226.

Nei, M., 1987. Molecular Evolutionary Genetics. Columbia University Press, New York, p. 512.

Norris, J.L., Pollock, K.H., 1996. Nonparametric MLE under two closed capture-recapture models with heterogeneity. Biometrics 52, 639–649.

Notarbartolo di Sciara, G., 2016. Marine mammals in the Mediterranean Sea: An overview. In: Notarbartolo di Sciara, G., Podestà, M., Curry, B.E. (Eds.), Mediterranean Marine Mammals Ecology and Conservation. Advances in Marine Biology, vol. 75. Elsevier, Amsterdam, pp. 1–36.

Notarbartolo di Sciara, G., Bearzi, G., 2010. National Strategy and Action Plan for the Conservation of Cetaceans in Greece, 2010–2015. Initiative for the Conservation of Cetaceans in Greece, Athens, Greece.

Otis, D.L., Burnham, K.P., White, G.C., Anderson, D.R., 1978. Statistical inference from capture data on closed animal populations. Wildl. Monogr. 62, 3–135.

Pollock, K.H., 1975. A K-sample tag-recapture model allowing for unequal survival and catchability. Biometrika 62, 577–583.

Pollock, K.H., 1982. A capture–recapture design robust to unequal probability of capture. J. Wildl. Manage. 46, 757–760.

Read, A.J., Urian, K.W., Wilson, B., Waples, D.M., 2003. Abundance of bottlenose dolphins in the bays, sounds and estuaries of North Carolina. Mar. Mamm. Sci. 19, 59–73.

Rees, A.F., Margaritoulis, D., Newman, R., Riggall, T.E., Tsaros, P., Zbinden, J.A., Godley, B.J., 2013. Ecology of loggerhead marine turtles *Caretta caretta* in a neritic foraging habitat: movements, sex ratios and growth rates. Mar. Biol. 160, 519–529.

Reeves, R.R., Smith, B.D., Crespo, E., Notarbartolo di Sciara, G., 2003. Dolphins, Whales, and Porpoises: 2000–2010 Conservation Action Plan for the World's Cetaceans. IUCN, Gland, Switzerland.

Richards, V.P., Greig, T.W., Fair, P.A., McCulloch, S.D., Politz, C., Natoli, A., Driscoll, C.A., Hoelzel, A.R., David, V., Bossart, G.D., Lopez, J.V., 2013. Patterns of population structure for inshore bottlenose dolphins along the eastern United States. J. Hered. 104, 765–778.

Romanić, S.H., Holcer, D., Lazar, B., Klinčić, D., Mackelworth, P., Fortuna, C.M., 2014. Organochlorine contaminants in tissues of common bottlenose dolphins *Tursiops truncatus* from the northeastern part of the Adriatic Sea. Environ. Toxicol. Pharmacol. 38, 469–479.

Shane, S.H., 1990. Behavior and ecology of the bottlenose dolphin at Sanibel Island, Florida. In: Reeves, R., Leatherwood, S. (Eds.), The Bottlenose Dolphin. Academic Press, New York, pp. 245–265.

Shoham-Frider, E., Kress, N., Wynne, D., Scheinin, A., Roditi-Elsar, M., Kerem, D., 2009. Persistent organochlorine pollutants and heavy metals in tissues of common bottlenose dolphin (*Tursiops truncatus*) from the Levantine Basin of the Eastern Mediterranean. Chemosphere 77, 621–627.

Storelli, M.M., Marcotrigiano, G.O., 2003. Levels and congener pattern of polychlorinated biphenyls in the blubber of Mediterranean bottlenose dolphins *Tursiops truncatus*. Environ. Int. 28, 559–565.

Storelli, M.M., Barone, G., Piscitelli, G., Storelli, A., Marcotrigiano, G.O., 2007. Tissue-related polychlorinated biphenyls accumulation in Mediterranean cetaceans: assessment of toxicological status. Bull. Environ. Contam. Toxicol. 78, 206–210.

Struntz, D.J., McLellan, W.A., Dillaman, R.M., Blum, J.E., Kucklick, J.R., Pabst, D.A., 2004. Blubber development in bottlenose dolphins (*Tursiops truncatus*). J. Morphol. 259, 7–20.

Tanabe, S., Tatsukawa, R., Maruyama, K., Miyazaki, N., 1982. Transplacental transfer of PCBs and chlorinated hydrocarbon pesticides from the pregnant striped dolphin (*Stenella coeruleoalba*) to her fetus. Agric. Biol. Chem. 46, 1249–1254.

Tezanos-Pinto, G., Baker, C.S., Russell, K., Martien, K., Baird, R.W., Hutt, A., Stone, G., Mignucci-Giannoni, A.A., Caballero, S., Endo, T., 2009. A worldwide perspective on the population structure and genetic diversity of bottlenose dolphins (*Tursiops truncatus*) in New Zealand. J. Hered. 100, 11–24.

Tsabaris, C., Patiris, D.L., Fillis-Tsirakis, E., Kapsimalis, V., Pilakouta, M., Pappa, F.K., Vlastou, R., 2015. Vertical distribution of 137Cs activity concentration in marine sediments at Amvrakikos Gulf, western of Greece. J. Environ. Radioact. 144, 1–8.

Tsangaris, C., Cotou, E., Papathanassiou, E., Nicolaidou, A., 2010. Assessment of contaminant impacts in a semi-enclosed estuary (Amvrakikos Gulf, NW Greece): bioenergetics and biochemical biomarkers in mussels. Environ. Monit. Assess. 161, 259–269.

Urian, K., Read, A., Gorgone, A., Balmer, B., Wells, R.S., Berggren, P., Durban, J., Eguchi, T., Rayment, W., Hammond, P.S., 2015. Recommendations for photo-identification methods used in capture-recapture models with cetaceans. Mar. Mamm. Sci. 31, 298–321.

Viaud-Martinez, K.A., Brownell Jr., R.L., Komnenou, A., Bohonak, A.J., 2008. Genetic isolation and morphological divergence of Black Sea bottlenose dolphins. Biol. Conserv. 141, 1600–1611.

Wafo, E., Sarrazin, L., Diana, C., Dhermain, F., Schembri, T., Lagadec, V., Pecchia, M., Rebouillon, P., 2005. Accumulation and distribution of organochlorines (PCBs and DDTs) in various organs of *Stenella coeruleoalba* and a *Tursiops truncatus* from Mediterranean littoral environment (France). Sci. Total Environ. 348, 115–127.

Waples, R.S., Gaggiotti, O., 2006. What is a population? An empirical evaluation of some genetic methods for identifying the number of gene pools and their degree of connectivity. Mol. Ecol. 15, 1419–1439.

Weir, B.S., Cockerham, C.C., 1984. Estimating F-statistics for the analysis of population structure. Evolution 38, 1358–1370.

Wells, R.S., Scott, M.D., 1999. Bottlenose dolphin *Tursiops truncatus* (Montagu, 1821). In: Ridgway, S.H., Harrison, R. (Eds.), Handbook of Marine Mammals. The Second Book of Dolphins and Porpoises, vol. 6. Academic Press, San Diego, CA, pp. 137–182.

Wells, R.S., Scott, M.D., Irvine, A.B., 1987. The social structure of free-ranging bottlenose dolphins. In: Genoways, H.H. (Ed.), Current Mammalogy, vol. 1. Plenum, New York, pp. 247–305.

Wilson, B., Thompson, P.M., Hammond, P.S., 1997. Habitat use by bottlenose dolphins: seasonal distribution and stratified movement patterns in the Moray Firth, Scotland. J. Appl. Ecol. 34, 1365–1374.

Wilson, B., Hammond, P.H., Thompson, P.M., 1999. Estimating size and assessing trends in a coastal bottlenose dolphin population. Ecol. Appl. 9, 288–300.

World Health Organization (WHO), 1979. DDT and Its Derivatives, Environmental Health Criteria 9. WHO, Geneva. Available online athttp://www.inchem.org/documents/ehc/ehc/ehc009.htm.

Würsig, B., Jefferson, T.A., 1990. Methods of photo-identification for small cetaceans. In: Hammond, P.S., Mizroch, P., Donovan, G. (Eds.), Individual Recognition of Cetaceans: Use of Photo-Identification and Other Techniques to Estimate Population Parameters. Report to the International Whaling Commission, Special Issue 12, pp. 43–52.

Würsig, B., Würsig, M., 1979. Behaviour and ecology of the bottlenose dolphin, *Tursiops truncatus*, in the South Atlantic. Fish. Bull. 77, 399–412.

Yordy, J.E., Pabst, D.A., McLellan, W.A., Wells, R.S., Rowles, T.K., Kucklick, J.R., 2010. Tissue-specific distribution and whole-body burden estimates of persistent organic pollutants in the bottlenose dolphin (*Tursiops truncatus*). Environ. Toxicol. Chem. 29, 1263–1273.

CHAPTER TEN

Dolphins in a Scaled-Down Mediterranean: The Gulf of Corinth's Odontocetes

G. Bearzi[*,†,‡,1], S. Bonizzoni[*,†,‡], N.L. Santostasi[*,§], N.B. Furey[*,¶], L. Eddy[*,†], V.D. Valavanis[||], O. Gimenez[§]

[*]Dolphin Biology and Conservation, Oria, Italy
[†]OceanCare, Wädenswil, Switzerland
[‡]Texas A&M University at Galveston, Galveston, TX, United States
[§]Centre d'Ecologie Fonctionnelle et Evolutive, Montpellier, France
[¶]University of British Columbia, Vancouver, BC, Canada
[||]Marine Geographic Information Systems Lab, Hellenic Centre for Marine Research, Heraklion, Greece
[1]Corresponding author: e-mail address: giovanni.bearzi@gmail.com

Contents

Advances in Marine Biology, Volume 75
ISSN 0065-2881
http://dx.doi.org/10.1016/bs.amb.2016.07.003

Abstract

The Gulf of Corinth is a 2400-km^2 semi-enclosed inland system (a *mediterraneus*) in central Greece. Its continental shelf areas, steep bottom relief, and waters up to 500–900 m deep offer suitable habitat to neritic and pelagic species. We used photographic capture–recapture, distribution modelling, and direct observations to investigate the abundance, status, habitat preferences, movements, and group size of four odontocete species regularly observed in the Gulf, based on five years (2011–2015) of survey effort from small boats. Striped dolphins (*Stenella coeruleoalba*) are more abundant (1324 individuals, 95%CI 1158–1515) than was determined from previous estimates. Striped dolphins appear to be confined to the Gulf, where they favour deep and oligotrophic waters, and were encountered in single-species and mixed-species groups. Short-beaked common dolphins (*Delphinus delphis*) (22 individuals, 95%CI 16–31), individuals with intermediate pigmentation (possibly striped/common dolphin hybrids) (55, 95%CI 36–83), and a single Risso's dolphin (*Grampus griseus*) were only encountered in mixed-species groups with striped dolphins. Short-beaked common dolphins constitute a discrete conservation unit (subpopulation), and based on the current estimate, would qualify as Critically Endangered according to International Union for the Conservation of Nature (IUCN) Red List criteria. Common bottlenose dolphins (*Tursiops truncatus*) (39 animals, 95%CI 33–47) occur in single-species groups; they prefer continental shelf waters and areas near fish farms in the northern sector, and several animals appear to move into and out of the Gulf. Additionally, we contribute records of marine fauna and an assessment of the fishing fleet operating in the Gulf. Our study shows that the importance of this vulnerable marine environment has been underestimated, and management action must be taken to mitigate human impact and ensure long-term protection.

1. INTRODUCTION

In the Mediterranean Sea, formal commitments to protect cetaceans clash with geopolitical complexity, socio-economic or naval interests, and a generally poor political resolve, resulting in inaction. Beyond the political arena, a poor understanding of ecological components in the region leads to yet another reason for inertia (Portman et al., 2013). Though action may not be taken even when extensive information *does* become available (Bearzi, 2007), lack of information will invariably undermine marine conservation efforts. Therefore, contributing rigorous data on cetacean population abundance, status, distribution, and movements is a fundamental first step.

The Gulf of Corinth (GOC) is a semi-enclosed embayment in central Greece. Its intracontinental rift—one of Earth's most active and fastest spreading (Beckers et al., 2015; Taylor et al., 2011)—generated a fault along the Gulf's southern margin, where waters 500–900 m deep are found close to the coast. The occurrence of continental shelf areas, steep bottom relief, and deep waters within a 2400-km^2 inland basin gave rise to a scaled-down Mediterranean (literally meaning 'in the middle of the land') that offers suitable habitat to a variety of neritic and pelagic species. Striped dolphins (*Stenella coeruleoalba*)—a typically pelagic odontocete (Aguilar and Gaspari, 2012; Archer, 2009)—occur near the coast and are abundant (Frantzis and Herzing, 2002). Research conducted since the mid-1990s has documented mixed-species groups including striped and short-beaked common dolphins (*Delphinus delphis*) (Azzolin et al., 2010; Bearzi et al., 2011a; Frantzis and Herzing, 2002; Frantzis et al., 2003; Gkafas et al., 2007; Mardikis et al., 1999; Podiadis et al., 2003; Zafiropoulos et al., 1999). Animals with intermediate striped-common dolphin pigmentation were also observed and are suspected to be hybrids (Bearzi et al., 2011a; Frantzis and Herzing, 2002). Additionally, between 1997 and 2001 two Risso's dolphins (*Grampus griseus*) were found within groups of striped and common dolphins (Frantzis and Herzing, 2002). A fourth odontocete encountered in the GOC is the common bottlenose dolphin (*Tursiops truncatus*). Bottlenose dolphins were never observed in mixed-species groups and some individuals were documented to range as far as 265 km, moving into and out of the GOC (Bearzi et al., 2011b).

Ensuring conservation of odontocetes occurring in the GOC is crucial, particularly when one considers the status of these species in the broader Mediterranean region (Notarbartolo di Sciara and Birkun, 2010). Striped dolphins are the most abundant cetaceans in the Mediterranean Sea (Aguilar, 2000), including in the waters of Greece (Frantzis et al., 2003). However, they have been classified as Vulnerable on the IUCN Red List of Threatened Species, primarily because of past mortality events caused by viral infections (related to contamination by xenobiotics), incidental capture in pelagic driftnets, and decreased food availability caused by overfishing (Aguilar and Gaspari, 2012). Common dolphins were once abundant, but since the 1960s they have been declining dramatically throughout the Mediterranean and are currently classified as Endangered (Bearzi, 2003). The causes of their decline include historical culling and, more recently, prey depletion by overfishing, incidental mortality in fishing gear, and sea temperature shifts related to climate change (Bearzi et al., 2003, 2008a; Cañadas and Vázquez, in press). The status of Risso's dolphins is poorly known and their Mediterranean subpopulation is classified as Data Deficient (Bearzi et al., 2011c; Gaspari and Natoli, 2012). Finally, bottlenose dolphins in the Mediterranean are classified as Vulnerable due to declines as a result of culling, overfishing of their prey, mortality in fishing gear, and health effects caused by pollution (Bearzi et al., 2008b, 2012).

Several management measures to protect cetaceans in the GOC have been proposed, but no action has been taken to date. In 2007, the Agreement on the Conservation of Cetaceans of the Black Sea, Mediterranean Sea, and Contiguous Atlantic Area (ACCOBAMS) (also ratified by Greece) listed the GOC as an area of special importance for common dolphins and other cetaceans, and called for the creation of a Marine Protected Area (MPA) (resolution 3.22; ACCOBAMS, 2007). In the same year, Greenpeace proposed the creation of a marine reserve (Greenpeace, 2007). A National Strategy and Action Plan for the Conservation of Cetaceans in Greece, 2010–2015, granted high conservation priority to the GOC (Notarbartolo di Sciara and Bearzi, 2010). More recently, the European Union (EU) funded project, Monitoring and Evaluation of Spatially Managed Areas (MESMA) proposed a network of MPAs including parts of the GOC (Giakoumi et al., 2012; Issaris et al., 2012; Stelzenmüller et al., 2013; Vassilopoulou et al., 2012). Though praiseworthy, these efforts were based on limited and preliminary information (e.g. regarding population sizes of individual species, habitat preferences, movement patterns, and current anthropogenic stressors).

Here, we present methodology and results from five years (2011–2015) of field research in the GOC, to provide insight on the biology and ecology of these four odontocete species. Data from boat surveys that encompassed the entire Gulf are combined with oceanographic datasets and measures of human impact to describe cetacean distribution and habitat preferences. Photographic sampling and individual photo-identification are used to obtain abundance estimates for each species, and for animals of intermediate pigmentation (thought to be hybrids). In addition, we provide ancillary information on dolphin group size and behaviour, records of marine fauna, and an assessment of the active fishing fleet.

2. METHODS

2.1 Study Area

The GOC, between the Peloponnese and mainland Greece, is a semi-enclosed basin of 2400 km^2 marked by broad bays (Fig. 1). The 1.9-km-wide Strait of Rion separates the GOC from the outer Gulf of Patras. The 6.4-km long Corinth Canal, cutting through the narrow Isthmus of Corinth, connects the GOC to the Saronic Gulf and the Aegean Sea; it is

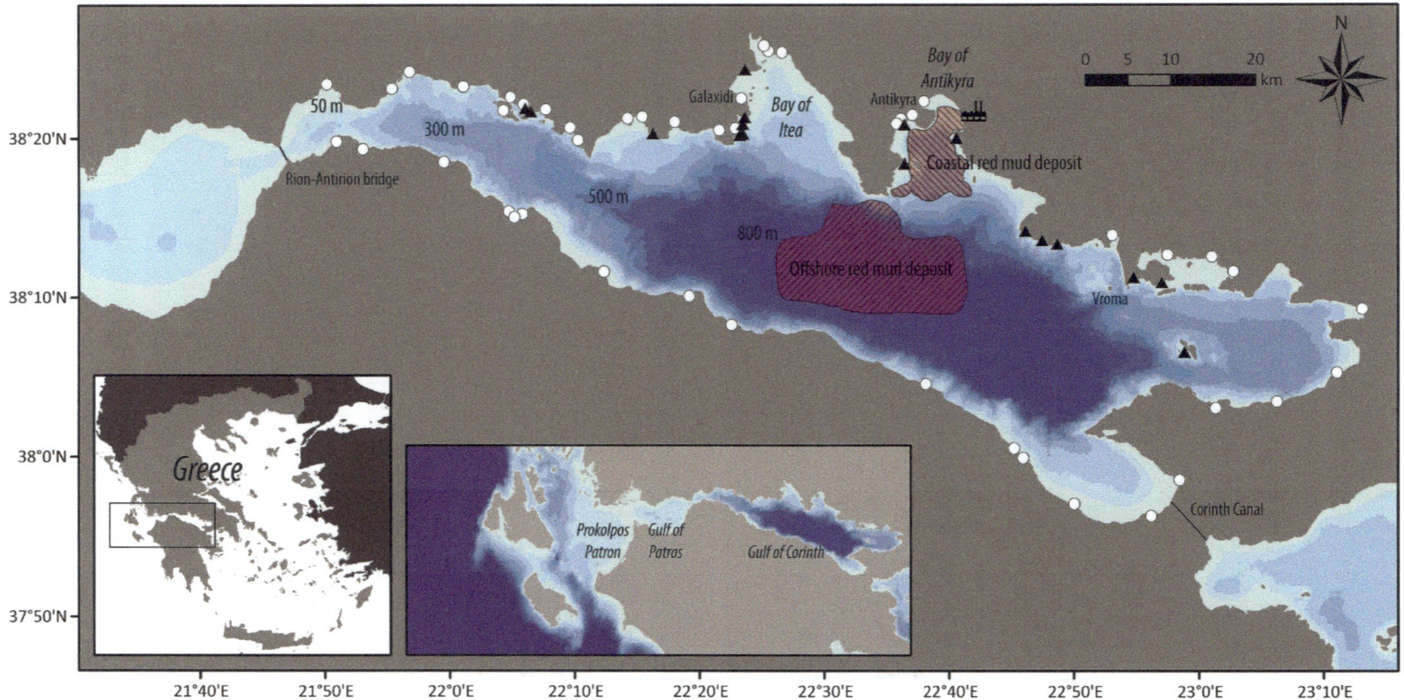

Fig. 1 The Gulf of Corinth. Important locations for the current study, 50–800 m isobaths, perimeter of coastal and offshore red mud deposits, positions of 17 active fish farms (*black triangles*), and 47 ports, as well as other shelters where fishing boats were recorded (*white dots*), are shown. *Insets* show the position of the Gulf in Greece and its inland deep-water basin separated from open Ionian Sea waters by the shallow Gulf of Patras and Prokolpos Patron.

only 21.4 m wide and 8 m deep, making it impassable for most modern large ships. The canal is also largely inaccessible to cetaceans (Frantzis and Herzing, 2002). The central portion of the GOC includes a vast extension of waters 500–900 m deep, encompassing 900 km^2. The Gulf's western quarter is shallower (200–400 m), with a constriction less than 2 km wide and a maximum depth of 65 m at the Rion-Antirion Bridge (Fig. 1).

The Gulf waters are generally oligotrophic, with rivers that are mostly seasonal/temporal and limited water intake from small streams. Input of pollutants comes primarily from industrial discards, city sewage, and agriculture runoff (Botsou and Hatzianestis, 2012). A large factory processing bauxite for aluminium production has been operating since 1966, near the city of Antikyra (Fig. 1). The industrial residual after extraction of aluminium (the Bayer process used for refining bauxite results in a by-product called 'red mud') has been dumped into the GOC for decades. Red mud, primarily composed of iron oxides, aluminium, and titanium, has a fine granulometry leading to great dispersion in water (Tsakiridis et al., 2004). Until 1969, red mud was discarded by barges in waters less than 50 m deep. Increased industrial production led to the construction of a series of underwater pipes, discarding red mud at depths between 120 and 265 m (Iatrou, 2013). Discards of red mud at sea have ranged between 500,000 and 700,000 tonnes annually (Papatheodorou et al., 1999; Pontikes, 2007; Varnavas and Achilleopoulos, 1995; Varnavas et al., 1986). In 2006, the factory introduced filter presses as a mitigation measure, and in 2011 it was reported to have stopped discarding red mud at sea (Issaris et al., 2012; and see http://www.alhellas.com). Two main deposits of red mud have been located on the seafloor: a coastal deposit covering 36 km^2 (with an estimated volume of 41 million m^3), and an offshore deposit covering 288 km^2 (with an estimated volume of 2 million m^3) (Iatrou, 2013; Iatrou et al., 2010a,b). Fig. 1 shows the estimated expanse of red mud deposits (based on Iatrou, 2013).

2.2 Survey and Photo-Identification Effort

We used 5.8 m long inflatable crafts with rigid hulls, powered by 100 HP four-stroke outboard engines, from May to October between the years 2011 and 2015, totalling 211 survey days, 1344 h at sea, and 21,435 km of navigation (Table 1; Fig. 2). Navigation was conducted under the following conditions: (1) daylight and long-distance visibility, (2) sea state ≤ 2 Douglas, (3) at least two experienced observers scanning the sea surface, (4) eye elevation of 1.6–1.8 m, and (5) survey speeds between 26 and

Table 1 Summary of Research Effort Surveying for Delphinid Species in the Gulf of Corinth (2011–2015)

Year	2011	2012	2013	2014	2015	Total
Survey days	31	28	49	52	51	211
Survey effort (km)	4171	3362	4243	4514	5145	21,435
Days with striped dolphin encounters	21	20	30	31	30	132
Days with bottlenose dolphin encounters	7	7	9	13	9	45
Tracking: striped dolphins (km)	316	342	450	382	383	1873
Tracking: bottlenose dolphins (km)	17	53	84	104	77	335
Observation time: striped dolphins	51 h 49 min	52 h 30 min	76 h 22 min	64 h 45 min	65 h 02 min	310 h 28 min
Observation time: bottlenose dolphins	3 h 08 min	9 h 13 min	16 h 28 min	21 h 59 min	16 h 22 min	67 h 10 min
Photos: striped dolphins (total)	9979	11,167	14,254	11,563	13,629	60,592
Photos: bottlenose dolphins (total)	281	1246	1032	1809	1134	5502

Numbers for striped dolphins (*Stenella coeruleoalba*) include counts from both single-species and mixed-species (with short-beaked common dolphins, *Delphinus delphis*, and/or with the single Risso's dolphin, *Grampus griseus*, encountered) groups. Common bottlenose dolphins (*Tursiops truncatus*) were only encountered in single-species groups.

30 km/h. Navigation was interrupted as soon as dolphins and other marine fauna were observed. Whenever possible, we attempted to obtain photographs or videos of marine fauna (sea turtles, fishes, cephalopods, etc.) to assist in identifying the species and estimating body size.

We spent a total of 378 h observing and tracking dolphins (Fig. 3), following them with the boat and recording the boat's position with a GPS at 1 min intervals. Striped dolphins and mixed-species groups were followed for 310 h 28 min (mean = 41 min per encounter, SD = 44.2, n = 457 encounters), across 1873 km, and on a total of 132 days; bottlenose dolphin groups were followed for 67 h 10 min (mean = 76 min per encounter, SD = 77.3, n = 53 encounters), across 335 km, on 45 days. The size and composition of dolphin groups were estimated using counts performed at

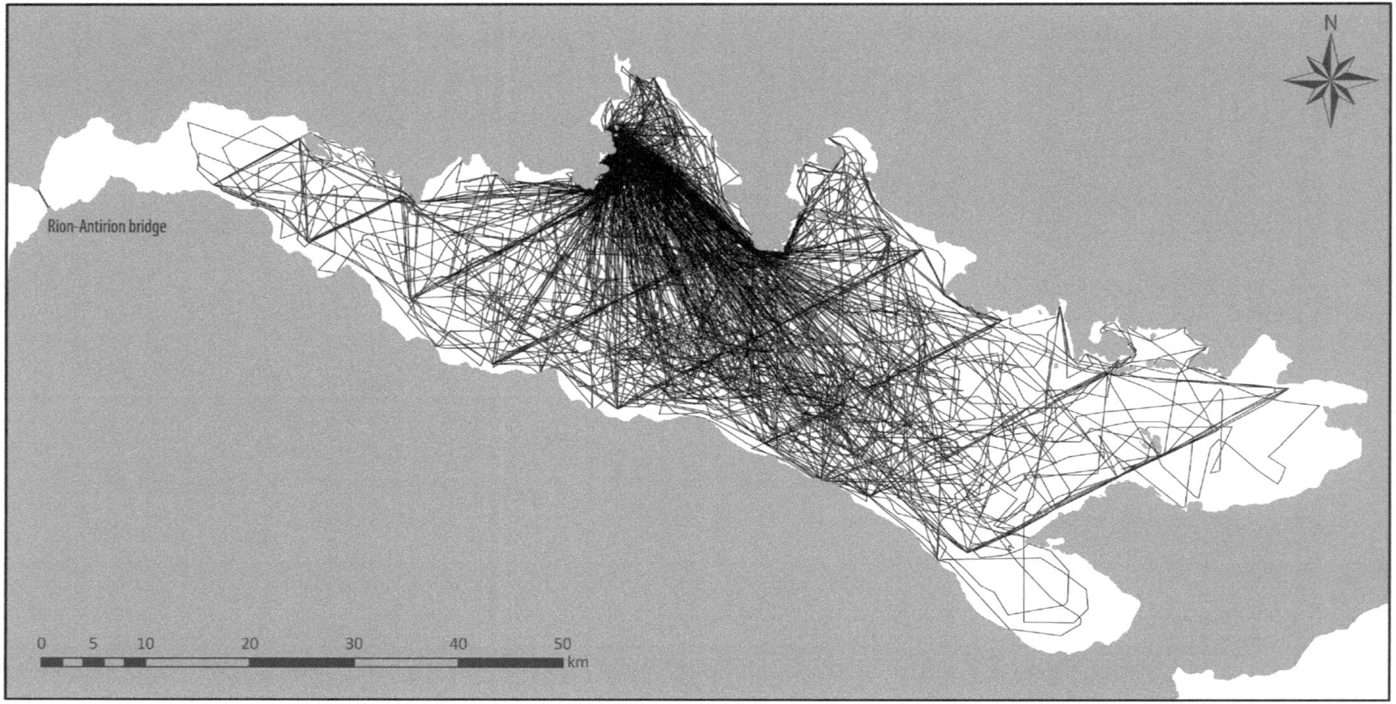

Fig. 2 Survey effort (navigation tracks) in the Gulf of Corinth, during the five-year study period (2011–2015).

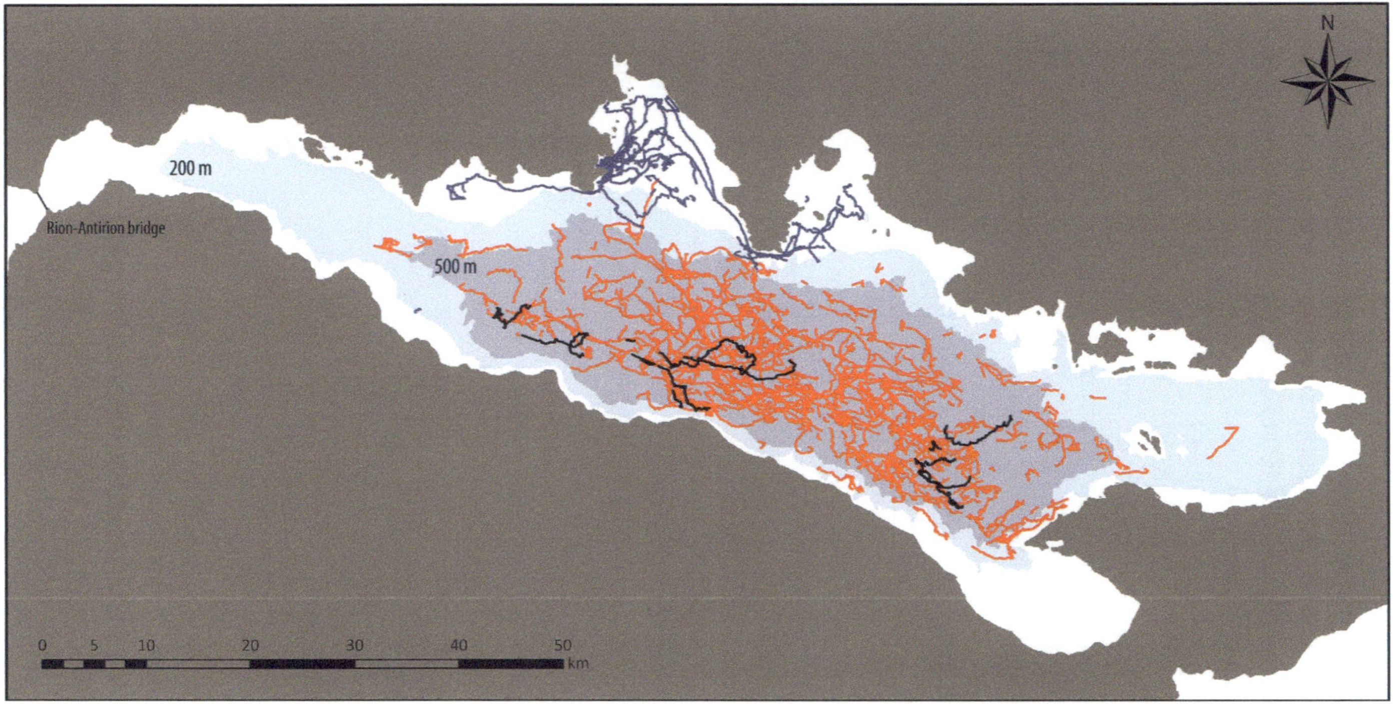

Fig. 3 Dolphin movements plotted based on GPS positions during 378 h of tracking in the Gulf of Corinth. *Red*: striped dolphins (*Stenella coeruleoalba*) (occasionally with short-beaked common dolphins, *Delphinus delphis*, and with individuals of intermediate pigmentation); *black*: Risso's dolphin (*Grampus griseus*; always in mixed-species groups with striped dolphins); *blue*: common bottlenose dolphins (*Tursiops truncatus*). *Shading* indicates 200 and 500 m isobaths.

15-min intervals. Groups were defined as 'dolphins observed in apparent association, moving in the same direction and often, but not always, engaged in the same activity' (Shane, 1990). Members of the focal group usually remained within approximately 100 m of each other and were all potentially photo-identifiable. To account for the difficulty of estimating the size of large, dispersed groups, counts above 15 individuals were attributed to categories (16–20, 21–30, 31–50, 51–75, 76–100, 101–150, etc.), using mid values within each category as best estimates. Distant (outlying) animals other than focal group members ('dolphins in sight'; Bearzi et al., 1997) were not included in the count but their occurrence was always recorded.

While tracking dolphins we attempted to photograph each individual in the focal group, irrespective of species, dorsal fin markings, or body size. Photo-identification was performed following Würsig and Jefferson (1990), using 18-megapixel digital cameras equipped with 70–200 mm f2.8 AF zoom lenses. Photographs suitable for individual identification were obtained on 158 days. Of 66,094 digital photos taken, 50,391 were selected based on recommendations by Read et al. (2003). These were cropped around the dorsal fin and visible parts of the body, and further subset using consistent criteria independent of fin markings (high sharpness, entire dorsal fin visible, and perpendicular to camera, no water spray masking fins), resulting in 26,704 high-quality, high-resolution photographs of single individuals. Images were then matched based on conspicuous dorsal fin markings (Bearzi et al., 2011a; Würsig and Würsig, 1977). Animals with absent or inconspicuous markings were considered unidentified. Strict selection, scoring, and matching criteria helped meet the 'mark recognition assumption' in capture–recapture analyses (Bearzi et al., 2011a; Wilson et al., 1999). Much attention was devoted to detecting possible mark changes over time (Gowans and Whitehead, 2001; Wilson et al., 1999).

2.3 Mixed Groups

Because striped and common dolphins (as well as individuals of pigmentation that appears to be intermediate between the two species) have similar size and morphology, in mixed groups one cannot discriminate between the two species based on photographs showing only their dorsal fins. Therefore, these individuals were considered together in capture–recapture analyses. The ratio of each species was then estimated based on a subset of 1175 photographs of animals showing sufficiently large body portions during aerial

Fig. 4 Examples of photographs showing sufficiently large body portions during aerial behaviour or conspicuous surfacing, used to estimate the proportion of striped dolphins *(Stenella coeruleoalba)* (A), short-beaked common dolphins (*Delphinus delphis*) (B), and individuals with intermediate pigmentation (C), in the Gulf of Corinth. If assessors were unable to attribute a category A, B, or C, or they attributed different categories, the photo was scored as 'controversial' (D).

behaviour or conspicuous surfacing, based on the method described by Bearzi et al. (2011a). This subset of photographs, extracted from a dataset of 23,995 photographs of striped dolphin groups, was also used to assess the proportion of common dolphins and animals of intermediate pigmentation (Fig. 4), which were quite distinctive from the known pattern and pigmentation variability of either species (see Rosso et al., 2008).

Criteria for selecting and scoring photographs in this study were considerably more rigorous than those used in the study by Bearzi et al. (2011a), which relied on a comparatively smaller sample size. Retained images had 100% agreement between two independent assessors with 15+ years of experience. If assessors were unable to attribute a species category (exemplified by photos A, B, and C in Fig. 4), or they attributed different categories, the image was considered 'controversial' (e.g. photo D in Fig. 4) and was discarded from the analysis. The final abundance estimate was corrected using the proportion of photographs of each species, and a coefficient of variation was calculated and incorporated in the final abundance estimate of each species, following Bearzi et al. (2011a).

2.4 Capture–Recapture Analyses

Striped, common, and intermediate dolphins were indistinguishable using dorsal fin photos alone, and they were treated as a single dataset in capture–recapture analyses. Therefore, we worked on two separate datasets: one for striped, common, and intermediate dolphins collectively, and one for bottlenose dolphins. We created a capture history (in our case, a photographic record) for each individual dolphin; a detection event in a sampling occasion was denoted as 1 and a nondetection as 0. To obtain a combined abundance estimate of striped, common, intermediate, and bottlenose dolphins, we used capture–recapture models within Pollock's robust design (Kendall et al., 1997; Pollock, 1982). These models relied on maximum likelihood estimation procedures (White and Burnham, 1999) expressing the probability of the observed data (capture history frequencies) as a function of population parameters such as population abundance and detection probability. These parameters were then estimated as the values that maximise the likelihood function (Lebreton et al., 1992). Analyses were carried out using the package RMark (Laake, 2013) in program R (R Core Team, 2014) to construct models from program MARK (White and Burnham, 1999).

Estimates obtained from capture–recapture models relate only to the marked proportion of the population (Read et al., 2003; Wilson et al., 1999). To estimate total population size, we scaled the estimates based on the proportion of identified individuals (Bearzi et al., 2011a; Williams et al., 1993). The variance of the abundance estimates was calculated following Wilson et al. (1999) as:

$$\mathrm{Var}(N_{\mathrm{tot}}) = N_{\mathrm{tot}}^2\left(\frac{\mathrm{var}N}{N^2} + \frac{1-\theta}{n\theta}\right)$$

where n = number of animals captured, N = estimate of number of marked animals, θ = proportion of identifiable animals, N_{tot} = estimate of total population size after correcting for proportion of identifiable individuals, and $\mathrm{var}N$ = variance of marked animals.

2.5 Distribution Modelling

To assess factors influencing dolphin distribution, survey data were linked to a variety of geographic, bathymetric, oceanographic, and anthropogenic variables. During surveys and dolphin group follows, GPS positions were recorded every 1 min. Only data collected under sea state S1 (flat), S2 (calm but rippled), and S3 (nonbreaking wavelets less than 20 cm high) were used

(Bonizzoni et al., 2014), corresponding to sea states up to approximately 2 Beaufort. Each position was associated with the following variables: sea state, sampling effort index (to account for variable survey effort across the Gulf), latitude, longitude, sea surface temperature (SST), chlorophyll *a* (Chl-*a*), upwelling areas, bottom depth, bottom slope, distance to coast, distance to fish farms, distance to coastal red mud deposit, and distance to offshore red mud deposit. Sampling effort index was calculated following Bonizzoni et al. (2014). Sea surface temperature and Chl-*a* satellite data were obtained from NASA OceanColor (oceancolor.gsfc.nasa.gov) as monthly averaged MODIS-SMI products. Using monthly averaged data, upwelling areas were the regions with negative SST anomalies and positive Chl-*a* anomalies based on a five-year climatology (2011–2015) for both environmental variables (Valavanis et al., 2004). Bottom depth was obtained from EMODNET (http://www.emodnet-hydrography.eu). All datasets were converted to a gridded surface of points and interpolated to a spatial resolution of 220 m within ArcGIS (http://www.arcgis.com) to match the mean resolution of field-sampled data. Bottom slope was calculated via spatial analyst tools using GIS software (ESRI ArcMap 10). Active fish farms ($n = 17$) were located through direct observations at sea, nautical charts, and information from Google Earth. Each farm's geometry was mapped with GPS by circumnavigating the farm. Perimeter of red mud deposits was obtained by georeferencing a map in Iatrou (2013). All distances (m) were calculated as minimum distance between the survey point and the corresponding polygon edge of the feature of interest, taking into account coastal profiles, by using the cost distance function within ArcGIS.

The relationships between dolphin presence and these variables were assessed within a modelling framework combining visual survey data and dolphin group follows via generalised additive models (GAM) (Hastie and Tibshirani, 1990; Wood, 2006) and generalised estimation equations (GEE) (Liang and Zeger, 1986), within a framework that has been used previously (Bonizzoni et al., 2014; Pirotta et al., 2011). Briefly, binomial GAM with a logit link was employed. To prevent overfitting by GAM, each explanatory variable was given a maximum number of degrees of freedom (df) to restrict flexibility (Ciannelli et al., 2008). GEE allow for relaxed assumptions of independence among GPS positions within autocorrelated blocks (assigned as in Bonizzoni et al., 2014), but maintain independence among blocks. A simple working independence model was identified as the most appropriate correlation structure for GEE, as advised by Pan (2001). To construct models, GEE-generalised linear models (GEE-GLM) were first constructed in R with the 'geepack' package (Højsgaard et al., 2016) in R (R Core Team, 2014), and then the package 'splines' was used to allow for smoothing splines within the GEE-GLM, resulting in GAM–GEE (Bonizzoni et al., 2014; Pirotta et al., 2011).

A framework of four submodels (geographic, bathymetric, oceanographic, and anthropogenic) was used to describe dolphin presence by incorporating subsets of the 13 variables in GAM and GEE (as in Bonizzoni et al., 2014). This framework of multiple submodels was used because a single global model with all variables suffered from collinearity, as demonstrated by variance inflation factors (VIF) of multiple variables exceeding four. Latitude and longitude were entered within the geographic submodel; bottom depth, bottom slope, and distance to the coast within the bathymetric submodel; SST, Chl-*a*, and distance to nearest upwelling area within the oceanographic submodel; distance to nearest fish farm and distance to red mud deposits within the anthropogenic submodel. This framework allows for submodels to be complementary rather than competing (Planque et al., 2011), and for the identification of a variety of factors correlated to dolphin occurrence. Effort index and sea state were included in all submodels to account for sampling bias. Prior to stepwise selection, multicollinearity was investigated for each submodel via VIF. Explanatory variables with $\mathrm{VIF} \geq 4$ (distance to offshore red mud deposit in the bottlenose dolphin dataset, and distance to coastal red mud deposit in the striped dolphin dataset) were removed from the analyses. The importance of variables was investigated by using a manual backward stepwise selection procedure based on minimising the quasi-likelihood under the independence model criterion (QIC) value (Pan, 2001).

2.6 Assessment of Fishing Fleets

Information on the composition of regional fishing fleets is essential for management and it can help estimate important information such as biomass of fish landed and food-web impacts on dolphins and other species (Bearzi et al., 2010; Piroddi et al., 2011b). Because little information existed on fishing effort in the GOC, and use of official fleet registers may be misleading (Gonzalvo et al., 2011), in September 2013 we catalogued all the fishing boats in 47 ports, docks, and roadsteads in the GOC (Fig. 1). Concealed fishing shelters were located by using nautical charts and Google Earth. Fishing boats were photographed and filed in a database that includes boat length, fishing gear, and activity (see Bearzi et al., 2010).

3. RESULTS

3.1 Striped and Short-Beaked Common Dolphins

During the five years of the study, we photo-identified a total of 393 striped and common dolphins that were categorised as marked individuals. The proportion of marked animals was 0.25 (95%CI 0.22–0.27). Our five-year average of the annual estimates of abundance obtained through capture–recapture methods, combining groups of striped dolphins only and mixed-species groups, was 1401 animals (95%CI 1241–1582; CV 0.06), with annual point estimates ranging between 1288 and 1534 and insignificant interannual variability ($p > 0.05$ for all pairwise z-test between years with Bonferroni corrections for multiple testing; Lebreton et al., 1992). Based on photographs of animals showing sufficiently large body portions during aerial behaviour or conspicuous surfacings, the proportion of striped dolphins was estimated as 0.9453 (CV 0.0191), common dolphins 0.0158 (CV 0.2830), and individuals of intermediate pigmentation 0.0389 (CV 0.3668). Considering these proportions and all the associated uncertainties (Bearzi et al., 2011a), we obtained final abundance estimates of 1324 striped dolphins (95%CI 1158–1515; CV 0.07), 22 common dolphins (95%CI 16–31; CV 0.17), and 55 intermediates (95%CI 36–83; CV 0.22). Across the five years of the study, the ratio of common dolphins versus intermediate animals showed a markedly negative trend, dropping from 1.0 in 2011 down to 0.38 in 2015. However, a linear regression failed to support such hypothesis ($R^2 = 0.77$, $p = 0.052$), likely due to the short duration of our study ($n = 5$).

The mean size of striped dolphin and mixed-species groups was 37 animals (SD = 33.1, $n = 1170$, range 1–225). Occurrence of distant individuals that were not included in the count was recorded during 639 samples (55%). When occurrence of common dolphins could be confirmed, groups were significantly larger (mean = 45 animals, SD = 31.1, $n = 232$, range 5–125; Mann–Whitney $U = 67{,}754$, $p < 0.001$).

Distribution models (GAM–GEE) for striped dolphins retained latitude and longitude in the geographic submodel, bottom depth in the bathymetric submodel, Chl-*a* in the oceanographic submodel, and distance to nearest fish farm as well as distance to offshore red mud deposit in the anthropogenic submodel (Fig. 5). Striped dolphin occurrence was positively affected by bottom depth, and negatively by Chl-*a*: deep oligotrophic waters were clearly preferred. Occurrence was higher in the central and southern sectors

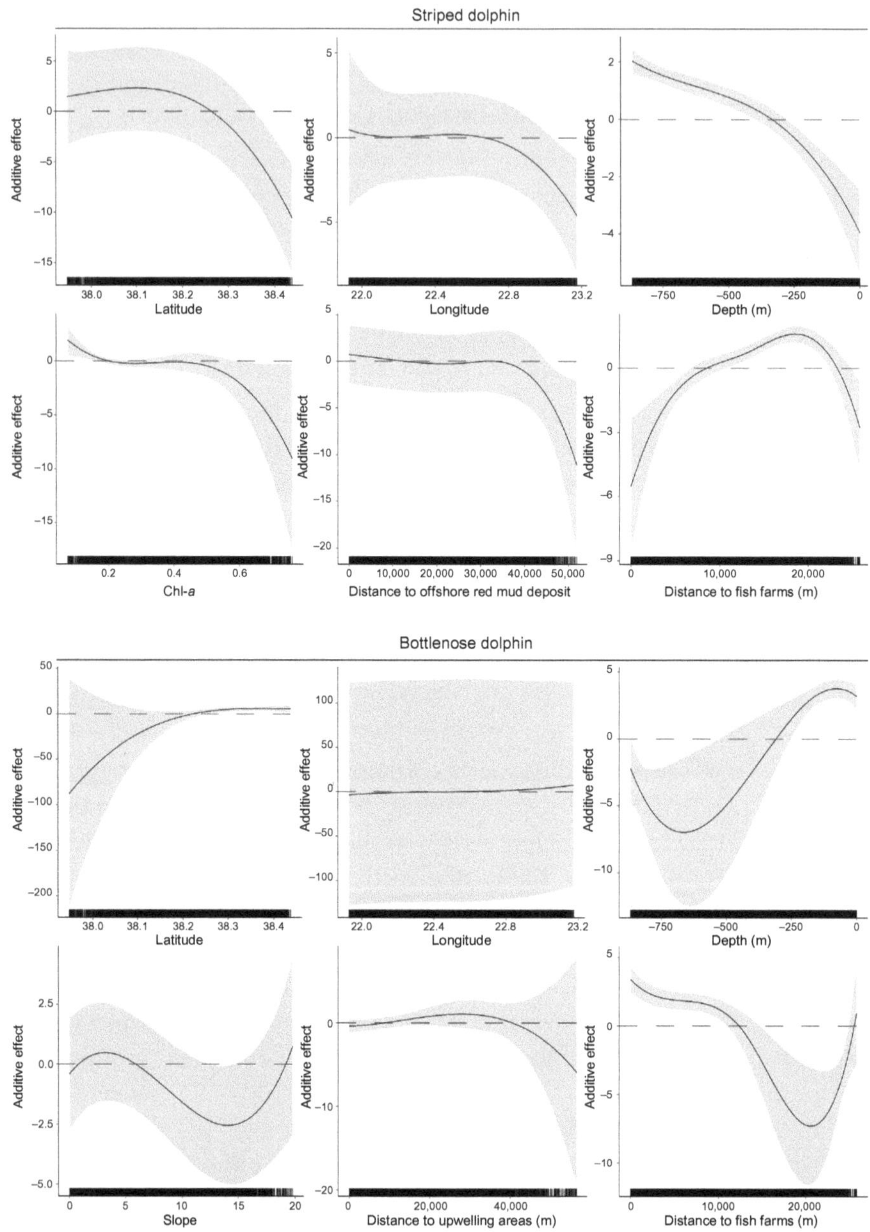

Fig. 5 Response curves of the relationships among explanatory variables and dolphin occurrence in the Gulf of Corinth. Zero on the vertical axes corresponds to no effect of the covariate on the estimated response. Positive values on the vertical axes correspond to a positive relationship between the covariate and presence of dolphins, while negative values correspond to a negative relationship. *Shaded areas* represent 95%CI as calculated by GEE.

of the Gulf. Although retained within the geographic submodel, the response curve for longitude shows wide confidence intervals, indicating the modelled increased occurrence in the west is misleading (no observations ever occurred west of 22°08′30″E). Lower occurrence of striped dolphins at distances greater than 40 km from the offshore red mud deposit likely relates to their strong preference for deep waters (where red mud naturally settles). A subtotal of 119 h 02 min of striped and common dolphin tracking (38.3% of the total) was conducted in areas over red mud deposits.

3.2 Risso's Dolphin

One of the two Risso's dolphins previously photo-identified in the GOC (Frantzis and Herzing, 2002) was never encountered. The only Risso's dolphin encountered was an adult female with characteristic dorsal fin and body markings, including a healed shark bite. This animal was encountered in each year between 2011 and 2015 (a total of 12 days). Consistent with previous findings (Frantzis and Herzing, 2002), the Risso's dolphin was observed within groups of striped and common dolphins, and never alone. Based on movements tracked across 114 km (24 h 27 min), habitat encompassed waters 296–859 m deep, 0.7–10.7 km off the nearest coast. Mean size of mixed-species groups when the Risso's dolphin was present was 44 animals (SD = 37.1, range 4–225), distant dolphins being recorded in 71% of group size samples. Occurrence of striped and/or common dolphin calves was recorded in 45% of samples, and the Risso's dolphin was often tightly associated with one or more of those calves. Surfacing intervals recorded across eight continuous timing sessions, totalling 10 h 48 min of timing, yielded dives between 2 and 139 s (mean = 23 s, SD = 17.8, $n = 1628$). During daytime observations the Risso's dolphin never performed long dives suggestive of feeding behaviour (Bearzi et al., 2011c). Observed behaviours included travelling, milling, physical contact with either of the other dolphin species, occasional playing with jellyfish and plastic bags, wave riding, and bow-riding small boats as well as a 95 m cargo ship.

3.3 Common Bottlenose Dolphins

Across the five years of the study, we photo-identified a total of 48 individual animals that were considered as marked. The proportion of marked animals was 0.704 (95%CI 0.650–0.757). The 2011 estimate (14 animals, 95%CI 5–39; CV 0.57) was considered unreliable due to insufficient sample size (see Table 1). Our final abundance estimate for the years 2012–2015 was 39 animals (95%CI 33–47; CV 0.10), with annual point estimates ranging between 31 and 56 and insignificant interannual variability ($p > 0.05$ for all pairwise z-test between years with Bonferroni corrections for multiple testing; Lebreton et al., 1992).

Number of encounters with marked individuals varied widely; only three were encountered in every year of this study, 23 in three or four of the years, 10 in two of the years, and 12 were only encountered in one of the years. Six of 48 marked individuals observed in the GOC between 2011 and 2015 were previously identified in other parts of Greece (Bearzi et al., 2005, 2010, 2011b). Bottlenose dolphin groups were composed of a mean of eight individuals (SD = 4.5, $n = 259$, range 1–28), 16% of samples including distant (outlying) dolphins not included in the count.

Distribution models (GAM–GEE) for bottlenose dolphins retained latitude and longitude in the geographic submodel, bottom depth and bottom slope in the bathymetric submodel, distance to upwelling areas in the oceanographic submodel, and distance to nearest fish farm in the anthropogenic submodel (Fig. 5). Bottlenose dolphins occurred predominantly in the northern sector of the Gulf. The response curve for longitude shows wide confidence intervals and no influence is apparent. The response curve for bottom depth indicates higher occurrence in waters shallower than approximately 300 m, whereas the response curve for slope suggests preference for gentler bottom contours. Confidence intervals around steeper slope values are wider and the exact trend of the estimated relationship should be interpreted with caution. Distance to upwelling areas suggests lower occurrence away from productive areas, though past 40 km the negative relationship has wide confidence intervals. Bottlenose dolphin occurrence was higher in areas within approximately 10 km of fish farms, and increased in their immediate proximity. A subtotal of 6 h 20 min of bottlenose dolphin tracking (9.4% of the total) was conducted in areas over red mud deposits.

3.4 Other Marine Fauna

Here, we provide a brief account of the marine fauna encountered during boat surveys conducted during the five-year study period.

*3.4.1 Mediterranean Monk Seal (*Monachus monachus*)*

We encountered a single monk seal on three occasions (June and July 2014, July 2015) in and around the Bay of Itea. Photographs indicated it was the same adult female, approximately 2 m long.

3.4.2 Sea Turtles

We encountered 64 loggerhead sea turtles (*Caretta caretta*). Of these, one had longline protruding from its mouth (we cut the line and released the animal), and three were dead (one also hooked on a longline). One adult was observed feeding on the Mediterranean jelly (*Cotylorhiza tuberculata*). Mean carapace length was 56 cm (SD = 11.9, range 20–80 cm). In August 2012, we encountered a leatherback sea turtle (*Dermochelys coriacea*) that was approximately 1.5–2 m long; this is the first reported observation of a living individual in the GOC (Bearzi et al., 2015).

3.4.3 Fishes

Scombridae schools were frequently observed feeding at the surface. Of 313 encounters, 35 were 'large tuna' 80–150 cm long (mode = 100 cm), 249 'small tuna' 30–70 cm long (mean = 43 cm, SD = 7.5), and 29 tuna of undetermined size. Based on photographs, 'small tuna' included juvenile Atlantic bluefin tuna (*Thunnus thynnus*), little tunny (*Euthynnus alletteratus*), and Atlantic bonito (*Sarda sarda*) (in frequent association with flocks of Scopoli's shearwater, *Calonectris diomedea*), whereas 'large tuna' were mostly bluefin. Myliobatidae (likely giant devil ray, *Mobula mobular*) were observed once in July 2015: two individuals of 2 m disc width. Swordfish (*Xiphias gladius*) totalled 70 individuals (mean = 138 cm, SD = 39.2, range 40–250). Other fish included sunfish (*Mola mola*; 10 individuals) and unidentified flyingfish (14 individuals 5–30 cm long). Under floating debris we observed juveniles of common dolphinfish (*Coryphaena hippurus*), greater amberjack (*Seriola dumerili*), wreckfish (*Polyprion americanus*), imperial blackfish (*Schedophilus ovalis*), grey triggerfish (*Balistes capriscus*), and pilotfish (*Naucrates ductor*). No sharks were observed during the five-year study period.

3.4.4 Invertebrates

Dead cephalopods or parts thereof were found floating adrift in deep pelagic waters on 24 occasions. Species included 19 specimens of the long-armed squid (*Chiroteuthis veranyi*), three umbrella squid (*Histioteuthis bonnellii*) and two unidentified parts. Thirteen findings, of which 12 were *C. veranyi*, occurred while tracking striped dolphins, suggesting that these squids—known prey of striped dolphins (Öztürk et al., 2007)—could have been killed by them (all squids were fresh and most carried lesions or amputations). *Cotylorhiza tuberculata* had a massive bloom during August and September 2012, when the entire surface of the Gulf was dotted by these jellyfish. Striped and common dolphins were often observed rubbing *Cotylorhiza* with their bodies as well as blowing them to pieces with their rostra.

3.5 Fishing Fleet

Assessments during windy (no-fishing) days can reasonably estimate the number of vessels of a small-scale fishing fleet, considering that most small-scale fishing in Greece is conducted near home ports to reduce operating costs (Tsitsika and Maravelias, 2008). The GOC small-scale fishing fleet included 361 boats: 301 active and 60 inactive. Mean boat size for the active small-scale fleet was 7.6 m (SD = 1.67, $n = 301$, range 5–12). Fishing gear was identified for 262 of the fishing boats. Of these, 77% ($n = 202$) used set nets (either trammel nets, gillnets, or nylon 'Japanese' nets), 12% ($n = 32$) used longlines, 6% ($n = 15$) a combination of set nets and longlines, and 5% ($n = 13$) were beach seiners that also deployed set nets.

The intermediate scale (industrial) fishing fleet included six purse seiners of 15–18 m, one bottom trawler of 20 m, and one longliner of 15–18 m. Purse seiners were occasionally observed fishing illegally in shallow near-shore areas, including in the Bay of Itea where purse seine fishing is banned. Information obtained during this study indicated that a few nonresident bottom trawlers routinely entered the GOC when the winter fishing season opened; the size and impact of this fleet of nomadic trawlers need to be properly assessed.

4. DISCUSSION

4.1 Geographic Isolation and Genetic Differentiation

Information on geographic and genetic isolation is relevant for management of local populations and can contribute to assessments of their conservation status (IUCN, 2012a). Cetaceans that have restricted ranges and disjunct distributions can become isolated, and are especially vulnerable to anthropogenic impacts. Further divergence can occur as groups become resident within discrete and geographically separated subareas. Studies on common dolphins show a clear population boundary between the western and the eastern Mediterranean Sea, with additional differentiation in the Ionian Sea (Moura et al., 2013; Natoli et al., 2008). Studies on striped and bottlenose dolphins also show evidence of fine-scale population genetic structure within the Mediterranean Basin (Gaspari et al., 2007, 2015a,b).

Geographic isolation of striped dolphins and common dolphins in the GOC has been proposed based on absence of records in the western quarter of the Gulf and in the adjacent Gulf of Patras (Bearzi et al., 2011a; Frantzis, 2009; Frantzis et al., 2003). Consistent with the hypothesis of geographic isolation, our study yielded no occurrence west of 22°08′E. Such geographic isolation may have led to genetic differentiation. Genetic evidence from 25 striped dolphins and three common dolphins sampled in the GOC suggested significant differentiation for both species from individuals in the Ionian Sea and other Mediterranean areas (Gkafas, 2011; Gkafas et al., 2007; Moura, 2010; Moura et al., 2013). The GOC's western quarter is comparatively shallower, with a depth of 65 m under the Rion-Antirion Bridge. Shallow waters continue west of the bridge due to the continental shelf of the Gulf of Patras and Prokolpos Patron, and thus any pelagic species would need to cross 80+ km of waters less than 50–100 m deep to pass between the Ionian Sea and the GOC (Fig. 1). This physiography might have contributed to the apparent fragmentation of striped dolphins, considering that shallow waters have the potential of inhibiting movements of pelagic species, for instance because key prey may not be found in shallow continental shelf areas (Cañadas et al., 2002). The habitat and range of common dolphins, never observed in single-species groups in the GOC, may be influenced by the habitat preferences and movement patterns of striped dolphins.

4.2 Striped Dolphins

Striped dolphins are abundant in the GOC and we found their numbers to be one order of magnitude higher than those inferred from early observations (Frantzis, 2009), and also substantially higher than those estimated by Bearzi et al. (2011a). The increase in numbers of individuals encountered in the current study likely relates to insufficient sampling effort in early studies and does not reflect dramatic shifts in striped dolphin abundance (Santostasi et al., 2016). Indeed, our population estimates did not show significant differences across the five years of this study, but research over longer periods of time is needed to detect trends. Even though the population is 'large', we lack a true baseline of historical striped dolphin abundance. In the past, rapidly spreading epizootics, potentially triggered by pollutants and decreased food availability, have decimated Mediterranean striped dolphins (Aguilar, 2000). We would expect the apparently isolated GOC population to be especially vulnerable if an outbreak were to occur. Vulnerability resulting from isolation within the deep-water portion of the GOC requires careful monitoring and precautionary management action for this population.

4.3 Short-Beaked Common Dolphins and Mixed-Species Groups with Striped Dolphins

Mixed-species groups are thought to occur due to foraging advantages and predator avoidance, as well as social and reproductive benefits (Herzing and Elliser, 2013; Stensland et al., 2003). Frantzis and Herzing (2002) noted that when numbers of Mediterranean common dolphins decline, the animals tend to associate with striped dolphins due to their tendency of staying in large groups. The authors suggested that as the decline continues, common dolphins start 'depending' on striped dolphins, becoming progressively more scattered across mixed-species groups. The two species, however, have different ecological needs and diets (Aguilar, 2000; Bearzi et al., 2003) and while common dolphins may adapt their behaviour to coexist with striped dolphins, changes in habitat and prey likely come at a cost.

Detecting common dolphin abundance trends in the GOC is hampered by insufficient analytical power resulting *inter alia* from the low number of animals—a classic problem with small populations (Taylor and Gerrodette, 1993). The steep negative trend ($R^2 = 0.77$) in the proportion of common dolphins versus dolphins with intermediate pigmentation, suggesting a continuous decline of the former species, was statistically inconclusive given the short duration of this study. Still, the proportion of common dolphins relative to animals of intermediate pigmentation may represent a valuable proxy to assess relative abundance over time.

Based on the available information, common dolphins in the GOC constitute a geographically distinct conservation unit, and likely have little demographic and genetic exchange. These dolphins face a high risk of extinction due to their small population size (22 animals estimated during the 2011–2015 study period), limited distribution, and suspected hybridisation with a 60-fold larger subpopulation of striped dolphins. We suggest that, given the low number of individuals, this subpopulation is likely nonviable (Traill et al., 2010). Under standard criteria provided by the IUCN Red List to assess extinction risk (IUCN, 2012a,b), the subpopulation would qualify as Critically Endangered (Bearzi et al., 2016).

4.3.1 Possible Striped Dolphin–Short-Beaked Common Dolphin Hybrid Individuals

Hybridisation, including that among small delphinids (e.g. Amaral et al., 2014; Brown et al., 2014; Silva et al., 2005), is relatively widespread in the order Cetacea, suggesting incomplete postmating barriers to interbreeding (Crossman et al., 2016). Species pairs that share a greater number of traits have a higher propensity to hybridise (Crossman et al., 2016). Because striped and common dolphins are closely related (Amaral et al., 2012, 2014; LeDuc et al., 1999), intermating may be likely to occur in mixed-species groups, possibly resulting in hybridisation (Bérubé, 2009). Hybridisation and introgression are significant threats for rare species coexisting with more abundant species (Allendorf et al., 2001; Levin, 2002), as is the case for common dolphins in the GOC. Hybridisation may lead to local eradication through genetic swamping, where 'pure' species are progressively replaced by hybrids (as suggested by the markedly negative trend in the ratio of common dolphins versus intermediate animals), or by demographic swamping, where population growth rates are reduced due to the expression of deleterious alleles and production of maladaptive hybrids (Rojas-Bracho and Taylor, 1999; Todesco et al., 2016). Population viability analyses taking into account hybridisation would be a valuable tool to assess survival under different scenarios (Allendorf et al., 2013; Van Dyke, 2008; Wolf et al., 2001). Collection of DNA samples from individuals with intermediate patterns of pigmentation, and subsequent molecular genetic analysis will be necessary to determine whether hybridisation has occurred among the striped and short-beaked common dolphins in the GOC.

4.3.2 Risso's Dolphin Individual in Mixed-Species Groups of Striped Dolphins and Short-Beaked Common Dolphins

Only one Risso's dolphin was observed in the GOC: an adult female observed in mixed-species groups throughout this study. Lack of historical information prevents understanding of whether Risso's dolphins (Frantzis and Herzing, 2002) used to be regular in the Gulf and have declined to the single individual observed today. Further, no information exists on Risso's dolphin distribution, movements, and abundance in the Ionian Sea, and no sighting records were reported from the Gulf of Patras and adjacent waters.

4.4 Common Bottlenose Dolphins

At least some of the bottlenose dolphins observed in the GOC can cross the strait leading to open Ionian Sea waters. Bearzi et al. (2011b) showed that eight of a total of 31 individuals photo-identified in the Gulf in 2009 were sighted in other areas of western Greece, up to 265 km apart. Though about half of the bottlenose dolphins photo-identified in the GOC showed moderate degrees of site fidelity (having been encountered in the Gulf repeatedly), many were only observed sporadically and are unlikely to permanently inhabit the Gulf, where they may return on opportunistic basis. Dispersion and roaming typically increase as a response to low density and patchy distribution of prey, or as a strategy to reduce the likelihood of inbreeding (Silva et al., 2008).

Bottlenose dolphins in the GOC appear to be strongly attracted to fish farms on the northern coast (see Fig. 1). This finding is consistent with observations in other coastal areas of Greece, where fish farms were described as 'a new trophic resource for bottlenose dolphins' (Piroddi et al. 2011a) and these dolphins were described as 'fish farm specialists' (Bonizzoni et al., 2014, 2015). Fish farms can attract wild fish by providing structure, refuge from predators, and food resources (Dempster et al., 2002), with influences extending beyond the immediate vicinity of the farmed area (Machias et al., 2005; Weir and Grant, 2005). Productive waters around fish farms have become important feeding spots for bottlenose dolphins, which seem to travel from one fish farm cluster to the next in search of prey. Such behaviour, possibly a response to prey depletion (Bearzi et al., 2008b) and low prey availability away from fish farms, may prompt movements to distant areas outside of the Gulf. While bottlenose dolphins observed during this study regularly interacted with fish farms and closely approached fish cages (often within 1 m), we recorded no evidence of depredation or conflict with farmers.

4.5 Other Species

The Mediterranean monk seal—the world's most endangered seal species (Karamanlidis et al., 2016)—was historically common in the GOC. The northern portion of the Gulf reportedly hosted the 'only important population known for continental Greece', particularly in and around the Bay of Antikyra and between Galaxidi and the island of Vroma (see Fig. 1), where approximately 20 monk seals—including young animals and a breeding colony—were regularly observed between 1969 and 1976 (Marchessaux and Duguy, 1977a,b, 1978). Later on, monk seal populations declined dramatically (Karamanlidis and Dendrinos, 2015), including in the GOC. For instance, Azzolin et al. (2014) reported two encounters with an individual in July 2012, described as the 'first sightings of the century'. Present numbers throughout the GOC are likely to be low, as we only observed a single animal between 2011 and 2015. Mediterranean monk seals, however, have recently shown unsuspected resilience and numbers have the potential to increase, given appropriate management (Notarbartolo di Sciara, 2010; Notarbartolo di Sciara and Kotomatas, 2016).

In addition to marine mammals, the GOC hosts a variety of protected species listed in the EU Habitats Directive and other international conservation conventions (Issaris et al., 2012). Our study documents a relatively high occurrence of loggerhead sea turtles, tuna (including juvenile bluefin), and swordfish, as well as of other charismatic and threatened species. To support high numbers of pelagic dolphins, the GOC must host a considerable biomass of epi- and mesopelagic prey, but formal assessments of these communities are lacking. Oceanic cephalopods with circadian vertical movements are likely important prey (Öztürk et al., 2007). Striped dolphin dives indicative of feeding behaviour were infrequent during daytime, but long diving typically increased before sunset, suggesting night-time feeding. Studies on dolphin (acoustic) behaviour at night, and investigations on distribution, abundance, and vertical movements of epi- and mesopelagic fauna (particularly deep-water cephalopods) could provide important data to increase our understanding of delphinid biology and ecology in the Gulf.

4.6 Anthropogenic Impacts

Though a colossal amount of red mud was discarded into the GOC for over half a century, effects on local marine food webs are unknown. Environmental concerns caused by disposal of red mud relate to its high alkalinity and sodicity (Paramguru et al., 2005; Power et al., 2011) and its documented hazards to sea life (Blackman and Wilson, 1973; Dauvin, 2010; Dethlefsen and Rosenthal, 1973; Pagano et al., 2002). Because red mud is a valuable resource that can be reused (e.g. Paramguru et al., 2005; Pontikes and Angelopoulos, 2013), systematic dumping of millions of tonnes at sea, let alone within a semienclosed Gulf such as the GOC, is not only hazardous but also wasteful. High levels of metals were found in seagrass (*Posidonia oceanica*) from the Bay of Antikyra (Malea et al., 1994), an area where levels of polycyclic aromatic hydrocarbons in sediments, and concentrations of polychlorinated biphenyls (PCBs) and dichlorodiphenyl trichloroethane and its metabolites (DDTs) in Mediterranean mussels (*Mytilus galloprovincialis*) were among the highest sampled throughout Greece (Botsou and Hatzianestis, 2012; Tsangaris et al., 2010, 2011). In the present study, distribution modelling showed no strong correlation between red mud deposits and dolphin occurrence. Because striped dolphins tend to be epi- and mesopelagic feeders and red mud dumping impacts predominantly the seafloor, avoidance would not be expected. However, indirect effects such as contamination up the food web are possible (Jepson et al., 2016). Bottlenose dolphins are primarily benthic feeders, and thus any use of the area is likely to result in direct and indirect exposure to toxic contaminants, with unknown health effects. On one occasion, bottlenose dolphins in the Bay of Antikyra were observed surfacing covered by red mud, indicating bottom feeding on the coastal deposit. Bottlenose dolphins are opportunistic feeders and they can occur in areas heavily impacted by human activities as long as prey is available (Bearzi et al., 2008b; Bonizzoni et al., 2014). Overlap between dolphin habitat and red mud deposits in the GOC raises concern, considering the immunotoxic and other detrimental effects of environmental pollutants (Desforges et al., 2016; Jepson et al., 2016).

The fishing fleet operating in the GOC is predominantly small scale. Though illegal fishing by purse and beach seiners was observed, current fishing bans and other regulations would contribute to mitigating overfishing if properly enforced. A scarcity of information and lack of baseline data prevent understanding of the past and present impacts of fishing in this area, including potential depletion of dolphin prey. During our study, we interviewed 104 fishers operating in the GOC, and 46 (44%) mentioned beach seiners, purse seiners, bottom trawlers, or overfishing in general as negatively influencing their catch (S. Bonizzoni and G. Bearzi, unpublished data). Overall, seiners and trawlers scored as the main anthropogenic factor perceived as a threat to fish stock viability in the GOC (also see Bearzi et al., 2008a, 2010). Future studies of fishing capacity should consider an appropriate assessment of the year-round industrial fishing effort and landings, also taking into account instances of illegal fishing and occurrence of incidental mortality in fishing gear of dolphins and other protected species (Macías López et al., 2012; Marçalo et al., 2015). Ecosystem modelling would be a valuable tool to investigate trophic interactions and fisheries-related ecological perturbations (Piroddi et al., 2010, 2011b).

Underwater noise and disturbance are known threats to cetaceans and marine life generally (Nowacek et al., 2007; Würsig and Richardson, 2009). The GOC is an area of great interest for geophysical research, and seismic surveys are not infrequent (e.g. Beckers et al., 2015; Taylor et al., 2011). The impact on cetaceans of noise generated by seismic research is of concern, and its effects should be carefully assessed. Further, cargos and ships up to approximately 100 m, as well as motor yachts of all sizes, have been regularly observed crossing dolphin habitat, sometimes at high speeds. Our photographs included striped dolphins with fresh propeller wounds and cut-off dorsal fins, suggesting occurrence of collisions. High-speed sport contests overlapping dolphin critical habitat (e.g. the jet-ski race held in the GOC in 2013; http://www.hjsba.gr, http://www.jetraidgreece.com) pose a high risk of collision and disturbance, and they should be banned.

4.7 Conclusions

We demonstrate that much of the GOC's biological wealth has been underestimated. The fate of such a rich natural heritage depends on the enforcement of legislation consistent with commitments to protect marine biodiversity—one of Greece's main treasures. We recommend that new pieces of evidence provided by this study be appropriately incorporated into management plans and spatial planning efforts, and we hope that such evidence will bring about timely action, ensuring mitigation of anthropogenic impacts and long-term protection of a vulnerable inland sea and its cetacean fauna.

ACKNOWLEDGEMENTS

This study was largely funded by OceanCare. Additional support was received from Andrea Sacchi and Texas A&M University. Special thanks to Sigrid Lüber, who greatly helped us develop this project. Our thanks also go to Silvia Frey, Vera Bürgi, Fabienne McLellan, and everybody at OceanCare. Thank you Bernd Würsig for opportunities and open doors. We are grateful to our collaborators Mariana Ferreira Da Silveira, Dagmar Knäbel, Sarah Piwetz, Eva Greene, Ana Catarina Cardoso Fonseca, Valeria Senigaglia, Annalucia Cantafaro, Philippa Dell, Riccardo Grigoletto, Manon Roucaute, Nicola Stoppelli, and all the volunteers who contributed to data collection and management. Many thanks to Margarita Iatrou, Paolo Guidetti, Dimitris Andressakis, Thomas Siarmpas, Spiros Giannoulatos, Danila Cuccu, and Patrick Louisy for contributing important information. Thank you Chrisoula Papalexi, Kostas Papalexis, Christos Goultidis, and Alexandra Kaliva for valuable support. We are grateful to two anonymous reviewers whose comments helped improve our manuscript. Field work in Greece, coordinated by the Tethys Research Institute until July 2011, was conducted under research permits issued by the Hellenic Ministry of Environment, Energy, and Climate Change. N.L.S. was supported by a research grant by La Sapienza University of Rome; N.B.F. by a Vanier Graduate Scholarship and a Travel Research Grant provided by the Canadian Society of Zoologists.

REFERENCES

ACCOBAMS, 2007. Report of the 3rd Meeting of the Contracting Parties to ACCOBAMS, the UNEP/CMS Agreement on the Conservation of Cetaceans of the Black Sea, Mediterranean Sea and Contiguous Atlantic Area, Dubrovnik, Croatia, 22–25 October. .

Aguilar, A., 2000. Population biology, conservation threats and status of Mediterranean striped dolphins (*Stenella coeruleoalba*). J. Cetacean Res. Manag. 2, 17–26.

Aguilar, A., Gaspari, S., 2012. *Stenella coeruleoalba* (Mediterranean subpopulation). The IUCN Red List of Threatened Species 2012. e.T16674437A16674052.

Allendorf, F.W., Leary, R.F., Spruell, P., Wenburg, J.K., 2001. The problems with hybrids: setting conservation guidelines. Trends Ecol. Evol. 16, 613–622.
Allendorf, F.W., Luikart, G., Aitken, S.N., 2013. Conservation and the Genetics of Populations, second ed. Wiley-Blackwell, Chichester, UK.
Amaral, A.R., Jackson, J.A., Möller, L.M., Beheregaray, L.B., 2012. Species tree of a recent radiation: the subfamily Delphininae (Cetacea, Mammalia). Mol. Phylogenet. Evol. 64, 243–253.
Amaral, A.R., Lovewell, G., Coelho, M.M., Amato, G., Rosenbaum, H.C., 2014. Hybrid speciation in a marine mammal: the clymene dolphin (*Stenella clymene*). PLoS One 9, e83645.
Archer, F.I., 2009. Striped dolphin *Stenella coeruleoalba*. In: Perrin, W.F., Würsig, B., Thewissen, J.G.M. (Eds.), Encyclopedia of Marine Mammals, second ed. Academic Press, Amsterdam, pp. 1127–1129.
Azzolin, M., Anichini, M., Galli, A., Papale, E., Giacoma, C., 2010. Evidence of a striped dolphin (*Stenella coeruleoalba*) nursery zone in the waters adjacent to the terrestrial Natura 2000 area 'Gerania Mount' (Gulf of Corinth). In: Proceedings 24th Conference European Cetacean Society, Stralsund, Germany, 22–24 March.
Azzolin, M., Papale, E., Giacoma, C., 2014. Mediterranean monk seal (*Monachus monachus*) in the Gulf of Corinth (Greece): first sightings of the century. In: Proceedings 28th Conference European Cetacean Society, Liège, Belgium, 5–9 April.
Bearzi, G., 2003. Short-beaked common dolphin *Delphinus delphis* (Mediterranean subpopulation). The IUCN Red List of Threatened Species 2003. e.T41762A10557372.
Bearzi, G., 2007. Marine conservation on paper. Conserv. Biol. 21, 1–3.
Bearzi, G., Notarbartolo di Sciara, G., Politi, E., 1997. Social ecology of bottlenose dolphins in the Kvarneric (northern Adriatic Sea). Mar. Mamm. Sci. 13, 650–668.
Bearzi, G., Reeves, R.R., Notarbartolo di Sciara, G., Politi, E., Cañadas, A., Frantzis, A., Mussi, B., 2003. Ecology, status and conservation of short-beaked common dolphins (*Delphinus delphis*) in the Mediterranean Sea. Mammal Rev. 33, 224–252.
Bearzi, G., Politi, E., Agazzi, S., Bruno, S., Costa, M., Bonizzoni, S., 2005. Occurrence and present status of coastal dolphins (*Delphinus delphis* and *Tursiops truncatus*) in the eastern Ionian Sea. Aquat. Conserv. 15, 243–257.
Bearzi, G., Agazzi, S., Gonzalvo, J., Costa, M., Bonizzoni, S., Politi, E., Piroddi, C., Reeves, R.R., 2008a. Overfishing and the disappearance of short-beaked common dolphins from western Greece. Endanger. Species Res. 5, 1–12.
Bearzi, G., Fortuna, C.M., Reeves, R.R., 2008b. Ecology and conservation of common bottlenose dolphins *Tursiops truncatus* in the Mediterranean Sea. Mammal Rev. 39, 92–123.
Bearzi, G., Agazzi, S., Gonzalvo, J., Bonizzoni, S., Costa, M., Petroselli, A., 2010. Biomass removal by dolphins and fisheries in a Mediterranean Sea coastal area: do dolphins have an ecological impact on fisheries? Aquat. Conserv. 20, 549–559.
Bearzi, G., Bonizzoni, S., Agazzi, S., Gonzalvo, J., Currey, R.J.C., 2011a. Striped dolphins and short-beaked common dolphins in the Gulf of Corinth, Greece: abundance estimates from dorsal fin photographs. Mar. Mamm. Sci. 27, E165–E184.
Bearzi, G., Bonizzoni, S., Gonzalvo, J., 2011b. Mid-distance movements of common bottlenose dolphins in the coastal waters of Greece. J. Ethol. 29, 369–374.
Bearzi, G., Reeves, R.R., Remonato, E., Pierantonio, N., Airoldi, S., 2011c. Risso's dolphin *Grampus griseus* in the Mediterranean Sea. Mamm. Biol. 76, 385–400.
Bearzi, G., Fortuna, C.M., Reeves, R.R., 2012. *Tursiops truncatus* (Mediterranean subpopulation). The IUCN Red List of Threatened Species 2012. e.T16369383A16369386.
Bearzi, G., Casale, P., Margaritoulis, D., Bonizzoni, S., Santostasi, N.L., 2015. Observation of a leatherback sea turtle *Dermochelys coriacea* in the Gulf of Corinth, Greece. Mar. Turt. Newsl. 146, 6–9.

Bearzi, G., Bonizzoni, S., Santostasi, N.L., Eddy, L., Gimenez, O., 2016. Short-beaked common dolphins in the Gulf of Corinth are Critically Endangered. In: 1st International Workshop 'Conservation and Research Networking on Short-Beaked Common Dolphin (*Delphinus delphis*) in the Mediterranean Sea', Ischia Island, Italy, 13–15 April, 2016, p. 42.

Beckers, A., Hubert-Ferrari, A., Beck, C., Bodeux, S., Tripsanas, E., Sakellariou, D., De Batist, M., 2015. Active faulting at the western tip of the Gulf of Corinth, Greece, from high-resolution seismic data. Mar. Geol. 360, 55–69.

Bérubé, M.B., 2009. Hybridism. In: Perrin, W.F., Würsig, B., Thewissen, J.G.M. (Eds.), Encyclopedia of Marine Mammals, second ed. Academic Press, Amsterdam, pp. 588–592.

Blackman, R.A.A., Wilson, K.W., 1973. Effects of red mud on marine animals. Mar. Pollut. Bull. 4, 169–171.

Bonizzoni, S., Furey, N.B., Pirotta, E., Valavanis, V.D., Würsig, B., Bearzi, G., 2014. Fish farming and its appeal to common bottlenose dolphins: modelling habitat use in a Mediterranean embayment. Aquat. Conserv. 24, 696–711.

Bonizzoni, S., Eddy, L., Würsig, B., Bearzi, G., 2015. Fish farm specialists: bottlenose dolphins in the Southern Evoikos Gulf, Greece. In: Proceedings 29th Conference European Cetacean Society, St. Julians, Malta, 23–25 March.

Botsou, F., Hatzianestis, I., 2012. Polycyclic aromatic hydrocarbons (PAHs) in marine sediments of the Hellenic coastal zone, eastern Mediterranean: levels, sources and toxicological significance. J. Soils Sediments 12, 265–277.

Brown, A.M., Kopps, A.M., Allen, S.J., Bejder, L., Littleford-Colquhoun, B., Parra, G.J., Cagnazzi, D., Thiele, D., Palmer, C., Frère, C.H., 2014. Population differentiation and hybridisation of Australian snubfin (*Orcaella heinsohni*) and Indo-Pacific humpback (*Sousa chinensis*) dolphins in north-western Australia. PLoS One 9, e101427.

Cañadas, A., Vázquez, J.A., in press. Common dolphins in the Alborán Sea: facing a reduction in their suitable habitat due to an increase in sea surface temperature. Deep-Sea Res. Pt. II.

Cañadas, A., Sagarminaga, R., García-Tiscar, S., 2002. Cetacean distribution related with depth and slope in the Mediterranean waters off southern Spain. Deep-Sea Res. Pt. I 49, 2053–2073.

Ciannelli, L., Fauchald, P., Chan, K.S., Agostini, V.N., Dingsor, G.E., 2008. Spatial fisheries ecology: recent progress and future prospects. J. Mar. Syst. 71, 223–236.

Crossman, C.A., Taylor, E.B., Barrett-Lennard, L.G., 2016. Hybridization in the Cetacea: widespread occurrence and associated morphological, behavioral, and ecological factors. Ecol. Evol. 6, 1293–1303. http://dx.doi.org/10.1002/ece3.1913.

Dauvin, J.C., 2010. Towards an impact assessment of bauxite red mud waste on the knowledge of the structure and functions of bathyal ecosystems: the example of the Cassidaigne canyon (north-western Mediterranean Sea). Mar. Pollut. Bull. 60, 197–206.

Dempster, T., Sanchez-Jerez, P., Bayle-Sempere, J.T., Giménez-Casalduero, F., Valle, C., 2002. Attraction of wild fish to sea-cage fish farms in the south-western Mediterranean Sea: spatial and short-term temporal variability. Mar. Ecol. Prog. Ser. 242, 237–252.

Desforges, J.P., Sonne, C., Levin, M., Siebert, U., De Guise, S., Dietz, R., 2016. Immunotoxic effects of environmental pollutants in marine mammals. Environ. Int. 86, 126–139.

Dethlefsen, V., Rosenthal, H., 1973. Problems with dumping of red mud in shallow water: a critical review of selected literature. Aquaculture 2, 267–280.

Frantzis, A., 2009. Cetaceans in Greece: present status of knowledge. Initiative for the Conservation of Cetaceans in Greece. Athens, Greece. 94 p. http://thalassaproject.mom.gr/LH2Uploads/ItemsContent/46/Cetaceans-in-Greece–Present-status-of-knowledge.pdf.

Frantzis, A., Herzing, D., 2002. Mixed-species associations of striped dolphins (*Stenella coeruleoalba*), short-beaked common dolphins (*Delphinus delphis*) and Risso's dolphins (*Grampus griseus*) in the Gulf of Corinth (Greece, Mediterranean Sea). Aquat. Mamm. 28, 188–197.

Frantzis, A., Alexiadou, P., Paximadis, G., Politi, E., Gannier, A., Corsini-Foka, M., 2003. Current knowledge of the cetacean fauna of the Greek Seas. J. Cetacean Res. Manag. 5, 219–232.

Gaspari, S., Natoli, A., 2012. *Grampus griseus* (Mediterranean subpopulation). The IUCN Red List of Threatened Species 2012. e.T16378423A16378453.

Gaspari, S., Azzellino, A., Airoldi, S., Hoelzel, A.R., 2007. Social kin associations and genetic structuring of striped dolphin populations (*Stenella coeruleoalba*) in the Mediterranean Sea. Mol. Ecol. 16, 2922–2933.

Gaspari, S., Holcer, D., Mackelworth, P., Fortuna, C., Frantzis, A., Genov, T., Vighi, M., Natali, C., Rako, N., Banchi, E., Chelazzi, G., Ciofi, C., 2015a. Population genetic structure of common bottlenose dolphins (*Tursiops truncatus*) in the Adriatic Sea and contiguous regions: implications for international conservation. Aquat. Conserv. 25, 212–222.

Gaspari, S., Scheinin, A., Holcer, D., Fortuna, C., Natali, C., Genov, T., Frantzis, A., Chelazzi, G., Moura, A.E., 2015b. Drivers of population structure of the bottlenose dolphin (*Tursiops truncatus*) in the Eastern Mediterranean Sea. Evol. Biol. 42, 177–190.

Giakoumi, S., Katsanevakis, S., Vassilopoulou, V., Panayotidis, P., Kavadas, S., Issaris, Y., Kokkali, A., Frantzis, A., Panou, A., Mavrommati, G., 2012. Could European marine conservation policy benefit from systematic conservation planning? Aquat. Conserv. 22, 762–775.

Gkafas, G.A., 2011. Molecular Ecology and Fitness of Striped (*Stenella coeruleoalba*, Meyen, 1833) and Common (*Delphinus delphis*, Linnaeus, 1758) Dolphins. Ph.D. thesis, Durham University, Durham, UK. http://etheses.dur.ac.uk/743/.

Gkafas, A., Podiadis, V., Tsigenopoulos, K., Reid, B., Karpouzli, E., Verriopoulos, G., Mamuris, Z., Magoulas, A., Exadactylos, A., 2007. Genetic polymorphism and population structure of the striped dolphin, *Stenella coeruleoalba*, and the common dolphin, *Delphinus delphis*, within the ACCOBAMS area. In: Proceedings 10th International Conference on Environmental Science and Technology, Kos Island, Greece, 5–7 September, pp. 426–431.

Gonzalvo, J., Moutopoulos, D.K., Bearzi, G., Stergiou, K.I., 2011. Fisheries mismanagement in a Natura 2000 area in western Greece. Fish. Manag. Ecol. 18, 25–38.

Gowans, S., Whitehead, H., 2001. Photographic identification of northern bottlenose whales (*Hyperoodon ampullatus*): sources of heterogeneity from natural marks. Mar. Mamm. Sci. 17, 76–93.

Greenpeace, 2007. Gulf of Corinth: A Small Haven of Life. Greenpeace, Greece. (in Greek), 30 p.

Hastie, T.J., Tibshirani, R.J., 1990. Generalised Additive Models. Monographs on Statistics and Applied Probabilities, vol. 46. Chapman & Hall, London.

Herzing, D.L., Elliser, C.R., 2013. Directionality of sexual activities during mixed-species encounters between Atlantic spotted dolphins (*Stenella frontalis*) and bottlenose dolphins (*Tursiops truncatus*). Int. J. Comp. Psychol. 26, 124–134.

Højsgaard, S., Halekoh, U., Yan, J., 2016. Generalized estimating equation package version 1.2-0.2. http://CRAN.R-project.org/package=geepack.

Iatrou, M., 2013. The Study of the Dispersion of the Red Mud Slurry in the Central Corinth Gulf Using Geophysical, Sedimentological and Geochemical Approaches. Ph.D. thesis (in Greek), University of Patras, Greece. http://www.didaktorika.gr/eadd/readonline?handle=10442/37557.

Iatrou, M., Papatheodorou, G., Geraga, M., Ferentinos, G., 2010a. The study of heavy metal concentrations in the red mud deposits at the Gulf of Corinth, using multivariate techniques. Bull. Geol. Soc. Greece 18, 1018–1028.

Iatrou, M., Papatheodorou, G., Geraga, M., Ferentinos, G., 2010b. Submarine bauxitic tailings disposal: a case study in Corinth Gulf, central Greece. Rapp. Comm. Int. Mer. Médit. 39, 260.

Issaris, Y., Katsanevakis, S., Pantazi, M., Vassilopoulou, V., Panayotidis, P., Kavadas, S., Kokkali, A., Salomidi, M., Frantzis, A., Panou, A., Damalas, D., Klaoudatos, D., Sakellariou, D., Drakopoulou, V., Kyriakidou, C., Maina, I., Fric, J., Smith, C., Giakoumi, S., Karris, G., 2012. Ecological mapping and data quality assessment for the needs of ecosystem-based marine spatial management: case study Greek Ionian Sea and the adjacent gulfs. Mediterr. Mar. Sci. 13, 297–311.

IUCN, 2012a. Guidelines for Application of IUCN Red List Criteria at Regional Levels: Version 4.0. IUCN, Gland.

IUCN, 2012b. Red List Categories and Criteria: Version 3.1, second ed. IUCN, Gland.

Jepson, P.D., Deaville, R., Barber, J.L., Aguilar, A., Borrell, A., Murphy, S., Barry, J., Brownlow, A., Barnett, J., Berrow, S., Cunningham, A.A., ten Doeschate, M., Esteban, R., Ferreira, M., Foote, A.D., Genov, T., Gimenez, J., Loveridge, J., Llavona, A., Martin, V., Maxwell, D.L., Papachlimitzou, A., Penrose, R., Perkins, M.W., Smith, B., de Stephanis, R., Tregenza, N., Verborgh, P., Fernandez, A., Law, R.J., 2016. PCB pollution continues to impact populations of orcas and other dolphins in European waters. Sci. Rep. 6, 18573. http://dx.doi.org/10.1038/srep18573.

Karamanlidis, A., Dendrinos, P., 2015. *Monachus monachus*. The IUCN Red List of Threatened Species 2015. e.T13653A45227543.

Karamanlidis, A.A., Dendrinos, P., de Larrinoa, P.F., Gücü, A.C., Johnson, W.M., Kiraç, C.O., Pires, R., 2016. The Mediterranean monk seal *Monachus monachus*: status, biology, threats, and conservation priorities. Mammal Rev. 46, 92–105. http://dx.doi.org/10.1111/mam.12053.

Kendall, W.L., Nichols, J.D., Hines, J.E., 1997. Estimating temporary emigration using capture-recapture data with Pollock's robust design. Ecology 78, 563–578.

Laake, J.L., 2013. RMark: an R interface for analysis of capture-recapture data with MARK: AFSC Processed Rep. 2013-01. Alaska Fish. Sci. Cent., NOAA, Natl. Mar. Fish. Serv., 7600 Sand Point Way, NE, Seattle, WA 98115.

Lebreton, J.D., Burnham, K.P., Clobert, J., Anderson, D.R., 1992. Modeling survival and testing biological hypotheses using marked animals: a unified approach with case studies. Ecol. Monogr. 62, 67–118.

LeDuc, R.G., Perrin, W.F., Dizon, A.E., 1999. Phylogenetic relationships among the delphinid cetaceans based on full cytochrome b sequences. Mar. Mamm. Sci. 15, 619–648.

Levin, D.A., 2002. Hybridization and extinction. Am. Sci. 90, 254–261.

Liang, K.Y., Zeger, S.L., 1986. Longitudinal data analysis using generalized linear models. Biometrika 73, 13–22.

Machias, A., Karakassis, I., Somarakis, S., Giannoulaki, M., Papadopoulou, K.N., Smith, C., 2005. The response of demersal fish communities to the presence of fish farms. Mar. Ecol. Prog. Ser. 288, 241–250.

Macías López, D., García Barcelona, S., Báez, J.C., De la Serna, J.M., Ortiz de Urbina, J.M., 2012. Marine mammal bycatch in Spanish Mediterranean large pelagic longline fisheries, with a focus on Risso's dolphin (*Grampus griseus*). Aquat. Living Resour. 25, 321–331.

Malea, P., Haritonidis, S., Kevrekidis, T., 1994. Seasonal and local variations of metal concentrations in the seagrass *Posidonia oceanica* (L.) Delile in the Antikyra Gulf, Greece. Sci. Total Environ. 153, 225–235.

Marçalo, A., Feijó, D., Katara, I., Araújo, H., Oliveira, I., Santos, J., Ferreira, M., Monteiro, S., Pierce, G.J., Silva, A., Vingada, J., 2015. Quantification of interactions between the Portuguese sardine purse seine fishery and cetaceans. ICES J. Mar. Sci. 72, 2438–2449.

Marchessaux, D., Duguy, R., 1977a. Le phoque moine, *Monachus monachus* (Hermann, 1779), en Grèce. Mammalia 41, 419–439.

Marchessaux, D., Duguy, R., 1977b. Note sur l'observation du phoque moine, *Monachus monachus*, en Grèce. In: 25th Congrès-Assemblée plénière, Comité des Vertébrés marins et Céphalopodes, Split, 22–30 October, 1976.

Marchessaux, D., Duguy, R., 1978. Le phoque moine, *Monachus monachus*, en Grèce. In: Ronald, K., Duguy, R. (Eds.), The Mediterranean Monk Seal. UNEP Technical Series, vol. 1. Pergamon Press, Oxford, pp. 65–84.

Mardikis, I., Podiadis, V., Verriopoulos, G., 1999. High sighting frequency of the pelagic species striped dolphin *Stenella coeruleoalba* in a closed sea area. Eur. Res. Cetaceans, 13, pp, 244–247.

Moura, A.E., 2010. Investigating the Relative Influence of Genetic Drift and Natural Selection in Shaping Patterns of Population Structure in Delphinids (*Delphinus delphis*; *Tursiops* spp.). Ph.D. thesis, Durham University, Durham, UK. http://etheses.dur.ac.uk/755/.

Moura, A.E., Natoli, A., Rogan, E., Hoelzel, A.R., 2013. Atypical panmixia in a European dolphin species (*Delphinus delphis*): implications for the evolution of diversity across oceanic boundaries. J. Evol. Biol. 26, 63–75.

Natoli, A., Cañadas, A., Vaquero, C., Politi, E., Fernandez-Navarro, P., Hoelzel, A.R., 2008. Conservation genetics of the short-beaked common dolphin (*Delphinus delphis*) in the Mediterranean Sea and in the eastern North Atlantic Ocean. Conserv. Genet. 9, 1479–1487.

Notarbartolo di Sciara, G., 2010. The word's two remaining monk seal species: how many different ways are there of being Critically Endangered? Monachus Guardian 13, 39–41.

Notarbartolo di Sciara, G., Bearzi, G., 2010. National Strategy and Action Plan for the conservation of cetaceans in Greece, 2010–2015. Initiative for the Conservation of Cetaceans in Greece, Athens. 55 p.

Notarbartolo di Sciara, G., Birkun, A., 2010. Conserving Whales, Dolphins and Porpoises in the Mediterranean and Black Seas: An ACCOBAMS Status Report. ACCOBAMS, Monaco.

Notarbartolo di Sciara, G., Kotomatas, S., 2016. Are Mediterranean monk seals, *Monachus monachus*, being left to save themselves from extinction? In: Notarbartolo di Sciara, G, Podestà, M., Curry, B.E. (Eds.), Mediterranean Marine Mammals Ecology and Conservation. Advances in Marine Biology, vol. 75. Elsevier, Amsterdam, pp. 359–386.

Nowacek, D.P., Thorne, L.H., Johnston, D.W., Tyack, P.L., 2007. Responses of cetaceans to anthropogenic noise. Mammal Rev. 37, 81–115.

Öztürk, B., Salman, A., Öztürk, A.A., Tonay, A., 2007. Cephalopod remains in the diet of striped dolphins (*Stenella coeruleoalba*) and Risso's dolphins (*Grampus griseus*) in the eastern Mediterranean. Vie Milieu 57, 53–59.

Pagano, G., Meric, S., De Biase, A., Iaccarino, M., Petruzzelli, D., Tünay, O., Warnau, M., 2002. Toxicity of bauxite manufacturing by-products in sea urchin embryos. Ecotoxicol. Environ. Saf. 51, 28–34.

Pan, W., 2001. Akaike's information criterion in generalized estimating equations. Biometrics 57, 120–125.

Papatheodorou, G., Lyberis, E., Ferentinos, G., 1999. Use of factor analysis to study the distribution of metalliferous bauxitic tailings in the seabed of the Gulf of Corinth, Greece. Nat. Resour. Res. 8, 277–286.

Paramguru, R.K., Rath, P.C., Misra, V.N., 2005. Trends in red mud utilization: a review. Miner. Process. Extr. Metall. Rev. 26, 1–29.
Piroddi, C., Bearzi, G., Christensen, V., 2010. Effects of local fisheries and ocean productivity on the northeastern Ionian Sea ecosystem. Ecol. Model. 221, 1526–1544.
Piroddi, C., Bearzi, G., Christensen, V., 2011a. Marine open cage aquaculture in the eastern Mediterranean Sea: a new trophic resource for bottlenose dolphins. Mar. Ecol. Prog. Ser. 440, 255–266.
Piroddi, C., Bearzi, G., Gonzalvo Villegas, J., Christensen, V., 2011b. From common to rare: the case of the Mediterranean common dolphin. Biol. Conserv. 144, 2490–2498.
Pirotta, E., Matthiopoulos, J., MacKenzie, M., Scott-Hayward, L., Rendell, L., 2011. Modelling sperm whale habitat preference: a novel approach combining transect and follow data. Mar. Ecol. Prog. Ser. 436, 257–272.
Planque, B., Loots, C., Petitgas, P., Lindstrom, U., Vaz, S., 2011. Understanding what controls the spatial distribution of fish populations using a multi-model approach. Fish. Oceanogr. 20, 1–17.
Podiadis, V., Dosi, A., Verriopoulou, A., Trivourea, M., Reichelt, M., Hayes Tieken, R., Brooke, B., Sini, M., Trigonis, V., Mumme, C., Genov, T., Zlatar, V., Büeler, D., Fourniotis Pavlatos, P., Prichard, D., Verriopoulos, G., 2003. The striped dolphins in the Gulf of Korinth, 1996–2002: a comparative study. Eur. Res. Cetaceans, 17, p. 263.
Pollock, K.H., 1982. A capture-recapture design robust to unequal probability of capture. J. Wildl. Manag. 46, 752–757.
Pontikes, Y.T., 2007. Utilisation of Red Mud in the Heavy Clay Industry. Ph.D. thesis (in Greek), University of Patras, Greece.
Pontikes, Y., Angelopoulos, G.N., 2013. Resources, conservation and recycling bauxite residue in cement and cementitious applications: current status and a possible way forward. Resour. Conserv. Recycl. 73, 53–63.
Portman, M., Notarbartolo di Sciara, G., Agardy, T., Katsanevakis, S., Possingham, H., Di Carlo, G., 2013. He who hesitates is lost: why conservation in the Mediterranean Sea is necessary and possible now. Mar. Policy 42, 270–279.
Power, G., Gräfe, M., Klauber, C., 2011. Bauxite residue issues: current management, disposal and storage practices. Hydrometallurgy 108, 33–45.
R Core Team, 2014. R: A Language and Environment for Statistical Computing. R Foundation for Statistical Computing, Vienna, Austria. http://www.R-project.org/.
Read, A.J., Urian, K.W., Wilson, B., Waples, D.M., 2003. Abundance of bottlenose dolphins in the bays, sounds and estuaries of North Carolina. Mar. Mamm. Sci. 19, 59–73.
Rojas-Bracho, L., Taylor, B.L., 1999. Risk factors affecting the vaquita (*Phocoena sinus*). Mar. Mamm. Sci. 15, 974–989.
Rosso, M., Moulins, A., Würtz, M., 2008. Colour patterns and pigmentation variability on striped dolphin *Stenella coeruleoalba* in north-western Mediterranean Sea. J. Mar. Biol. Assoc. U.K. 88, 1211–1219.
Santostasi, N.L., Bonizzoni, S., Bearzi, G., Eddy, L., Gimenez, O., 2016. Revised abundance estimates of striped and short-beaked common dolphins in the Gulf of Corinth, Mediterranean Sea. In: Proceedings 30th Annual Conference European Cetacean Society, Funchal, Madeira, 14–16 March.
Shane, S.H., 1990. Comparison of bottlenose dolphin behavior in Texas and Florida, with a critique of methods for studying dolphin behavior. In: Leatherwood, S., Reeves, R.R. (Eds.), The Bottlenose Dolphin. Academic Press, San Diego, pp. 541–558.
Silva, J.M., Silva, F.J.L., Sazima, I., 2005. Two presumed interspecific hybrids in the genus *Stenella* (Delphinidae) in the Tropical West Atlantic. Aquat. Mamm. 31, 468–472.
Silva, M.A., Prieto, R., Magalhães, S., Seabra, M.I., Santos, R.S., Hammond, P.S., 2008. Ranging patterns of bottlenose dolphins living in oceanic waters: implications for population structure. Mar. Biol. 156, 179–192.

Stelzenmüller, V., Breen, P., Stamford, T., Thomsen, F., Badalamenti, F., Borja, Á., Buhl-Mortensen, L., Carlstöm, J., D'Anna, G., Dankers, N., Degraer, S., Dujin, M., Fiorentino, F., Galparsoro, I., Giakoumi, S., Gristina, M., Johnson, K., Jones, P.J.S., Katsanevakis, S., Knittweis, L., Kyriazi, Z., Pipitone, C., Piwowarczyk, J., Rabaut, M., Sørensen, T.K., van Dalfsen, J., Vassilopoulou, V., Vega Fernandez, T., Vincx, M., Vöge, S., Weber, A., Wijkmark, N., Jak, R., Qiu, W., ter Hofstede, R., 2013. Monitoring and evaluation of spatially managed areas: a generic framework for implementation of ecosystem based marine management and its application. Mar. Policy 37, 149–164.

Stensland, E., Angerbjörn, A., Berggren, P., 2003. Mixed species groups in mammals. Mammal Rev. 33, 205–223.

Taylor, B.L., Gerrodette, T., 1993. The uses of statistical power in conservation biology: the vaquita and the northern spotted owl. Conserv. Biol. 7, 489–500.

Taylor, B., Weiss, J.R., Goodliffe, A.M., Sachpazi, M., Laigle, M., Hirn, A., 2011. The structures, stratigraphy and evolution of the Gulf of Corinth rift, Greece. Geophys. J. Int. 185, 1189–1219.

Todesco, M., Pascual, M.A., Owens, G.L., Ostevik, K.L., Moyers, B.T., Hübner, S., Heredia, S.M., Hahn, M.A., Caseys, C., Bock, D.G., Rieseberg, L.H., 2016. Hybridization and extinction. Evol. Appl. 9, 892–908. http://dx.doi.org/10.1111/eva.12367.

Traill, L.W., Brook, B.W., Frankham, R.R., Bradshaw, C.J.A., 2010. Pragmatic population viability targets in a rapidly changing world. Biol. Conserv. 143, 28–34.

Tsakiridis, P.E., Agatzini-Leonardou, S., Oustadakis, P., 2004. Red mud addition in the raw meal for the production of Portland cement clinker. J. Hazard. Mater. B116, 103–110.

Tsangaris, C., Kormas, K., Strogyloudi, E., Hatzianestis, I., Neofitou, C., Andral, B., Galgani, F., 2010. Multiple biomarkers of pollution effects in caged mussels on the Greek coastline. Comp. Biochem. Physiol. C Toxicol. Pharmacol. 151, 369–378.

Tsangaris, C., Hatzianestis, I., Catsiki, V.A., Kormas, K.A., Strogyloudi, E., Neofitou, C., Andral, B., Galgani, F., 2011. Active biomonitoring in Greek coastal waters: application of the integrated biomarker response index in relation to contaminant levels in caged mussels. Sci. Total Environ. 412–413, 359–365.

Tsitsika, E.V., Maravelias, C.D., 2008. Fishing strategy choices of purse seines in the Mediterranean: implications for management. Fish. Sci. 74, 19–27.

Valavanis, V.D., Kapantagakis, A., Katara, I., Palialexis, A., 2004. Critical regions: a GIS-based model of marine productivity hotspots. Aquat. Sci. 66, 139–148.

Van Dyke, F., 2008. Conservation Biology: Foundations, Concepts, Applications, second ed. Springer, Berlin.

Varnavas, S.P., Achilleopoulos, P.P., 1995. Factors controlling the vertical and spatial transport of metal-rich particulate matter in seawater at the outfall of bauxitic red mud toxic waste. Sci. Total Environ. 175, 199–205.

Varnavas, S.P., Ferentinos, G., Collins, M., 1986. Dispersion of bauxite red mud in the Gulf of Corinth. Mar. Geol. 70, 211–222.

Vassilopoulou, V., Katsanevakis, S., Panayotidis, P., Anagnostou, C., Damalas, D., Dogrammatzi, A., Drakopoulou, V., Giakoumi, S., Haralabous, J., Issaris, Y., Kavadas, S., Klaoudatos, D., Kokkali, A., Kyriakidou, C., Maina, I., Mavromati, G., Palialexis, A., Sakellariou, D., Salomidi, M., 2012. Application of the MESMA Framework, Case Study: Inner Ionian Archipelago and Adjacent Gulfs: MESMA Report. 240 p.

Weir, L.K., Grant, J.W.A., 2005. Effects of aquaculture on wild fish populations: a synthesis of data. Environ. Rev. 13, 145–168.

White, G.C., Burnham, K.P., 1999. Program MARK: survival estimation from populations of marked animals. Bird Study 46, 120–139.

Williams, J.A., Dawson, S.M., Slooten, E., 1993. The abundance and distribution of bottlenosed dolphins (*Tursiops truncatus*) in Doubtful Sound, New Zealand. Can. J. Zool. 71, 2080–2088.
Wilson, B., Hammond, P.H., Thompson, P.M., 1999. Estimating size and assessing trends in a coastal bottlenose dolphin population. Ecol. Appl. 9, 288–300.
Wolf, D.E., Takebayashi, N., Rieseberg, L.H., 2001. Predicting the risk of extinction through hybridization. Conserv. Biol. 15, 1039–1053.
Wood, S.N., 2006. Generalized Additive Models: An Introduction with R. Chapman and Hall/CRC, Boca Raton.
Würsig, B., Jefferson, T.A., 1990. Methods of photo-identification for small cetaceans. Rep. Int. Whal. Comm. Spec. Issue 12, 43–52.
Würsig, B., Richardson, W.J., 2009. Effects of noise. In: Perrin, W.F., Würsig, B., Thewissen, J.G.M. (Eds.), Encyclopedia of Marine Mammals, second ed. Academic Press, Amsterdam, pp. 765–773.
Würsig, B., Würsig, M., 1977. The photographic determination of group size, composition, and stability of coastal porpoises (*Tursiops truncatus*). Science 198, 755–756.
Zafiropoulos, D., Verriopoulos, G., Merlini, L., 1999. Geographical distribution of small cetaceans in several Greek coastal areas. Eur. Res. Cetaceans, 13, pp. 282–284.

CHAPTER ELEVEN

Harbour Porpoises, *Phocoena phocoena*, in the Mediterranean Sea and Adjacent Regions: Biogeographic Relicts of the Last Glacial Period

M.C. Fontaine[1]

Groningen Institute for Evolutionary Life Sciences (GELIFES), University of Groningen, Groningen, The Netherlands

[1]Corresponding author: e-mail address: mikafontaine@gmail.com

Contents

Abstract

The harbour porpoise, *Phocoena phocoena*, is one of the best studied cetacean species owing to its common distribution along the coastal waters of the Northern Hemisphere. In European waters, strandings are common and bycatch mortalities in commercial fisheries reach alarming numbers. Lethal interactions resulting from human activities together with ongoing environmental changes raise serious concerns about population viability throughout the species' range. These concerns foster the need to fill critical gaps in knowledge of harbour porpoise biology, including population structure, feeding ecology, habitat preference and evolutionary history, that are critical information for planning effective management and conservation efforts. While the species is distributed fairly continuously in the North Atlantic Ocean, it becomes fragmented in the south-eastern part with isolated populations occurring along the Atlantic coasts of the Iberian Peninsula, Northwest Africa and the Black Sea. The latter population is separated from Atlantic populations by the Mediterranean Sea, where the species is almost entirely absent. Understanding the evolutionary history of these populations occurring

Advances in Marine Biology, Volume 75
ISSN 0065-2881
http://dx.doi.org/10.1016/bs.amb.2016.08.006

in marginal habitats holds the potential to reveal fundamental aspects of the species' biology such as the factors determining its distribution, ecological niche, and how past and recent environmental variation have shaped the current population structure. This information can be critical for understanding the future evolution of the species in consideration of ongoing environmental changes. This chapter summarizes the recent advances in our knowledge regarding the populations bordering the Mediterranean Sea with a special emphasis on their ecological and evolutionary history, which has recently been reconstructed from genetic analyses.

1. INTRODUCTION

The harbour porpoise (*Phocoena phocoena*) (Fig. 1) is one of the smallest and most abundant odontocete cetaceans, and is widely distributed in cold to temperate shelf waters of the Northern Hemisphere (Fig. 2A). The species' distribution is circumpolar and typical of boreal-temperate species; harbour porpoises are common in shallow continental shelf waters, frequenting bays, estuaries and tidal channels (Gaskin, 1984; Gaskin et al., 1974; Read, 1999). Given their small size and relatively rapid reproductive output (Kanwisher and Sundnes, 1965; Lockyer, 2003), harbour porpoises apparently favour waters that are rich enough to sustain their demanding and energetically challenging life cycle (Kanwisher and Sundnes, 1965; Lockyer, 2007; Read and Hohn, 1995; Wisniewska et al., 2016). Therefore, harbour porpoise distribution is generally restricted to coastal continental shelf waters, usually shallower than 200 m (Read, 1999), where primary production is high enough to support a rich trophic network dominated by top predators (Benke et al., 2014; Fontaine et al., 2007; Johnston et al., 2005; Lambert et al., in press; Sveegaard et al., 2012).

Fig. 1 A surfacing harbour porpoise (*Phocoena phocoena*). *Photograph: Ari S. Friedlaender.*

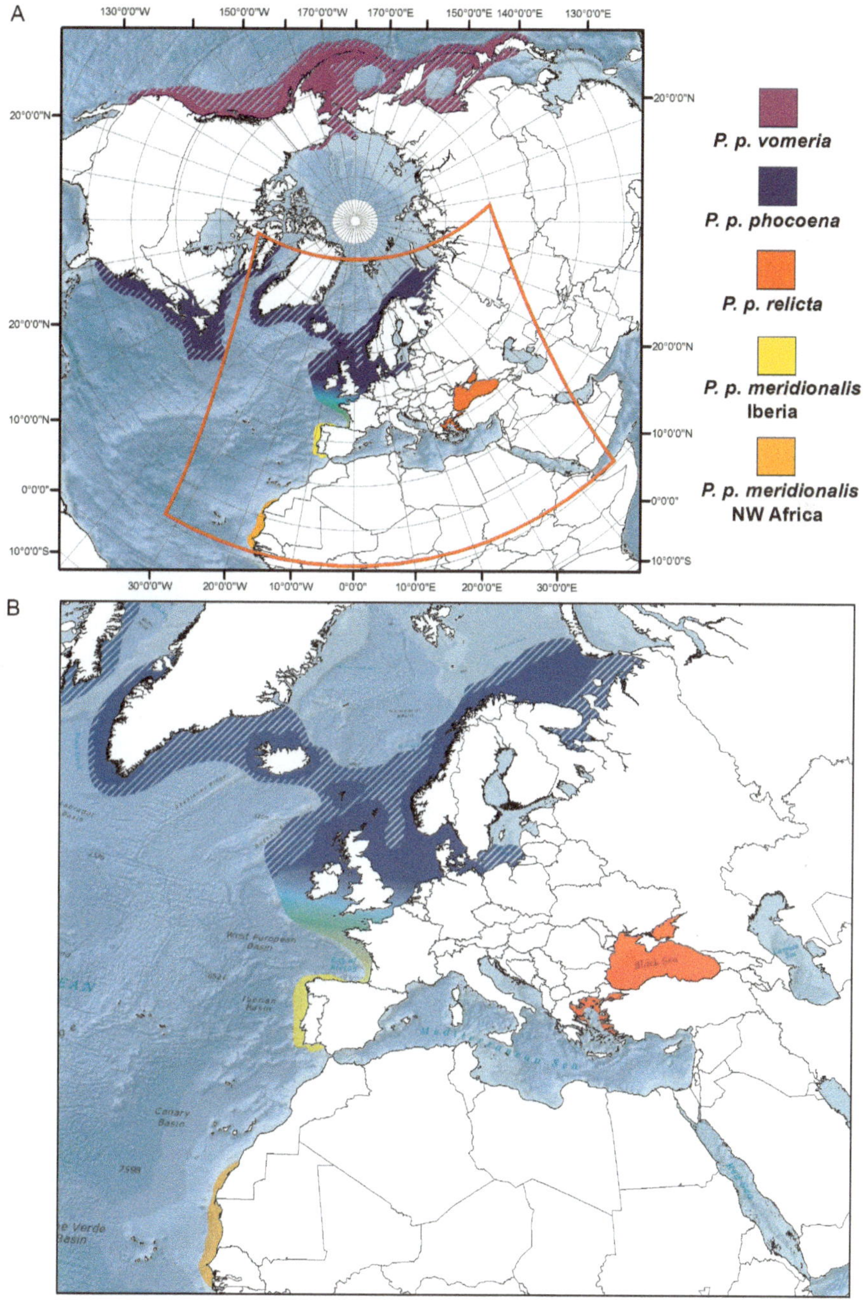

Fig. 2 See legend on next page.

With these ecological preferences, and because of their tendency to feed on fish such as herring, sprat and bottom-dwelling species, habitats and resources exploited by harbour porpoises largely overlap with commercial fisheries and lead to extremely high levels of casualties related to incidental captures in gillnet fisheries throughout the species' distribution (Jefferson and Curry, 1994; Read et al., 2006; Rosel, 1997). Harbour porpoises are the most affected species in the North Atlantic, especially in bottom set gill and tangle net fisheries (Bjørge et al., 2013; Hammond et al., 2013; Read et al., 2006; Stenson, 2003; Vinther and Larsen, 2004). In the early 1990s, for example, an estimated 2200 harbour porpoises were taken annually by English and Irish hake fisheries in the Celtic Sea (Tregenza et al., 1997), and between 6000 and 7000 were bycaught annually in Danish gillnet fisheries in the central and southern North Sea (Vinther and Larsen, 2004). These high mortality levels raised serious concern about population viability and sustainability among several nations and the International Whaling Commission (IWC). The Agreement on the Conservation of Small Cetaceans of the Baltic and North Sea (ASCOBAMS) in the Northeast Atlantic Ocean and the Agreement on the Conservation of Cetaceans in the Black Sea, Mediterranean Sea and Contiguous Atlantic Area (ACCOBAMS) emphasize the protection of the harbour porpoise and encourage continuous efforts to assess and mitigate these casualties.

While the species is distributed fairly continuously in the North Atlantic, it becomes fragmented in the south-eastern region, with isolated populations

Fig. 2 Distribution of the harbour porpoise (*Phocoena phocoena*), showing ranges for the different subspecies, including the proposed *P. p. meridionalis*, and genetically differentiated populations. (A) The global circumpolar distribution of the harbour porpoise. *Polygon* (in *red*) indicates the enlarged view displayed in (B). (B) The Northeast Atlantic distribution of the species with the different subspecies and genetically distinct populations bordering the Mediterranean Sea. The four subspecies are displayed with different colours: *P. p. phocoena* in *blue*; *P. p. meridionalis*, with two genetically differentiated populations, in *yellow* and *orange*; *P. p. relicta* in *red*; and *P. p. vomerina* in *purple*. The known and possible species distribution is shown with *plain and hashed surfaces*, respectively. A coloured gradient (*yellow* to *dark blue*) in the south of the British Isles and Bay of Biscay shows the approximate geographic distribution of the admixture zone between the Iberian population of *P. p. meridionalis* and the population of *P. p. phocoena* north of the Bay of Biscay. Maps were generated after Gaskin (1984), IWC (1996), and Read (1999) and were updated with recent observations from the Black Sea and northern Aegean Sea. *Figure prepared using ESRI ArcGIS™ v.10.3.*

along the Atlantic coasts of Iberia and Northwest Africa, and in the Black Sea, separated from the Atlantic populations by the Mediterranean Sea, where the species is almost entirely absent (see Fig. 2). Understanding the evolutionary history of these populations that occur in marginal habitats holds the potential to reveal fundamental aspects of harbour porpoise biology, including the factors that determine distribution, ecological niches, and how past and recent environmental variation have shaped current population structure (Hampe and Jump, 2011). The evolutionary history of these populations can potentially be informative regarding the future evolution of harbour porpoises and population differentiation as it may be affected by ongoing environmental changes. Previous studies have thoroughly reviewed the distribution, abundance, stock structure and status of the harbour porpoise in the North Atlantic (Andersen, 2003; Donovan and Bjørge, 1995; Gaskin, 1984; Read, 1999; Rosel, 1997). This chapter summarizes the recent advances in our knowledge regarding the populations bordering the Mediterranean Sea, including their distribution and ecology with a special emphasis on evolutionary history.

2. ECOLOGY AND OVERVIEW OF DISTRIBUTION RANGE

The global distribution of the harbour porpoise has been reviewed comprehensively by Gaskin (1984) and later updated by Read (1999). This species occurs fairly continuously throughout the coastal waters of the North Pacific and North Atlantic, with a relict population in the Black Sea (Fig. 2A). The species' distribution in the North Atlantic has received the largest amount of attention from the scientific community (Andersen, 2003; Donovan and Bjørge, 1995; Rosel, 1997). In this section, I provide an updated overview of harbour porpoise distribution in the North Atlantic Ocean and Black Sea, to outline the specificities of the populations bordering the Mediterranean Sea.

In the Northwest Atlantic (see Fig. 2), exceptional stranding records of harbour porpoises have been reported as far south as northern Florida along the east coast of the United States; however, they are considered scarce south of Cape Hatteras, North Carolina (Polacheck et al., 1995). Northward, porpoises become more common until Upernavik, West Greenland (Gaskin, 1984; Read, 1999), and are considered rare again on the eastern coasts of Greenland (Teilmann and Dietz, 1998). Harbour porpoises also inhabit the coastal waters of Iceland in the central North Atlantic. In the Northeast Atlantic (Fig. 2B), they have been reported continuously from the Barents Sea (Novaja Zemlja, 73°N), with some exceptional observations off Svalbard (Bjørge and Øien, 1995), southward to the European continental shelf. The highest abundances are found in the North Sea and adjacent waters including the Atlantic waters of Great Britain and Ireland, and the English Channel, with seasonal and yearly variation in abundance, as reported from field surveys and stranding records (Camphuysen and Siemensma, 2011; Haelters and Camphuysen, 2009; Hammond et al., 2002, 2013; Peltier et al., 2013; Scheidat et al., 2012). Harbour porpoises occupy the western side of the Baltic Sea and the northern side of the Bay of Biscay, but their abundance decreases towards the southern Bay of Biscay (Hammond et al., 2002, 2013).

Harbour porpoises also occur further south, with a small population along the Atlantic coast of the Iberian Peninsula, from Galicia to the Bay of Cadiz (Hammond et al., 2002, 2013), and another population off the coast of Northwest Africa. Although not much is known about this African population, it seems to be separated from the Iberian population by a gap in the distribution of about 1000 km (see Fig. 2B). African porpoises have been reported from Agadir, Morocco, south to Joal-Fadiouth (14°09′N, 16° 49′W, Senegal) and are considered most common off the Mauritanian coasts; however, no abundance estimates are available in that area (Jefferson et al., 1997; Robineau and Vely, 1998; Smeenk et al., 1992; Van Waerebeek, 2007; Van Waerebeek et al., 2000, 2003). The harbour porpoise populations off Iberia and Northwest Africa inhabit the Eastern Central Atlantic Upwelling system, with its northern limit encompassing the Atlantic coasts of the Iberian Peninsula (Arístegui et al., 2009). Despite the great distance separating them, Iberian and Mauritanian porpoises both inhabit an interconnected upwelling system, where cold nutrient-rich deep waters are brought to the surface by the winds and sustain a rich trophic network. The ecological upwelling conditions in which these southern porpoises thrive are quite distinct from the ecological conditions found further north on the European continental shelves north of the Bay of Biscay. General ecological and morphological similarities between Iberian and Mauritanian porpoises suggest that they inhabit the same kind of environment and rely on similar resources. However, stomach content analyses, stable isotopes and field surveys suggest that their habitat and diet differ from the porpoises living on the European continental shelf north of the Bay of Biscay (Méndez-Fernandez et al., 2013; Pierce et al., 2010; Pinela et al., 2010). This differentiation is further supported by morphological differences, with the body sizes of Iberian and Mauritanian porpoises often exceeding ~200 cm in contrast to the 150 cm commonly observed in the individuals north of the Bay of Biscay on the European continental shelf (Donovan and Bjørge, 1995; Smeenk et al., 1992).

The harbour porpoise also occurs in the Black Sea, and this population is completely isolated today from the Atlantic portion of the species' range by the Mediterranean Sea, from which the species is almost entirely absent, except at its extremities (Donovan and Bjørge, 1995; Frantzis et al., 2001; Gaskin, 1984; IWC, 1996; Read, 1999). In the Western Mediterranean Sea, only some accidental incursions of porpoises from the neighbouring population in the Atlantic have been reported, with the strandings of two porpoises in Malaga, Spain, one in 1981 (Rey and Cendrero, 1982) and another one in 2006 (Notarbartolo di Sciara and Birkun, 2010), and two strandings and one sighting in Cadiz close to the Strait of Gibraltar in 1999 (Frantzis et al., 2001). These porpoises are believed to be stragglers venturing into the Mediterranean Sea from the neighbouring Atlantic populations (Gaskin, 1984; Rosel, 1997), and no established population has ever been reported in the western Basin. In the Eastern Mediterranean no established population was reported, except for in the northern Aegean Sea, where a total of 19 strandings have been recorded for this species over the last two decades from Greek and Turkish coastlines (Frantzis, 2009; Frantzis et al., 2001; Tonay and Dede, 2013). Furthermore, the presence of free-ranging porpoises in Greek and Turkish waters was recently confirmed during field surveys, particularly in waters close to the Gulf of Saros, north-western Aegean Sea, Turkey (Ryan et al., 2014). Genetic analyses of porpoises stranded along the northern Aegean Sea coastlines showed that they came from the same genetic pool as the porpoises from the Black Sea (Fontaine et al., 2012; Rosel et al., 2003; Viaud-Martínez et al., 2007). These Aegean porpoises likely follow the outflowing surface waters from the Black Sea into the Marmara Sea and northern Aegean Sea. The northern Aegean basin is indeed under the influence of cold, low salinity waters that pour out of the Black Sea, which is entrained into a cyclonic circulation affecting the northern and western parts of the Aegean, causing an ecological isolation of the northern basin from the southern basin (Theocharis et al., 1993). In the southern Aegean basin the continental shelf becomes more reduced and the waters quickly become warm and oligotrophic, a Mediterranean characteristic (Longhurst, 2010; Moraitou-Apostolopoulou, 1985); as such, they are less suitable than the cold-water habitat preferred by harbour porpoises.

3. TAXONOMY AND ECOTYPE DELIMITATION

Allopatric distribution (see Fig. 2) as well as morphological and genetic differences have justified the recognition of three subspecies of harbour porpoises: *P. p. vomerina* in the North Pacific Ocean, *P. p. phocoena* in the North Atlantic Ocean and *P. p. relicta* in the Black Sea (Gaskin et al., 1974; Read, 1999; Rosel et al., 1995; Tolley and Rosel, 2006; Viaud-Martínez et al., 2007). Recently porpoises from the southern waters of the Northeast Atlantic off the coasts of Iberia and Mauritania (Fig. 2) have been proposed as belonging to a fourth subspecies, *P. p. meridionalis* (Fontaine et al., 2014).

These southern porpoises were already known to be distinct with respect to their unusually large body size, often exceeding 200 cm, compared to the 150 cm of harbour porpoises found further north in the Atlantic and in the Black Sea (Donovan and Bjørge, 1995; Read, 1999; Smeenk et al., 1992). Such morphological differences are likely related to the ecologically differentiated habitat in which they are successful, and to genetic differentiation (Fontaine et al., 2016). These porpoises inhabit a distinct environment (Arístegui et al., 2009), relying on the upwelling-related trophic network (Méndez-Fernandez et al., 2013; Pierce et al., 2010; Pinela et al., 2010), which contrasts with the predominantly shallow habitat and demersal feeding habits of porpoises from the European continental shelf (e.g. Santos and Pierce, 2003; Spitz et al., 2006). The evidence solving this puzzle actually came from new genetic analyses showing that the extent of genetic differentiation in the mitochondrial genome between the porpoises from southern and northern Northeast Atlantic, was as large as the genetic differentiation observed between the porpoises from the Black Sea and those from the European waters north of the Bay of Biscay (Fontaine et al., 2014). Given this level of genetic differentiation, Fontaine et al. (2014) proposed that the southern porpoises from Iberia and Northwest Africa should be raised to the level of subspecies—the same taxonomic level as the porpoises from the Black Sea (Gaskin, 1984; Rosel et al., 1995).

Beyond these taxonomic considerations, harbour porpoises from the more southern waters can be considered as a genetically, ecologically and morphologically differentiated ecotype that is distinct from the population occurring north of the Bay of Biscay, and from the relict population inhabiting the Black Sea. Recent genetic results also showed that harbour porpoises from Iberia and Mauritania are related to each other as they descend from a recent common ancestor (Fig. 3A), but that they form

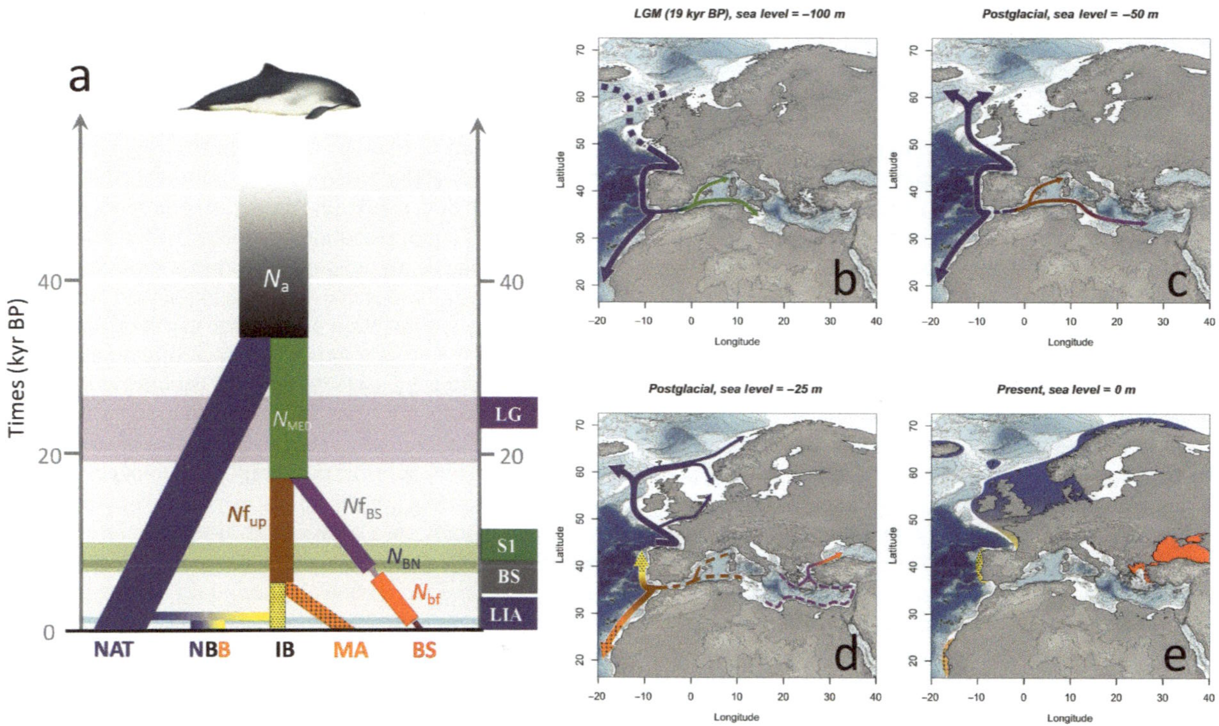

Fig. 3 See legend on next page.

Fig. 3 Evolutionary history inferred from analyses of genetic variation in the harbour porpoise (*Phocoena phocoena*). (A) A diagram depicting the evolutionary history of each differentiated group with a schematic population tree. Each contemporaneous group is shown at the bottom with a colour coding following Fig. 2 (*BS*, population from the Black Sea *P. p. relicta*; *IB*, Iberian population of the southern ecotype, the proposed *P. p. meridionalis*; *MA*, Mauritanian–Northwest African population of the southern ecotype; *NAT*, North-eastern Atlantic ecotype (*P. p. phocoena*) inhabiting the European continental shelf north of the Bay of Biscay; *NBB*, admixed population between IB and NAT in the northern side of the Bay of Biscay). Each group coalesced backward in time (upward) as indicated by the *Y*-axis which provides the timeline in thousands of years before present (kyr BP). Changes in population size are depicted by line width. Major environmental changes related to the demographic history of harbour porpoises are also plotted: Last Glacial Maximum (LGM period, ca. 23–19 kyr BP) (Clark et al., 2009), Mediterranean Sapropel S1 period (ca. 9.5–6.5 kyr BP) (Roberts et al., 2011; e.g. Spötl et al., 2010), flooding of the Black Sea (BS, ca. 8.4–9.4 kyr BP) (Giosan et al., 2009; Major et al., 2006) and Little Ice Age (LIA, ca. 250–700 yr BP) (Osborn and Briffa, 2006). (B–E) Geographic landscape of the porpoise evolutionary history at four time steps determined by historical change in sea level. During the LGM (glacial sea level ~100 m lower than today), porpoises from the Atlantic likely colonized the Mediterranean Sea and lead to a divergent group that formed the ancestral Mediterranean population(s) (B; also N_{MED} in A); during the postglacial Holocene warming and the sea level rise (C), these ancestral populations in the Mediterranean Sea split into an eastern (*purple*, Nf_{BS} in A) and western lineages (*brown*, Nf_{UP}) from which descended the porpoise of the Black Sea and the two populations of the upwelling waters of Iberia and Mauritania–Northwest Africa. The Mediterranean conditions became unsuitable for the harbour porpoise at the end of the African Humid Period and Mediterranean Sapropel episodes (D). Porpoises were thus forced out of the Mediterranean taking refuge where conditions were still suitable for the species: into the Black Sea for the Eastern Mediterranean lineage, reconnected to the Mediterranean ca. 8.4 kyr BP, and for the Western Mediterranean lineage back to the Atlantic waters of Iberia and Northwest Africa, reaching their present distribution (E). Maps were drawn using MARMAP v0.9.5 package (Pante and Simon-Bouhet, 2013) for R (R Core Team, 2016) using the ETOPO1 dataset available on the United States National Geophysical Data Center (Amante and Eakins, 2009). Sea level was adjusted to account for the historical variation during the LGM and post-LGM period. *Panel (A): Modified from Fontaine, M.C., Snirc, A., Frantzis, A., Koutrakis, E., Öztürk, B., Oztürk, A.A., Austerlitz, F., 2012. History of expansion and anthropogenic collapse in a top marine predator of the Black Sea estimated from genetic data. Proc. Natl. Acad. Sci. U.S.A. 109, E2569–E2576. doi:10.1073/pnas.1201258109; Fontaine, M.C., Roland, K., Calves, I., Austerlitz, F., Palstra, F.P., Tolley, K.A., Ryan, S., Ferreira, M., Jauniaux, T., Llavona, Á., Öztürk, B., Oztürk, A.A., Ridoux, V., Rogan, E., Sequeira, M., Siebert, U., Vikingsson, G.A., Borrell, A., Michaux, J.R., Aguilar, A., 2014. Postglacial climate changes and rise of three ecotypes of harbour porpoises,* Phocoena phocoena, *in western Palearctic waters. Mol. Ecol. 23, 3306–3321. doi:10.1111/mec.12817.*

genetically differentiated populations (Fontaine et al., 2014) (see Fig. 2). This suggests that the populations from Iberia and Mauritania have been evolving separately from each other for some time, which is consistent with the distribution gap of about a thousand kilometres between the two.

Little information is available on the ecological and behavioural differentiation of these southern porpoises beyond differences in feeding ecology and morphology that can readily be observed from stranded and bycaught individuals. Estimates of local abundance along the Iberian coasts are not available. Compared to those in the northern waters of the Bay of Biscay and the North Sea (see Hammond et al., 2013), the population size of the Iberian population seems to be quite low. No estimates are available for harbour porpoise abundance in Northwest African waters (Jefferson et al., 1997; Robineau and Vely, 1998; Smeenk et al., 1992; Van Waerebeek, 2007; Van Waerebeek et al., 2000, 2003). The level of genetic diversity of the Iberian and Northwest African populations was comparable, but lower than the genetic diversity observed in the Black Sea population and much lower than the values reported in the harbour porpoises inhabiting the shelf waters north of the Bay of Biscay (Fontaine et al., 2007, 2010, 2014). This also supports the small population size of the populations from the southern ecotype.

4. WHY DO HARBOUR PORPOISES NO LONGER OCCUR IN THE MEDITERRANEAN SEA?

The generally accepted explanation is that the present-day Mediterranean Sea is unsuitable for a cold-water species such as the harbour porpoise. The region is considered to be a warm-temperate sea that is becoming increasingly warmer and more oligotrophic, with an average sea surface temperature increasing from the west to the east (Rohling et al., 2009). However, such qualitative argumentation deserves some more details and quantitative considerations. Kaschner et al. (2006) recently used a modelling approach to assess the potential relative environmental suitability of a given area for a species throughout its range and applied it to marine mammals including the harbour porpoise. This approach predicts the relative occurrence of a species in geographic space by investigating the relationship between known species' occurrence and environmental parameters in ecological space, that are selected to represent most of the key physical factors structuring the habitat of marine species (Longhurst, 2010). The variables

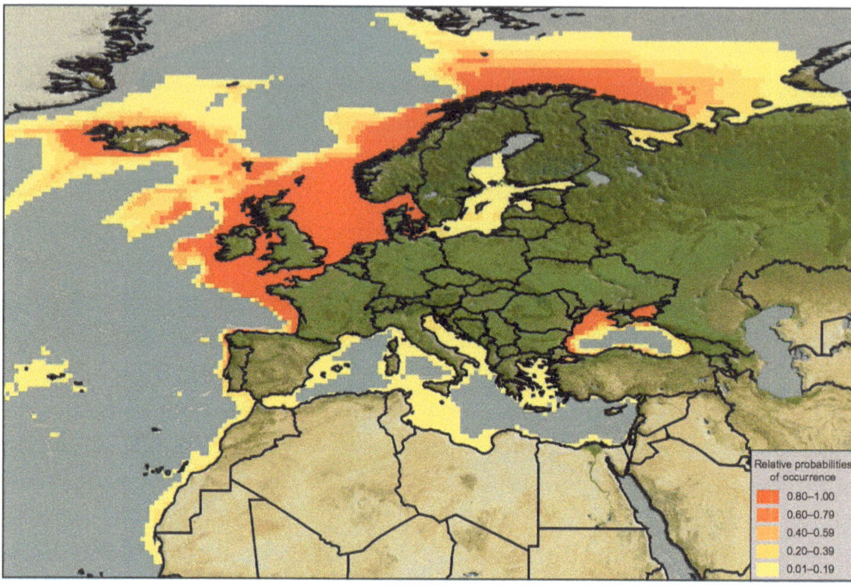

Fig. 4 Distribution of suitable habitats for the harbour porpoise (*Phocoena phocoena*) in the Northeast Atlantic Ocean, Mediterranean Sea and Black Sea, modelled from the reported species occurrence (Kaschner et al., 2006, 2011, 2013). Distribution range *colours* indicate degree of suitability of habitat, which can be interpreted as probabilities of occurrence. Map produced using AquaMaps (Kaschner et al., 2011, 2013).

included by the authors for the harbour porpoise were the bathymetry, average annual sea surface temperature, surface salinity, primary production and sea ice concentration (Kaschner et al., 2006, 2011, 2013). The resulting map of predicted suitable habitat for the harbour porpoise (Fig. 4) shows a relatively good match with the known occurrence of the species in the Northeast Atlantic, especially in areas of highest abundance, such as the Black Sea, the Atlantic coasts of Iberia and the more northern Northeast Atlantic waters. However, the prediction deviated from the known distribution in the southern part of the range, such as off the Northwest African coast where the porpoise occurrence was underestimated by the model. In contrast, the model predicted occurrences of marginal suitable habitats in the Azores, Canary Islands and Mediterranean Sea. The occurrence probability reached up to 40% in some areas of the Mediterranean Sea, such as the Gulf of Lyon, France, and in the northern Adriatic Sea (Kaschner et al., 2006, 2011, 2013) (see Fig. 4). Such discrepancy between predictions and observations may have arisen from inaccurate "presence" records that were not properly curated, or from an imperfect representation of the ecological niche of the harbour porpoises. This predictive modelling approach assumed that the species is monotypic, which was the common belief in 2006. However, Kaschner et al. (2006, 2011, 2013) did not incorporate subsequent results from genetic and ecological studies that indicated harbour porpoises from the southern waters are genetically and ecologically differentiated from the more northern populations, and actually belong to a distinct ecotype with distinct ecological preferences (Fontaine et al., 2007, 2010, 2014). Therefore, it may not be surprising that deviations of the model from known occurrence were mainly noted for the southern populations. Future research should model the distribution of suitable habitats separately for each ecotype in order to account for their differing ecological niche. In addition to refining the predicted distribution of the suitable habitats for each ecotype, modelling could also provide a hypothesis testing framework to advance our knowledge of the ecological separation between these ecotypes.

Despite the caveats of such a modelling approach, the model predictions offered some interesting observations that some areas of the Mediterranean Sea may still be suitable for the harbour porpoise, as shown by the 40% occurrence probability of porpoise occurrences in the Gibraltar Strait, the Gulf of Lions, the northern Adriatic Sea and the northern Aegean Sea (Kaschner et al., 2006, 2011, 2013) (see Fig. 4). As noted in Section 2, no established population was ever reported in the western or eastern Mediterranean basins, with the exception of the northern Aegean Sea, which displays oceanographic and ecological properties in many respects similar to the Black Sea. Even if suitable habitats may be potentially available for the harbour porpoise in the Mediterranean Sea, these model-based predictions more closely approximate a species' fundamental niche than its realized niche. Other factors, such as competitive exclusion by other species, including humans, may be involved in reducing the realized niche of a species, especially in places where food availability is a limiting factor, which is the case of the oligotrophic Mediterranean Sea. Supporting this line of thought are the lethal interactions reported between common bottlenose dolphins (*Tursiops truncatus*) and harbour porpoises in the waters of Great Britain (MacLeod et al., 2007; Patterson et al., 1998; Ross and Wilson, 1996), and (indirectly) recently reported interactions between grey seals (*Haliochoerus grypus*) and harbour porpoises in the southern North Sea (Haelters et al., 2012; Leopold et al., 2015). Dolphin-induced mortalities have been reported along coastlines of the United Kingdom and were a primary cause of mortality in some localities, such as Wales and the south-western United Kingdom (Jepson, 2006). Harbour porpoises are pursued and killed, but crucially, not eaten by bottlenose dolphins. MacLeod et al. (2007) analysed the process of mass-dependent predation risk in this system and showed that where high levels of dolphin-induced mortality occurs, porpoises carry significantly less energy reserves than would otherwise be expected. This lack of reserves equated to reducing the length of time that a porpoise could survive without feeding by approximately 37%. A similar situation could have occurred in the Mediterranean Sea, where bottlenose dolphins are common. Dolphins could have caused a decline in the remnant Mediterranean porpoises and might also have pushed them away from potentially suitable habitat. If so, the Mediterranean porpoises may still remain in areas where food availability is high enough to mitigate interspecific competition, as is true for the upwelling waters off Iberia and Northwest Africa in the Western Mediterranean, and at the Eastern Mediterranean entrance to the Black Sea.

5. EVOLUTIONARY HISTORY OF THE HARBOUR PORPOISES: THE FATE OF THE MEDITERRANEAN POPULATIONS

The presence of a relict population in the Black Sea is a living record that harbour porpoises were present in the Mediterranean Sea at a time in the past when environmental conditions were very different from those of the present day, with surface water far colder and more productive than the current conditions (Frantzis et al., 2001). However, no fossil records are available to confirm historical occurrence of the harbour porpoise in the Mediterranean (Frantzis et al., 2001). Speculations about the evolutionary history of the harbour porpoise, and in particular the colonization of the Black Sea, have been debated since at least the early 1980s (Gaskin, 1982). Harbour porpoises are assumed to have colonized the Black Sea from the Atlantic Ocean. This assumption was made initially based on the proximity of porpoise populations in the Northeast Atlantic and by phylogenetic analyses showing a closer genetic relationship between Black Sea individuals with the North Atlantic congeners than with any other populations in the North Pacific (Rosel et al., 1995).

In addition to enabling the identification of genetically differentiated groups, analyses of DNA variation and inferences of gene and individual genealogies using statistical modelling can be used to delineate the evolutionary history and past demography of populations up to the most recent common ancestor (Schraiber and Akey, 2015). Such genetic inferences about the demographic history of the differentiated harbour porpoise populations in the Northeast Atlantic and Black Sea were recently conducted by Fontaine et al. (2014; and see also Fontaine et al., 2010, 2012). The analyses of genetic variation suggested that harbour porpoises from the upwelling zones—the Iberian Peninsula and Northwest Africa—and the Black Sea, shared a common ancestor prior to splitting from the porpoises currently living on the European continental shelf north of the Bay of Biscay (N_{MED} in Fig. 3A). This common ancestor to the populations bordering the Mediterranean Sea, supports the suspected past existence of now extinct populations in the Mediterranean Sea (Frantzis et al., 2001; Gaskin, 1982). A divergence time estimate between the Northeast Atlantic porpoises and a hypothetical ancestral Mediterranean group(s) (N_{MED}, Fig. 3A) would have occurred in the Mediterranean Sea between 26,500 and 19,000 years before present (yr BP). This corresponds to the Last Glacial Maximum (LGM) of the last Ice Age (Clark et al., 2009). At that time (see Fig. 3B), the glacial sea-level averaged about 100 m below the present-day level, while global temperatures averaged 10°C colder, and the Mediterranean Sea was closer to a subarctic boreal sea than it is presently (Rohling et al., 2009). With the sea level dropping during the glacial period, the Mediterranean Sea remained connected to the Atlantic Ocean due to the depth, of at least 300 m, of the Gibraltar Strait. On the other hand, the connection between the Black Sea and Mediterranean Sea was interrupted given the shallow sill depth of the Bosporus and Dardanelle Straits, turning the Black Sea into a brackish interior sea (Aksu et al., 1999; Georgievski and Stanev, 2006; Rohling et al., 2009; Ryan et al., 1997). Cold-water species, such as the harbour porpoise, could have thus found the glacial conditions of the Mediterranean Sea quite suitable and likely colonized first the western Basin and then the eastern Basin (Fig. 3A and B).

Given the size of the Mediterranean Sea and the very different oceanographic conditions existing between the western and eastern basins (Rohling et al., 2009), it is likely that population subdivisions occurred during, or after, the eastward colonization process (see Fig. 3A and C). The divergence between the western and eastern populations in the Mediterranean Sea likely occurred during the postglacial period, around ca. 14 kyr BP according to the genetic data (Fontaine et al., 2014); thus, it occurred before the reopening of the Black Sea into the Mediterranean Sea, which occurred ca. 8.4 kyr BP (Aksu et al., 1999; Georgievski and Stanev, 2006; Rohling et al., 2009; Ryan et al., 1997). The genetic evidence of postglacial presence of the harbour porpoises in the Mediterranean Sea further correlates with major environmental transitions in the Mediterranean that occurred during the Holocene (Roberts et al., 2011). The period from the early to mid-Holocene (14.8–6 kyr BP) corresponds to the 'African Humid Period' during which abrupt climatic and hydrologic changes driven by monsoon-like effects occurred in the Mediterranean Sea and North Africa (deMenocal et al., 2000; Kropelin et al., 2008; Kuper and Kröpelin, 2006; Liu et al., 2006; Shanahan et al., 2015; Tjallingii et al., 2008). In the Mediterranean Sea, these environmental changes resulted in interspersed nutrient-rich episodes known as the Mediterranean 'Sapropel episodes', characterized by the deposition of organic-rich sediments on the seafloor, which formed as a result of increased primary productivity and rearrangements of water masses (Calvert et al., 1992; Rohling et al., 2009). This period was suitable for the development of a rich and productive marine food web and favourable to the occurrence of top marine predators, such as the harbour porpoises. While these events were particularly intense in the Eastern Mediterranean Sea, in the Western region there were contemporaneous oceanographic and biological shifts (i.e. plankton fauna), around 8.4 kyr BP, as a result of increased inflow of Atlantic waters (Jimenez-Espejo et al., 2007). These environmental conditions may have created a rich and productive trophic network favourable to top marine predators like the harbour porpoises.

After the termination of the African Humid Period and disappearance of the nutrient-rich Sapropel episodes during the second half of the Holocene period (5.5 kyr BP to present; see Fig. 3A and D), the Mediterranean progressively shifted towards warm and oligotrophic conditions (Rohling et al., 2009) unsuitable for the harbour porpoise. Mediterranean porpoise populations would have been forced to retreat to areas where suitable habitats were available. The Black Sea reconnected to the Mediterranean ~8.4 kyr BP (Aksu et al., 1999; Georgievski and Stanev, 2006; Rohling et al., 2009; Ryan et al., 1997), which offered a suitable refuge for Eastern Mediterranean harbour porpoise populations. This timescale is consistent with genetic data suggesting a drastic founder event (N_{BN}, Fig. 3A) followed by a demographic expansion (N_{bf}, Fig. 3A) of the Black Sea population from a single "mitochondrial eve" (Fontaine et al., 2010, 2012, 2014). In contrast, harbour porpoises from the Western Mediterranean would have migrated out of the Mediterranean Sea back to the Atlantic, consistent with the results of Fontaine et al. (2014), which suggest that the Iberian and Mauritanian populations are descended from extinct Western Mediterranean populations. The Iberian and Northwest African populations would have diverged from each other ~3.1 kyr BP (90% confidence interval: 1.4–14.1 kyr BP), which would have been contemporaneous with the end of the Mediterranean Sapropel episodes and the end of the favourable conditions in the Mediterranean Sea (Fontaine et al., 2014). Thus, overall genetic inference suggests that the Iberian and Mauritanian populations were part of an extended Western Mediterranean group, which has been forced to a relict distribution within the "local" upwelling habitats that remained productive enough to sustain the elevated energetic requirements of the demanding harbour porpoise life style (Lockyer, 2007; Read and Hohn, 1995; Wisniewska et al., 2016).

The two subspecies of harbour porpoise—*P. p. relicta* and the proposed *P. p. meridionalis*—are thus suggested to be relicts of the Last Ice Age derived from a palaeo-Mediterranean ancestral group(s) which split from the Atlantic populations of *P. p. phocoena* presently living north of the Bay of Biscay. The three subspecies (or ecotypes) have been following independent evolutionary trajectories and correspond to three distinct Evolutionary Significant Units (ESUs) (Moritz, 2002) diverging from each other with respect to their ecology, morphology and genetic make-up, and specializing and adapting to the local specificities of their respective environments (Fontaine et al., 2014, 2016). Such divergent evolution is a primary step of a speciation process, which can reach completion with the instalment of reproductive isolation mechanisms (Coyne and Orr, 2004). However, such reproductive isolation can only be tested when the two diverging groups are crossed under control conditions in a laboratory, which is impossible to conduct with cetacean species, or if the two groups meet again in natural conditions, forming a secondary contact zone. Fortunately, harbour porpoises present this situation in the Northeast Atlantic, where the Iberian population of the southern ecotype came back into contact with the northern ecotype during the last millennium (Fontaine et al., 2010, 2014). A secondary contact zone that is geographically restricted to the northern part of the Bay of Biscay, Celtic Sea, Irish Sea, South-western coasts of Britain and the western side of the English Channel (Fig. 2B) was described by Fontaine et al. (2014, 2016). In this zone, harbour porpoises have an admixed (hybrid) genetic ancestry from the two ecotypes. This shows that reproduction and gene flow between the two populations is still possible. However, Fontaine et al. (2010, 2014) showed that such gene flow was fully asymmetrical and directed from the southern to the northern ecotype. This admixture zone quickly vanishes northwards with a sharp transition to the pure northern ecotype. The current delimitation of the admixture zone raises the question of what environmental and ecological factors determine the distributions of the ecotypes and extent of the contact zone, and whether the distributions are stable or dynamic. Previous work has shown that the structure and distribution of harbour porpoise populations have been influenced by changes in oceanographic conditions which affect food resources (Fontaine et al., 2007, 2014). Therefore, the location and extent of the Biscay admixture zone is likely to be similarly dynamic and sensitive to past and future changes in climate, which influence shifts in oceanographic and ecological conditions. For instance, warming waters may facilitate a northward expansion of the southern ecotype, which would be detectable by a shift in the extent of the admixture zone.

6. CONCLUSIONS AND PERSPECTIVES

Although harbour porpoises do not presently occupy the Mediterranean Sea (except at the extremities; e.g. in the Aegean Sea, and as vagrants in the Alborán Sea), the evolutionary history of harbour porpoise populations and ecotypes surrounding the region have been deeply shaped by the palaeoenvironmental changes of the Mediterranean Sea. The last glaciations had a major impact on the terrestrial and marine environments (e.g. Hewitt, 2000), and the different subspecies or ecotypes of harbour porpoises are living records of such rich history (Fontaine et al., 2010, 2012, 2014). The evolutionary history of the harbour porpoise reinforces the role of past variation in marine primary production as an important driver of cetacean evolution (e.g. Marx and Uhen, 2010). Past environmental variation likely triggered the porpoise range expansion into the Mediterranean Sea during a cold nutrient-rich period followed by its contraction and fragmentation with the postglacial Holocene warming, leaving behind relict populations in fragmented habitats. These processes led to ecologic, morphologic and genetic divergence of three ecotypes with potentially also some ecological specialization. The relatively narrow secondary contact zone between the two Atlantic ecotypes in the Bay of Biscay might suggest that speciation process has been initiated (Nosil et al., 2009). However, additional investigation is required to fully understand the extent of their divergence and the dynamics of the contact zone in face of the ongoing climate change.

Harbour porpoise evolutionary history may have also revealed how the species is sensitive to stochastic environmental variation and could help to identify which populations are the most threatened by such stochastic processes. This is important as locally adapted populations can be brought to extinction if the environmental conditions become unsuitable or if resources becomes unavailable. Genetic analyses revealed the uniqueness of the populations of the southern ecotype inhabiting the upwelling waters off the Iberian Peninsula and Northwest Africa, and how small are their population sizes and genetic diversity. While harbour porpoises from the Black Sea are already recognized as an Endangered "subpopulation" on the Red List of the International Union for Conservation of Nature (IUCN) (Birkun and Frantzis, 2008), the status and threats faced by the southern ecotype still require a proper evaluation. The new information obtained over the last decade underscores the need for immediate assessment of the southern ecotype populations. Given that ecological differences among populations may lead to differential responses to climate change, the identification of a new distinct ecotype in the upwelling zones, comprised of two populations locally restricted to the coasts of Iberia and Mauritania, highlights the importance of adequate genetic sampling to correctly characterize the population genetic structure of a species. Further information on basic aspects of their life history, demography and ecology is greatly needed to assess the impact of incidental exploitation and the ecological process underlying their divergence.

This chapter tentatively showed that accurate characterization of the evolutionary history of a species is critical for improving predictions about how populations of marine mammals will respond to future changes in climate and to adapt conservation and management initiatives. Genetic approaches have been instrumental for revealing the evolutionary history of the harbour porpoises and clarify the timing and modalities of the divergence process between ecotypes (subspecies). The revolution in sequencing technologies and democratization of the costs to sequence whole genomes for nonmodel species hold great promise to advance further our knowledge about the evolutionary history of populations and species and for better understanding the processes that lead to adaptations and specializations (Cammen et al., 2016; Ellegren, 2014; Garner et al., 2016; Kelley et al., 2016; Shafer et al., 2015). Examples of such potentials are already available for some species such as killer whales (*Orcinus orca*) (e.g. Foote et al., 2016) and will likely expand dramatically to other species in the forthcoming years (Cammen et al., 2016; Kelley et al., 2016). Such an initiative is currently underway for the species that comprise the genus *Phocoena*, including the harbour porpoise (Fontaine et al., 2015), and may prove useful to future conservation and management efforts in the face of continued climate change.

ACKNOWLEDGEMENTS

I thank Frederic Labbé for his assistance in preparing the figures, Ari S. Friedlaender for contributing the photograph of a harbour porpoise and Ayaka A. Öztürk for sharing useful literature. The manuscript was greatly improved by the comments of the editors, Giuseppe Notarbartolo di Sciara and Barbara E. Curry.

REFERENCES

Aksu, A.E., Hiscott, R.N., Yaşar, D., 1999. Oscillating quaternary water levels of the Marmara Sea and vigorous outflow into the Aegean Sea from the Marmara Sea–Black Sea drainage corridor. Mar. Geol. 153, 275–302.

Amante, C., Eakins, B.W., 2009. ETOPO1 1 Arc-Minute Global Relief Model: procedures, data sources and analysis. NOAA Technical Memorandum NESDIS NGDC-24. National Geophysical Data Center, NOAA. http://dx.doi.org/10.7289/V5C8276M available at https://www.ngdc.noaa.gov/mgg/global/global.html (accessed 08.18.2016).

Andersen, L.W., 2003. Harbour porpoises (*Phocoena phocoena*) in the North Atlantic: distribution and genetic population structure. NAMMCO Scientific Publications vol. 5. NAMMCO, Tromsø, Norway, p. 11.

Arístegui, J., Barton, E.D., Álvarez-Salgado, X.A., Santos, A.M.P., Figueiras, F.G., Kifani, S., Hernández-León, S., Mason, E., Machú, E., Demarcq, H., 2009. Sub-regional ecosystem variability in the Canary Current upwelling. Prog. Oceanogr. 83, 33–48.

Benke, H., Bräger, S., Dähne, M., Gallus, A., Hansen, S., Honnef, C.G., Jabbusch, M., Koblitz, J.C., Krügel, K., Liebschner, A., Narberhaus, I., Verfuß, U.K., 2014. Baltic Sea harbour porpoise populations: status and conservation needs derived from recent survey results. Mar. Ecol. Prog. Ser. 495, 275–290. http://dx.doi.org/10.3354/meps10538.

Birkun Jr., A., Frantzis, A., 2008. *Phocoena phocoena* ssp. *relicta*. available at http://dx.doi.org/10.2305/IUCN.UK.2008.RLTS.T17030A6737111.en (accessed 08.18.2016).

Bjørge, A., Øien, N., 1995. Distribution and abundance of harbour porpoise, *Phocoena phocoena*, in Norwegian waters. Rep. Int. Whal. Commn. (spec. issue vol. 16), 89–98.

Bjørge, A., Skern-Mauritzen, M., Rossman, M.C., 2013. Estimated bycatch of harbour porpoise (*Phocoena phocoena*) in two coastal gillnet fisheries in Norway, 2006–2008. Mitigation and implications for conservation. Biol. Conserv. 161, 164–173. http://dx.doi.org/10.1016/j.biocon.2013.03.009.

Calvert, S.E., Nielsen, B., Fontugne, M.R., 1992. Evidence from nitrogen isotope ratios for enhanced productivity during formation of eastern Mediterranean sapropels. Nature 359, 223–225. http://dx.doi.org/10.1038/359223a0.

Cammen, K.M., Andrews, K.R., Carroll, E.L., Foote, A.D., Humble, E., Khudyakov, J.I., Louis, M., McGowen, M.R., Olsen, M.T., Van Cise, A.M., 2016. Genomic methods take the plunge: recent advances in high-throughput sequencing of marine mammals. J. Hered. 107, 481–495. http://dx.doi.org/10.1093/jhered/esw044.

Camphuysen, K.C.J., Siemensma, M.L., 2011. Conservation Plan for the Harbour Porpoise *Phocoena phocoena* in The Netherlands: Towards a Favourable Conservation Status: NIOZ Report 2011–07. Royal Netherlands Institute for Sea Research, Texel, The Netherlands.

Clark, P.U., Dyke, A.S., Shakun, J.D., Carlson, A.E., Clark, J., Wohlfarth, B., Mitrovica, J.X., Hostetler, S.W., McCabe, A.M., 2009. The last glacial maximum. Science 325, 710–714. http://dx.doi.org/10.1126/science.1172873.

Coyne, J.A., Orr, H.A., 2004. Speciation. Sinauer Associates Inc., Sunderland, MA.

deMenocal, P., Ortiz, J., Guilderson, T., Adkins, J., Sarnthein, M., Baker, L., Yarusinsky, M., 2000. Abrupt onset and termination of the African Humid Period: rapid climate responses to gradual insolation forcing. Quat. Sci. Rev. 19, 347–361.

Donovan, G.P., Bjørge, A., 1995. Harbour porpoises in the North Atlantic: edited extract from the report of the IWC Scientific Committee, Dublin 1995. Rep. Int. Whal. Commn. (spec. issue vol. 16), 3–26.

Ellegren, H., 2014. Genome sequencing and population genomics in non-model organisms. Trends Ecol. Evol., 29, 51–63. http://dx.doi.org/10.1016/j.tree.2013.09.008.

Fontaine, M.C., Baird, S.J.E., Piry, S., Ray, N., Tolley, K.A., Duke, S., Birkun, A., Ferreira, M., Jauniaux, T., Llavona, Á., Öztürk, B., Öztürk, A.A., Ridoux, V., Rogan, E., Sequeira, M., Siebert, U., Vikingsson, G.A., Bouquegneau, J.-M., Michaux, J.R., 2007. Rise of oceanographic barriers in continuous populations of a cetacean: the genetic structure of harbour porpoises in Old World waters. BMC Biol. 5, 30. http://dx.doi.org/10.1186/1741-7007-5-30.

Fontaine, M.C., Tolley, K.A., Michaux, J.R., Birkun, A., Ferreira, M., Jauniaux, T., Llavona, Á., Öztürk, B., Oztürk, A.A., Ridoux, V., Rogan, E., Sequeira, M., Bouquegneau, J.-M., Baird, S.J.E., 2010. Genetic and historic evidence for climate-driven population fragmentation in a top cetacean predator: the harbour porpoises in European water. Proc. R. Soc. B 277, 2829–2837. http://dx.doi.org/10.1098/rspb.2010.0412.

Fontaine, M.C., Snirc, A., Frantzis, A., Koutrakis, E., Öztürk, B., Oztürk, A.A., Austerlitz, F., 2012. History of expansion and anthropogenic collapse in a top marine predator of the black Sea estimated from genetic data. Proc. Natl. Acad. Sci. U. S. A. 109, E2569–E2576. http://dx.doi.org/10.1073/pnas.1201258109.

Fontaine, M.C., Roland, K., Calves, I., Austerlitz, F., Palstra, F.P., Tolley, K.A., Ryan, S., Ferreira, M., Jauniaux, T., Llavona, Á., Öztürk, B., Oztürk, A.A., Ridoux, V., Rogan, E., Sequeira, M., Siebert, U., Vikingsson, G.A., Borrell, A., Michaux, J.R., Aguilar, A., 2014. Postglacial climate changes and rise of three ecotypes of harbour porpoises, *Phocoena phocoena*, in western Palearctic waters. Mol. Ecol. 23, 3306–3321. http://dx.doi.org/10.1111/mec.12817.

Fontaine, M.C., Chehida, Y.B., Roland, K., Calves, I., Austerlitz, F., Palstra, F.P., Tolley, K.A., Ferreira, M., Jauniaux, T., Llavona, Á., Öztürk, B., Oztürk, A.A., Ridoux, V., Rogan, E., Sequeira, M., Siebert, U., Vikingsson, G.A., Borrell, A., Aguilar, A., Morin, P.A., 2015. Postglacial rise of three ecotypes of harbor porpoises in western Palearctic waters and genomic perspective of adaptive divergence. In: Poster Presented at the 21st Biennial Conference on the Biology of Marine Mammals. 13–18 December, San Francisco, CA, USA.

Fontaine, M.C., Thatcher, O., Ray, N., Piry, S., Brownlow, A., Davinson, N.J., Jepson, P.D., Deaville, R., Goodman, S.J., 2016. Mixing of porpoise ecotypes in South Western UK waters revealed by genetic profiling. bioRxiv. http://dx.doi.org/10.1101/045807.

Foote, A.D., Vijay, N., Ávila-Arcos, M.C., Baird, R.W., Durban, J.W., Fumagalli, M., Gibbs, R.A., Hanson, M.B., Korneliussen, T.S., Martin, M.D., Robertson, K.M., Sousa, V.C., Vieira, F.G., Vinar, T., Wade, P., Worley, K.C., Excoffier, L., Morin, P.A., Gilbert, M.T.P., Wolf, J.B.W., 2016. Genome-culture coevolution promotes rapid divergence of killer whale ecotypes. Nat. Commun. 7, 11693. http://dx.doi.org/10.1038/ncomms11693.

Frantzis, A., 2009. Cetaceans in Greece: Present Status of Knowledge. Initiative for the Conservation of Cetaceans in Greece, Athens, Greece.

Frantzis, A., Gordon, J., Hassidis, G., Komnenou, A., 2001. The enigma of harbor porpoise presence in the Mediterranean Sea. Mar. Mamm. Sci. 17, 937–944.

Garner, B.A., Hand, B.K., Amish, S.J., Bernatchez, L., Foster, J.T., Miller, K.M., Morin, P.A., Narum, S.R., O'brien, S.J., Roffler, G., Templin, W.D., Sunnucks, P., Strait, J., Warheit, K.I., Seamons, T.R., Wenburg, J., Olsen, J., Luikart, G., 2016. Genomics in conservation: case studies and bridging the gap between data and application. Trends Ecol. Evol. 31, 81–83. http://dx.doi.org/10.1016/j.tree.2015.10.009.

Gaskin, D.E., 1982. The Ecology of Whales and Dolphins. Heinemann, London and Exeter.

Gaskin, D.E., 1984. The harbour porpoise *Phocoena phocoena*: regional populations, status and information on direct and indirect catches. Rep. Int. Whal. Commn. 34, 569–586.

Gaskin, D.E., Arnold, P.W., Blair, B.A., 1974. *Phocoena phocoena*. Mamm. Species 42, 1–8. http://dx.doi.org/10.2307/42.1.

Georgievski, G., Stanev, E.V., 2006. Paleo-evolution of the Black Sea watershed: sea level and water transport through the Bosporus Straits as an indicator of the Lateglacial–Holocene transition. Clim. Dyn. 26, 631–644. http://dx.doi.org/10.1007/s00382-006-0123-y.

Giosan, L., Filip, F., Constatinescu, S., 2009. Was the Black Sea catastrophically flooded in the early Holocene? Quat. Sci. Rev. 28, 1–6. http://dx.doi.org/10.1016/j.quascirev.2008.10.012.

Haelters, J., Camphuysen, K.C.J., 2009. The Harbour Porpoise in the Southern North Sea: abundance, threats and research & management proposals: Report to the IFAW (International Fund for Animal Welfare), Brussels, Belgium.

Haelters, J., Kerckhof, F.-M., Jauniaux, T., Degraer, S., 2012. The Grey Seal (*Halichoerus grypus*) as a predator of Harbour Porpoises (*Phocoena phocoena*)? Aquat. Mamm. 38, 343–353. http://dx.doi.org/10.1578/AM.38.4.2012.343.

Hammond, P.S., Berggren, P., Benke, H., Borchers, D.L., Collet, A., Heide-Jørgensen, M.P., Heimlich, S., Hiby, A.R., Leopold, M.F., Øien, N., 2002. Abundance of harbour porpoise and other cetaceans in the North Sea and adjacent waters. J. Appl. Ecol. 39, 361–376. http://dx.doi.org/10.1046/j.1365-2664.2002.00713.x.

Hammond, P.S., Macleod, K., Berggren, P., Borchers, D.L., Burt, L., Cañadas, A., Desportes, G., Donovan, G.P., Gilles, A., Gillespie, D., 2013. Cetacean abundance and distribution in European Atlantic shelf waters to inform conservation and management. Biol. Conserv. 164, 107–122.

Hampe, A., Jump, A.S., 2011. Climate relicts: past, present, future. Annu. Rev. Ecol. Evol. Syst. 42, 313–333. http://dx.doi.org/10.1146/annurev-ecolsys-102710-145015.

Hewitt, G., 2000. The genetic legacy of the Quaternary ice ages. Nature 405, 907–913.

International Whaling Commission (IWC), 1996. Report of the sub-committee on the small cetaceans. Rep. Int. Whal. Commn. 46, 160–179.

Jefferson, T.A., Curry, B.E., 1994. A global review of porpoise (Cetacea: Phocoenidae) mortality in gillnets. Biol. Conserv. 67, 167–183.

Jefferson, T.A., Curry, B.E., Leatherwood, S., Powell, J.A., 1997. Dolphins and porpoises of West Africa: a review of records (Cetacea: Delphinidae, Phocoenidae). Mammalia 61, 87–108.

Jepson, P.D., 2006. Cetacean strandings investigation and co-ordination in the UK 2000–2004: Final report to Department for Food and Rural Affairs.

Jimenez-Espejo, F.J., Martinez-Ruiz, F., Sakamoto, T., Iijima, K., Gallego-Torres, D., Harada, N., 2007. Paleoenvironmental changes in the western Mediterranean since the last glacial maximum: high resolution multiproxy record from the Algero–Balearic basin. Palaeogeogr. Palaeoclimatol. Palaeoecol. 246, 292–306. http://dx.doi.org/10.1016/j.palaeo.2006.10.005.

Johnston, D.W., Westgate, A.J., Read, A.J., 2005. Effects of fine-scale oceanographic features on the distribution and movements of harbour porpoises *Phocoena phocoena* in the Bay of Fundy. Mar. Ecol. Prog. Ser. 295, 279–293. http://dx.doi.org/10.3354/meps295279.

Kanwisher, J., Sundnes, G., 1965. Physiology of a small cetacean. Hvalrad. Skr. 48, 45–53.
Kaschner, K., Watson, R., Trites, A.W., Pauly, D., 2006. Mapping world-wide distributions of marine mammal species using a relative environmental suitability (RES) model. Mar. Ecol. Prog. Ser. 316, 1–26.
Kaschner, K., Ready, J.S., Agbayani, E., Kesner-Reyes, K., Rius-Barile, J., Eastwood, P.D., South, A.B., Kullander, S.O., Rees, T., Watson, R., Pauly, D., Froese, R., 2011. Using "Aquamaps" for representing species distribution in Regional Seas. In: Christensen, V., Lai, S., Palomares, M.L., Zeller, D., Pauly, D. (Eds.), The State of Biodiversity and Fisheries in Regional Seas. Fisheries Centre, University of British Columbia, Vancouver, pp. 17–21. http://dx.doi.org/10.14288/1.0074734.
Kaschner, K., Rius-Barile, J., Kesner-Reyes, K., Garilao, C., Kullander, S.O., Rees, T., Froese, R., 2013. AquaMaps: Predicted Range Maps for Aquatic Species. World Wide Web Electronic Publication. www.aquamaps.org. version 08/2013.
Kelley, J.L., Brown, A.P., Therkildsen, N.O., Foote, A.D., 2016. The life aquatic: advances in marine vertebrate genomics. Nat. Rev. Genet. 17, 523–534. http://dx.doi.org/10.1038/nrg.2016.66.
Kropelin, S., Verschuren, D., Lezine, A.M., Eggermont, H., Cocquyt, C., Francus, P., Cazet, J.P., Fagot, M., Rumes, B., Russell, J.M., Darius, F., Conley, D.J., Schuster, M., Suchodoletz von, H., Engstrom, D.R., 2008. Climate-driven ecosystem succession in the Sahara: the past 6000 years. Science 320, 765–768. http://dx.doi.org/10.1126/science.1154913.
Kuper, R., Kröpelin, S., 2006. Climate-controlled Holocene occupation in the Sahara: motor of Africa's evolution. Science 313, 803–807. http://dx.doi.org/10.1126/science.1130989.
Lambert C., Laran S., David L., Dorémus G., Pettex E., Van Canneyt O. and Ridoux V., How does ocean seasonality drive habitat preferences of highly mobile top predators? Part I: the north-western Mediterranean Sea, *Deep Sea Res. Pt. II*, in press. http://dx.doi.org/10.1016/j.dsr2.2016.06.012.
Leopold, M.F., Begeman, L., van Bleijswijk, J.D.L., IJsseldijk, L.L., Witte, H.J., Gröne, A., 2015. Exposing the grey seal as a major predator of harbour porpoises. Proc. R. Soc. B 282, 20142429. http://dx.doi.org/10.1098/rspb.2014.2429.
Liu, Z., Wang, Y., Gallimore, R., Notaro, M., Prentice, I.C., 2006. On the cause of abrupt vegetation collapse in North Africa during the Holocene: climate variability vs. vegetation feedback. Geophys. Res. Lett. 33, L22709. http://dx.doi.org/10.1029/2006GL028062.
Lockyer, C., 2003. Harbour porpoises (*Phocoena phocoena*) in the North Atlantic: biological parameters. NAMMCO Scentific Publications, vol. 5. NAMMCO, Tromsø, Norway, pp. 71–78.
Lockyer, C., 2007. All creatures great and smaller: a study in cetacean life history energetics. J. Mar. Biol. Assoc. U. K. 87, 1035. http://dx.doi.org/10.1017/S0025315407054720.
Longhurst, A.R., 2010. Ecological Geography of the Sea, second ed. Academic Press, San Diego, CA.
MacLeod, R., MacLeod, C.D., Learmonth, J.A., Jepson, P.D., Reid, R.J., Deaville, R., Pierce, G.J., 2007. Mass-dependent predation risk and lethal dolphin-porpoise interactions. Proc. R. Soc. B 274, 2587–2593. http://dx.doi.org/10.1098/rspb.2007.0786.
Major, C.O., Goldstein, S.L., Ryan, W.B., Lericolais, G., Piotrowski, A.M., Hajdas, I., 2006. The co-evolution of Black Sea level and composition through the last deglaciation and its paleoclimatic significance. Quat. Sci. Rev. 25, 2031–2047.
Marx, F.G., Uhen, M.D., 2010. Climate, critters, and cetaceans: Cenozoic drivers of the evolution of modern whales. Science 327, 993–996. http://dx.doi.org/10.1126/science.1185581.
Méndez-Fernandez, P., Pierce, G.J., Bustamante, P., Chouvelon, T., Ferreira, M., González, A.F., Lopez, A., Read, F.L., Santos, M.B., Spitz, J., Vingada, J.V.,

Caurant, F., 2013. Ecological niche segregation among five toothed whale species off the NW Iberian Peninsula using ecological tracers as multi-approach. Mar. Biol. 160, 2825–2840. http://dx.doi.org/10.1007/s00227-013-2274-9.

Moraitou-Apostolopoulou, M., 1985. The Zooplankton communities of the eastern Mediterranean (Levantine Basin, Aegean Sea); influence of man-made factors. Medterranean Marine Ecosystems. Springer, Boston, MA, pp. 303–331. http://dx.doi.org/10.1007/978-1-4899-2248-9_14.

Moritz, C., 2002. Strategies to protect biological diversity and the evolutionary processes that sustain it. Syst. Biol. 51, 238–254. http://dx.doi.org/10.1080/10635150252899752.

Nosil, P., Harmon, L.J., Seehausen, O., 2009. Ecological explanations for (incomplete) speciation. Trends Ecol. Evol. 24, 145–156. http://dx.doi.org/10.1016/j.tree.2008.10.011.

Notarbartolo di Sciara, G., Birkun Jr., A., 2010. Conserving Whales, Dolphins and Porpoises in the Mediterranean and Black Seas: An ACCOBAMS Status Report. ACCOBAMS and Department of External Relations, Principality of Monaco. 212 pp.

Osborn, T.J., Briffa, K.R., 2006. The spatial extent of 20th-century warmth in the context of the past 1200 years. Science 311, 841–844. http://dx.doi.org/10.1126/science.1120514.

Pante, E., Simon-Bouhet, B., 2013. marmap: a package for importing, plotting and analyzing bathymetric and topographic data in R. PLoS One 8, e73051. http://dx.doi.org/10.1371/journal.pone.0073051.

Patterson, I.A.P., Reid, R.J., Wilson, B., Grellier, K., Ross, H.M., Thompson, P.M., 1998. Evidence for infanticide in bottlenose dolphins: an explanation for violent interactions with harbour porpoises? Proc. R. Soc. B 265, 1167–1170. http://dx.doi.org/10.1098/rspb.1998.0414.

Peltier, H., Baagøe, H.J., Camphuysen, K.C.J., Czeck, R., Dabin, W., Daniel, P., Deaville, R., Haelters, J., Jauniaux, T., Jensen, L.F., Jepson, P.D., Keijl, G.O., Siebert, U., Van Canneyt, O., Ridoux, V., 2013. The stranding anomaly as population indicator: the case of harbour porpoise *Phocoena phocoena* in North-Western Europe. PLoS One 8, e62180. http://dx.doi.org/10.1371/journal.pone.0062180.

Pierce, G.J., Caldas, M., Cedeira, J., Santos, M.B., Llavona, Á., Covelo, P., Martinez, G., Torres, J., Sacau, M., Lopez, A., 2010. Trends in cetacean sightings along the Galician coast, north-west Spain, 2003–2007, and inferences about cetacean habitat preferences. J. Mar. Biol. Assoc. U. K. 90, 1547–1560. http://dx.doi.org/10.1017/S0025315410000664.

Pinela, A.M., Borrell, A., Cardona, L., Aguilar, A., 2010. Stable isotope analysis reveals habitat partitioning among marine mammals off the NW African coast and unique trophic niches for two globally threatened species. Mar. Ecol. Prog. Ser. 416, 295–306. http://dx.doi.org/10.3354/meps08790.

Polacheck, T., Wenzel, F.W., Early, G., 1995. What do stranding data say about harbor porpoises (*Phocoena phocoena*)? Rep. Int. Whal. Commn. (spec. issue vol. 16), 169–179.

R Core Team, 2016. R: A Language and Environment for Statistical Computing. R Foundation for Statistical Computing. Vienna, Austria. https://www.R-project.org.

Read, A.J., 1999. Harbour porpoise (*Phocoena phocoena*). In: Ridgway, S., Harrison, R. (Eds.), Handbook of Marine Mammals. Academic Press, London, San Diego, pp. 323–350.

Read, A.J., Hohn, A.A., 1995. Life in the fast lane: the life history of harbor porpoises from the Gulf of Maine. Mar. Mam. Sci. 11, 423–440.

Read, A.J., Drinker, P., Northridge, S., 2006. Bycatch of marine mammals in U.S. and global fisheries. Conserv. Biol. 20, 163–169.

Rey, J.C., Cendrero, O., 1982. Les mammifères marins trouvés sur les cotes espagnoles (No. ICES Council Meeting 1982/N:6). Conseil International pour l'Exploitation de la Mer (CIEM/ICES).

Roberts, N., Brayshaw, D., Kuzucuoglu, C., Perez, R., Sadori, L., 2011. The mid-Holocene climatic transition in the Mediterranean: causes and consequences. The Holocene 21, 3–13. http://dx.doi.org/10.1177/0959683610388058.

Robineau, D., Vely, M., 1998. Les cétacés des cotes de Mauritanie (Afrique du Nord-Ouest). Particularités et variations spatio-temporelles de répartition: roles des facteurs océanographiques. Rev. Ecol. (Terre Vie) 53, 123–152.

Rohling, E.J., Abu-Zeid, R., Casford, C., Hayes, A., Hoogakker, B., 2009. The Mediterranean Sea: present and past. In: Woodward, J. (Ed.), Physical Geography of the Mediterranean Basin. Oxford University Press, Oxford, UK, pp. 33–68.

Rosel, P.E., 1997. A review and assessment of the status of the harbor porpoise (*Phocoena phocoena*) in the North Atlantic. In: Dizon, A.E., Chivers, S.J., Perrin, W.F. (Eds.), Molecular Genetics of Marine Mammals. The Society for Marine Mammalogy, San Diego, CA, pp. 209–226.

Rosel, P.E., Dizon, A.E., Haygood, M.G., 1995. Variability of the mitochondrial control region in populations of the harbour porpoise, *Phocoena phocoena*, on interoceanic and regional scales. Can. J. Fish. Aquat. Sci. 52, 1210–1219.

Rosel, P.E., Frantzis, A., Lockyer, C., Komnenou, A., 2003. Source of Aegean Sea harbour porpoises. Mar. Ecol. Prog. Ser. 247, 257–261.

Ross, H.M., Wilson, B., 1996. Violent interactions between bottlenose dolphins and harbour porpoises. Proc. R. Soc. B 263, 283–286. http://dx.doi.org/10.1098/rspb.1996.0043.

Ryan, W.B.F., Pitman III, W.C., Major, C.O., Shimkus, K., Moskalenko, V., Jones, G.A., Dimitrov, P., Gorür, N., Sakinç, M., Yüce, H., 1997. An abrupt drowning of the Black Sea shelf. Mar. Geol. 138, 119–126. http://dx.doi.org/10.1016/S0025-3227(97)00007-8.

Ryan, C., Cucknell, A.C., Romagosa, M., Boisseau, O., Moscorop, A., Frantzis, A., McLachlan, R., 2014. A visual and acoustic survey for marine mammals in the Eastern Mediterranean Sea during summer 2013. Final report prepared for the Marine Conservation Research International, Essex, UK.

Santos, M., Pierce, G., 2003. The diet of harbour porpoise (*Phocoena phocoena*) in the northeast Atlantic. Oceanogr. Mar. Biol. 41, 355–390.

Scheidat, M., Verdaat, H., Aarts, G., 2012. Using aerial surveys to estimate density and distribution of harbour porpoises in Dutch waters. J. Sea Res. 69, 1–7. http://dx.doi.org/10.1016/j.seares.2011.12.004.

Schraiber, J.G., Akey, J.M., 2015. Methods and models for unravelling human evolutionary history. Nat. Rev. Genet. 16, 727–740. http://dx.doi.org/10.1038/nrg4005.

Shafer, A.B.A., Wolf, J.B.W., Alves, P.C., Bergström, L., Bruford, M.W., Brännström, I., Colling, G., Dalén, L., De Meester, L., Ekblom, R., Fawcett, K.D., Fior, S., Hajibabaei, M., Hill, J.A., Hoezel, A.R., Höglund, J., Jensen, E.L., Krause, J., Kristensen, T.N., Krützen, M., McKay, J.K., Norman, A.J., Ogden, R., Österling, E.M., Ouborg, N.J., Piccolo, J., Popović, D., Primmer, C.R., Reed, F.A., Roumet, M., Salmona, J., Schenekar, T., Schwartz, M.K., Segelbacher, G., Senn, H., Thaulow, J., Valtonen, M., Veale, A., Vergeer, P., Vijay, N., Vilà, C., Weissensteiner, M., Wennerström, L., Wheat, C.W., Zieliński, P., 2015. Genomics and the challenging translation into conservation practice. Trends Ecol. Evol. 30, 78–87. http://dx.doi.org/10.1016/j.tree.2014.11.009.

Shanahan, T.M., McKay, N.P., Hughen, K.A., Overpeck, J.T., Otto-Bliesner, B., Heil, C.W., King, J., Scholz, C.A., Peck, J., 2015. The time-transgressive termination of the African Humid Period. Nat. Geosci. 8, 140–144. http://dx.doi.org/10.1038/ngeo2329.

Smeenk, C., Leopold, M.F., Addink, M.J., 1992. Note on the harbour porpoise *Phocoena phocoena* in Mauritania, West Africa. Lutra 35, 98–104.

Spitz, J., Rousseau, Y., Ridoux, V., 2006. Diet overlap between harbour porpoise and bottlenose dolphin: an argument in favour of interference competition for food? Estuar. Coast. Shelf Sci. 70, 259–270. http://dx.doi.org/10.1016/j.ecss.2006.04.020.
Spötl, C., Nicolussi, K., Patzelt, G., Boch, R., team, D., 2010. Humid climate during deposition of sapropel 1 in the Mediterranean Sea: assessing the influence on the Alps. Glob. Planet. Chang. 71, 242–248. http://dx.doi.org/10.1016/j.gloplacha.2009.10.003.
Stenson, G.B., 2003. Harbour porpoise (*Phocoena phocoena*) in the North Atlantic: abundance, removals, and sustainability of removals. NAMMCO Scentific Publications, vol. 5. NAMMCO, Tromsø, Norway, pp. 271–302. http://dx.doi.org/10.7557/3.2830.
Sveegaard, S., Andreasen, H., Mouritsen, K.N., Jeppesen, J.P., Teilmann, J., Kinze, C.C., 2012. Correlation between the seasonal distribution of harbour porpoises and their prey in the Sound, Baltic Sea. Mar. Biol. 159, 1029–1037. http://dx.doi.org/10.1007/s00227-012-1883-z.
Teilmann, J., Dietz, R., 1998. Status of the harbour porpoise in Greenland. Polar Biol. 19, 211–220. http://dx.doi.org/10.1007/s003000050237.
Theocharis, A., Georgopoulos, D., Lascaratos, A., Nittis, K., 1993. Water masses and circulation in the central region of the Eastern Mediterranean: eastern Ionian, South Aegean and Northwest Levantine, 1986–1987. Deep Sea Res. Pt. II 40, 1121–1142. http://dx.doi.org/10.1016/0967-0645(93)90064-T.
Tjallingii, R., Claussen, M., Stuut, J.-B.W., Fohlmeister, J., Jahn, A., Bickert, T., Lamy, F., Röhl, U., 2008. Coherent high- and low-latitude control of the northwest African hydrological balance. Nat. Geosci. 1, 670–675. http://dx.doi.org/10.1038/ngeo289.
Tolley, K.A., Rosel, P.E., 2006. Population structure and historical demography of eastern North Atlantic harbour porpoises inferred through mtDNA sequences. Mar. Ecol. Prog. Ser. 327, 297.
Tonay, A.M., Dede, A., 2013. First stranding record of a harbour porpoise (*Phocoena phocoena*) in the Southern Aegean Sea. J. Black Sea/Mediterr. Environ. 19, 132–137.
Tregenza, N., Berrow, S.D., Hammond, P.S., Leaper, R., 1997. Harbour porpoise (*Phocoena phocoena* L.) by-catch in set gillnets in the Celtic Sea. ICES. J. Mar. Sci. 54, 896–904.
Van Waerebeek, K., 2007. Conservation status of the Northwest African population of the harbour porpoise (No. CMS/ScC14/Doc.7, Agenda item 6(a)). In: 14th Meeting of the CMS Scientific Council. CMS/UNEP, Bonn, Germany.
Van Waerebeek, K., Ndiaye, E., Djiba, A., Diallo, M., 2000. A Survey of the Conservation Status of Cetaceans in Senegal, the Gambia and Guinea-Bissau. UNEP/CMS Secretariat, Bonn, Germany.
Van Waerebeek, K., Barnett, L., Camara, A., Cham, A., Diallo, M., Djiba, A., Jallow, A., Ndiae, E., Samba Ould-Bilal, A.O., Bamy, I.L., 2003. Conservation of Cetaceans in The Gambia and Senegal, 1999-2001, and Status of the Atlantic Humpback Dolphin. UNEP/CMS Secretariat, Bonn, Germany.
Viaud-Martínez, K.A., Vergara, M.M., Gol'din, P.E., Ridoux, V., Ozturk, A.A., Ozturk, B., Rosel, P.E., Frantzis, A., Komnenou, A., Bohonak, A.J., 2007. Morphological and genetic differentiation of the Black Sea harbour porpoise *Phocoena phocoena*. Mar. Ecol. Prog. Ser. 338, 281–294.
Vinther, M., Larsen, F., 2004. Updated estimates of harbour porpoise (*Phocoena phocoena*) bycatch in the Danish North Sea bottom-set gillnet fishery. J. Cetac. Res. Manage. 6, 19–24.
Wisniewska, D.M., Johnson, M., Teilmann, J., Rojano-Doñate, L., Shearer, J., Sveegaard, S., Miller, L.A., Siebert, U., Madsen, P.T., 2016. Ultra-High foraging rates of harbor porpoises make them vulnerable to anthropogenic disturbance. Curr. Biol. 26, 1441–1446. http://dx.doi.org/10.1016/j.cub.2016.03.069.

CHAPTER TWELVE

Are Mediterranean Monk Seals, *Monachus monachus*, Being Left to Save Themselves from Extinction?

G. Notarbartolo di Sciara*[,1], S. Kotomatas[†]
*Tethys Research Institute, Acquario Civico, Milano, Italy
[†]WWF Greece, Athens, Greece
[1]Corresponding author: e-mail address: disciara@gmail.com

Contents

Abstract

Mediterranean monk seals (*Monachus monachus*), amongst the most endangered marine mammals, are showing localised signs of recovery warranting their recent down-listing, from Critically Endangered to Endangered, on the International Union for the Conservation of Nature (IUCN) Red List. This, however, cannot be taken as a reason for complacency, as the species' condition is still very critical, having been extirpated from most of its historical range. Monk seals within the Mediterranean, a 'unit to conserve' separate from Atlantic conspecifics, were once widely distributed throughout the Mediterranean Sea, with their range also extending into the Sea of Marmara and the Black Sea. Today breeding nuclei persist only in the northeastern portion of the region, in Greek, Turkish and Cypriot waters. The main reasons for their decline include deliberate killing and human encroachment of their critical habitat. Past conservation efforts have mostly failed due to the inability of implementing institutional commitments, lack of coordination and continuity of efforts and insufficient consideration of the socioeconomic implications of conserving monk seals. Yet the small reversal of the species' decline that has been observed in Greece may have resulted from conservation efforts by civil society, combined with ongoing societal change within the local communities coexisting with the seals. The inaccessibility of large portions of monk seal habitat in the Eastern Mediterranean Sea may also have contributed, by offering to the monk seals a

Advances in Marine Biology, Volume 75
ISSN 0065-2881
http://dx.doi.org/10.1016/bs.amb.2016.08.004

refuge from persecution and encroachment. Despite continued threats to monk seals, conservation activities at the local scale that utilise lessons learned from previous failures and successes could secure the survival of the largest Mediterranean colony of monk seals, while also providing a model to support the species' recovery in other portions of its former range.

1. INTRODUCTION

Mediterranean monk seals (*Monachus monachus*) have been considered to be on the brink of extinction for about half a century. A telling narrative of their plight can begin with a 1965 International Union for the Conservation of Nature (IUCN) Red List assessment that told of the species being 'very rare and believed to be decreasing in numbers'; *M. monachus* later became assessed as Endangered in 1986, in 1988, in 1990 and in 1994, and successively as Critically Endangered in 1996, and again in 2008 and 2013.

Today there are hints that the species' condition may be changing. The status of Mediterranean monk seals is still extremely precarious overall; however, in specific locations in Madeira (Portugal), Mauritania and Greece, local colonies or breeding nuclei have halted their decreasing trends and are actually increasing in number. Based on this, the species' status on the 2015 IUCN Red List was down-listed to Endangered, after having been Critically Endangered for the previous 19 years (Karamanlidis and Dendrinos, 2015).

Unfortunately these signs of recovery, albeit encouraging, should not lead people into thinking that the conservation of Mediterranean monk seals has succeeded. In the specific locations, within the countries mentioned above, numbers have been stable or increasing because conditions are characterised by very low levels of human interaction. Although the observed localised improvements certainly provide renewed stimulus and encouragement to strive to consolidate the species' recovery, they cannot be taken as a reason for complacency. Outside these well-circumscribed locations, little has been achieved in terms of actions to effectively protect monk seals anywhere, nor has institutional neglect diminished significantly. The species' condition is still very critical, having been extirpated from the greater portion of its historical range. Localised increases in breeding colony size do not warrant a relaxation of urgent attention and effort to conserve this iconic species. The consolidation of monk seal recovery, particularly in the Mediterranean, will demand hard work and uncompromising determination from all the concerned parties.

A critical examination of what might have been, and still are, the factors at play in causing a sign of reversal in the conservation conditions of parts of the population strengthens strategies to conserve the species. Considering that the countless conservation measures invoked in endless official meetings during the past 50 years (Table 1) cannot be seen as responsible for any improvement, what other factors have played, and what factors are currently playing, a role in Mediterranean monk seal conservation?

Table 1 Main Meetings Dedicated to the Conservation of Mediterranean Monk Seals, *Monachus monachus*

Year	Location (Date)	Meeting	References
1972	Guelph, Canada (18–19 August)	IUCN working meeting of seal specialists on threatened and depleted seals of the world	Israëls (1992)
1974	London (5 October)	Monk seal meeting	Israëls (1992)
1976	Rome (May)	Meeting 'The monk seal along the Italian coasts: problems and perspectives for its positive protection'	Israëls (1992)
1978	Rhodes, Greece (2–5 May)	First International Conference on the Mediterranean monk seal	Ronald and Duguy (1979)
1979	Athens (11–13 October)	Conference on the protection of Greek flora–fauna biotopes	Israëls (1992)
1984	La Rochelle (5–6 October)	Second International Conference on the Mediterranean Monk Seal	Ronald and Duguy (1984)
1985	Port-Cros, France (13–14 June)	'Séminaire Intérnational sur la stratégie de conservation du phoque moine'	Israëls (1992)
1986	Strasbourg (15–16 September)	First meeting of the monk seal Expert Group convened by the Council of Europe	
1986	Bruxelles (30 October)	Meeting of experts on the Mediterranean monk seal held under the auspices of the Directorate of the Environment, Consumer Protection and Nuclear Safety Commission of the European Communities	
1987	Antalya, Turkey (2–6 November)	Third International Conference on the Mediterranean monk seal	

Continued

Table 1 Main Meetings Dedicated to the Conservation of Mediterranean Monk Seals, *Monachus monachus*—cont'd

Year	Location (Date)	Meeting	References
1988	Athens (11–12 January)	Joint expert consultation on the conservation of the Mediterranean monk seal, organised by UNEP/MAP in cooperation with IUCN (UNEP/MAP and IUCN 1988)	
1988	Port-Cros, France (26 May)	Meeting of the International Scientific Committee on the monk seal	Israëls (1992)
1988	Strasbourg, France (30–31 May)	Second meeting of the monk seal Expert Group convened by the Council of Europe	Israëls (1992)
1989	Madeira (20–22 September)	Meeting of coordination of national and international programmes on the conservation of the Mediterranean monk seal. Organised by the Council of Europe in coordination with UNEP-MAP-RAC/SPA, IUCN, CMS, the Portuguese Government and the Regional Government of Madeira	Israëls (1992)
1990	Bruxelles (6 November)	Sixth Meeting of the monk seal Specialist Group	Israëls (1992)
1990	Texel, The Netherlands (10–11 December)	'Urgent action meeting for safeguarding the Mediterranean monk seal as a species'	Israëls (1992)
1991	Antalya, Turkey (1–4 May)	Seminar on the conservation of the Mediterranean monk seal	Council of Europe (1991)
1994	Rabat, Morocco (7–9 October)	Meeting of experts on the evaluation of the implementation of the Action plan for the management of Mediterranean monk seals	UNEP-MAP-RAC/SPA (1994)
1998	Monaco (19–20 January)	The World Marine Mammal Science Conference. Workshop on the biology and conservation of the world's endangered monk seals, Monaco, 19–20 January 1998. The Society for Marine Mammalogy and The European Cetacean Society	

Table 1 Main Meetings Dedicated to the Conservation of Mediterranean Monk Seals, *Monachus monachus*—cont'd

Year	Location (Date)	Meeting	References
1998	Arta, Greece (29–31 October)	Meeting of Experts on the Implementation of the Action Plans for Marine Mammals (monk seal and cetaceans) adopted within MAP	UNEP-MAP-RAC/SPA (1998)
2002	Lattakia, Syria (29–30 September)	Meeting of experts on the conservation of the Mediterranean monk seal: proposal of priority activities to be carried out in the Mediterranean Sea	Bayed et al. (2003)
2006	Antalya, Turkey (17–19 September)	International Conference on monk seal conservation	UNEP-MAP-RAC/SPA (2006)
2008	Monaco (14 November)	First meeting of the Working Group: 'Reintroduction of the monk seal to the Western Mediterranean', organised by the Foundation Albert II, Prince of Monaco	
2009	Monaco (30 January)	Second meeting of the Working Group: 'Reintroduction of the monk seal to the Western Mediterranean', organised by the Foundation Albert II, Prince of Monaco	
2009	Istanbul (28 February)	'Who are our seals? Moving towards a standardised population estimate approach for *Monachus monachus*'. Workshop conducted within the framework of the European Cetacean Society Annual Conference, sponsored by the RAC/SPA and the Principality of Monaco	
2009	Maui, Hawaii (30 March–3 April)	First International Conference on Marine Mammal Protected Areas. Workshop on MMPAs and MMPA networks for monk seal conservation	Reeves (2009)
2010	Monaco (10 June)	Third meeting of the Working Group: 'Reintroduction of the monk seal to the Western Mediterranean', organised by the Foundation Albert II, Prince of Monaco	

Continued

Table 1 Main Meetings Dedicated to the Conservation of Mediterranean Monk Seals, *Monachus monachus*—cont'd

Year	Location (Date)	Meeting	References
2011	Martinique, French Antilles (9 November)	Second International Conference on Marine Mammal Protected Areas. Workshop on the conservation of monk seals	Hoyt (2012)
2014	Adelaide (9–11 November)	Third International Conference on Marine Mammal Protected Areas. Workshop on synergies between marine mammal conservation and Marine Spatial Planning	Hoyt (2015)
2015	Hawaii (6–17 April)	Workshop on the International Collaboration for the Conservation of Monk Seals, organised by the Joint Institute for Marine and Atmospheric Research JIMAR of the University of Hawaii in collaboration with the NOAA Pacific Island Fisheries Science Center (PIFSC)	
2016	Madeira, Portugal (12 March)	30th International Conference of European Cetacean Society. Workshop on the development of a conservation status surveillance system for monk seals	

This analysis of factors leading to conservation success for the species concentrates on monk seals in the Mediterranean. First, basic background on species status and main threats is provided. Past conservation efforts are then critically considered, outlining the many failures and the few successes. In addition to evaluating conservation efforts, other unanticipated factors which may have played an important role in slowing down, and even reversing, the condition of local populations are examined. Concluding remarks attempt to point out promising directions that may contribute to stabilising a budding recovery process in some areas, hopefully providing the stimulus and paving the way for expanding such recovery to other portions of the Mediterranean monk seal's former range.

2. BACKGROUND AND SPECIES STATUS

Monk seals (Phocidae: Monachinae) are a subfamily of pinnipeds living in tropical and warm temperate climates. Three species are recognised, recently separated into two genera: *Monachus* with one species, the Mediterranean monk seal, *M. monachus* (Hermann, 1779) (Fig. 1); and *Neomonachus* with two species, the Hawaiian monk seal, *N. schauinslandi* (Matschie, 1905) and the now extinct Caribbean monk seal, *N. tropicalis* (Gray, 1850).

Mediterranean monk seals are distributed in the Mediterranean Sea and in parts of the Northeast Atlantic Ocean, where two separate populations (a larger one along the west African coast occurring between Sahara and Mauritania, and a smaller nucleus in the Madeira Archipelago) are apparently increasing (Karamanlidis et al., 2015a). Monk seals in the Atlantic have been geographically separated from their Mediterranean conspecifics for sufficient time to develop noticeable morphological (Van Bree, 1979) and genetic (Karamanlidis et al., 2015b; Pastor et al., 2007) differences. Accordingly, monk seals in the Mediterranean are treated here as a 'unit to conserve' (Taylor, 2005), the conservation of which must be addressed independently from the populations living in the North Atlantic, with a separate assessment in the IUCN Red List.

Historically, the species was widely distributed throughout the Mediterranean Sea, the Sea of Marmara and the Black Sea. In the Sea of Marmara monk seals are limited today to a handful of individuals (Inanmaz et al., 2014), whereas in the Black Sea they are considered extirpated (Kiraç, 2001). The species' presence in the Mediterranean is currently severely depleted, with individuals largely confined to the region's northeastern corner.

Two countries, Greece and Turkey, stand out as the most important regional monk seal repositories today, where breeding nuclei still persist

Fig. 1 A Mediterranean monk seal, *Monachus monachus*, observed in the waters of Kalamos, Greek Ionian Sea, on 12 June 2015. *Photograph by Joan Gonzalvo/Tethys Research Institute.*

as the last remaining significant assets of the species in the Mediterranean. These sites should therefore be given the highest priority as far as regional conservation action is concerned. Due to the geography of the Aegean Sea, monk seals found in Greek and Turkish waters can be presumed to be intermixing as part of a single transboundary population. In contrast, low levels of mitochondrial heterogeneity (mtDNA diversity) have been detected between Aegean and Ionian monk seals (Karamanlidis et al., 2015b). Despite ongoing high-level concern regarding low total population numbers for monk seals in the Mediterranean, it is encouraging that numbers in north-eastern Mediterranean may have increased, since Marchessaux (1989) provided estimates more than a quarter century ago.

Based on Notarbartolo di Sciara et al. (2009b), and supplemented by more recent data, significant breeding concentrations of monk seals are known to exist in Greece in the following locations: Gyaros island (65–70 indiv., with a mean annual pup production of 9.75: Dendrinos et al., 2008; MOm, 2015); Northern Sporades islands (>50 individuals, with a mean annual pup production of >8); Kimolos and Polyaigos islands (<50 indiv., mean pups/year <8); North Karpathos and Saria islands (>20 indiv., mean pups/year <4); Ionian Islands: Kefallinia, Lefkada, Ithaca and Zakynthos (about 20 indiv.: Panou, 2009). In addition to the above locations, monk seals are widely, albeit in small numbers, distributed over the entire maritime territory of Greece, with pupping occurring in many places. This makes it extremely hard to produce a realistic total population estimate of monk seals in Greece.

In Turkey, monk seals are scattered along both the Aegean and Levantine coasts, from the Dardanelles to the border with Syria, with three main breeding concentrations (Güçlüsoy et al., 2004; Kıraç et al., 2011): Northern Aegean (35 indiv.); Southern Aegean (<30 indiv.) and Mediterranean coast (Levantine Sea) (>40 indiv.: Gücü et al., 2009b).

Cyprus is also known to host reproducing monk seals, although the relative importance of the contribution from the Republic of Cyprus is now minimal, with probably <10 individuals left (Demetropoulos, 2011); however, there is evidence that pupping still occurs (Marcou, 2015). In Northern Cyprus, 3–17 individuals were estimated in 2006–2007; a young seal observed there was likely to have been born locally (Gücü et al., 2009a). The recent evidence of breeding in Cyprus requires close attention, considering the very small and likely declining number of seals still present on that island.

Evidence that monk seals occur episodically elsewhere in the Mediterranean—although with no sign of breeding success—is provided by a remarkable number of recent sightings. This is powerful testimony to the species' potential for recolonising its former habitat in several countries, if given the chance. Sightings include (listed clockwise from the west):

- *Spain*. Reliable information exists of an individual photographed in 2008 at Isla del Toro, Majorca, Baleares; the first documented presence in European Spain in 50 years (Monachus Guardian, 2008). More sightings in the area were reported by Font and Mayol (2009) and summarised by Gazo and Mo (2012). By contrast, the small colony of seals known to have been surviving in the Chafarinas Archipelago, along the African coast, is presumed extinct (Monachus Guardian, 2004).
- *Italy*. Mo (2011) presented information on 81 observations documented between 1998 and 2010, corresponding to a minimum of 35 distinct sighting events. During the last two decades monk seals were sighted in Liguria, Tuscany, Sardinia, Latium, Sicily, Calabria and Apulia.
- *Croatia*. Based on numerous sighting reports, Antolovic et al. (2007) considered that monk seals were still present in Croatian coastal waters from 2000 to 2005, particularly around the offshore islands of the Dalmatian Archipelago. Gomerčić et al. (2011) listed 31 sightings of monk seals in Croatia since 2005, including an older mature adult female that was repeatedly photographed and filmed in the Kamenjak Natural Reserve, near the southern tip of the Istria peninsula, where the individual died (cause of death was diagnosed as chronic pneumonia) in August 2014 (Huber, 2014).
- *Albania*. Although very little information exists about the status of monk seal habitat in the country (Gazo and Mo, 2012; White et al., 2005), a recent documented sighting in the area south of Vlore on 4 August 2012 testifies to the presence of the species in the nation's coastal waters (Monachus Guardian, 2012).
- *Syria*. The continued presence of the species is mentioned by Mo et al. (2003) and Gücü (2004). Documented proof was provided by Jony and Ibrahim (2006), with a sighting 10 km north of Latakia in April 2005, and several reports by local fishermen. More recently, in 2013 a near-term pregnant female was shot by the coast guard in a non-specified location along the Syrian coast (Abou-Zahra, 2013).
- *Lebanon*. Two separate monk seal encounters were filmed underwater in Northern Lebanon, on 15 August and 4 September 2010, likely involving the same individual seal (Monachus Guardian, 2010).
- *Israel*. After an absence from the country of more than 50 years, monk seals were reported along the Israeli coast 45 times between November 2009

and September 2010; one report included photographs of a young female resting inside the breakwater of Herziliya Marina (Scheinin et al., 2011). Although it is uncertain if all the sightings were of a single individual or whether there were multiple animals present, Scheinin et al. (2011) suggested that there likely were at least two seals in the area at the time.

- *Egypt*. The species was considered as having disappeared from the country for about 20 years, but the presence of at least one monk seal was documented from Marsa Matrouh, western Egypt, in March 2011 (Notarbartolo di Sciara and Fouad, 2011a,b).
- Libyan Arab Jamahiriya, particularly Cyrenaica, apparently had an estimated 20 individuals around the 1970s, as reported by Sergeant et al. (1979). Although current numbers are unknown in spite of the considerable effort invested in surveys (Hamza et al., 2003), the recent finding (25 March 2012) of a dead young female in the area of Ain El Ghazala, near the border with Egypt, testifies to the continued presence of the species in that country (Alfaghi et al., 2013; Gazo and Mo, 2012).

Other Mediterranean countries where monk seals are presumed to still occasionally occur, although no recent sightings have been reported to our knowledge, include Tunisia (Ouerghi et al., 2001), Algeria (Boutiba, 2006; Gazo and Mo, 2012) and Morocco (Mo et al., 2011). However, and in stark contrast with the situation in the Eastern Mediterranean, the decline of the species has been particularly dramatic in north-west African countries, considering that only three decades ago estimates of monk seal numbers from that area probably exceeded 140 individuals, of which about 20 were in Tunisia (Marchessaux, 1986), 100 in Algeria (Marchessaux, 1977) and 20 in Morocco (Avella and Gonzalez, 1984; Marchessaux, 1989).

3. THREATS TO MEDITERRANEAN MONK SEALS

Threats to monk seal survival in the Mediterranean Sea have been listed in the past in minute detail, without engaging in an effort of ranking them in order of importance (e.g. Bayed et al., 2003; Israëls, 1992; Ronald, 1984; Ronald and Duguy, 1979; UNEP-MAP-RAC/SPA, 1994, 1998). For example, an expert meeting held in Latakia, Syria, in September 2002 listed no less than 21 types of different threats to monk seals (Bayed et al., 2003). Here, we suggest that it would be more effective to resist the doctrinaire temptation of making long lists of pressure factors, real or hypothetical, that may or may not threaten monk seal survival; such exercises, while perhaps academically interesting, carry dubious value in real-world management because they distract from concentrating on the heart of the matter. To a great extent, Mediterranean monk seals have declined because: (a) they are deliberately killed (mostly by fishermen); (b) their critical habitat, particularly during the breeding season, is being encroached by highly disruptive human presence; and (c) seals (mostly young individuals) drown when accidentally caught in fishing gear (Karamanlidis and Dendrinos, 2015).

Deliberate killing, notably by artisanal fishermen, is considered the single most important source of mortality for monk seals in the Mediterranean, as was first documented almost two decades ago by Androukaki et al. (1999). However, while killing of monk seals still unquestionably occurs, it is unclear whether this continues to be the most important threat, and a reassessment of its relative importance is recommended. Also, illegal fishing with dynamite (Fotis et al., 1988) still frequently takes place in monk seal habitat in the Eastern Mediterranean Sea, and although it is not necessarily occurring with the deliberate purpose of killing the seals, the technique is likely to have significant direct and indirect consequences on the species' status.

The second category of threats that significantly impact monk seals in the Mediterranean includes habitat deterioration, destruction and fragmentation, such as that caused by the construction of power plants, marine terminals, hotels and housing in pristine monk seal habitat, particularly in Turkey (Kiraç, 2011), or caused by tourists entering breeding caves during the reproductive season (e.g. Notarbartolo di Sciara, 2014). A third threat causing monk seal mortality is entanglement in fishing gear, which particularly affects subadult, inexperienced individuals (Karamanlidis et al., 2008; Kıraç et al., 2011; Veryeri et al., 2001).

The only historical pressure that no longer affects Mediterranean monk seals is exploitation for subsistence or commercial harvest for skins and oil (Israëls, 1992). Conversely, a number of other potential threats, such as disease outbreaks (e.g. morbillivirus), toxic algal blooms, prey depletion, contamination by pollutants and oil spills in critical habitat, have the potential of affecting monk seal survival. These potential threats could devastate monk seal conservation status, given the minuscule sizes of current monk seal nuclei scattered across a vast marine region.

Inbreeding of Mediterranean monk seals also raises concern. Based on levels of mtDNA diversity (haplotypic diversity: 0.03; variable sites: 0.6%), Karamanlidis et al. (2015b) found the Mediterranean monk seal to be 'one of the most genetically depauperate mammals on Earth', a condition thought to substantially increase the probability of extinction. However many mammal species, particularly pinnipeds such as northern elephant seals (*Mirounga angustirostris*) and Guadalupe fur seals (*Arctocephalus townsendi*), were capable of spectacular population recoveries despite having undergone through substantial genetic bottlenecks (Weber et al., 2004). The continued observed capability of monk seals in the Eastern Mediterranean for breeding and producing viable pups, able to grow and face the many threats the population is subjected to suggests that Mediterranean monk seals might have managed, for now, to escape from the Allee effect, whereby population decline below a critical size leads to recruitment failure and puts the species on an irreversible path towards extinction (Courchamp et al., 1999).

The relative importance of threats is not evenly distributed across the species' range, emphasising the need for tailoring strategic actions to local conditions on the basis of careful, site-specific analyses. For example, in Greece deliberate killings used to rank very high (Androukaki et al. 1999), whereas habitat degradation is naturally mitigated by the large extent of area made inaccessible (or of limited access), due to the geography and wind conditions of the widely scattered insular environment. Conversely, in Turkey the highest ranking threat is habitat degradation, which takes many different forms, from coastal development irreversibly destroying pristine coasts (Kiraç, 2011; Savas, 1999) to disturbance occurring within monk seal critical habitat, including overfishing and illegal fishing such as with dynamite, recreational boating, swimming, snorkelling and diving in prime habitat including breeding caves. Deliberate killing of monk seals was reported as significant in Turkey decades ago (Berkes et al., 1979); however, the dearth of similar reports in more recent times could be an artefact of the law being harshly enforced in that country.

In conclusion, regardless of threat types and their relative importance across the species range, the obvious common thread to all pressures is their derivation from human activities. This emphasises the importance of addressing with the greatest care the socioeconomic aspects of monk seal conservation.

4. WHY HAS MONK SEAL CONSERVATION WORKED SO POORLY IN THE PAST

It was only in the very recent past that observations of localised monk seal nuclei have showed a halt in decline, and even some signs of recovery. And yet, international concern for Mediterranean monk seal conservation dates from at least the 1970s, as testified by the impressive list of meetings dedicated to the cause, during which all the possible resolutions and recommendations on monk seal conservation had been excogitated and formally, repeatedly adopted (see Table 1). Monk seal action plans and conservation strategies have been drafted and adopted in international fora (e.g. GFCM, 2011; IUCN, 2009; UNEP-MAP-RAC/SPA, 2005; UNEP-MAP-RAC/SPA and IUCN, 1988) as well as at the national level; for example in Algeria (UNEP-MAP-RAC/SPA, 2006), Cyprus (Demetropoulos, 2011), Egypt (Notarbartolo di Sciara and Fouad, 2011a), Greece (MOm, 1996, superseded by Notarbartolo di Sciara et al., 2009a; MOm, 2009), and Turkey (Kıraç et al., 2011). Unfortunately, declarations, action plans, resolutions and recommendations, year after year, are now collecting dust without the surviving monk seals being able to take much notice. Even a resolution formally adopted by the Contracting Parties to the Barcelona Convention at their 18th Meeting (Istanbul, 3–6 December 2013), requesting the Parties themselves to 'take the necessary measures for the implementation of the' ... 'Regional Strategy for the conservation of Monk Seals in the Mediterranean' (Notarbartolo di Sciara, 2013), 'and report on their implementation according to the cycle and format of the MAP reporting system' still awaits implementation.

As is the case in many similar failures in environmental conservation, there has been a widespread inability by the concerned institutions to close the gap between 'knowing' and 'doing'. In fact, knowledge gaps are no longer a problem, and the old pretext of stalling action on the basis of 'not knowing enough' about the species' ecology and status no longer stands. Ecological, biological and veterinary medical knowledge, although still leaving ample room for improvement, is today substantive and helpful enough to implement successful conservation measures. Threats are well identified, and the measures to address them understood. We know beyond any possible doubt the factors that are causing the monk seals' decline, and the path to addressing them is evident. Not even regulatory insufficiency can be blamed, given that legal provisions at all possible levels (national, regional, European and international) could not be more complete. The problem solely concerns the willingness and ability to enact the recommended conservation measures all the stakeholders were resolved to take (Notarbartolo di Sciara, 2010). We suggest that the resounding failure in securing monk seal survival in the Mediterranean can be explained by three main reasons.

First, the governmental unwillingness or inability to implement legal commitments in terms of conservation and sustainable use of marine resources certainly remains at the forefront of the Mediterranean monk seal conservation debacle. Too much attention has been devoted to extensively listing pressure factors, and too little to addressing factors central to monk seal decline, to endeavouring to affect the human behaviours that are at the source of decline, and to understanding how can the problems be addressed and solved in practice. For example, conferring effective protection to the last 'pristine' remnants of monk seal critical habitat through the establishment of Marine Protected Areas (MPAs) has been in many cases grossly inadequate, and in some of the very few places where such protection was decreed on paper (in a typical 'top-down' fashion), effective protection has remained in the realm of unfulfilled intentions due to lack of management, management capacity, allocation of human and financial resources, monitoring and enforcement, and most relevantly due to the absence of local interest and support (Karamanlidis et al., 2015a; Oikonomou and Dikou, 2008). Management failures, within and outside protected areas, include even simple and apparently straightforward actions such as enforcing the prohibition of carrying firearms aboard fishing vessels or of fishing illegally with explosives, still rampant in core monk seal habitat. In spite of deliberate killing of seals being recognised as a major cause of the species' impending extinction, and considering that killing a monk seal is a criminal offence in the legislation of most Mediterranean nations, to the best of our knowledge no one in the Mediterranean was ever convicted for this crime. Even more importantly, many pressure factors on monk seals, such as those deriving from fishing and fish farming activities or from the degradation of the marine environment, are elements that all coastal nations are supposed to address within their recognised commitment of ensuring that human activities at sea be sustainably managed. Failure to effectively pursue the sustainability of fisheries and the good health of the seas is still a serious flaw in Mediterranean marine governance, and has dire socioeconomic implications. The loss of species, even charismatic ones such as monk seals, is just one of the many consequences of this malaise. Therefore, while combating overfishing, illegal fishing and marine pollution remains an action of paramount importance in terms of monk seal conservation concerns, this should be implemented as part of each nation's coastal and marine management and conservation duties rather than as part of a monk seal conservation strategy.

Second, efforts to conserve the Mediterranean monk seal, a natural asset shared by all the region's coastal states, have sorely lacked in coordination and continuity, in spite of the commendable prescriptions contained in the Monk Seal Action Plan adopted within the framework of the Barcelona Convention (UNEP-MAP-RAC/SPA, 2003). While conserving the marine environment and the protected species living in it is an individual duty recognised by all single nations, international cooperation is of foremost importance particularly in the case of species having transboundary ranges. Yet, too many action plans have been produced that have remained on paper instead of becoming the backbone of a concerted effort, seeing the active involvement and cooperation of all the components of Mediterranean civil society at large, public and private, national and international. Funds for monk seal conservation have been allocated in a piecemeal fashion instead of being invested to support a science-based, long-term, region-wide strategy. The need for coordination is particularly acute in an arena that sees so many players, as well as many international bodies, taking interest in highly mobile animals that are not necessarily confined to waters within the jurisdiction of any single nation (e.g. Adamantopoulou et al., 2011). Monk seals offer an exemplary case in which conservation needs cooperation amongst range states and concerned international bodies, which include, in addition to the Barcelona Convention, the Convention on Migratory Species (which lists Mediterranean monk seals in its Appendix I), the General Fisheries Commission for the Mediterranean (e.g. GFCM, 2011), the Bern Convention (Mediterranean monk seals listed in its Appendix II) and the European Union (which lists Mediterranean monk seals as *priority species* in Annex II of Council Directive 92/43/EEC, known as the 'Habitats Directive'). Although the greatest achievements in monk seal conservation in the Mediterranean during the recent decades, largely in the realms of scientific knowledge and public awareness, were secured thanks to the laudable commitment of a handful of non-governmental organizations (NGOs) often benefiting from international (in most cases, European) support, in the end the dearth of institutional interest and leadership from within the concerned nations has failed to create a common, coordinated reference base. The lack of institutional commitment, that could have been conducive to a climate of constructive cooperation towards a shared goal and might have capitalised on a wide popular basis of expertise, passion and enthusiasm, ultimately eroded civil society's goodwill and squandered priceless human capital.

Third, until now the overwhelming emphasis of monk seal conservation actions has focused on ecology and biology rather than on the human interactions. However, the root of monk seal conservation has a social rather than an ecological nature, because problems for the species are derived solely from its devastating interactions with people. Those who have long been involved in monk seal conservation—naturalists, biologists, ecologists and veterinarians—urgently need to collaborate with social scientists, economists, as well as policy, legal, media and education experts, if actions are to become more incisive where the problems are most acute. Solving monk seal conservation problems must be perceived as residing in, and fully coinciding with, the solution of the wider environmental and socioeconomic problems of the human communities involved. However, merely advocating greater stakeholder participation is insufficient, because participation will not happen by itself. It is, to a large extent, from within such communities that a viable solution to monk seal conservation problems must be consolidated, but for this to happen an intense effort must be implemented to raise awareness amongst local stakeholders on the advantages of a healthy marine environment, and on the need for a proactive engagement to keep that environment, with all of its natural components, in a healthy state.

5. FACTORS THAT HAVE HELPED MONK SEAL CONSERVATION

Against the background of major shortcomings in the half century-long efforts to conserve monk seals in the Mediterranean, considering that the species not only has not gone extinct yet, but has even recently managed in a few locations to reverse its decline (Karamanlidis and Dendrinos, 2015), one wonders what factors might have been at play (in the face of generalised institutional inertia), to cause this surprising turn of events.

Some of these factors are the result of tireless activism by committed individuals and organisations that have surrogated institutional duties. Key actions include provision of basic scientific knowledge, spearheading of effective establishment and management of MPAs, rehabilitation of monk seal pups, and building public awareness. Instigated by civil society's pressure, some of the MPAs that have been established in the Eastern Mediterranean (e.g. Northern Sporades and Karpathos in Greece, Foça in Turkey) are concrete conservation measures that are producing results, at a minimum contributing to the conservation of important parcels of critical monk seal habitat. Indicative of the success of a concerted effort at a local scale, which included contributions from local fishermen and the NGO community, is the surveillance system that operated continuously for more than a decade in the Northern Sporades, actively enforcing a 'no take zone' in the core area of the park and allowing the local monk seal population to breed undisturbed (Karamanlidis et al., 2004; MOm, 2005b). Similarly, a programme of rescue, rehabilitation and release of orphaned pups in Greece has led to the recovery of more than 20 seals. This may have significantly supported the species' survival considering the very small size of the population of origin (Androukaki et al., 2003; MOm, 2005a). Although no long-term data on released pup survival exist to demonstrate the programme's ecological conservation value, at a minimum, it has contributed significantly to popular awareness in Greece. Actions were further strengthened in Greece by a long series of awareness programmes over the last two decades, culminating with recent nationwide campaign that resulted in a significant increase in the general public's appreciation of the plight of the monk seal and of other marine mammals, and on the need to conserve these species (MOm, 2014).

However, awareness and *in situ* conservation actions by elements of civil society may have also been greatly facilitated by other factors that are totally independent from the conservation efforts discussed earlier. First, the relative importance of artisanal fishing in many island economies has been declining in recent years (Gonzalvo et al., 2014; Moutopoulos and Koutsikopoulos, 2014), in large part due to a combination of dwindling revenues and increasing costs that is inducing younger generations to abandon their parents' traditional professions in favour of employment in the tourism industry. This is causing a significant decrease in opportunities for fishers to interact with monk seals, and consequently a decrease in impact on monk seal survival.

Second, positive changes in public opinion regarding environmental stewardship and animal welfare, coupled with the socioeconomic changes affecting even small island communities, have likely caused a greater resistance to killing monk seals. Langford et al. (2001) investigated the use and nonuse values for the Mediterranean monk seals using a survey conducted in Mytilene, on the island of Lesvos, Greece. The research showed that even people with potential economic motives for self-interest regarding the monk seals were able to consider wider issues and to contemplate the ethical and temporal implications of their choices in relation to the natural environment; the analysis also showed that younger people were more likely to pledge 'willingness to pay' amounts to enable them to continue to experience the monk seals personally.

Third, and most relevantly, large portions of monk seal habitat are generally inaccessible to humans, and this may have offered the species refuge that has been essential for the population to avoid extinction. Such refuge has been offered by: (a) areas made inaccessible by law because of the presence of prisons or military establishments (e.g. the island of Gyaros), and (b) the geographic characteristics of the monk seal habitat, particularly in the Aegean

Sea, where thousands of remote, uninhabited islets—becoming particularly inaccessible during the windy summers—offer at least partial refuge from human encroachment and disturbance. In reality the Aegean environment has acted as a natural reservoir for monk seal conservation at the species level, creating a buffer to maintain a core of the species waiting for other conditions to favourably change. Now, perhaps, that time has come.

6. THE WAY FORWARD THROUGH LESSONS LEARNED

Here we present a local scale case study, taking stock of the main reasons why Mediterranean monk seal conservation has failed in the past, to implement innovative monk seal conservation. The case study example provided here is particularly relevant because it addresses conservation of the monk seal colony of the island of Gyaros, in the Cyclades Archipelago in Greece, the largest in the Mediterranean. A four-year project known as CYCLADES LIFE was launched in 2013, adopting a new and challenging conservation approach that includes the implementation and effective management of a newly established MPA around Gyaros. Of critical importance, conservation strategies and MPA design are being implemented based on experience and knowledge gained from MPAs established elsewhere in the world (e.g. Chuenpagdee et al., 2013; Gleason et al, 2013; Portman et al., 2013). Key elements of the CYCLADES LIFE project include:

1. *Mobilisation of, and commitment from, the relevant entities at the international, European, national, regional and local levels from the start of the initiative.* The CYCLADES LIFE project is jointly implemented by several public and private partners, including Greek national agencies, regional authorities, local development agencies, environmental NGOs and academic institutions, and is directly supported by the European Union and other international environmental foundations. Most relevantly, in addition to the formal engagement of authorities at all levels, the area's users and stakeholders are also directly involved in key conservation activities such as research and monitoring, surveillance and enforcement, ecosystem-based and marine spatial planning processes and education and awareness.
2. *Continuity and coordination of the conservation measures.* In spite of the formal four-year duration of the CYCLADES LIFE project, the overall scope of the endeavour is based on a continuous long-term effort (initiated in 2001) to effectively manage the MPA that includes the island of Gyaros and adjacent waters. Conservation and management activities planned

and implemented within the project's framework envisage their sustainability in the future. For instance, new surveillance and monitoring technologies (e.g. land-mounted radar, remote sensing, infrared cameras, unmanned aerial vehicles) tested and deployed during the course of the project will remain as an endowment to future MPA management, and new approaches in management are being utilised to minimise the resources required, thereby ensuring that conservation activities are sustained at least in the medium term. By framing the future MPA within the Aegean seascape, the project also envisages the adoption of ecosystem-based management, and future inclusion of the MPA within a wider marine spatial planning process requiring the involvement of various expertise and stakeholders.

3. *Inclusion of the human component and the social aspect of the wider Gyaros area in the CYCLADES LIFE initiative since its inception.* Surveys and analyses of the concerned local communities conducted by social scientists are a backbone of the project, and have identified and inventoried human uses of the project area, recorded existing and potential conflicts, interests and aspirations and assessed local attitudes towards nature conservation in general and towards monk seals in particular. The results of the social analyses create the centrepiece for the design of the future MPA and of the conservation measures to be implemented. Most importantly, MPA design and its future management is being foreseen by the project with a strong participatory character through the establishment of a Consortium of Stakeholders, thereby avoiding the traditional MPA top-down approach that has consistently alienated local communities in Greece and in the wider Mediterranean. As a collective public/private body, with membership from state, regional, and local authorities, local users, NGOs and scientists, the Consortium will be involved in the design of the MPA boundaries and zoning, and in the development of management actions supporting both marine biodiversity conservation and sustainable local development.

If the approach adopted and the governance scheme proposed in the CYCLADES LIFE project will succeed in protecting the local monk seal population and the overall marine environment of the area, while at the same time promoting and supporting the livelihoods of the neighbouring local societies through the development of alternative sources of income, it is likely that the conservation of the most important monk seal concentration in the Mediterranean will be secured in the long term. Furthermore, in parallel to continuing the actions in Gyaros and to strengthening what was accomplished until now, most notably in terms of stakeholder involvement, lessons learned in Gyaros could inform monk seal conservation efforts in other Mediterranean locations containing suitable monk seal habitat, where the species has been extirpated from in historical times. In many of such locations, such as the Egadi islands and north-eastern Sardinia in Italy (Mo, 2011), and the numerous islands and islets bordering on the Croatian coast (Gomerčić et al., 2011), sporadic appearances of monk seal individuals still occur to the present day, and targeted efforts could accelerate ongoing attitudinal changes by the local human communities, thereby facilitating recolonisation by the monk seals of their former habitats.

7. WHY CONSERVING MONK SEALS TRANSCENDS THE SPECIES' IMPORTANCE

Emboldened by the first significant signs that the species is capable of recovery, even though improvement is currently limited to a handful of circumscribed examples, we concur with Redford et al. (2011) that successful conservation of monk seals in the Mediterranean should extend beyond the simple goal of avoiding its extinction. For a real recovery a condition should be reached whereby multiple self-sustaining, healthy and genetically robust monk seal populations are maintained across the species' regional range, so that these populations will become resilient to environmental change, as well as to a human encroachment, which is not expected to be significantly abating in the foreseeable future. Furthermore, facilitating the recovery of monk seals in the Mediterranean Sea would go far beyond the already important imperative of conserving a charismatic element of Mediterranean marine biodiversity.

Faced with the prospect of investing the considerable amount of time, effort and resources needed to reverse the critical conservation status of monk seals in the Mediterranean, many could find it legitimate to question the ethical aspects of dedicating far greater attention to a single species than is given to most of the region's other marine organisms. By contrast, we suggest that dedicating extraordinary attention and resources to monk seal conservation is indeed legitimate for many important reasons, in addition to the simple fact that protecting monk seals is a legal requirement wherever they occur.

One reason in favour of protecting monk seals in the Mediterranean involves the species' nonconsumptive use value: (a) as an apex predator concurring to the maintenance of the ecological functionality of coastal ecosystems that are already heavily perturbed (Bianchi et al., 2014), (b) as a potential ally in combating the diffusion of noxious alien fish species such as siganids (Giakoumi, 2013), and (c) as a resource for a carefully managed ecotourism (Oikonomou and Dikou, 2008; Sakinan, 2004). Also option value—i.e., a means of assigning a value to risk aversion in the face of uncertainty—and existence value—i.e., value attached to the existence of a species or habitat by someone who may have no intention of ever visiting or using it (McNeely, 1988)—are at play in determining the importance of conserving monk seals. Langford et al. (2001) reported on surveys on nonuse values of the highly charismatic Mediterranean monk seals, where the significant presence of option and existence values was detected amongst respondents.

Furthermore, protecting monk seals can also be important because actions favouring monk seal conservation can extend their benefits to other co-occurring species and to the environment they are part of, given the monk seals' potential quality of umbrella species (Roberge and Angelstam, 2004). For instance, protected areas specifically designated to foster monk seal conservation are likely to support the maintenance or restoration of the good environmental status of concerned coastal and marine habitats with cascading benefits on many other species.

Finally, preventing the extinction of charismatic monk seals in the Mediterranean also carries symbolic and political significance because biodiversity is likely to become increasingly recognised as a public good (Rands et al., 2010), and within such frame of mind monk seal extinction would create a devastating loss of regional institutional credibility. This is why a forceful monk seal conservation strategy, embraced regionally as a best practice example, must be solidly integrated within a wider strategy for the conservation of the Mediterranean marine environment (Notarbartolo di Sciara, 2013). While conserving monk seals and their habitat in the Mediterranean is an obligation that the region's nations have explicitly committed to on the basis of many national, regional, international and, where appropriate, European legal instruments, the species' future will be secured only if (a) the region's extended civil society will attribute to the seals the value they deserve, and (b) saving monk seals from extinction will be seen as the epitome of the effort of reversing the devastating trend of loss of naturalness that is plaguing the Mediterranean Sea. Ideally, monk seals will become the symbol of a renewed effort towards Mediterranean marine conservation.

ACKNOWLEDGEMENTS

First we wish to acknowledge the contribution from three reviewers, who improved substantively the manuscript with their comments and suggestions. The CYCLADES LIFE project mentioned in this chapter is implemented with the contribution of the LIFE financial instrument of the European Union and with the support of the Prince Albert II of Monaco Foundation. Additional conservation actions within the CYCLADES initiative were supported by the Blue Marine Foundation and the Fondation Ensemble. We wish to thank UNEP MAP's Regional Activity Centre for specially protected areas for permitting the use of elements from the introductory part of the 'Regional strategy for the conservation of Monk Seals in the Mediterranean (2014–2019)' (Notarbartolo di Sciara, 2013).

REFERENCES

Abou-Zahra, A., 2013. Rare and Pregnant Monk Seal Falls Victim to Arab Instability. International Fund for Animal Welfare. Available from: http://www.ifaw.org/united-kingdom/news/rare-and-pregnant-monk-seal-falls-victim-arab-instability.

Adamantopoulou, S., Androukaki, E., Dendrinos, P., Kotomatas, S., Paravas, V., Psaradellis, M., Tounta, E., Karamanlidis, A.A., 2011. Movements of Mediterranean monk seals (*Monachus monachus*) in the eastern Mediterranean Sea. Aquat. Mamm. 37 (3), 256–261. http://dx.doi.org/10.1578/AM.37.3.2011.256.

Alfaghi, I.A., Abed, A.S., Dendrinos, P., Psaradellis, M., Karamanlidis, A.A., 2013. First confirmed sighting of the Mediterranean monk seal (*Monachus monachus*) in Libya since 1972. Aquat. Mamm. 39 (1), 81–84. http://dx.doi.org/10.1578/AM.39.1.2013.81.

Androukaki, E., Adamantopoulou, S., Dendrinos, P., Tounta, E., Kotomatas, S., 1999. Causes of mortality in the Mediterranean monk seal (*Monachus monachus*) in Greece. Contrib. Zoogeogr. Ecol. East. Medit. Reg. 1, 405–411.

Androukaki, E., Fatsea, E., Hart, L., Osterhaus, A.D.M.E., Tounta, E., Komnenou, N., 2003. Treating stranded Mediterranean monk seal pups in Greece. In: Proceedings of the 15th Biennial Conference on the Biology of Marine Mammals, Society for Marine Mammals, Greensboro, North Carolina, USA.

Antolovic, J., Antolovic, M., Antolovic, N., Furlan, B., Adamic-Antolovic, L., Antolovic, R., Cok, I., 2007. Monk seal (*Monachus monachus*) sightings in the Croatian part of the Adriatic with a special reference to the population of open-sea islands. Monachus Guardian 10 (1).

Avella, F.J., Gonzalez, L.M., 1984. Monk seal (*Monachus monachus*): a survey along the Mediterranean coast of Morocco. In: Ronald, K., Duguy, R. (Eds.), Les phoques moines—Monk Seals. Proceedings of the Second International Conference, La Rochelle, France, 5–6 October. Ann. Soc. Sci. Nat. Charente-Maritime, Suppl., décembre 1984, pp. 60–78.

Bayed, A., Gücü, A., Mo, G., Dendrinos, P., 2003. The conservation of the Mediterranean monk seal: proposal of priority activities to be carried out in the Mediterranean Sea. In: Sixth Meeting of National Focal Points for SPAs, Marseilles, 17–20 June. UNEP (DEC)/MED WG.232/Inf 6, 45 p.

Berkes, F., Anat, H., Kislalioglu, M., Esenel, M., 1979. Distribution and ecology of *Monachus monachus* on Turkish coasts. In: Ronald, K., Duguy, R. (Eds.), The Mediterranean Monk Seal. Proceedings of the First International Conference, Rhodes, Greece, 2–5 May 1978. In: UNEP Technical Series, vol. 1. Pergamon Press, Oxford, pp. 113–128.

Bianchi, C.N., Corsini-Foka, M., Morri, C., Zenetos, A., 2014. Thirty years after: dramatic change in the coastal marine ecosystems of Kos Island (Greece), 1981–2013. Mediterr. Mar. Sci. 15 (3), 482–497. http://dx.doi.org/10.12681/mms.678.

Boutiba, Z., 2006. Propositions d'actions concrètes pour la mise en oeuvre d'un plan de conservation et gestion des sites identifiés comme critiques potentiels pour le phoque moine sur le littoral ouest algérien: Report Submitted to the Regional Activity Centre for Specially Protected Areas (RAC/SPA), Tunis. 45 p.

Chuenpagdee, R., Pascual-Fernandez, J.J., Szelianszky, E., Alegret, J.L., Fraga, J., Jentoft, S., 2013. Marine protected areas: re-thinking their inception. Mar. Policy 39, 234–240.

Council of Europe, 1991. In: Seminar on the Conservation of the Mediterranean Monk Seal: Technical and Scientific Aspects, Antalya, Turkey, 1–4 May. T-PVS (91)25:1–94.

Courchamp, F., Clutton-Brock, T., Grenfell, B., 1999. Inverse density dependence and the Allee effect. Trends Ecol. Evol. 14 (10), 405–410.

Demetropoulos, A., 2011. National action plan for the conservation of the Mediterranean monk seal in Cyprus. Contract RAC/SPA: N20/RAC/SPA_2011, 24 p.

Dendrinos, P., Karamanlidis, A.A., Kotomatas, S., Paravas, V., Adamantopoulou, S., 2008. Report of a new Mediterranean monk seal (*Monachus monachus*) breeding colony in the Aegean Sea, Greece. Aquat. Mamm. 34 (3), 355–361. http://dx.doi.org/10.1578/AM.34.3.2008.355.

Font, A., Mayol, J., 2009. Mallorca's lone seal: the 2009 follow-up. Monachus Guardian 12 (2). Available from: http://www.monachus-guardian.org/library/mguard24.pdf.

Fotis, G., Paleokostas, A., Evagelidis, D., 1988. Fishing with dynamite and other toxic substances. In: Proceedings of the 4th Hellenic Ichthyological Congress, vol. 194, p. 7.

Gazo, M., Mo, G., 2012. Action Plan for the Conservation/Management of the Monk Seal in Low Density Areas of the Mediterranean. Regional Activity Centre for Specially Protected Areas (RAC/SPA), Tunis. MoU n. 34/RAC/SPA_2011, 29 p.

GFCM, 2011. Recommendation GFCM/35/2011/5 on fisheries measures for the conservation of the Mediterranean monk seal (*Monachus monachus*) in the GFCM Competence Area: Report of the General Fisheries Commission for the Mediterranean's 35th Session, Rome. 3 p.

Giakoumi, S., 2013. Distribution patterns of the invasive herbivore *Siganus luridus* (Rüppell, 1829) and its relation to native benthic communities in the central Aegean Sea, Northeastern Mediterranean. Mar. Ecol. 2013, 1–10. http://dx.doi.org/10.1111/maec.12059.

Gleason, M., Fox, E., Ashcraft, S., Vasques, J., Whiteman, E., Serpa, P., Saarman, E., Caldwell, M., Frimodig, A., Miller-Jenson, M., Kirlin, J., Ota, B., Pope, E., Weber, M., Wiseman, K., 2013. Designing a network of marine protected areas in California: achievements, costs, lessons learned, and challenges ahead. Ocean Coast. Manag. 74, 90–101.

Gomerčić, T., Huber, D., Đuras Gomerčić, M., Gomerčić, H., 2011. Presence of the Mediterranean monk seal (*Monachus monachus*) in the Croatian part of the Adriatic Sea. Aquat. Mamm. 37 (3), 243–247. http://dx.doi.org/10.1578/AM.37.3.2011.243.

Gonzalvo, J., Giovos, I., Moutopoulos, D.K., 2014. Fishermen's perception on the sustainability of small-scale fisheries and dolphin–fisheries interactions in two increasingly fragile coastal ecosystems in western Greece. Aquat. Conserv. 25 (1), 91–106. http://dx.doi.org/10.1002/aqc.2444.

Gray, J.E., 1850. Catalogue of the Specimens of Mammalia in the Collections of the British Museum. Part 2. Seals. British Museum of Natural History, London.

Güçlüsoy, H., Kýraç, C.O., Veryeri, N.O., Savas, Y., 2004. Status of the Mediterranean monk seal, *Monachus monachus* (Hermann, 1779) in the coastal waters of Turkey. E.U. J. Fish. Aquat. Sci. 21 (3–4), 201–210.

Gücü, A.C., 2004. Is the broken link between two isolated colonies in the Northeastern Mediterranean re-establishing? Monachus Guardian 7 (2). Available from: http://www.monachus-guardian.org/library/mguard14.pdf.

Gücü, A.C., Ok, M., Sakinan, S., 2009a. A survey of the critically endangered Mediterranean monk seal *Monachus monachus* (Hermann, 1779) along the coast of Northern Cyprus. Isr. J. Ecol. Evol. 55 (1), 77–82. http://dx.doi.org/10.1560/IJEE.55.1.77.

Gücü, A.C., Sakinan, S., Ok, M., 2009b. Occurrence of the critically endangered Mediterranean monk seal, *Monachus monachus* (Hermann, 1779), at Olympos-Beydaglarý National Park, Turkey. Zool. Middle East 46, 3–8.

Hamza, A., Mo, G., Tayeb, K., 2003. Results of a preliminary mission carried out in Cyrenaica, Libya, to assess monk seal presence and potential coastal habitat. Monachus Guardian 6 (1). Available from: http://www.monachus-guardian.org/library/mguard11.pdf.

Hermann, J., 1779. Beschreibung der Münchs-Robbe. Beschäft. Berlinischen Ges. Naturf. Fr. 4, 456–509.

Hoyt, E. (Ed.), 2012. Proceedings of the Second International Conference on Marine Mammal Protected Areas (ICMMPA 2), Fort-de-France, Martinique, 7–11 November 2011, p. 103.

Hoyt, E. (Ed.), 2015. Summary Proceedings of the Third International Conference on Marine Mammal Protected Areas (ICMMPA 3), Adelaide, Australia, 9–11 November 2014, p. 85.

Huber, D., 2014. Note on an old female monk seal that died in Croatia in 2014. Monachus Guardian. Available from: http://www.monachus-guardian.org/wordpress/2014/10/29/note-on-an-old-female-monk-seal-that-died-in-croatia-in-2014/. 29 October 2014.

Inanmaz, O.E., Degirmenci, O., Gücü, A.C., 2014. A new sighting of the Mediterranean Monk Seal, *Monachus monachus* (Hermann, 1779), in the Marmara Sea (Turkey). Zool. Middle East 60, 278–280. http://dx.doi.org/10.1080/09397140.2014.944438.

Israëls, L.D.E., 1992. Thirty years of Mediterranean monk seal conservation, a review. Mededelingen, vol. 28 Nederlandsche Commissie voor Internationale Natuurbescherming, Amsterdam, Netherlands, pp. 1–65.

IUCN, 2009. Resolution 4.023. Conservation and recovery of the Mediterranean monk seal *Monachus monachus*. In: Resolutions and Recommendations. IUCN, Gland, Switzerland, pp. 23–24. 158 p.

Jony, M., Ibrahim, A., 2006. The first confirmed record for Mediterranean monk seals in Syria. Abstract, p. 54, In: UNEP/MAP, RAC/SPA. Report of the International Conference on Monk Seal Conservation, Antalya, Turkey, 17–19 September, p. 69.

Karamanlidis, A., Dendrinos, P., 2015. *Monachus monachus*. In: The IUCN Red List of Threatened Species http://dx.doi.org/10.2305/IUCN.UK.2015-4.RLTS.T13653A45227543.en. Downloaded on 28 December 2015.

Karamanlidis, A.A., Dendrinos, P., Tounta, E., Kotomatas, S., 2004. Monitoring human activity in an area dedicated to the protection of the endangered Mediterranean monk seal. Coast. Manag. 32, 293–306.

Karamanlidis, A.A., Androukaki, E., Adamantopoulou, S., Chatzispyrou, A., Johnson, W.M., Kotomatas, S., Papadopoulos, A., Paravas, V., Paximadis, G., Pires, R., Tounta, E., Dendrinos, P., 2008. Assessing accidental entanglement as a threat to the Mediterranean monk seal *Monachus monachus*. Endanger. Species Res. 5, 205–213. http://dx.doi.org/10.3354/esr00092.

Karamanlidis, A.A., Dendrinos, P., Fernández de Larrinoa, P., Gücü, A.C., Johnson, W.M., Kiraç, C.O., Pires, R., 2015a. The Mediterranean monk seal *Monachus monachus*: status, biology, threats, and conservation priorities. Mammal Rev. 46 (2), 92–105. http://dx.doi.org/10.1111/mam.12053.

Karamanlidis, A.A., Gaughran, S., Aguilar, A., Dendrinos, P., Huber, D., Pires, R., Schultz, J., Skrbinšek, T., Amato, G., 2015b. Shaping species conservation strategies using mtDNA analysis: the case of the elusive Mediterranean monk seal (*Monachus monachus*). Biol. Conserv. 194, 71–79. http://dx.doi.org/10.1016/j.biocon.2015.11.014.

Kiraç, C.O., 2001. Witnessing the monk seal's extinction in the Black Sea. Monachus Guardian 4 (2), 1–3. Available from: http://www.monachus-guardian.org/library/mguard08.pdf.

Kiraç, C.O., 2011. Conservation of the Mediterranean monk seal *Monachus monachus* in Turkey and the role of coastal and marine protected areas. In: Hoyt, E. (Ed.), Proceedings of the Second International Conference on Marine Mammal Protected Areas, vol. 41. Agency des Aires Marines Protégés/NOAA, Fort-de-France, Martinique, France/USA.

Kıraç, C.O., Veryeri, N.O., Güçlüsoy, H., Savaş, Y., 2011. National Action Plan for the Conservation of Mediterranean Monk Seal *Monachus monachus* in Türkiye. UNEP-MAP-RAC/SPA, Tunis. 35 p.

Langford, I.H., Skourtos, M.S., Kontogianni, A., Day, R.J., Georgiou, S., Bateman, I.J., 2001. Use and nonuse values for conserving endangered species: the case of the Mediterranean monk seal. Environ. Plan. A 33, 2219–2233. http://dx.doi.org/10.1068/a348.

Marchessaux, D., 1977. Will the Mediterranean monk seal survive? Aquat. Mamm. 5 (3), 87.

Marchessaux, D., 1986. Etude de l'evolution du statut du phoque moine en Tunisie et dans l'archipel de la Galite. Regional Activity Centre for Specially Protected Areas (RAC/SPA)/IUCN, Tunis/Gland, Switzerland. 25 p.

Marchessaux, D., 1989. Distribution et statut des populations du phoque moine *Monachus monachus* (Hermann, 1799). Mammalia 53 (4), 621–642.

Marcou, M., 2015. The Mediterranean monk seal *Monachus monachus* in Cyprus. Monachus Guardian. Available from: http://www.monachus-guardian.org/wordpress/2015/05/21/the-mediterranean-monk-seal-monachus-monachus-in-cyprus/. 21 May 2015.

Matschie, P., 1905. Eine Robbe von Laysan. Sitz. Ber. Ges. Naturf. Fr. Berl. 1905, 254–262.

McNeely, J.A., 1988. Economics and Biological Diversity: Developing and Using Economic Incentives to Conserve Biological Resources. IUCN, Gland, Switzerland.

Mo, G., 2011. Mediterranean monk seal (*Monachus monachus*) sightings in Italy (1998–2010) and implications for conservation. Aquat. Mamm. 37 (3), 236–240. http://dx.doi.org/10.1578/AM.37.3.2011.236.

Mo, G., Gazo, M., Ibrahim, A., Ammar, I., Ghanem, W., 2003. Monk seal presence and habitat assessment: results of a preliminary mission carried out in Syria. Monachus Guardian 6 (1). Available from: http://www.monachus-guardian.org/library/mguard11.pdf.

Mo, G., Bazairi, H., Bayed, A., Agnesi, S., 2011. Survey on Mediterranean monk seal (*Monachus monachus*) sightings in Mediterranean Morocco. Aquat. Mamm. 37 (3), 248–255. http://dx.doi.org/10.1578/AM.37.3.2011.248.

MOm, 1996. Strategy for the Protection of the Mediterranean Monk Seal *Monachus monachus* in Greece. Archipelagos—Marine and Coastal Management and Mom/Hellenic Society for the Study and Protection of the Monk Seal (MOm), Athens, Greece. 10 p.

MOm, 2005a. Mediterranean Monk Seal Rehabilitation in Greece 1990–2004: 15 Years of Action. Hellenic Society for the Study and Protection of the Monk Seal (MOm), Athens, Greece. 31 p. Available from: http://www.monachus-guardian.org/library/mom05c.pdf.

MOm, 2005b. Operational Manual for the Surveillance of the National Marine Park of Alonnisos Northern Sporades (in Greek). Hellenic Society for the Study and Protection of the Monk Seal (MOm), Athens, Greece. 60 p.

MOm, 2009. Action plan for the mitigation of the negative effects of monk seal–fisheries interactions in Greece. LIFE Nature Project: "MOFI: Monk Seal and Fisheries: Mitigating the Conflict in Greek Seas" (LIFE05NAT/GR/000083). Hellenic Society for the Study and Protection of the Monk Seal (MOm), Athens, Greece. 11 p. Available from: http://www.monachus-guardian.org/library/mom09a.pdf.

MOm, 2014. Final technical report. LIFE-Info Project "Thalassa campaign: learn, act, protect/awareness, educational and participation campaign for marine mammals in Greece" (LIFE09INF/GR/000320). Hellenic Society for the Study and Protection

of the Monk Seal (MOm). 69 p. Available from: http://thalassaproject.mom.gr/LH2Uploads/ItemsContent/6/Final_Technical_Report_Thalassa%20Project.pdf.
MOm, 2015. Second Annual Conservation Status Report of the Mediterranean Monk Seal Population in the island of Gyaros. Publication prepared as part of the LIFE-Nature Project "CYCLADES LIFE: Integrated Monk Seal Conservation in the Northern Cyclades" (LIFE12 NAT/GR/000688). Hellenic Society for the Study and Protection of the Monk Seal (MOm). 43 p.
Monachus Guardian, 2004. Mediterranean news: Morocco. Monachus Guardian 7 (2). Available from: http://www.monachus-guardian.org/library/mguard14.pdf.
Monachus Guardian, 2008. Mediterranean news: Spain. Seal returns after 50-year absence. Monachus Guardian 11 (2). Available from: http://www.monachus-guardian.org/library/mguard22.pdf.
Monachus Guardian, 2010. Mediterranean news: Lebanon. Seal sightings in Lebanon. Monachus Guardian 13 (2). Available from: http://www.monachus-guardian.org/library/mguard26.pdf.
Monachus Guardian, 2012. Monk seal sighting in Albania. Monachus Guardian. Available from: http://www.monachus-guardian.org/wordpress/2012/08/23/monk-seal-sighting-in-albania/. 23 August 2012.
Moutopoulos, D.K., Koutsikopoulos, C., 2014. Fishing strange data in national fisheries statistics in Greece. Mar. Policy 48, 114–122. http://dx.doi.org/10.1016/j.marpol.2014.03.017.
Notarbartolo di Sciara, G., 2010. The world's two remaining monk seal species: how many different ways are there of being Critically Endangered? Monachus Guardian 13 (1). Available from:http://www.monachus-guardian.org/library/mguard25.pdf.
Notarbartolo di Sciara, G., 2013. Regional strategy for the conservation of monk seals in the Mediterranean (2014–2019). In: Action Plans under the Specially Protected Areas and Biological Diversity Protocol including Monk Seal, Marine Turtles, Birds, Cartilaginous Fishes, and Dark Habitats. Decision IG 21/4 Adopted by the Contracting Parties to the Barcelona Convention, 18th Conference of Parties, Istanbul, 3–6 December.
Notarbartolo di Sciara, G., 2014. Tourism Could Help Saving Mediterranean Monk Seals, but If Unmanaged It Will Only Accelerate Their Demise. Wave Action. Available from: http://www.disciara.org/tourism-help-saving-mediterranean-monk-seals-unmanaged-will-accelerate-demise/.
Notarbartolo di Sciara, G., Fouad, M., 2011a. National Action Plan for the Conservation of Marine Mammals in the Egyptian Mediterranean Sea—2012–2016. Regional Activity Centre for Specially Protected Areas, Tunis. 54 p. Available from: http://www.disciara.org/documents/NotarbartolodiSciara_2011b.pdf.
Notarbartolo di Sciara, G., Fouad, M., 2011b. Monk seal sightings in Egypt. Monachus Guardian. Online edition. 29 April 2011. Available from: http://www.monachus-guardian.org/wordpress/2011/04/29/monk-seal-sightings-in-egypt/.
Notarbartolo di Sciara, G., Adamantopoulou, S., Androukaki, E., Dendrinos, P., Karamanlidis, A.A., Paravas, V., Kotomatas, S., 2009a. National Strategy and Action Plan for the Conservation of the Mediterranean Monk Seal in Greece, 2009–2015. Hellenic Society for the Study and Protection of the Mediterranean Monk Seal (MOm), Athens, Greece, p. 19.
Notarbartolo di Sciara, G., Adamantopoulou, S., Androukaki, E., Dendrinos, P., Karamanlidis, A.A., Paravas, V., Kotomatas, S., 2009b. National strategy and action plan for the conservation of the Mediterranean monk seal in Greece, 2009–2015. Report on evaluating the past and structuring the future. Publication prepared as part of the LIFE-Nature Project: MOFI: Monk Seal and Fisheries: mitigating the conflict in Greek Seas, Hellenic Society for the Study and Protection of the Mediterranean Monk Seal (MOm). 71 p. Available from: http://www.monachus-guardian.org/library/notarb09b.pdf.

Oikonomou, Z.-S., Dikou, A., 2008. Integrating conservation and development at the National Marine Park of Alonissos, Northern Sporades, Greece: perception and practice. Environ. Manag. 42, 847–866. http://dx.doi.org/10.1007/s00267-008-9163-x.

Ouerghi, A., Mo, G., Di Domenico, F., Majhoub, H., 2001. Assessment of Mediterranean Monk Seal (*Monachus monachus*) Habitat at La Galite, Tunisia: Towards a Monk Seal Conservation Strategy in northern Tunisia and Nearby Waters. Regional Activity Centre for Specially Protected Areas/Istituto Centrale per la Ricerca Applicata al Mare (ICRAM)/Agence Nationale de la Protection de l'Environnement (ANPE), Tunis/Rome/Tunisia, p. 3.

Panou, A., 2009. Monk seal sightings in the central Ionian Sea: a network of fishermen for the protection of the marine resources. Archipelagos—Environment and Development, Greece, In: Presentation at the "Who Are Our Seals?" Workshop, European Cetacean Society Annual Conference, Istanbul, Turkey, 28 February, p. 6.

Pastor, T., Garza, J.C., Aguilar, A., Tounta, E., Androukaki, E., 2007. Genetic diversity and differentiation between the two remaining populations of the critically endangered Mediterranean monk seal. Anim. Conserv. 2007, 1–9. http://dx.doi.org/10.1111/j.1469-1795.2007.00137.x.

Portman, M.E., Notarbartolo di Sciara, G., Agardy, T., Katsanevakis, S., Possingham, H.P., Di Carlo, G., 2013. He who hesitates is lost: why conservation in the Mediterranean Sea is necessary and possible now. Mar. Policy 42, 270–279. http://dx.doi.org/10.1016/j.marpol.2013.03.004.

Rands, M.R.W., Adams, W.M., Bennun, L., Butchart, S.H.M., Clements, A., Coomes, D., Entwistle, A., Hodge, I., Kapos, V., Scharlemann, J.P.W., Sutherland, W.J., Vira, B., 2010. Biodiversity conservation: challenges beyond 2010. Science 329, 1298–1303. http://dx.doi.org/10.1126/science.1189138.

Redford, K.H., Amato, G., Baillie, J., Beldomenico, P., Bennett, E.L., Clum, N., Cook, R., Fonseca, G., Hedges, S., Launay, F., Lieberman, S., Mace, G.M., Murayama, A., Putnam, A., Robinson, J.G., Rosenbaum, H., Sanderson, E.W., Stuart, S.N., Thomas, P., Thorbjarnarson, J., 2011. What does it mean to successfully conserve a (vertebrate) species? BioScience 61, 39–48. http://dx.doi.org/10.1525/bio.2011.61.1.9.

Reeves, R.R. (Ed.), 2009. Proceedings of the First International Conference on Marine Mammal Protected Areas, Maui, Hawaii, USA, March 30–April 3. NOAA, p. 133.

Roberge, J.-M., Angelstam, P., 2004. Usefulness of the umbrella species concept as a conservation tool. Conserv. Biol. 18 (1), 76–85.

Ronald, K., 1984. Action for the conservation of monk seal. In: Ronald, K., Duguy, R. (Eds.), Les phoques moines—Monk Seals. Proceedings of the Second International Conference, La Rochelle, France, 5–6 October, pp. 109–112. Ann. Soc. Sci. Nat. Charente-Maritime, Suppl., décembre 1984.

Ronald, K., Duguy, R. (Eds.), 1979. The Mediterranean Monk Seal. Proceedings of the First International Conference, Rhodes, Greece, 2–5 May 1978. In: UNEP Technical Series, vol. 1. Pergamon Press, Oxford.

Ronald, K., Duguy, R. (Eds.), 1984. Les phoques moines—Monk Seals. Proceedings of the Second International Conference, La Rochelle, France, 5–6 October. Ann. Soc. Sci. Nat. Charente-Maritime, Suppl., décembre 1984, 120 p.

Sakinan, S., 2004. Mediterranean news. Turkey. Cilician Basin ecotourism project enters second phase. Monachus Guardian 7 (2). Available from: http://www.monachus-guardian.org/library/mguard14.pdf.

Savas, Y., 1999. How tourism has ruined the coastal habitats of the monk seal on the Bodrum Peninsula, Turkey. Monachus Guardian 2 (2), 21–27. Available from: http://www.monachus-guardian.org/library/mguard04.pdf.

Scheinin, A.P., Goffman, O., Elasar, M., Perelberg, A., Kerem, D.H., 2011. Mediterranean monk seal (*Monachus monachus*) resighted along the Israeli coastline after more than half a century. Aquat. Mamm. 37 (3), 241–242. http://dx.doi.org/10.1578/AM.37.3.2011.241.

Sergeant, D., Ronald, K., Boulva, J., Berkes, F., 1979. The recent status of *Monachus monachus* the Mediterranean monk seal. In: Ronald, K., Duguy, R. (Eds.), The Mediterranean Monk Seal. Proceedings of the First International Conference, Rhodes, Greece, 2–5 May 1978. In: UNEP Technical Series vol. 1. Pergamon Press, Oxford, pp. 31–54.

Taylor, B.L., 2005. Identifying units to conserve. In: Reynolds, I.E., Perrin, W.F., Reeves, R.R., Montgomery, S., Ragen, T.J. (Eds.), Marine Mammal Research: Conservation Beyond Crisis. The Johns Hopkins University Press, Baltimore, pp. 149–164.

UNEP-MAP-RAC/SPA, 1994. Present status and trend of the Mediterranean monk seal (*Monachus monachus*) populations. In: Meeting of Experts on the Evaluation of the Implementation of the Action Plan for the Management of the Mediterranean monk seal, Rabat, Morocco, 7–9 October. UNEP(OCA)/MED WG. 87/3. 44 p.

UNEP-MAP-RAC/SPA, 1998. Report of the meeting of experts on the implementation of the action plans for marine mammals (monk seal and cetaceans) adopted within MAP. In: Meeting of Experts on the Implementation of the Action Plans for Marine Mammals (Monk Seal and Cetaceans) Adopted within MAP, Arta, Greece, 29–31 October. UNEP (OCA)/MED WG. 146/5. 122 p. Available from: http://195.97.36.231/dbases/MEETING_DOCUMENTS/98WG146_5_Eng.pdf.

UNEP-MAP-RAC/SPA, 2003. Action Plan for the Management of the Mediterranean Monk Seal (*Monachus monachus*). Regional Activity Centre for Specially Protected Areas (RAC/SPA), Tunis. 12 p. Available from: http://www.monachus-guardian.org/library/unepma87.pdf.

UNEP-MAP-RAC/SPA, 2005. Information report on the status of the monk seal in the Mediterranean. In: Seventh Meeting of the National Focal Points for SPAs, Seville, 31 May–3 June. UNEP(DEC)/MED WG. 268/Inf 3. 45 p.

UNEP-MAP-RAC/SPA, 2006. Report of the International Conference on Monk Seal Conservation. Antalya, Turkey, 17–19 September 2006. UNEP. 69 p. Available from: http://www.rac-spa.org/sites/default/files/doc_monackus/report_int_monk_seal_conf_antalya_2006.pdf.

UNEP-MAP-RAC/SPA, IUCN, 1988. Report of the joint expert consultation on the conservation of the Mediterranean monk seal, Athens, 11–12 January. IUCN/UNEP/MEDU/MM-IC/5. 8 p.

Van Bree, P.J.H., 1979. Notes on the differences between monk seals from the Atlantic and the Western Mediterranean. In: Ronald, K., Duguy, R. (Eds.), The Mediterranean Monk Seal. Proceedings of the First International Conference, Rhodes, Greece, 2–5 May 1978. In: UNEP Technical Series, vol. 1. Pergamon Press, Oxford, p. 99.

Veryeri, O., Güçlüsoy, H., Savas, Y., 2001. Snared and drowned: are fishing nets killing off a new generation of monk seals in Turkey's protected areas? Monachus Guardian 4 (1). Available from: http://www.monachus-guardian.org/library/mguard07.pdf.

Weber, D.S., Stewart, B.S., Lehman, N., 2004. Genetic consequences of a severe population bottleneck in the Guadalupe fur seal (*Arctocephalus townsendi*). J. Hered. 95 (2), 144–153.

White, M., Haxhiu, I., Kouroutos, V., Gace, A., Vaso, A., Beqiraj, S., Plytas, A., Dedej, Z., 2005. Rapid Assessment Survey of important marine turtle and monk seal habitats in the coastal area of Albania, October–November 2005: Report to UNEP. p. 36. Available from: http://www.seaturtle.org/PDF/WhiteM_2006_GEFSGPTechReport.pdf.

CHAPTER THIRTEEN

The International Legal Framework for Marine Mammal Conservation in the Mediterranean Sea

T. Scovazzi[1]

Dipartimento giuridico delle istituzioni nazionali ed europee, Università degli Studi di Milano-Bicocca, Milan, Italy

[1]Corresponding address: e-mail address: tullio.scovazzi@unimib.it

Contents

Advances in Marine Biology, Volume 75
ISSN 0065-2881
http://dx.doi.org/10.1016/bs.amb.2016.07.006

Abstract

The paper reviews the international treaties that are today applicable for the protection of marine mammals in the Mediterranean Sea. They include instruments applicable at the world or the regional level. Emphasis is put on the International Whaling Commission, which is today affected by a confrontation between the two opposing groups of nonwhaling and whaling parties, the Agreement on the Conservation of Cetaceans of the Black Sea, Mediterranean Sea and Contiguous Atlantic Area (ACCOBAMS) and its implementation through the resolutions adopted by the Meeting of the Parties, as well as the subregional Agreement between France, Italy and Monaco for the establishment of a sanctuary for the protection of marine mammals (Pelagos Sanctuary). While several treaty provisions are in place to ensure the protection of marine mammals in the Mediterranean, specific mandatory provisions are still lacking for certain subjects, such as underwater noise, collisions with ships, by catch in fishing gear and ingestion of plastic litter.

1. INTRODUCTION

This paper will review the international treaties which are today applicable for the conservation of marine mammals in the Mediterranean Sea, with special emphasis placed on the Agreement on the Conservation of Cetaceans of the Black Sea, Mediterranean Sea and Contiguous Atlantic Area (ACCOBAMS), which is the treaty of regional scope specifically related to cetaceans. It will not consider the domestic legislation of the bordering countries and the instruments adopted by the European Union.[1]

2. OTHER TREATIES OF WORLD SCOPE THAT ARE APPLICABLE TO MARINE MAMMALS

A number of treaties having a world scope of geographical application and a general purpose are relevant for marine mammals in the Mediterranean.

[1] Also particularly relevant for marine mammals are Directive 92/43/EEC of 21 May 1992 on the conservation of natural habitats and of wild fauna and flora, and Directive 2008/56/EC of 17 June 2008 establishing a framework for community action in the field of marine environmental policy (Marine Strategy Framework Directive). Among the instruments applicable to fisheries relevant are Regulation 1005/2008 of 29 September 2008, establishing a community system to prevent, deter and eliminate illegal, unreported and unregulated fishing, Regulation 1224/2009 of 20 November 2009, establishing a community control system for ensuring compliance with the rules of the common fisheries policy, and Regulation No. 1239/98 of 8 June 1998, which bans the use of driftnets (*infra*, para. 5.F).

2.1 CITES

The purpose of the Convention on International Trade in Endangered Species of Wild Fauna and Flora (CITES; Washington, 1973) is to protect endangered species by strictly regulating their international trade. The CITES provides for a system based on the exchange of export and import permits and certificates. Such documentation is granted by States parties where certain conditions are met and in particular where the trade is not determined as detrimental to the survival of the species.

The CITES applies to specimens, whether alive or dead, and readily recognizable parts or derivatives thereof. It is completed by three Appendices which list the species whose international trade is regulated by the Convention. Inclusion in Appendix I, II or III depends on the different level of the endangerment of a species. Trade in specimens of species listed in Appendix I must be subject to particularly strict regulation in order not to further endanger their survival and must be authorized only in exceptional circumstances. The export of any specimen of species listed in Appendix II requires the prior presentation of an export permit which is granted only if certain conditions are met. Appendix III provides a list of species included at the request of a party that already regulates trade in these species and needs the cooperation of other countries to prevent unsustainable or illegal exploitation. Twenty-three cetacean species and two seal species are listed in Appendix I.

2.2 Bonn Convention

In the preamble of the Convention on the Conservation of Migratory Species of Wild Animals (Bonn, 1979) the Parties recognize 'that wild animals in their innumerable forms are an irreplaceable part of the earth's natural system which must be conserved for the good of mankind' and declare themselves aware 'that each generation of man holds the resources of the earth for future generations and has an obligation to ensure that this legacy is conserved and, where utilized, is used wisely'.

Migratory animals face several threats, perhaps especially during their travels, such as pollution of habitats, deterioration of their natural stop-over places, direct hazards from hunting or fishing. The Parties to the Bonn Convention acknowledge 'the need to take action to avoid any migratory species becoming endangered' (Art. II, paras 1 and 2).[2] Sixteen cetacean species are

[2] '"Migratory species" means the entire population or any geographical separate part of the population of any species or lower taxon of wild animals, a significant proportion of whose members cyclically and predictably cross one or more jurisdictional boundaries' (Art. I, para. 1, *a*). '"Habitat" means any area in the range of a migratory species which contains suitable living conditions for that species' (Art. I, para. 1, *g*).

listed in Appendix I (endangered migratory species) and several others in Appendix II (migratory species having an unfavourable conservation status).

The Bonn Convention calls for the conclusion of specific agreements for the protection of certain species worldwide or in particular regions. Art. IV, para. 4, encourages the Parties 'to take action with a view to concluding agreements for any population or any geographically separate part of the population of any species or lower taxon of wild animals, numbers of which periodically cross one or more national jurisdictional boundaries'.[3]

2.3 UNCLOS

The United Nations Convention on the Law of the Sea (UNCLOS; Montego Bay, 1982) provides that all States are under a general obligation 'to protect and preserve the marine environment' (Art. 192). Accordingly, under Art. 194, para. 5, the measures taken to protect and preserve the marine environment 'shall include those necessary to protect and preserve rare or fragile ecosystems as well as the habitat of depleted, threatened or endangered species and other forms of marine life'. This obligation has a general scope of application. It covers every kind of vulnerable marine ecosystems and species, wherever in the sea they are located, irrespective of the legal condition of the relevant waters. A typical measure to protect such ecosystems and species is the establishment of a marine protected area.

Two provisions of the UNCLOS directly relate to marine mammals:

> *Nothing in this Part [= Part V: Exclusive economic zone] restricts the right of a coastal State or the competence of an international organization, as appropriate, to prohibit, limit or regulate the exploitation of marine mammals more strictly than provided for in this Part. States shall cooperate with a view to the conservation of marine mammals and in the case of cetaceans shall in particular work through the appropriate international organizations for their conservation, management and study (Art. 65).*
>
> *Article 65 also applies to the conservation and management of marine mammals in the high seas (Art. 120).*

It follows that, under Arts. 65 and 120, the normal scheme of exploitation of marine living resources provided for by the UNCLOS, which is based on the objective of optimum utilization of the living resources and the determination of the total allowable catch, does not apply to marine mammals. Unlike other marine living resources of the sea, the exploitation of these animals can be prohibited, limited or regulated by States or a competent

[3] For the agreement applicable to the Mediterranean, see *infra*, para. 5.

international organization, irrespective of the fact that they are in danger of extinction or their stocks are being depleted.

2.4 The CBD

The United Nations Convention on Biological Diversity (CBD; Rio de Janeiro, 1992) aims at setting a balance between the ecological objective of nature conservation and the economic needs of development. It fulfils three general purposes, namely to conserve biological diversity, to sustainably use the components of biological diversity and to fairly and equitably share the benefits arising from the use of genetic resources.

The CBD sets out a series of measures for *in situ* conservation. *Inter alia*, parties are required, as far as possible and as appropriate, to 'establish a system of protected areas or areas where special measures need to be taken to conserve biological diversity' (Art. 8, *a*), to 'develop, where necessary, guidelines for the selection, establishment and management of protected areas where special measures need to be taken to conserve biological diversity' (Art. 8, *b*), and to 'regulate or manage biological resources important for the conservation of biological diversity whether within or outside protected areas, with a view to ensuring their conservation and sustainable use' (Art. 8, *c*). This provision has a broad scope, as it implies control of a number of activities which could affect the biological resources concerned. Under Art. 8, *d*, the Parties shall promote the protection of ecosystems, natural habitats and the maintenance of viable populations of species in natural surroundings. Under Art. 8, *f*, Parties shall rehabilitate and restore degraded ecosystems and promote the recovery of threatened species. Under Art. 8, *k*, Parties shall develop or maintain necessary legislation or other regulatory provisions for the protection of threatened species and populations.

3. TREATIES OF WORLD SCOPE SPECIFICALLY RELATED TO SPECIES OF MARINE MAMMALS: THE WHALING CONVENTION

The purpose of the International Convention for the Regulation of Whaling (Washington, 1946) is to provide for the proper conservation of whale stocks.[4] The preamble of the Convention puts emphasis on the need to prevent the depletion of whales and to safeguard them for future generations as a 'great natural resource':

4 The Convention does not provide a definition of 'whale'.

> *Recognizing the interest of the nations of the world in safeguarding for future generations the great natural resources represented by the whale stocks;*
> *Considering that the history of whaling has seen over-fishing of one area after another and of one species of whale after another to such a degree that it is essential to protect all species of whales from further over-fishing;*
> *Recognizing that the whale stocks are susceptible of natural increases if whaling is properly regulated, and that increases in the size of whale stocks will permit increases in the number of whales which may be captured without endangering these natural resources;*
> *Recognizing that it is in the common interest to achieve the optimum level of whales stocks as rapidly as possible without causing widespread economic and nutritional distress;*
> *Recognizing that in the course of achieving these objectives, whaling operations should be confined to those species best able to sustain exploitation in order to give an interval for recovery to certain species now depleted in numbers;*
> *Desiring to establish a system of international regulation for the whale fisheries to ensure proper and effective conservation and development of whale stocks (...);*
> *Having decided to conclude a convention to provide for the proper conservation of whale stocks and thus make possible the orderly development of the whaling industry.*[5]

Article III of the Convention establishes the International Whaling Commission (IWC), which is composed of one member from each State party. The main duty of the IWC is to keep under review and revise, as necessary, the measures laid down in the Schedule to the Convention which govern the conduct of whaling throughout the world and provide for the conservation and utilization of whale resources through the determination of

> (a) *protected and unprotected species;* (b) *open and closed seasons;* (c) *open and closed waters, including the designation of sanctuary areas;* (d) *size limits for each species;* (e) *time, methods, and intensity of whaling (including the maximum catch of whales to be taken in any one season);* (f) *types and specifications of gear and apparatus and appliances which may be used;* (g) *methods of measurement; and* (h) *catch returns and other statistical and biological records (Art. V, para. 1).*

The amendments to the Schedule, which require a three-fourths majority of parties voting,

> (a) *shall be such as are necessary to carry out the objectives and purposes of this Convention and to provide for the conservation, development, and optimum utilization of the whale resources;* (b) *shall be based on scientific findings (...) (Art. V, para. 2).*

[5] The early mentioning in the first preambular paragraph of 'future generations' is notable, considering that this expression is today linked to the concept of sustainable development, as embodied in the 1992 Rio Declaration on Environment and Development: 'The right to development must be fulfilled so as to equitably meet developmental and environmental needs of present and future generations' (Principle 3).

Sanctuaries where commercial whaling is prohibited were established by the IWC in vast areas, such as the Indian Ocean (1979) and the Southern Ocean (1994).

However, there are two evident loopholes in the Convention that can seriously affect the effectiveness of the measures adopted by the IWC.

First, under Art. III, para. 2, the decisions of the IWC to amend the Schedule become effective 90 days after their notification. However, if a State party presents an objection prior to the expiration of the 90-day period, the objected decision does not become effective with respect to that State until the objection is withdrawn. In a number of instances, the right to cast objections has been exercised by some parties to engage in whaling activities which would not be allowed under the amended Schedule.

Second, under Art. VIII, notwithstanding anything contained in the Convention, any State party may grant to its nationals special permits authorizing them to kill, take and treat whales 'for the purposes of scientific research', subject to the conditions that the State party thinks fit and in exemption from the operation of the Convention. This possibility is also fully exploited by some States parties. By the judgement rendered on 31 March 2014 on the *Whaling in the Antarctic* case (Australia vs Japan; New Zealand intervening), the International Court of Justice found that the special permits granted by Japan in connection with the Whale Research Programme called JARPA II did not fall within the provisions of Article VIII of the Whaling Convention. The Court also decided that Japan was bound to revoke any extant permit in relation to JARPA II and refrain from granting any further permits in pursuance of that programme.

While the original signatories of the Whaling Convention were 15 whaling countries, the parties now number 89 States. Far from being an instrument reserved to whaling countries, the Whaling Convention is open to the accession by any State (Art. X, para. 2). In 1946, the countries active in the hunting of marine mammals believed that conservation was a means to ensure the optimum utilization of a commercial resource and the only way to allow the continuation of whaling, which was their main concern. Today the situation has changed. The majority of the parties oppose whaling for commercial purposes. Most of the countries which previously engaged in whaling have ceased their activities, while whaling on a commercial basis is still carried out by a minority of parties to the Convention (Japan, Norway, Iceland, supported by other States). Several countries that were never active in whaling (for instance, Switzerland, Austria, Monaco or San Marino) have acceded to the Whaling Convention with the aim of strengthening a protectionist policy. The States belonging to the present nonwhaling majority prefer nonlethal ways of economic use of marine mammals, such as whale watching.

A remarkable achievement in the protectionist policy of the IWC occurred in 1982, when the IWC adopted a zero quota for commercial whaling with effect from the 1985–86 season:

> *Notwithstanding the other provisions of paragraph 10, catch limits for the killing for commercial purposes of whales from all stocks for the 1986 coastal and the 1985/86 pelagic seasons and thereafter shall be zero. This provision will be kept under review, based upon the best scientific advice, and by 1990 at the latest the Commission will undertake a comprehensive assessment of the effects of this decision on whale stocks and consider modification of this provision and the establishment of other catch limits (para. 10, e, of the Schedule).*

The moratorium thereby created, which does not apply to aboriginal subsistence whaling, is still in force today (June 2016).

The dilemma between a policy of exploitation or a policy of protection is the main reason for the critical situation which presently affects the IWC. In fact, the IWC is prevented from taking a number of substantive decisions because of the confrontation between two opposing groups of parties. Attempts to elaborate compromise solutions that would partially meet the concerns of both whaling and nonwhaling countries have not achieved successful results so far.

4. TREATIES OF REGIONAL SCOPE APPLICABLE ALSO TO MARINE MAMMALS

4.1 The Barcelona Convention and the SPA Protocol

The Convention on the Protection of the Mediterranean Sea against Pollution was opened to signature in 1976 in Barcelona. It is a framework treaty that is intended to be complemented by specific protocols. In subsequent years, the Barcelona Convention and some of its protocols have been amended and new protocols adopted. The present Barcelona system includes eight instruments, namely:

(a) the Barcelona Convention which, as amended in 1995, changed its name to the Convention for the Protection of the Marine Environment and the Coastal Region of the Mediterranean;

(b) the Protocol for the Prevention and Elimination of Pollution of the Mediterranean Sea by Dumping from Ships and Aircraft or Incineration at Sea (Barcelona, 1976; amended in 1995);

(c) the Protocol Concerning Cooperation in Preventing Pollution from Ships and, in Cases of Emergency, Combating Pollution of the Mediterranean Sea (Valletta, 2002);

(d) the Protocol for the Protection of the Mediterranean Sea against Pollution from Land-Based Sources and Activities (Athens, 1980; amended in 1996);

(e) the Protocol Concerning Specially Protected Areas and Biological Diversity in the Mediterranean (SPA Protocol; Barcelona, 1995);

(f) the Protocol Concerning Pollution Resulting from Exploration and Exploitation of the Continental Shelf, the Seabed and its Subsoil (Madrid, 1994);

(g) the Protocol on the Prevention of Pollution of the Mediterranean Sea by Transboundary Movements of Hazardous Wastes and their Disposal (Izmir, 1996);

(h) the Protocol on Integrated Coastal Zone Management in the Mediterranean (Madrid, 2008).

Particularly important for the protection of marine mammals is the SPA Protocol, which applies to all the maritime waters of the Mediterranean, irrespective of their legal condition, to the seabed and its subsoil and to the terrestrial coastal areas designated by each of the Parties. The extension of the application of the SPA Protocol to the high seas areas that still exist in the Mediterranean[6] was considered as necessary to protect those highly migratory marine species (such as marine mammals) which, because of their natural behaviour, do not respect the artificial boundaries drawn by man on the sea.

The SPA Protocol provides, *inter alia*, for the establishment of a List of Specially Protected Areas of Mediterranean Importance (SPAMI List). The SPAMI List may include sites that 'are of importance for conserving the components of biological diversity in the Mediterranean; contain ecosystems specific to the Mediterranean area or the habitats of endangered species; are of special interest at the scientific, aesthetic, cultural or educational levels' (Art. 8, para. 2). The existence of the SPAMI List does not exclude the right of each Party to create and manage protected areas that are not intended to be included among the SPAMIs, but deserve to be protected under the domestic legislation.

---

6 *Infra*, para. 5.2.

The procedures for the listing of SPAMIs are specified in detail in the Protocol:

Proposals for inclusion in the List may be submitted:

- ***(a)*** *by the Party concerned, if the area is situated in a zone already delimited, over which it exercises sovereignty or jurisdiction;*
- ***(b)*** *by two or more neighbouring Parties concerned if the area is situated, partly or wholly, on the high sea;*
- ***(c)*** *by the neighbouring Parties concerned in areas where the limits of national sovereignty or jurisdiction have not yet been defined (Art. 9, para. 2).*

The submission of a joint proposal may become a way to promote new forms of cooperation between the States concerned, irrespective of the fact that their maritime boundaries have not yet been defined.

In proposing a SPAMI, the Party or Parties concerned shall indicate the relevant protection and management measures, as well as the means for their implementation (Art. 9, para. 3). As paper areas would not comply with the purposes of the SPA Protocol, protection, planning and management measures 'must be adequate for the achievement of the conservation and management objectives set for the site in the short and long term, and take in particular into account the threats upon it' (Annex 1, para. D, 2).

Once areas are included in the SPAMI List, all the parties agree 'to recognize the particular importance of these areas for the Mediterranean', as well as 'to comply with the measures applicable to the SPAMIs and not to authorize nor undertake any activities that might be contrary to the objectives for which the SPAMIs were established' (Art. 8, para. 3). This gives to the SPAMIs, and to the measures adopted for their protection, an *erga omnes partes* effect, that is an effect that applies with respect to all of the Parties to the Protocol. As to the relationship with third countries, the Parties shall 'invite States that are not Parties to the Protocol and international organizations to cooperate in the implementation' of the Protocol (Art. 28, para. 1). The Parties also 'undertake to adopt appropriate measures, consistent with international law, to ensure that no one engages in any activity contrary to the principles and purposes' of the Protocol (Art. 28, para. 2). This provision aims at confronting potential problems arising from the fact that treaties, including the SPA Protocol itself, can produce rights and obligations among only the Parties.

The SPA Protocol is completed by three annexes, which were adopted in 1996 in Monaco. These are the Common Criteria for the Choice of Protected Marine and Coastal Areas that Could be Included in the SPAMI List (Annex I), the List of Endangered or Threatened Species (Annex II) and the List of Species Whose Exploitation is Regulated (Annex III). Nineteen species of marine mammals are listed in Annex II. Under Annex I, the sites included in the SPAMI List must be 'provided with adequate legal status, protection measures and management methods and means' (para. A, *e*) and must fulfil at least one of six general criteria (uniqueness, natural representativeness, diversity, naturalness, presence of habitats that are critical to endangered, threatened or endemic species, cultural representativeness). The SPAMIs must have a management body, a management plan and a monitoring programme (paras from D.6 to D.8).

However, only 1 (the Pelagos Sanctuary[7]) among the 34 SPAMIs so far established also covers areas beyond the limit of the territorial sea.

4.2 Bern Convention

Under the Convention on the Conservation of European Wildlife and Natural Habitats (Bern, 1979), concluded within the framework of the Council of Europe,[8] the parties are bound to take appropriate and necessary legislative and administrative measures to ensure the conservation of the habitats of the wild flora and fauna species, especially those specified in Appendices I (strictly protected flora species) and II (strictly protected fauna species), and the conservation of endangered natural habitats. They also undertake to give special attention to the protection of areas that are of importance for the migratory species listed in Appendices II and III (protected fauna species) and which are appropriately situated in relation to migration routes.

Thirty cetacean species and three seal species are listed in Appendix II. All other cetacean species are included in Appendix III. Also relevant is Appendix IV (prohibited means and methods of killing, capture and other forms of exploitation).

5. TREATIES OF REGIONAL SCOPE SPECIFICALLY RELATED TO SPECIES OF MARINE MAMMALS: ACCOBAMS

Particularly important among the agreements concluded within the framework of the Bonn Convention[9] is the Agreement on the Conservation of Cetaceans of the Black Sea, Mediterranean Sea and Contiguous Atlantic

---

7 *Infra*, para. 6.

8 Also four non-European States, namely Burkina Faso, Morocco, Senegal and Tunisia are parties to the Bern Convention.

9 *Supra*, para. 2.2.

Area (ACCOBAMS). This agreement was opened for signature in Monaco on 24 November 1996 and entered into force on 1 June 2001. It is now binding on 23 out of the 29 States that border the marine waters to which it applies.[10]

ACCOBAMS is composed of a preamble, 17 articles and two annexes (Indicative lists of cetaceans to which ACCOBAMS applies; Conservation plan). A number of related instruments (resolutions and recommendations) have been adopted at the five Meetings of Parties held so far.[11]

5.1 The Objectives

ACCOBAMS is based on the spirit of international cooperation for the protection of the environment, which is the result of the United Nations Conference on Environment and Development (Rio de Janeiro, 1992). The major environmental principles embodied in the Rio Declaration can be found in the text of ACCOBAMS. In the preamble, the Parties recognize that cetaceans must be conserved for the benefit of present and future generations (so-called principle of intergenerational equity) and that their conservation is a common concern. They acknowledge the importance of integrating actions to conserve cetaceans with activities related to socioeconomic development, including fishing and free circulation of vessels (principle of sustainable development). They stress the need to ensure cooperation among all the stakeholders, namely, States, regional economic integration organizations, intergovernmental organizations and the nongovernmental sector. They call for the provision of assistance, in a spirit of solidarity, to developing range States for research, training, monitoring and the establishment or improvement of scientific and administrative institutions.

Article II, para. 4, binds the Parties to apply the precautionary principle in implementing the measures prescribed under ACCOBAMS. This principle, which is not defined in ACCOBAMS, is the result of the general evolution of international environmental law. It provides in general that 'where there are threats of serious or irreversible damage, lack of full scientific certainty shall not be used as a reason for postponing cost-effective measures to prevent environmental degradation' (Principle 15 of the Rio Declaration on Environment and Development).

[10] The States of the region that are not yet parties to ACCOBAMS are Bosnia and Herzegovina, Israel, Palestine, the Russian Federation, Turkey and the United Kingdom. For the European Union, see *infra*, para. 5.2.

[11] *Infra*, para. 5.8.

According to para. 1, *c*, of Annex 2, an environmental impact assessment is required to allow activities that may affect cetaceans or their habitat, including fisheries, offshore exploration and exploitation, nautical sports, tourism and cetacean watching, as well as for establishing the conditions under which such activities may be conducted.

Several threats adversely affect the conservation status of cetaceans in the waters where ACCOBAMS applies, such as degradation and disturbance of their habitats, pollution, reduction of food resources, use and abandonment of nonselective fishing gear, as well as deliberate and incidental catches, as stated in the ACCOBAMS preamble. A more detailed list of threats to cetaceans includes the following:

> *competition between fishermen leading to culling operations and the use of these animals as bait, or as food for human or animal consummation;*
> *non-intentional catches, in fishing nets or even in broken nets, discarded at sea;*
> *reduction of their food resources due to over-fishing or deterioration of their environment;*
> *harmful disturbances caused by tourist observations upsetting certain phases of their life-cycle (this activity has a worthy impact on public awareness on cetaceans and on their conservation, but as an activity, it should be controlled and subject to strict ruling);*
> *pollution generated by harmful substances being discharged directly into the sea (tank cleaning, oil spills, etc.) or contamination from land-based sources (industry, overpopulated areas);*
> *menaces from ships, both from collisions and from the modern high speed vessels which today are becoming more frequent (these vessels make little underwater noise and are not easily detectable by the large cetaceans);*
> *acoustic disorder provoked by construction works along the coastal fringe or geo-physics prospection and other civilian or military activities;*
> *disturbances from aquatic sport activities, such as off-shore competitions, which accumulate the two risks just above mentioned.*[12]

ACCOBAMS binds the Parties to achieve and maintain a favourable conservation status for cetaceans.[13] The main obligations of the Parties are to prohibit any deliberate taking of cetaceans, to create and maintain a network of specially protected areas to conserve cetaceans (Art. II, para. 1) and to take the measures specified in the conservation plan (Annex 2).

[12] The list may be found in an introductory note on *ACCOBAMS—A Biodiversity Conservation Tool for the Mediterranean and Black Seas*, published in the booklet containing the official text of ACCOBAMS.

[13] Under Art. I, para. 3, ACCOBAMS, the expression favourable conservation status has to be defined as it is in Art. I, para. 1, *c*, of the Bonn Convention.

According to Art. I, para. 2, ACCOBAMS applies to all cetaceans that have a range[14] which lies entirely or partly within the Agreement area or those species that accidentally or occasionally frequent the Agreement area. In order to avoid ambiguity, a list of cetaceans covered by ACCOBAMS is given in Annex 1. It includes three species of the Black Sea and 18 species of the Mediterranean Sea and contiguous Atlantic waters. However, the list is only 'indicative' (Art. I, para. 2) and, consequently, also other species of cetaceans which are occasionally found in the Agreement area are covered by ACCOBAMS.

ACCOBAMS does not contain any provisions that exclude its application to military ships or State-owned ships in general. It may follow that also military activities fall in general under the scope of ACCOBAMS.

5.2 The Geographical Scope

During the negotiations for ACCOBAMS, some delegations supported an extension of the area of application of the agreement beyond the Black and Mediterranean Seas, to cover some Atlantic contiguous waters, which are part of the habitat of cetacean species during migration.

The original text of Art. I, para. 1, *a*, provides for a precise determination of three different areas of application (the Mediterranean Sea, the Black Sea and the contiguous Atlantic area):

> *The geographic scope of this Agreement, hereinafter referred to as the 'Agreement area', is constituted by all the maritime waters of the Black Sea and the Mediterranean and their gulfs and seas, and the internal waters connected to or interconnecting these maritime waters, and of the Atlantic area contiguous to the Mediterranean Sea west of the Strait of Gibraltar. For the purpose of this Agreement:*
>
> - *the Black Sea is bounded to the southwest by the line joining Capes Kelaga and Dalyan (Turkey);*
> - *the Mediterranean Sea is bounded to the east by the southern limits of the Straits of the Dardanelles between the lighthouses of Mehmetcik and Kumkale (Turkey) and to the west by the meridian passing through Cape Spartel lighthouse, at the entrance to the Strait of Gibraltar; and*
> - *the contiguous Atlantic area west of the Strait of Gibraltar is bounded to the east by the meridian passing through Cape Spartel lighthouse and to the west by the line joining the lighthouses of Cape St. Vicente (Portugal) and Casablanca (Morocco).*

[14] For the definition of 'range', see *infra*, para. 5.3.

As can be seen, ACCOBAMS applies in general to all 'maritime waters' within the 'Agreement area', irrespective of their legal condition, be they maritime internal waters, territorial seas, exclusive economic zones, fishing zones or ecological protection zones.[15]

The waters of the Strait of Gibraltar are included among the Mediterranean waters. The waters of the Strait of Dardanelles, of the Marmara Sea and of the Strait of Bosphorus are not included either in the Mediterranean or in the Black Seas, as the Mediterranean waters start beyond the southern limits of the Dardanelles and the Black Sea waters start beyond the eastern limits of the Bosphorus. However, also the Strait of Dardanelles, the Marmara Sea and the Strait of Bosphorus fall under the scope of geographic application of ACCOBAMS, insofar as they are included in the expression 'internal waters (...) interconnecting' the maritime waters of the Mediterranean and the Black Sea.

At the 1995 Meeting of the ACCOBAMS Parties, the Russian Federation and Ukraine made a statement relating to the Sea of Azov.[16]

The contiguous Atlantic area originally consisted of a relatively small portion of the Atlantic Ocean, delimited to the west by a line passing through Cape St. Vicente (Portugal) and the city of Casablanca (Morocco) and to the east by the meridian passing through Cape Spartel (Morocco). At their 2010 Meeting, on proposals submitted by Portugal and Spain, the ACCOBAMS Parties decided to extend the geographical scope of ACCOBAMS to cover a broader Atlantic area, consisting of the extent of the territorial seas and exclusive economic zones off the continental coasts of Portugal and Spain (Resolution A/4.1). The last sentence of Art. I, para. 1, *a*, was accordingly amended as follows:

> *the neighbouring Atlantic Area west of the Strait of Gibraltar is bounded to the east by the meridian passing through Cape Spartel lighthouse (Morocco); to the west by the line joining the lighthouses of Casablanca (Morocco) and Cape St. Vicente (Portugal) until this line reaches the parallel of latitude 36°N, then by the parallel*

[15] However, under Art. XV, a reservation may be entered in respect of a specifically delimited part of internal waters.

[16] (...) The delegations of the Russian Federation and the Ukraine deem (...) that, in regard of the Black Sea, the Sea of Azov is proposed to be included in the scope of the agreement. The Sea of Azov has been, up until the dissolution of the USSR, its internal waters. The present legal status of the Azov Sea has not yet been defined. The Russian Federation and the Ukraine, being the only coastal states of this sea, are going to hold negotiations to define its legal status in the nearest future. Until this crucial question has been resolved, the Russian Federation and the Ukraine cannot assume any legal responsibilities in regard of these waters in the framework of the Agreement being debated here.

of latitude 36°N until it reaches the external limit of marine waters covered by the sovereignty or jurisdiction of Portugal, then by the external limit of marine waters covered by the sovereignty or jurisdiction of Portugal and Spain until the land border between Spain and France.

The name of ACCOBAMS was also amended to become 'Agreement on the Conservation of Cetaceans of the Black Sea, Mediterranean Sea and neighbouring Atlantic Area'. However the 2010 amendments have not yet entered into force, since, under Art. III, para. 8, *d*, they also need to be subsequently approved by two-thirds of the Parties.[17]

Under Art. I, para. 3, *j*, the Agreement area is divided in two subregions, one comprising the coastal States of the Black Sea and the other the coastal States of the Mediterranean Sea and the contiguous Atlantic area.[18]

An important element from the legal point of view is the quite elaborate disclaimer provision contained in ACCOBAMS (Art. I, para. 1, *b* and *c*):

Nothing in this Agreement nor any act adopted on the basis of this Agreement shall prejudice the rights and obligations, the present and future claims or legal views of any State relating to the law of the sea or to the Montreux Convention of 20 July 1936 (Convention concernant le régime des détroits), in particular the nature and the extent of marine areas, the delimitation of marine areas between States with opposite or adjacent coasts, freedom of navigation on the high seas, the right and the modalities of passage through straits used for international navigation and the right of innocent passage in territorial seas, as well as the nature and extent of the jurisdiction of the coastal State, the flag State and the port State. No act or activity undertaken on the basis of this Agreement shall constitute grounds for claiming, contending or disputing any claim to national sovereignty or jurisdiction.

In fact, while all the States bordering the Black Sea and the contiguous Atlantic area have established an exclusive economic zone beyond a 12-mile territorial sea, there is at present in the Mediterranean a great variety of coastal zones subject to national jurisdiction beyond the territorial sea.[19]

---

17 So far the amendments have been approved by 6 Parties out of the 13 required.

18 One bordering State (Turkey) could for geographical reasons belong to either subregion. Non-bordering parties belong to the subregion in which their activities may affect the conservation of cetaceans (Art. I, para. 3, *j*).

19 Most Mediterranean States have established a 12-mile territorial sea. The exceptions are the United Kingdom (3 nm for Gibraltar and the Sovereign Base Areas of Akrotiri and Dhekelia in the island of Cyprus), Greece (6 nm) and Turkey (6 nm in the Aegean Sea, but 12 nm elsewhere). Several Mediterranean States (Albania, Algeria, Croatia, Cyprus, Egypt, France, Italy, Libya, Malta, Morocco, Montenegro, Spain, Tunisia and Turkey) apply legislation measuring the breadth of the territorial sea from straight baselines joining specific points located on the mainland or islands. The waters between the straight baselines and the coastline are considered as maritime internal waters. Historic bays are claimed by Italy (Gulf of Taranto) and Libya (Gulf of Sidra).

Four States (Algeria, Libya, Malta and Tunisia) have established a fishing zone; two States (Italy and Slovenia) have established an ecological protection zone; one State (Croatia) has established a fishing and ecological zone; eight States have established an exclusive economic zone (Cyprus, Egypt, France, Israel, Lebanon, Morocco, Spain and Syria); two States (Libya and Tunisia) have enacted framework legislation to enable the future establishment of an exclusive economic zone. If all the Mediterranean coastal States were to establish their own exclusive economic zone, as they are entitled to do, the high seas will disappear, as no point in the Mediterranean is located at a distance of more than 200 nm (corresponding to the maximum breadth of the exclusive economic zone) from the nearest land or island. As regards maritime boundaries, only a limited number of treaties have so far been concluded by Mediterranean coastal States for the delimitation of national maritime zones beyond the territorial sea. Several boundaries are still to be agreed upon by the States concerned.

In order to overcome these and other problems, the ACCOBAMS disclaimer provision is intended to convey two messages. First, the establishment of intergovernmental cooperation in the field of conservation of cetaceans does not affect a number of legal and political questions which have a very different nature and are still pending, such as those relating to the nature and extent of national coastal zones or to the drawing of marine boundaries between adjacent or opposite States. Second, the existence of such questions should not jeopardize or delay the adoption of measures necessary for the conservation of cetacean species.

5.3 Participation

As provided for in Art. XIII, ACCOBAMS is open for signature and ratification 'by any range State, whether or not areas under its jurisdiction lie within the Agreement area'. 'Range State' is defined as 'any State that exercises sovereignty and/or jurisdiction over any part of the range of a cetacean population covered by this Agreement, or a State, flag vessels of which are engaged in activities in the Agreement area which may affect the conservation of cetaceans' (Art. I, para. 3, *g*). In turn, 'range' is defined as 'all areas of water that a cetacean inhabits, stays in temporarily or crosses at any time on its normal migration route within the Agreement area' (Art. I, para. 3, *f*).

From these two definitions, it may be inferred that States which do not border the waters of the Agreement area can also become parties to

ACCOBAMS, provided that they exercise sovereignty or jurisdiction[20] over waters falling within the range of cetaceans that stay temporarily or cross the waters falling under the Agreement area. In addition, other States are also entitled to become parties to ACCOBAMS, insofar as ships flying their flag are engaged in activities which may affect the conservation of cetaceans (for example, shipping entailing the risk of collisions with cetaceans or naval exercises producing underwater noise[21]). Participation by these non-regional States concerned would only strengthen the forms of cooperation envisaged under ACCOBAMS.[22]

ACCOBAMS is also open to the participation by regional economic integration organizations of which at least one member is a range State, such as the European Union. This international organization is entitled to exercise an exclusive competence in the field of fisheries and a competence shared with its now 28 member States in the field of the protection of the environment. The European Union (at that time European Community) did participate in 1995 and 1996 to the negotiations for ACCOBAMS and pointed out that it is

> *(...) fully committed to the conservation of these species. As a matter of fact, all cetaceans are fully protected under the EU Directive for the conservation of natural habitats, and of wild flora and fauna.*
>
> *Furthermore, this draft agreement has some important aspects in relation with fisheries regulations, a matter of full Community competence. This implies that any disposition dealing specifically with the regulation of fisheries will have to be agreed upon within the framework of the Common Fisheries Policy, by all the Member States of the European Union, before it can be agreed upon by those Member States who will become Parties to the proposed agreement.*

Despite the entitlement to do so, the European Union has not yet decided to become a party to ACCOBAMS. However, 10 (out of 11) Member States that have a coast along the Agreement area are parties to it.

5.4 The Interdiction of Deliberate Takings

One of the main obligations of the Parties to ACCOBAMS is to

> *prohibit and take all necessary measures to eliminate, where this is not already done, any deliberate taking of cetaceans (Art. II, para. 1).*

[20] The definition of the terms 'sovereignty' and 'jurisdiction' is far from being clear. In practice, 'sovereignty' is used for the rights that the coastal State exercises over the territorial sea and 'jurisdiction' for some of the rights that the coastal State exercises over the exclusive economic zone.

[21] In this regard it would be conceivable, even if highly unlikely, that a State having an interest in whaling in the ACCOBAMS area decided to become a party to ACCOBAMS.

[22] For the time being, only States of the ACCOBAMS region are Parties to ACCOBAMS.

Under Art. I, para. 3, the term 'taking' is to be intended in the very broad meaning as it is defined in Art. I, para. 1, *i*, of the Bonn Convention, that is:

> *(...) taking, hunting, fishing, capturing, harassing, deliberate killing, or attempting to engage in any such conduct.*

Not only whaling, in any of its forms, but also all activities which may harass cetaceans are consequently banned in the Agreement area. As already remarked,[23] this kind of ban is in full conformity with the UNCLOS.

Two exceptions to the interdiction of 'taking' are envisaged by Art. II, para. 2, namely in emergency situations and for research purposes:

> *Any Party may grant an exception to the prohibition set out in the preceding paragraph only in emergency situations as provided for in Annex 2, paragraph 6, or, after having obtained the advice of the Scientific Committee, for the purpose of non-lethal* in situ *research aimed at maintaining a favourable conservation status for cetaceans. The Party concerned shall immediately inform the Bureau and the Scientific Committee, through the Agreement secretariat, of any such exception that has been granted. The Agreement secretariat shall inform all Parties of the exception without delay by the most appropriate means.*

The exceptions are defined in very strict terms. The emergency situations are only those 'where exceptionally unfavourable or endangering conditions' for the conservation status of cetaceans occur, as indicated in Annex 2, para. 6. Research activities which may entail the 'taking' of cetaceans are only those aimed at maintaining their favourable conservation status and must be 'nonlethal', that is not resulting in the killing of these animals, and '*in situ*', that is carried out in their natural habitat and not in other places (laboratories, dolphinaria, etc.).

Due to the broad meaning of 'taking', some activities which have taken place in the ACCOBAMS area, such as the use of dolphins for military exercises to detect underwater minefields or the keeping of cetaceans in captivity for exhibition or amusement, are to be considered as violations of the obligations arising from ACCOBAMS.

5.5 The Creation of Marine Protected Areas

Cetaceans use vast spaces and require specific environments for their natural needs and behaviours. Another main obligation of Parties to ACCOBAMS is the establishment of a network of Marine Protected Areas (MPAs) that would contribute to achieve and maintain a favourable conservation status for cetaceans:

---

[23] *Supra*, para. 2.3.

> *Parties (...) shall co-operate to create and maintain a network of specially protected areas to conserve cetaceans (Art. II, para. 1).*

This obligation reflects what is in general already provided under Art. 194, para. 5, of UNCLOS,[24] and Art. 8, para. *a*, of the CBD.[25]

At the Mediterranean regional level, the relevant instrument in this regard is the already mentioned SPA Protocol.[26] Para. 3 of Annex 2 to ACCOBAMS makes a specific reference to the Barcelona Convention and its relevant protocol as the appropriate framework within which can be established specially protected areas that serve as habitats for cetacean or provide important food resources for them.[27]

It appears that the ACCOBAMS Parties have still to achieve the objective of creating and maintaining a network of specially protected areas to conserve cetaceans. While also other areas have been proposed, so far only the Pelagos Sanctuary[28] has been established. Already in Resolution 4.15 (Marine Protected Areas of Importance for Cetaceans Conservation), adopted in 2010, the ACCOBAMS Meeting of the Parties identified 10 'areas of special importance for the common dolphin and other cetaceans', three 'areas of special importance for Black Sea cetaceans', five 'areas of special importance for the bottlenose dolphin', one 'area of special importance for the sperm whale' and three 'areas of special importance and diversity for various cetacean species'. The resolution encouraged the States concerned to promote the institution of areas of special importance for cetaceans, as identified in the resolution, and to ensure their effective management.

5.6 The Application of the Conservation Plan

Most of the conservation measures, instead of being included in the main text of ACCOBAMS, appear in Annex II. This does not detract from their

---

24 *Supra*, para. 2.3.

25 *Supra*, para. 2.4.

26 *Supra*, para. 4.1.

27 Para. 3 of Annex 2 to ACCOBAMS leaves open the possibility to use for this purpose 'other appropriate instruments'. This could be considered today as a reference to other instruments covering some parts of the ACCOBAMS areas, namely Annex V concerning the Protection and Conservation of the Ecosystems and Biological Diversity of the Maritime Area, added in 1998 to the Convention for the Protection of the Marine Environment of the North East Atlantic, known as the OSPAR Convention (Paris, 1992), and the Protocol on Black Sea Biodiversity and Landscape Protection (Sofia, 2002), concluded within the framework of the Convention on the Protection of the Black Sea against Pollution (Bucharest, 1992).

28 *Infra*, para. 6.

binding effect, as the annexes to ACCOBAMS form an integral part thereof (Art. I, para. 5). The only substantive difference between the main text and an annex is that the latter may be amended according to a simplified procedure (Art. X, paras 4 and 5).[29]

Annex 2 (conservation plan) to ACCOBAMS deals with the conservation, research and management measures that, as provided for in Art. 2, para 3, the Parties 'shall apply'. This wording confirms that in principle the Parties are under a legal obligation to apply the measures in question. However, the *chapeau* of Annex 2, allows the Parties to comply with this obligation 'to the maximum extent of their economic, technical and scientific capacities'. In this regard, the *chapeau* sets the following scale of priorities: 'giving priority to conserving those species or populations identified by the Scientific Committee as having the least favourable conservation status, and to undertaking research in areas or for species for which there is a paucity of data'.

The measures in question address a number of matters which present particular importance for cetaceans, namely 'adoption and enforcement of national legislation', 'assessment and management of human–cetacean interactions', 'habitat protection, research and monitoring', 'capacity building, collection and dissemination of information, training and education' and 'responses to emergency situations'.

For instance, in the case of fishing, which is one of the main activities leading to interactions between man and cetaceans, the original text of para. 1, *a*, of Annex 2 to ACCOBAMS bound the Parties not to allow any vessel to keep on board, or use for fishing, one or more driftnets whose individual or total length is more than 2.5 km.[30] This was, at the time of the negotiations for ACCOBAMS, the limit applied in the legislation of the European Union (Regulation No. 345/92 of 27 January 1992). Subsequently Regulation 1239/98 of 8 June 1998 banned the use of all driftnets for catching tunas, billfish, sharks, dolphin fishes, sea bream, sauries and cephalopods in all waters falling within the jurisdiction of Member States, as well as outside those waters. This development, together with others going in the same direction,[31] led in 2007 the Meeting of the Parties to ACCOBAMS to

[29] An amendment to Annex 2 has been adopted in 2007 by the Meeting of the Parties as regards the prohibition of driftnets (see, *infra*, in this paragraph).

[30] Under Art. II, para. 3, Parties shall apply measures concerning fisheries activities 'in all waters under their sovereignty and/or jurisdiction and outside these waters in respect of any vessel under their flag or registered within their territory'.

[31] Measures on driftnets applying to the Mediterranean have also been adopted within the framework of the General Fisheries Commission for the Mediterranean and the International Commission for the Conservation of Atlantic Tunas.

amend para. 1, *a*, of Annex 2, in order to provide that 'no vessels will be authorized to keep on board or to use any driftnets'.

5.7 The Adoption of National Legislation

ACCOBAMS provides that the Parties, besides cooperating among themselves and coordinating their action, shall adopt and enforce national legislation (Art. II, para. 3, *a*) that is the necessary legislative, regulatory or administrative measures to give full protection to cetaceans (Annex 2, para. 1).[32] Such measures apply in waters under the sovereignty or jurisdiction of the Parties, as well as on the high seas with respect to vessels flying the flag of the Parties or registered within their territory.

The analysis of ACCOBAMS shows that several provisions are not self-executing, in the sense that, due to their general character, they need to be implemented through the enactment of specific laws and regulations or the taking of administrative action by the Parties.[33] Among the obligations arising from ACCOBAMS provisions that do not seem self-executing, the following can be pointed out:

- As regards fishing activities, the obligations to adopt measures to minimize adverse effect on the conservation status of cetaceans (Annex 2, para. 1, *a*), to prevent fishing gear from be discarded or left adrift at sea (Annex 2, para. 1, *b*), to require the immediate release of cetaceans caught incidentally in fishing gears (Annex 2, para. 1, *b*).[34]
- As regards activities in general which may affect cetaceans, the obligation to carry out an environmental impact assessment as a basis for allowing or prohibiting the continuation or the future development of activities such as fishing, offshore exploration and exploitation, nautical sports, tourism and cetacean watching, as well as establishing conditions under which such activities may be conducted (Annex 2, para. 1, *c*).
- As regards pollution, the obligations to regulate the discharge at sea of pollutants believed to have adverse effects on cetaceans and to adopt within the framework of other appropriate legal instruments stricter standards for such pollutants (Annex 2, para. 1, *d*).

[32] Whether primary (laws) or secondary (regulations) legislation is needed is a question to be addressed according to the constitutional systems of the Parties.

[33] Obviously, no enactment of new legislation is needed if the appropriate national legislation was already in force before a State became a party to ACCOBAMS.

[34] The prohibition of driftnets could be considered as a self-executing provision. Domestic legislation is however needed to establish appropriate sanctions for transgressors.

- As regards the assessment and management of human–cetaceans interactions, the obligations set forth in para. 2 of Annex 2 (for example, to take appropriate remedial measures).
- As regards habitat protection, the obligations set forth in para. 3 of Annex 2 (for example, to establish and manage specially protected areas for cetaceans).
- As regards research and monitoring, the obligations set forth in para. 4 of Annex 2 (for example, to develop systematic research programmes on dead, stranded, wounded or sick cetaceans).
- As regards capacity building, collection and dissemination of information, training and education, the obligations set forth in para. 5 of Annex 2 (for example, to cooperate to develop systems for collecting data on observations, incidental catches, strandings, epizootics and other phenomena).
- As regards responses to emergency situations, the obligations set forth in para. 6 of Annex 2 (for example, to prepare emergency plans).
- As regards institutions, the obligation to create or strengthen national institutions with a view to furthering the implementation of ACCOBAMS (Annex 2, para. 1, *e*).
- As regards the functioning of ACCOBAMS, the obligations to designate a focal point and prepare reports on the implementation of this agreement (Art. VIII).

As it happens also for other treaties applying in the field of protection of the environment, the review of domestic legislation enacted by ACCOBAMS Parties shows a great variety of regimes. A number of instances of lack of regulation and, to a lesser extent, overlapping national regulations may be noticed.

As an appropriate step towards achieving an appropriate level of approximation and consistency in the applicable domestic regimes, it is always advisable that the Parties to the same treaty review how the relevant treaty provisions have been implemented in their domestic legislation and elaborate the content of uniform guidelines or model legislation, wherever appropriate. This task is, *inter alia*, performed by the ACCOBAMS Meetings of the Parties, which on the basis of drafts prepared by the Secretariat, have frequently adopted guidelines covering a broad range of subjects related to cetacean protection.

In any case, the provisions of ACCOBAMS cannot affect the right of any Party to maintain or adopt more stringent measures for the conservation of cetaceans and their habitats (Art. XI, para. 1). In other words, if there is a conflict between the provisions in ACCOBAMS and the provisions in national legislation, those that are more favourable for cetaceans prevail (criterion of the better protection).

5.8 The Resolutions Adopted by the Meeting of the Parties

Among the organs established to facilitate the implementation of ACCOBAMS—namely the Meeting of the Parties, the Secretariat,[35] the Coordination Units, the Bureau and the Scientific Committee—the Meeting of the Parties is the decision-making body (Art. III, para. 1). It is entrusted with a broad range of competences, as listed in Art. III, paras 7 and 8.

The provision on voting by the Meeting of the Parties was the subject of lengthy discussion during the negotiations for ACCOBAMS. As in the drafting of other treaties, the main question was whether to require consensus or to allow decision making by majority. As a compromise solution, it was finally agreed that the decisions of the Meeting of the Parties are adopted by consensus. However, if consensus cannot be achieved on matters covered by the annexes to ACCOBAMS, a decision may be adopted by a two-thirds majority of the Parties present and voting.[36] In the event of a vote, any Party may, within 150 days, notify the depositary in writing of its intention not to apply what has been decided by the Meeting of the Parties (Art. III, para. 6).[37]

As regards the content, the only limit to the decision-making power of the Meeting of the Parties is that the instrument adopted must fall within the competences granted to it by Art. III, paras 7 and 8. These competences include, *inter alia*, all matters relating to the implementation of ACCOBAMS (subpara. *i*).

As regards the question of the binding force of the instruments adopted, Art. III, para. 8, *c*, provides that the Meeting of the Parties can 'make recommendations to the Parties as it deems necessary or appropriate and adopt specific actions to improve the effectiveness' of ACCOBAMS. In fact, the Meeting of the Parties is granted a broad margin of discretion with respect to the legal effects of what is the subject of its decisions. It is entitled to adopt instruments which have only a hortatory value (properly called 'recommendations') or to adopt instruments which are the basis for specific

[35] In 2002 the Meeting of the Parties accepted the offer of the Government of Monaco to host the Permanent Secretariat of ACCOBAMS (Resolution 1.2). In 2010 a headquarter agreement was concluded between Monaco and the ACCOBAMS Parties.

[36] This implies that decisions on matters covered by the main text of ACCOBAMS must always be adopted by consensus.

[37] Even a Party that has voted in favour of a decision may avail itself of this opting-out provision.

actions and, as such, can have a binding character. The same instrument, however named, may also contain both binding and nonbinding provisions.

To determine whether or not an instrument adopted by the Meeting of the Parties has a binding force is a question more of content than of name. The legal effects of any instrument are to be carefully ascertained on the basis of its actual wording, the negotiations that have led to its adoption, as well as all other relevant circumstances.

The conclusion that the instruments adopted by the Meeting of the Parties can, in certain cases, have a binding force is also strengthened by the use in the text of ACCOBAMS of the word 'decision'.[38] According to an interpretation based on the ordinary meaning of the terms used in a treaty (see Art. 31, para. 1, of the 1969 Vienna Convention on the Law of Treaties), the word 'decision' is clearly intended to refer to binding instruments. The entitlement of the Meeting of the Parties to adopt binding instruments is confirmed by Art. III, para. 6. It gives a Party the right to notify in writing the depositary of its intention not to apply a decision. This means that the 'decision' in question has been given a binding nature by the Meeting of the Parties and that, exceptionally, a Party may avail itself of a so-called 'opting out' procedure. It would be useless to notify in writing the intention not to apply an instrument that does not entail any obligation to be applied.

In the practice of the five Meetings of the Parties held so far, the instruments taken are often called 'resolutions' and only seldom 'recommendations'. While by definition a 'recommendation' cannot have a mandatory effect, the term 'resolution' does not appear in the text of ACCOBAMS and no precision is given as to the legal effect of such an act. The binding force, if any, of the provisions contained in a resolution has consequently to be determined case by case, on the basis of the elements specified earlier.

From the practice developed so far, it appears that the Meetings of the Parties have in most cases preferred the adoption of instruments that, because of their very wording, can only have a hortatory character.[39] However, resolutions including provisions which can be considered as binding on the Parties may be found, even if in a minority of cases.

As regards their content, the resolutions adopted by the Meeting of the Parties cover a broad range of subjects. For instance, the Meeting of the Parties held in 2010 approved resolutions relating to, *inter alia*, the work

---

[38] 'The Meeting of the Parties shall be the decision-making body of this Agreement' (Art. III, para. 1). 'All decisions of the Meeting of the Parties shall be adopted by consensus (...)' (Art. III, para. 6).

[39] In some cases, the Meeting of the Parties has taken note in a resolution of 'guidelines' that had little legal content, but were in fact scientific studies and included also a bibliography. See, for example, Resolution 3.29, adopted in 2007, on 'guidelines for a coordinated cetacean stranding response'.

programme 2011–12, the format for national implementation reports, commercial cetacean-watching activities, the contribution from ACCOBAMS to the implementation of the European Union Marine Strategy Framework Directive, fisheries interactions with cetaceans, ship strikes on large cetaceans in the Mediterranean Sea, population structure studies, comprehensive cetacean population estimates and distribution, conservation of the Mediterranean short-beaked common dolphin (*Delphinus delphis*), climate change, marine protected areas of importance for cetacean conservation, guidelines for a coordinated cetacean stranding response, guidelines to address the impact of anthropogenic noise on cetaceans, guidelines for the granting of exceptions for the purpose of nonlethal *in situ* research, model measures for the conservation of cetaceans, strengthening the status of ACCOBAMS partners, conditions for use of the ACCOBAMS logo, ACCOBAMS strategy (period 2013–23).

The Meeting of the Parties held in 2013 adopted, *inter alia*, Resolution 5.4 on the ACCOBAMS Follow up Procedure. This instrument establishes a Follow up Procedure (also called in other instruments a compliance procedure) that has a nonconfrontational character and aims at determining the facts and causes of cases of noncompliance and assisting the Party concerned in furthering the implementation of the ACCOBAMS provisions. Resolution 5.4 presents a very advanced character. It creates a Follow up Committee, which is entrusted with the procedure and is composed of three members elected by the Parties and two members elected by the nongovernmental organizations and institutions having the status of ACCOBAMS Partner. The right to present submissions is attributed not only to States Parties and the ACCOBAMS Secretariat, but also to ACCOBAMS Partners. This ensures great transparency and the involvement of the civil society in the implementation of the Agreement. The Follow up Committee has the power to make recommendations that can be endorsed by the Meeting of the Parties. The first two submissions were presented in 2016 by an ACCOBAMS Partner and will in the near future be examined by the Follow up Committee.

5.9 The Relationship with Other Treaties

As already remarked,[40] ACCOBAMS has a very close connection with the Bonn Convention, being an agreement concluded pursuant to Art. IV, para. 4, of this treaty. For instance, under Art. I, para. 3, some terms used in ACCOBAMS, such as ‘migratory species’, ‘conservation status of

[40] *Supra*, para. 2.2.

migratory species', 'favourable' or 'unfavourable' conservation status, 'endangered' and 'taking', are intended according to the same meaning that they have in the Bonn Convention.

The Preamble of ACCOBAMS recalls a lengthy list of global and regional treaties and institutions. While most of the instruments in question are inspired by similar general principles and objectives, the overlapping of provisions contained in different treaties may create very complex legal problems where successive treaties relating to the same subject matter are in principle applicable and contain provisions that are not fully consistent. Article 30 of the 1969 Vienna Convention on the Law of Treaties provides for the following regime:

> **1.** *(...) the rights and obligations of States parties to successive treaties relating to the same subject-matter shall be determined in accordance with the following paragraphs.*
> **2.** *When a treaty specifies that it is subject to, or that it is not to be considered as incompatible with, an earlier or later treaty, the provisions of that other treaty prevail.*
> **3.** *When all the parties to the earlier treaty are parties also to the later treaty but the earlier treaty is not terminated or suspended in operation under article 59, the earlier treaty applies only to the extent that its provisions are compatible with those of the later treaty.*
> **4.** *When the parties to the later treaty do not include all the parties to the earlier one:*
> **(a)** *as between States parties to both treaties the same rule applies as in paragraph 3;*
> **(b)** *as between a State party to both treaties and a State party to only one of the treaties, the treaty to which both States are parties governs their mutual rights and obligations.*[41]

Accordingly, from a logical point of view, the questions to be addressed in order to determine the rights and obligations of States in cases of successive treaties are the following:

(a) whether the States concerned are parties to both the earlier and the later treaty (a *ratione personae* question);

(b) if they are parties to both treaties, whether a provision of the earlier treaty and a provision of the later one relate to the same subject matter and have the same scope of territorial application (a *ratione materiae* and a *ratione loci* question);

[41] Art. 59 of the Vienna Convention, referred to in Art. 30, para. 3, deals with the termination or suspension of the operation of a treaty implied by the conclusion of a later treaty.

(c) if so, whether one of the two treaties specifies that it is subject to the other;
(d) if not, whether and to what extent the two provisions in question are incompatible; in this respect and whenever possible, the provisions should be interpreted according to a meaning that leads to their reconciliation; for instance, a special provision contained in a treaty may be compatible with a more general provision contained in another treaty (*lex specialis derogat legi generali*);
(e) finally, if reconciliation between the two provisions is not possible, it must be determined which one is the later treaty[42] and, consequently, which is the prevailing provision according to Art. 30, para. 3, of the Vienna Convention.[43]

In the case of ACCOBAMS, Art. XI, para. 1, provides that

> *the provisions of this Agreement shall not affect (...) the rights or obligations of any Party deriving from any existing treaty, convention or agreement to which it is a party, except where the exercise of those rights and obligations would threaten the conservation of cetaceans.*

In consideration of the main concern for the conservation of cetaceans, this provision is an application of the already mentioned criterion of the better protection,[44] which is relevant also for the relationship between ACCOBAMS and national legislation.

As regards customary international law, the participants to the negotiations for ACCOBAMS reached consensus on a formulation according to which the Parties shall implement the Agreement 'consistently with their rights and obligations arising under the law of the sea' (Art. XI, para. 2). This wording was sufficiently broad to be accepted also by States that were not parties to the UNCLOS.

During the negotiations for ACCOBAMS, delegations strongly emphasized the necessity of encouraging interinstitutional cooperation by using, whenever appropriate, the existing subregional institutions, in particular the Secretariats of the Barcelona Convention, in the case of the Mediterranean Sea, and the Convention for the Protection of the Black Sea against Pollution (Bucharest, 1992), in the case of the Black Sea.

---

[42] Incidentally, this is a very problematic question. It may be asked whether the concepts of 'earlier' and 'later' are to be determined according to the dates of signature of the two treaties or to the dates of their entry into force on the international level or to the dates on which the treaties have become binding in the relationship between the States concerned. The answer is far from being clear.

[43] Under Art. 44, para. 3, of the Vienna Convention on the Law of Treaties, the provisions of a treaty are separable, as far as termination is concerned.

[44] *Supra*, para. 5.7.

6. TREATIES OF SUBREGIONAL SCOPE SPECIFICALLY RELATED TO SPECIES OF MARINE MAMMALS: PELAGOS SANCTUARY

In 1999 France, Italy and Monaco signed in Rome an Agreement for the establishment of a sanctuary for the protection of marine mammals (so-called Pelagos Sanctuary). The sanctuary extends for about 87,500 km^2 of waters located between the continental coasts of the three countries and the islands of Corsica (France) and Sardinia (Italy). These waters are inhabited by the eight cetacean species regularly found in the Mediterranean, namely the fin whale (*Balaenoptera physalus*), the sperm whale (*Physeter macrocephalus*), the Cuvier's beaked whale (*Ziphius cavirostris*), the long-finned pilot whale (*Globicephala melas*), the striped dolphin (*Stenella coeruleoalba*), the short-beaked common dolphin, the common bottlenose dolphin (*Tursiops truncatus*) and the Risso's dolphin (*Grampus griseus*). In the area of the sanctuary, the water currents create conditions favouring phytoplankton growth and abundance of northern krill (*Meganyctiphanes norvegica*), a small euphausiid that is preyed upon by pelagic vertebrates. Depending on their location, the waters included in the Pelagos Sanctuary have the legal status, of marine internal waters (in the case of France and Italy), territorial sea, ecological protection zone (in the case of Italy), exclusive economic zone (in the case of France)[45] or high seas.[46]

The parties undertake to adopt measures to ensure a favourable state of conservation for every species of marine mammal and to protect them and their habitat from negative impacts, both direct and indirect (Art. 4). They are bound to prohibit in the sanctuary any deliberate 'taking' (defined as 'hunting, catching, killing or harassing of marine mammals, as well as the attempting of such actions') or disturbance of mammals. Nonlethal catches may be authorized in urgent situations or for *in situ* scientific research purposes (Art. 7, *a*).

As regards to the crucial question of driftnet fishing, the parties are bound to comply with the relevant international and European Union regimes (Art. 7, *b*). This is an implicit reference to the already mentioned Regulation No. 1239/98 which banned driftnets.[47] The parties also undertake to exchange their views, if appropriate, in order to promote, in the competent

[45] After the establishment by France of an exclusive economic zone in the Mediterranean (Decree No. 2012-1148 of 12 October 2012), the high seas area within the sanctuary is restricted to the waters that would become the exclusive economic zone of Monaco, if this State were to establish such a zone.

[46] From a strictly legal point of view, the most interesting aspect of the Agreement is Art. 14, dealing with the enforcement on the high seas of the measures agreed upon by the parties.

[47] *Supra*, para. 5.6.

fora and after scientific evaluation, the adoption of regulations concerning the use of new fishing methods that could involve the incidental catch of marine mammals or endanger their food resources, taking into account the risk of loss or discard of fishing instruments at sea (Art. 7, *c*).

The parties commit themselves to exchange their views with the objective to regulate and, if appropriate, prohibit high-speed offshore races in the sanctuary (Art. 9), as well as to regulate whale watching activities for purposes of tourism (Art. 8).[48]

The parties are bound to hold regular meetings to ensure the application and follow up of the Agreement (Art. 12, para. 1).[49] In this framework, they are required to encourage national and international research programmes, as well as public awareness campaigns directed at professional and other users of the sea and nongovernmental organizations, relating *inter alia* to the prevention of collisions between vessels and marine mammals and the communication to the competent authorities of the presence of dead or distressed marine mammals (Art. 12, para. 2).

7. CONCLUSIONS

As it appears from the review made earlier, there are several treaties, starting from ACCOBAMS, which are in force to ensure in general the protection of marine mammals in the Mediterranean Sea. Provisions for the establishment of marine protected areas also exist, although they have been implemented so far only once as far as marine mammals are concerned (Pelagos Sanctuary).

What is probably lacking are specific provisions having a legally binding nature that directly address a number of threats affecting Mediterranean marine mammals, such as underwater noise, collisions with ships, bycatch in fishing gear and ingestion of plastic litter.[50]

[48] Whale watching for commercial purposes is carried out in the sanctuary by a certain number of vessels. There are promising prospects for the development in the sanctuary of this kind of activity, which, if well regulated, is a benign way of exploiting marine mammals.

[49] Criticism has been raised with respect to the lack of a proper management body for the Pelagos Sanctuary.

[50] At the European Union level, Annex I to the already quoted (*supra*, note a) Marine Strategy Framework Directive lists among the 'qualitative descriptors for determining good environmental status' also the fact that 'properties and quantities of marine litter do not cause harm to the coastal and marine environment' and 'introduction of energy, including underwater noise, is at levels that do not adversely affect the marine environment'.

SUBJECT INDEX

Note: Page numbers followed by "*f*" indicate figures, "*t*" indicate tables, and "*np*" indicate footnotes.

F

G

N

O

P

R

S

TAXONOMIC INDEX

Note: Page numbers followed by "*f*" indicate figures, and "*t*" indicate tables.

O

P

S

T

X

Z

FSC
www.fsc.org
MIX
Papier | Fördert gute Waldnutzung
FSC® C083411

Kolibri 360 GmbH
Ferdinand-Jühlke-Straße 7
99095 Erfurt, Deutschland
produktsicherheit@kolibri360.de